SEDIMENTATION CONSOLIDATION MODELS

Predictions and Validation

Proceedings of a Symposium sponsored by the
ASCE Geotechnical Engineering Division
in conjunction with the ASCE Convention in
San Francisco, California
October 1, 1984

Raymond N. Yong and Frank C. Townsend, Editors

AMERICAN
SOCIETY OF
CIVIL
ENGINEERS
FOUNDED
1852
®

Published by the
American Society of Civil Engineers
345 East 47th Street
New York, New York 10017-2398

The Society is not responsible for any statements made or opinions expressed in its publications.

TABLE OF CONTENTS

iii

Part II: Validation and Field Studies

SEDIMENTATION/CONSOLIDATION MODELS - PREDICTIONS AND VALIDATION

Raymond N. Yong[*] and Frank C. Townsend[**]

PROBLEM DEVELOPMENT

The standard methods generally used as a basis for development of analyses and predictions of the settling performance of saturated soils have traditionally addressed the problem in terms of either (a) sedimentation (e.g. Stokesian and Kynchian models), or (b) consolidation (e.g. Terzaghi and Biot models). Success in matching predictions with field performances has been obtained when field situations have conformed closely to the initial and boundary conditions used in the analyses, - provided that the input soil properties properly represent the field conditions.

In very soft sediments, questions have arisen concerning the capability of standard computational techniques used to implement calculations. The problems lie not as much in the viability of the models themselves, - although there are indeed some limiting constraints in most of the models - but in the manner in which the models are used. The problems become more apparent when settling predictions need to be made for solids-suspensions such as those obtained in slime and sludge ponds. Since the solids concentrations of the initially discharged slimes or sludges into the containment ponds are generally very low, - in the range of 2 to 5% by weight - the total settling performance of the suspended solids which combines (a) sedimentation at the initial and intermediate stages with (b) consolidation at the later stages of settling, requires judicious application of sedimentation/consolidation models for evaluation and prediction.

In the evaluation of sedimentation in containment ponds, the overall solids-settling performance has been identified as a process which can be characterized and modelled by at least three different physical models representing various stages of the settling process. These are:

(a) Unhindered "free-fall" discrete solids-settling.

(b) Hindered settling of discrete solids where physical interaction through contact with neighbouring solids occurs.

(c) Self-weight consolidation or subsidence where effective stresses and pore pressures are recorded.

---

[*]William Scott Professor of Civil Engineering and Applied Mechanics, and Director, Geotechnical Research Centre, McGill University, Montreal.
[**]Professor of Civil Engineering, University of Florida, Gainesville.

The above notwithstanding, studies on settling characteristics of suspended solids in containment ponds show that in the midlayer of the ponds, the solids concentrations do not significantly increase after the initial settling period of from 6 to 12 months. A somewhat stagnant solids concentration appears to exist for many types of discharge slurries - hovering from 18 to 30 percent - dependent obviously on the composition of the slurry. Decisions in regard to the range of applicability of the various models proposed to handle predictions of settling performance of the solids hinge around the establishment of relevant initial conditions and pertinent material "sedimentation/consolidation" properties.

PREDICTION OF SEDIMENTATION/CONSOLIDATION

Of the several choices available in seeking predictive methods for application to the solids-settling problem, the most common ones center around the classical sedimentation and consolidation models. Theories relying on Stokes' principles and solids concentrations appear to have been widely used to model the initial sedimentation phase of the settling process. These theories or models rely on solids concentration for derivation of sedimentation relationships, and accountability for hindrances is made through appropriate changes in the viscosity and density of the system.

Application of thickening theories as alternatives has also been considered. These depend on local concentrations of solids and a knowledge of the distribution of solids. The settling processes however are assumed to be controlled by gravitational forces, akin to the Stokesian model.

Developments in self-weight consolidation models and associated computational techniques are a step forward in meeting the requirements in modelling the physical performance of very high initial void ratio situations (void ratios of from 3 to 5) such as those faced in the sediment-formation stage. However, as noted previously, it is not immediately clear where or how one draws the boundary between the three separate types of models presently needed to handle the prediction and evaluation of the total sedimentation/consolidation phenomenon.

However, if proper evaluation and predictions of the settling performance of solids-suspensions and very soft sediments are to be made, it is clear that several questions and/or problems have to be addressed. Some of the more immediate ones that come easily to mind include:

1. The higher degree of reluctance of certain solids-suspensions (in contrast to others) to settle in accord with expectations from gravitational settling models: Why? How to characterize? Model? Measure?

2. Excess hydrostatic pressure and "effective" stresses: Where? When? How to measure? Limiting void ratios? How to identify?

3. Initial conditions and values: What? Where? How to determine?

4. Properties: What? How to measure?

5. Sampling: How? What? Slurry extraction? Disturbance?

6. Field measurements and validation: Solids-liquid interface? Turbidity? Pore pressures? Viscosity? Profile? Rates?

7. Boundary conditions, pond geometry, drainage conditions, and loading conditions.

In addition to the required critical scrutiny of model development and application, we note from the above that two other areas need to be examined. These include: (a) laboratory and/or field tests for determination of pertinent and specific properties of the material under study, and (b) validation procedures. Experiences with conventional laboratory and/or field tests for measurement of relevant soil properties by and large fail to properly account for or assess the physical and chemical interactive conditions which appear to dominate the actual performance of slimes/sludges and very soft sediments. Since the material under study has heretofore not been considered as "regular" soil material insofar as laboratory testing is concerned, new procedures and techniques need to be devised to provide a proper sensing of the kinds of "property" information relevant to the problem at hand.

RESPONSE TO PROBLEM IDENTIFICATION

Responding to the above challenges, the ASCE/GED Soil Properties Committee undertook to develop a Symposium which would (a) address the various aspects of the problem, (b) indicate the present state of knowledge, and (c) seek input and exchange of ideas/information with the geotechnical community. The Task Committee formed included:

H.Y. Ko	H.W. Olsen	A.S. Saada
R.L. Schiffman	E.T. Selig	F.C. Townsend (co-Chairman)
		R.N. Yong (Chairman)

In addition to the organization of the ASCE Symposium, titled SEDIMENTATION/CONSOLIDATION MODELS, the co-chairmen of the Task Committee also organized an ASTM Symposium titled CONSOLIDATION BEHAVIOUR OF SOILS, which was designed to address the laboratory testing problems and issues. This ASTM Symposium is scheduled for January 1985, in Ft. Lauderdale Florida, and the Proceedings will be published in the regular ASTM STP.

The ASCE Symposium program organized to meet the aims of the Soil Properties Committee is as follows:

Session 1

SEDIMENTATION/CONSOLIDATION: THEORIES AND PREDICTIONS

Chairman: H.E. Wahls

Moderator: H.Y. Ko

Status Report: *The Theory of One-dimensional Consolidation*
                     of Saturated Clays. IV. An Overview of non-
                     linear Finite Strain Sedimentation and Con-
                     solidation:

                     R.L. Schiffman, V. Pane and R.E. Gibson

Status Report: *Particle Interaction and Stability of Sus-*
                     pended Solids:

                     R.N. Yong

General Reporters: F.C. Townsend
                   G.A. Nystrom

Session 2

SEDIMENTATION/CONSOLIDATION: VALIDATION AND FIELD STUDIES

Chairman: J.T. Christian

Moderator: D. van Zyl

Status Report: *Consolidation of Mining Wastes:*

                     L.G. Bromwell

Status Report: *Perspectives on Modelling Consolidation of*
                     Dredged Materials:

                     R.J. Krizek and F. Somogyi

General Reporters: J. Garlanger
                   J.H.A. Crooks

GENERAL DISCUSSIONS

    The Status Reports which are in essence Position Papers, represent
the view of the respective researchers on specific issues identified in
the problem under study. They are designed to provide a basic appreciation of the status of development and solution of the aspects of the pro-
blem.addressed by the researchers and practitioners. Also included in
the Symposium are the General Reports which summarize and comment on the
various papers submitted (and peer reviewed) to the Symposium.

It was felt that with both the ASTM and ASCE Symposia, sufficient
exposure to the problem would be obtained and that the attention gained
would assist the Researchers and Practitioners. Obviously, much more
work remains to be done. It is hoped that the Symposia will provide
the basic platform to mount further work in this complex area of Geotechnical Engineering.

PUBLICATION AND ASCE/GED PEER REVIEW

Except for the Reporters' Reports, all the papers published in
this Special Publication were reviewed in accordance with standard
referee procedures: i.e. papers received two positive peer reviews as
per the regular ASCE/GED Geotechnical Engineering Journal procedure.
As such, all these papers are eligible for discussion in the GED Geotechnical Engineering Journal, and are also eligible for ASCE awards.

ACKNOWLEDGEMENTS

The success of this Symposium and the success in the development of
this Special Publication Volume obviously lie with the many contributions
made by the participants: (a) the Chairmen and the Moderators of the
two sessions, (b) the Status Reporters, (c) the General Reporters, (d)
the Panelists, (e) the many Authors who submitted valuable contributed
papers to the Symposium, and (f) the general audience. To these participants, the Profession owes a great debt. Without their combined input and work, the contributions reported herein would not be available.

The very fine assistance provided by (a) the Reviewers, and (b) the
members of the ASCE Soil Properties Committee, especially the chairman
of the Committee, Pro. H.Y. Ko, must also be acknowledged.

We wish to thank (a) the Task Committee members for their efforts
in the organization of the Symposium, (b) Mrs. M.L. Powell for her
valuable assistance and efforts in administrative details and preparation of this Special Publication, and (c) Ms. Shiela Menaker, Manager,
Book Production, ASCE, for her work in the arrangements for production
of the Publication.

THE THEORY OF ONE-DIMENSIONAL CONSOLIDATION
OF SATURATED CLAYS

IV. AN OVERVIEW OF NONLINEAR FINITE STRAIN
SEDIMENTATION AND CONSOLIDATION

by

Robert L. Schiffman[1], Vincenzo Pane[1], Robert E. Gibson[2]

ABSTRACT

This paper presents an overview of the theory and practical
application of the one-dimensional theory of nonlinear finite strain
consolidation. The paper has three parts. First, the theory of
nonlinear finite strain consolidation is reviewed. A modification of
the effective stress equation, to account of settling and consolidation as a unified phenomenon, is proposed. Second, nonlinear finite
strain effects are presented in the context of a series of examples
taken from practice. The examples relate to marine geotechnology,
marine geology, and mine waste planning. Finally, a discussion is
presented of the limiting assumptions and postulates of the current
theory. Suggestions are made of problem areas in which further
progress can be made in the development of a more complete theory.

INTRODUCTION

Modern geotechnical engineering was founded by the publication
of the theory of consolidation [27]. This theory first defines the
fundamental relationships governing the response of a soil system to
imposed loads and from this basis predicts the stresses and displacements of a loaded soil as functions of space and time. This theory,
in concept, is fundamental to the practice of geotechnical engineering
and enters every area where water and soil interact. It is invaluable

---

[1] University of Colorado, Boulder, Colorado, U.S.A.

[2] Golder Associates, Maidenhead, and University of Oxford, United
Kingdom.

1

when analyzing the settlement of structures, the stability of slopes, the design of pile foundations, the conduct of laboratory tests, and so on.

Although considered as the mathematical base for geotechnical engineering, the theory of consolidation which has been used in general practice until very recently [28] has sometimes shown significant discrepancies to exist between prediction and observation. In fact, it is widely recognized that the Terzaghi-Fröhlich theory is based on simplifying assumptions which are, in practice, only approximately satisfied.

One set of concerns revolves around the hypotheses inherent in the theory of consolidation as originally formulated. First, it is noted that the conventional theory is a one-dimensional representation of a three-dimensional process, and that coupled analysis theory [3,4] provides a more consistent formulation and its use is generally to be preferred in place of the Terzaghi-Fröhlich one-dimensional or the Terzaghi-Rendulic uncoupled three-dimensional theory [25].

A second approach to the improvement of the theory cites the linearity of the constitutive relationships in both the Terzaghi-Fröhlich and Biot theories; both the permeability and compressibility of the soil are assumed to remain constant during consolidation under a particular increment of load. Clearly, the errors arising from the assumption of linearity will depend on the magnitude of the void ratio changes during the consolidation process. Studies have sought to extend classical theory to take account of variations of permeability and compressibility with depth [24]. These factors are likely to be of real importance only if the void ratio changes and strains are appreciable; yet these studies have all been based essentially on the infinitesimal strain theory.

The third suggested improvement, namely that concerning the magnitude of the strains, is therefore inextricably linked to this nonlinear soil behavior: the softer the soil, the greater the nonlinearity of the consolidation properties. The development of finite strain consolidation theory has, as is natural, started with

one-dimensional geometries. A rigorous formulation has been
established [11,18]. These two theories are identical in principle
[21], the latter, however, having been derived on the assumption that
the initial condition is uniform. No such restriction was placed on
the derivation of the former.

Another important aspect of the theory, which is sometimes
neglected in conventional analysis, is the influence of the
self-weight of the consolidating layer. The weight of the soil must
be included in the analysis of thick natural clay strata, in which the
externally applied stresses are comparable with those arising from the
self-weight of the deposit. This weight may itself be the sole agency
causing consolidation; examples of this kind are slow and rapid
sedimentation processes which occur in the offshore environment and
are also used in planning of mine waste disposal areas, land
reclamation, and dredging projects. The Gibson, England and Hussey
theory [11] is unrestricted in this regard. It applies equally to
rapid deposition [21], slow accumulation [23] and to the process of
consolidation of loaded thick clay layers [12]. It has recently been
pointed out [15] that in certain cases the general diffusion-advection
equation [11] which explicitly includes an advection term accounting
for self-weight can be reduced to an equation of diffusion alone.
This will, in these cases, simplify the solution procedure.

THEORETICAL CONSIDERATIONS

The governing equation of one-dimensional nonlinear finite strain
consolidation [11] is

$$\frac{\partial}{\partial z}\left[g(e)\frac{\partial e}{\partial z}\right] - f(e)\frac{\partial e}{\partial z} = \frac{\partial e}{\partial t} \qquad (1a)$$

where

$$g(e) = -\frac{k(e)}{\gamma_w(1+e)}\frac{d\sigma'}{de} \qquad (1b)$$

$$f(e) = \left(\frac{\gamma_s}{\gamma_w} - 1\right)\frac{d}{de}\left[\frac{k(e)}{(1 + e)}\right] \tag{1c}$$

where e is the void ratio, γ_s and γ_w are the weights of solid and fluid per unit of their own volume respectively, k is the coefficient of permeability, σ' is the vertical effective stress and z is a reduced coordinate. This coordinate is defined as the volume per unit area of solids lying between the datum plane and the Lagrangian (initial) coordinate point. Here, the reduced coordinate direction is measured against gravity.

Equations (1) are mildly nonlinear partial differential equations of the diffusion-advection type. They require as inputs, along with the boundary and initial conditions, continuous values of the compressibility and permeability (as monotonic) functions of the void ratio.

The governing equation in terms of the reduced coordinate is usually most convenient for solving problems of loaded clay layers. Since we are dealing with large displacements the customary use of Eulerian coordinates will lead to difficulties since in terms of these the location of the surface of the layer (upon which certain boundary conditions must be imposed) is not given and must be discovered as part of the solution. However, the volume of solids in a clay layer never changes. Thus equations (1) can be treated as related to fixed boundaries. Once a solution is developed in terms of z and t, the instantaneous Eulerian ξ coordinate can be found from

$$\xi(z,t) = \int_0^z \left[1+e(z_1't)\right]\, dz' \tag{2}$$

Even if the clay layer is accreting, equations (1) are useful; in these cases the rate of solids building the clay layer is known a priori and thus the boundary value problem is given on a clearly defined domain in (z,t) space.

The consolidation process formulated above is based upon the concept that a fully matured soil is reacting to externally applied or self-weight stresses. However, consolidation is only one part of the process of formation of a soil deposit. A simple genesis of formation, especially of marine sediments, begins with pelagic rain in a quiet water environment. This can and is often modeled as sedimentation by Stokes' law [14]. As particles come together their density of packing increases. The downward motion of this system of particles is referred to as "hindered settling." The process was first formulated by Kynch [13] by means of the equation

$$\frac{\partial c}{\partial t} + \frac{d}{dc}\left[c\ v_s(c)\right]\frac{\partial c}{\partial \xi} = 0 \tag{3}$$

where c is the concentration of particles and v_s is the velocity of the solid particles. As the concentration of solids tends to zero, Kynch's theory reduces to Stokes' theory. Allowing the vertical effective stress approach to zero, it has been found that equations (1) reduce to [1]

$$\frac{\partial e}{\partial t} + f(e)\frac{\partial e}{\partial z} = 0 \tag{4}$$

The similarity between equations (3) and (4) is apparent and in fact, as noted by Been [1], the nonlinear finite strain equation reduces exactly to Kynch's theory when the effective stress goes to zero.

Experimental studies [2,17] have shown that there is a transition zone between a soil (fully developed vertical effective stresses), and a dispersion (zero vertical effective stresses). A simple way of incorporating this into the theory would be to modify the effective stress equation [22] to the form

$$\sigma = \beta(e)\sigma' + u_w \tag{5}$$

where σ is the vertical total stress, u_w is the pore water pressure and β is a monotonic function of e which is supposed known.

In a dispersion β is zero; in a mature soil β becomes unity. The resulting governing equation can be written as

$$\frac{\partial}{\partial z} \left\{ a(e) \left[\beta \frac{\partial \sigma'}{\partial z} + \sigma' \frac{\partial \beta}{\partial z} \right] \right\} + f(e) \frac{\partial e}{\partial z} + \frac{\partial e}{\partial t} = 0 \qquad (6a)$$

$$a(e) = \frac{k}{\gamma_w(1+e)} \qquad (6b)$$

This relationship would then govern the process of sediment formation and consolidation. Thus it is seen that nonlinear finite strain consolidation, along with a modified effective stress equation (5), forms a general expression for the genesis of a soil formation starting with palegic deposition and terminating with consolidation.

In the remainder of this paper we confine ourselves to a discussion of the consolidation process and defer a detailed discussion of the general process of sediment formation to a later paper.

APPLICATIONS OF NONLINEAR FINITE STRAIN CONSOLIDATION

The theories presented above are general in their applicability to practical geotechnical engineering problems. In the discussion which follows we shall limit ourselves to an analysis of nonlinear finite strain consolidation. In this regard we will concentrate on comparisons between the results of nonlinear finite strain analyses and more conventional practice.

We consider the application of the theory presented here to a series of practical problems. These are divided into two classes. First, we consider the response of a clay layer, upon which a surcharge load is applied. Second, we analyze problems concerning the slow deposition of soft material. Here a deposit is built up over a period of time. In this interval the soil is increased in thickness

by depositional processes. Simultaneously consolidation occurs from
self-weight.

The materials considered will be of two types which are typical
of soft deposits, namely marine sediments and mine waste impoundment
materials (slimes).

The Loaded Clay Layer

Consolidation of a loaded clay layer occurs when the total
vertical stress is augmented by an imposed increment of load, Δq. In
the example which follows we consider a clay layer 15 meters thick.
The layer is in equilibrium prior to the instantaneous application of
a 200 kPa surcharge. The initial void ratio distribution is
calculated from the void ratio-effective stress relationship shown in
Figure 1 on the assumption that the layer is normally consolidated
under an effective overburden stress σ_0' given by

$$\sigma_0' = (\gamma_s - \gamma_w)(h_z - z) + q_0' \qquad (7)$$

where h_z is the total height of solids in the layer, and q_0' is the
vertical effective stress at the surface of the layer in the
equilibrium state. This leads to a nonuniform initial condition.

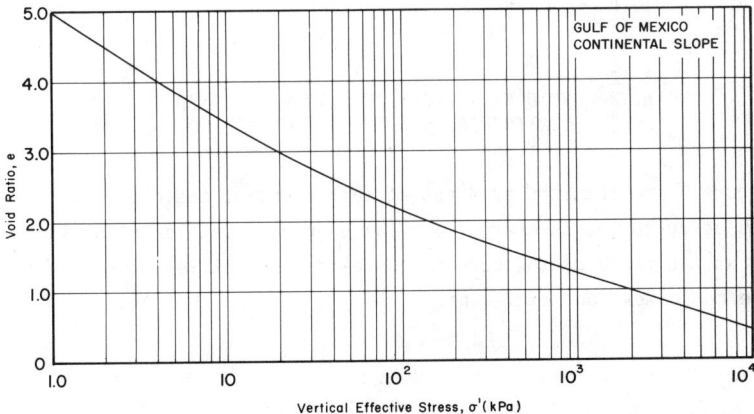

Fig. 1 - VOID RATIO EFFECTIVE STRESS RELATIONSHIP
CONTINENTAL SLOPE, GULF OF MEXICO

The data chosen for this example is based on test results of
recent sediments deposited on the Continental Slope in the Gulf of
Mexico [5].   These sediments are silty clays with high
plasticity.   The index properties are:

Liquid Limit: 137%

Plastic Limit: 40%

The natural water content at the surface of the sediment is 140%.
Figure 2 presents the void ratio-permeability relationships as
determined in the laboratory.

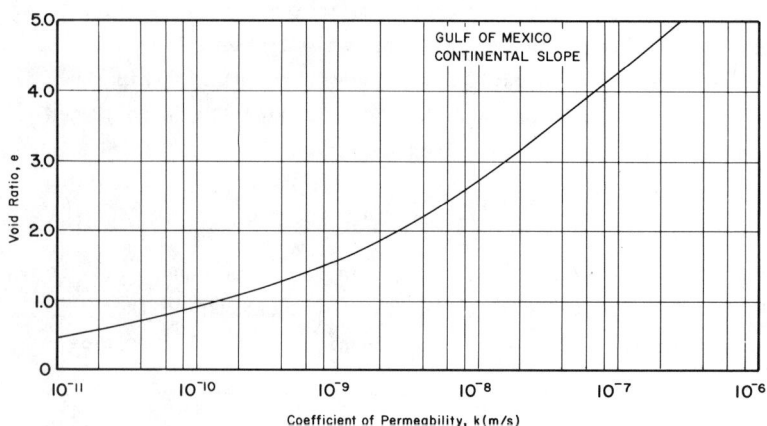

Fig. 2 - VOID RATIO-PERMEABILITY RELATIONSHIP
CONTINENTAL SLOPE, GULF OF MEXICO

Figure 3 presents the progress of settlement of the clay layer
when subjected to the surcharge Δq.   The degree of settlement U_s is
defined as the ratio of the current consolidation settlement to the
ultimate consolidation settlement.

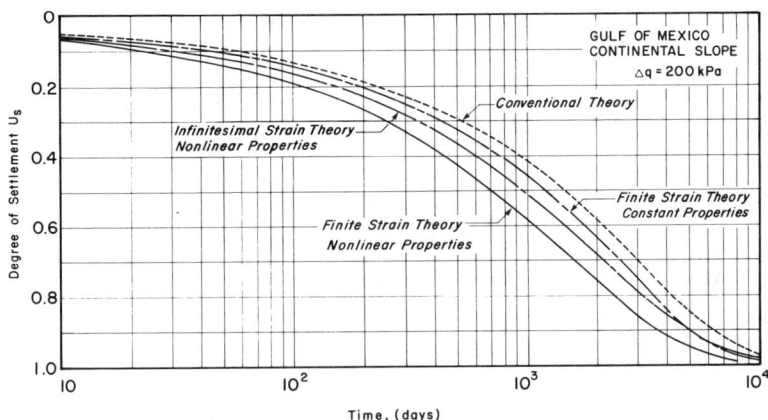

Fig. 3 - PROGRESS OF SETTLEMENT COMPARISONS
CONTINENTAL SLOPE, GULF OF MEXICO

In order to assess the effects of the theory we compare the
results of nonlinear finite strain consolidation to conventional
Terzaghi-Fröhlich theory and to nonlinear infinitesimal strain theory
and to finite strain consolidation with constant properties. We
assume average consolidation properties. The coefficient of
consolidation c_v, defined as

$$c_v = \frac{k(1+e)}{a_v \gamma_w} \qquad (8)$$

where a_v is the coefficient of compressibility, is 3.62 m²/s. This
is based on an average void ratio during the settlement process.

Four particular cases are shown in Figure 3. First, the progress
of nonlinear finite strain consolidation is presented. Second, finite
strain theory is applied, but a single average value of the compressibility and permeability are used in the calculations. Third, infinitesimal strain theory is applied using nonlinear consolidation
properties. Fourth, conventional theory is applied using the value of

c_v given above. That is, in addition to a constant permeability and compressibility, the strains are deemed to be small.

It is noted that nonlinear finite strain consolidation theory predicts a progress of settlement which is substantially faster than predicted when constant, average properties are used. This appears to be a general rule for loaded clay layers. It is consistent with field observations which show that conventional consolidation theory is an over-predictor of settlement times. This increase in the progress of settlement has, in the past, been attributed to two- and three-dimensional effects and the natural inhomogeneity and anisotropy of clay layers [24,25]. The effect of nonlinearity and realistic magnitudes of strain must also be considered.

Figure 3 also shows the importance of the nonlinearity and finite strain effects. It is seen that the effects of large strain can be comparable to the effects of nonlinear properties.

It is noted that a numerical procedure is used to determine the consolidation response. It is further noted that the algorithms for implementing nonlinear infinitesimal and finite strain theories are essentially the same. In fact, because of the use of reduced coordinates the numerical procedures for solving the finite strain governing equation are somewhat simpler than companion procedures for a small strain analysis.

Figure 4 presents excess pore water pressure isochrones at a particular time. It is noted that while finite strain theory predicts faster settlements, vis à vis conventional theory the predicted dissipation of excess pore water pressure is slower.

A degree of excess pore pressure dissipation U_e can be defined as the deviation from unity of the ratio of the average excess pore water pressure to the surcharge load intensity. If conventional theory is used, and the clay layer is homogeneous, U_s and U_e coincide. As shown in Figure 5, this is not the case when nonlinear finite strain theory is applied. It is shown here that for most of

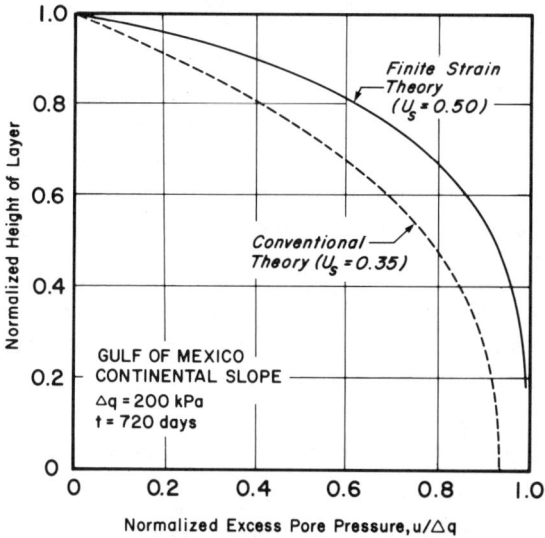

Fig. 4 - EXCESS PORE PRESSURE COMPARISONS
CONTINENTAL SLOPE, GULF OF MEXICO

the time involved the rate of settlement is substantially faster than
the rate of dissipation of excess pore water pressure.

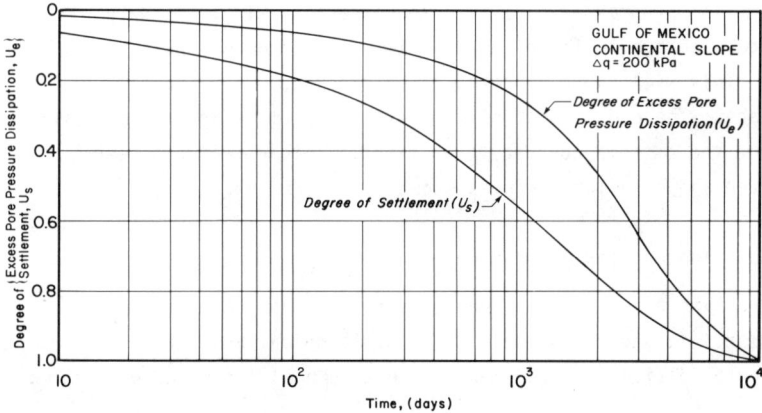

Fig. 5 - PROGRESS OF CONSOLIDATION COMPARISONS
CONTINENTAL SLOPE, GULF OF MEXICO

The results shown in Figure 5 are general in the sense that U_s and U_e are not coincidental. The exact nature of their difference depends on the specific properties of the clay being studied, and the extent of the nonlinearity. Clearly, conventional wisdom which holds that deformation and excess pore pressure dissipation are in a one-to-one correspondence is not valid. This can have a serious effect on the laboratory determination of strength parameters and on the results of stability analyses.

If the clay layer is not in a state of equilibrium when the surcharge Δq is applied, there will be two modes of consolidation: self-weight and surcharge. Then the effects described above will be more pronounced. The degree of the added self-weight consolidation effect will depend on the layer thickness, its properties, and the magnitude of Δq.

Conventional Terzaghi-Fröhlich theory provides a marked distinction in the progress of consolidation between free-draining and impervious boundaries. This stems from the assumption that the coefficients of permeability and compressibility are constant. Nonlinear finite strain consolidation, on the other hand, accommodates the decrease in both the permeability and compressibility as deformation progresses. As a consequence a pervious boundary will undergo a change in its properties immediately upon application of the load. This will cause a "cake" to be produced which inhibits drainage. The end product of this effect is to lessen the difference between the extreme boundary conditions. This effect has been observed and calculated and is shown elsewhere [8].

Accumulation by Slow Deposition

We turn our attention now to a problem that has substantial importance in geology as well as in geotechnical engineering. This is the consolidation of a sediment whose thickness increases with time.

As an example we first consider the accumulation and simultaneous consolidation of copper slimes placed in a tailings empoundment. For this purpose we will use the data for Bethlehem slimes [20]. The void ratio-effective stress and void ratio-permeability relationships are

$$e = 1.45 - 0.265 \log \sigma' \qquad (9a)$$

$$e = 1.42 + 0.56 \log (k/10^{-7}) \qquad (9b)$$

where σ' has units of kPa and k has units of m/s. We assume that the copper slimes are deposited at a void ratio of 1.3 over a period of 3000 days at a constant rate. After 3000 days the tailings empoundment remains in a quiescent state. Given reasonable estimates of the tonnages of slimes produced and the volume of solids (assume the specific gravity $G_s=2.60$) it is estimated that the Lagrangian height of the empoundment is 80 meters at 3000 days. The Lagrangian height h_a is the height the deposit would occupy if no consolidation took place. We assume that the water level is maintained at the top of the slimes and that the slimes rest on a impervious base. Given the Lagrangian height (and thus the Lagrangian rates of deposition) the material height h_z at any time can be calculated from

$$h_z = \int_0^{h_a} \frac{da'}{1+e(a',0)} \qquad (10)$$

where a is the Lagrangian coordinate. Figure 6 shows the Lagrangian and the calculated (Eulerian) height-time relationships. The latter data were calculated using nonlinear finite strain consolidation theory. It is noted that the height of the impoundment at the end of 3000 days is 71.7 meters. Thus, the consolidation settlement during filling is 8.3 meters. This amounts to an engineering strain of

11.6%.   The ultimate height of the impoundment is 64.6 meters, or a
total strain of 23.8%.

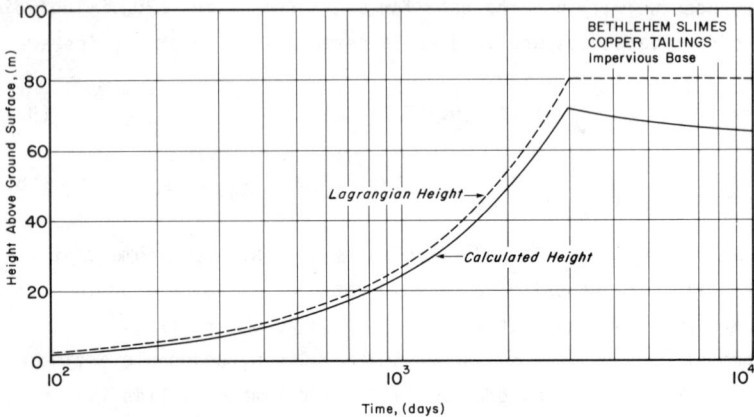

Fig. 6.- HEIGHT-TIME RELATIONSHIPS
BETHLEHEM SLIMES, COPPER TAILINGS

In order to put the results of the nonlinear finite strain
analysis into context we compare these with those of an analysis using
linear infinitesimal strain theory [10]. This theory, by its small
strain nature, does not account for changes in height due to consolidation. In order to compensate for this deficiency we used the finite
strain calculations as a guide. The use of linear consolidation
theory requires knowledge of the coefficient of consolidation c_v.
We chose to bracket this parameter using void ratio-coefficient of
consolidation data [20]. The bounds were based upon the void ratios
calculated by the finite strain analysis. The maximum and minimum
values of c_v are:

$$(c_v)_{max} = 3.40 \times 10^{-6} \ m^2/s$$

$$(c_v)_{min} = 2.25 \times 10^{-7} \ m^2/s$$

The pore water pressure profile at 3000 days is given in Figure 7.
This profile presents the static pore water pressure u_0, and the

pore water pressure u_w for the three cases analyzed. These terms
are related by

$$u_w = u_0 + u \qquad (11)$$

where u is the excess pore water pressure. It is noted that while the
finite and one of the small strain curves are in almost total
agreement this requires that one would have the foresight to choose
the maximum value of c_v. Even though the bounds of small strain
results are only about 14% apart it must be remembered that these
results are favorably biased by the use of finite strain results as
input to the small strain theory.

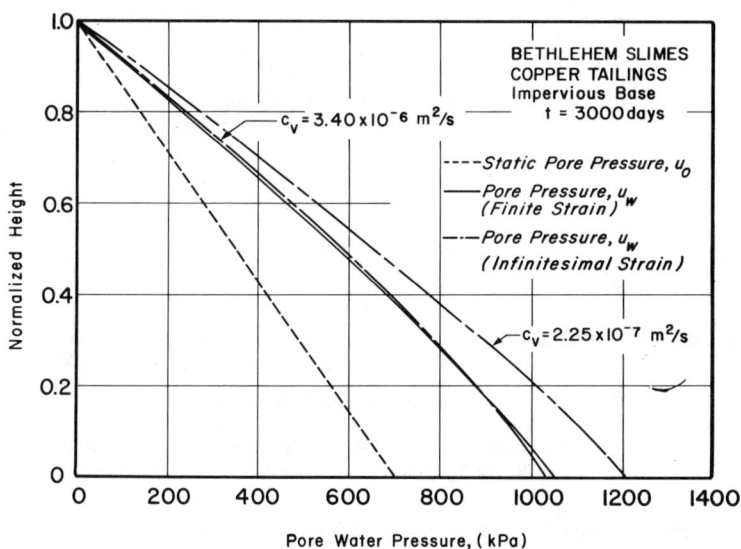

Fig. 7 - PORE WATER PRESSURE ISOCHRONES,
BETHLEHEM SLIMES, COPPER TAILINGS
IMPERVIOUS BASE

    Let us now assume that the tailings impoundment is underlain by a
dry permeable stratum. Then the persistent boundary condition at the
base of the slimes is one in which the pore water pressure u_w is

zero. For this condition the end-of-filling height is 66.9 meters and
the end-of-consolidation height is 62.2 meters. The pore water
pressure profiles at 3000 days are shown in Figure 8. In this case
the range of c_V values produces extremely large differences in the
pore water pressures calculated by linear small strain theory. They
bracket the nonlinear finite strain solution. An average of the two
linear small strain results is reasonably close to the nonlinear
curve, but this is fortuitous.

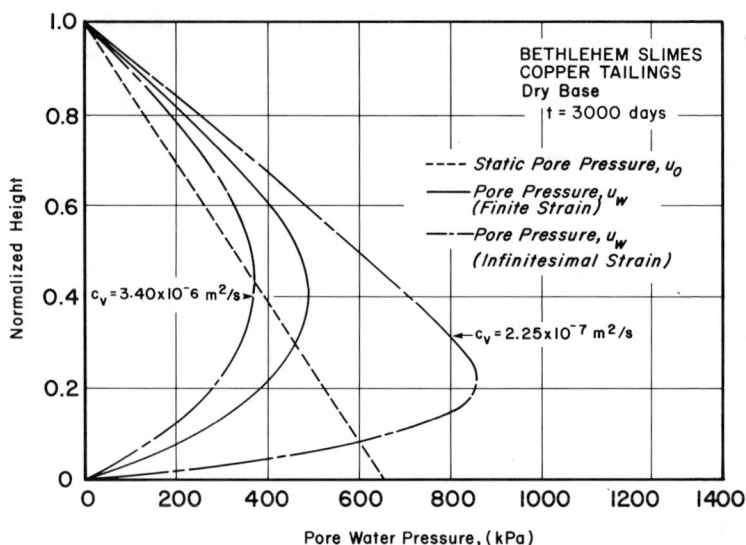

Fig. 8 - PORE WATER PRESSURE ISOCHRONES BETHLEHEM
SLIMES, COPPER TAILINGS DRY BASE

It is noted parenthetically that equations (1) were derived on
the basis that the excess pore water pressure u is defined as the excess over the static pore water pressure u_0. Using this definition,
the excess pore water pressures shown in Figure 8 can be negative. If
u were to be defined as the excess over the pressure associated with a
final state of steady uniform downward seepage (namely, atmospheric
pressure) then this would provide a positive value of u of the same
magnitude.

Conventional practice related to the design of tailings impoundments is based on approximations to linear, infinitesimal strain consolidation theory [29]. Nonlinear finite strain theory offers a consistent, potentially more accurate alternative.

Geotechnical Stratigraphy

We define geotechnical stratigraphy as the variation of geotechnical properties and processes of sediment formation as functions of space and time. In the geologic sense geotechnical stratigraphy uses geotechnical analyses to quantify the depositional history of a geologic formation. More about this later in this section. In the geotechnical sense the principles of geotechnical stratigraphy can be used to reconstruct aspects of the history of an earth structure. We use these principles to reconstruct the time-height relationship and pore pressure profiles of a copper tailings dam. The void ratio-effective stress and void ratio-permeability relationships of the tailings are:

$$e = 1.26 - 0.247 \log \sigma' \qquad (12a)$$

$$e = 1.27 + 0.288 \log (k/10^{-7}) \qquad (12b)$$

where σ' and k have units of kPa and m/s respectively. We also know, from field data, that at the end of 3390 days the copper slimes had reached a height of 49.5 meters above a dry, permeable base. The void ratio of the slimes at deposition e_0 was 1.45 and the average void ratio e_{av} of the slimes at 3390 days was 0.8. During the development of the empoundment a surcharge was placed on the surface. The commencement of surcharging was at 3030 days. The surcharge was placed at a constant rate of loading of 0.47 kPa/day for 360 days. Given a field monitoring of the actual height and the average void ratios as functions of time, the material height function $h_z(t)$ necessary as input to equations (1) could be determined directly, according to the relationship

$$h_z(t) = \frac{h(t)}{1 + e_{av}(t)} = \frac{h_a(t)}{1 + e_0} \qquad (13)$$

This procedure provides benchmarks that must be matched for the validation of the analysis. These benchmarks may be the measured current height, void ratio and pore pressure profiles, etc., depending on the availability of field data.

Figure 9 presents the history of the Lagrangian, calculated and measured heights of the slime empoundment. The calculated height is 51 meters, and represents a 3% difference with the measured benchmark height. The calculated average void ratio at 3390 days is 0.82.

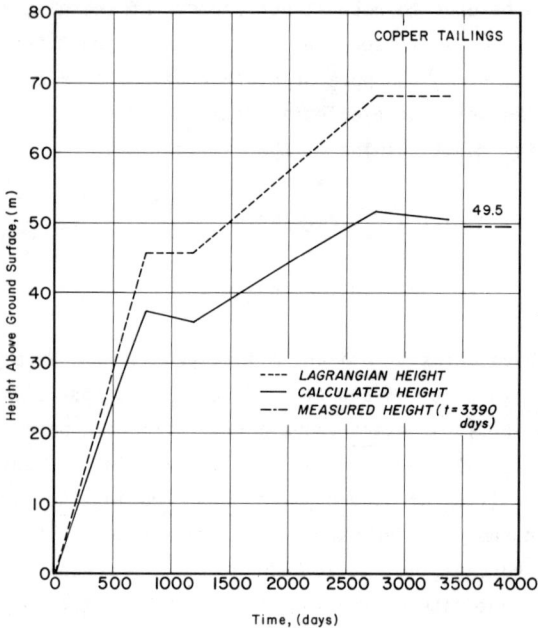

Fig. 9 - HEIGHT-TIME RELATIONSHIPS
COPPER TAILINGS

It is remarked that small strain theory, by the nature of its fundamental assumption, is incapable of reconstructing (or constructing) a rationally consistent height-time relationship.

In order to compare our nonlinear finite strain approach to a companion linear small strain analysis, pore water pressure profiles were calculated at 2760 and 3390 days. Figures 10 present the static u_0 and the pore water pressure profiles u_w for the slimes. The earliest date (2760 days) represents the end of filling. Note that this was 270 days before the commencement of surcharging. The second date represents the end of surcharging. It is noted that at 3390 days the water level is above the slimes, thus providing a non-zero static pore water pressure at the interface between the slimes and the surcharge. It is further noted that the linear analysis was based upon an average value of c_v of 2.46 x 10^{-7} m^2/s.

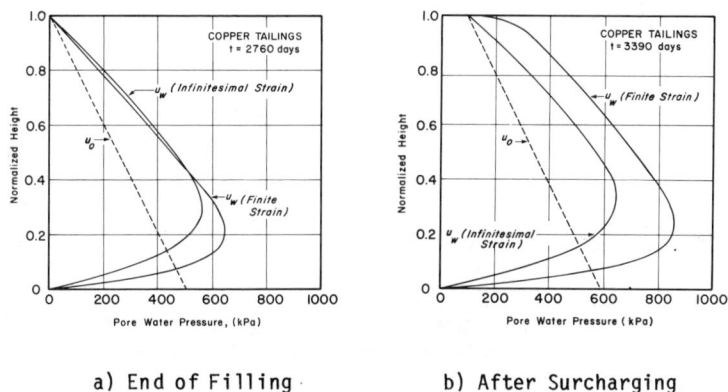

a) End of Filling   b) After Surcharging

Fig. 10 - PORE WATER PRESSURE ISOCHRONES
COPPER TAILINGS

We observe that prior to the placement of the surcharge (2760 days) there is little difference between the small and finite strain analyses in the upper 75% of the slime pond. On the other hand, a comparison with the Bethlehem slime profiles (Figures 7 and 8) indicates that a general observation of congruence cannot be made. We observe further the effect of the surcharge, shown in Figure 10b.

Here the use of linear theory would seriously underestimate the
pore water pressures developed.

Turning our attention to marine geology we reiterate a previously
published result of an analysis of the vertical effective stresses
developed in the Mississippi Delta sediments [23]. Figure 11 presents
a profile of the vertical effective stresses developed by use of the
two theories. We note the underconsolidated nature of the sediments.
This is typical of observations of soft marine sediments with
relatively high rates of accumulation. We remark that the nonlinear
finite strain analysis provides a markedly higher effective stress
profile than would be determined by the use of a linear analysis.
This more exact analysis can provide significant economic advantages
in the design of offshore structures.

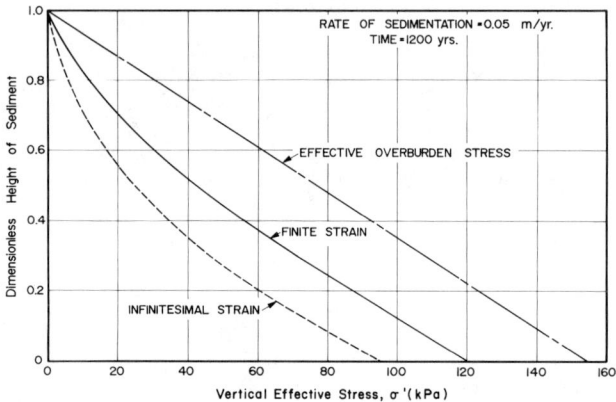

Fig. 11 - EFFECTIVE STRESS DURING SEDIMENTATION

Centrifuge Validation

Heretofore we have tacitly assumed that nonlinear finite strain
consolidation theory will be understood to be a more accurate
predictor of consolidation behavior than linear infinitesimal strain
theory. However, this assumption must be validated. One form of
validation is by means of geotechnical centrifuge testing. In this
regard we reproduce the results of a series of centrifuge tests
performed on a remoulded Georgia kaolin with an initial void ratio of

2.86. The kaolin had a liquid limit of 44% and an initial water content of 110% [8].

Figure 12 presents the results of these centrifuge tests. Similar results have been obtained for Speswhite kaolin [16] and for a Florida phosphatic clay slime [26]. It is seen that nonlinear finite strain theory is a good predictor of centrifuge consolidation compared with conventional Terzaghi-Fröhlich theory.

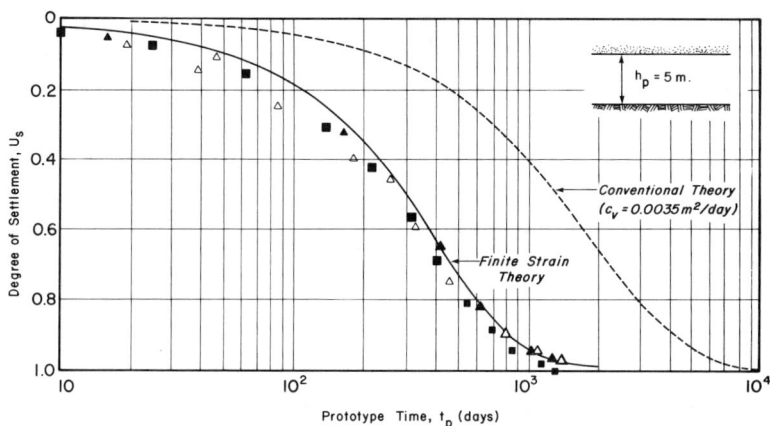

Fig. 12 - CENTRIFUGE VALIDATION OF NONLINEAR
FINITE STRAIN CONSOLIDATION THEORY

Field Verification

Moving from the laboratory to the field we report on a field tank test performed under the auspices of the Florida Phosphatic Clay Research Project, directed by Dr. L.G. Bromwell. The tests were conducted at the International Minerals and Chemical Company, Kingsford Mine in Florida.

A Florida phosphatic clay at a very high water content was permitted to settle in a tank for a period of approximately 400 days. Initial and final height measurements were taken. Furthermore, ports along the side of the tank permitted water content (void ratio) measurements. The initial height of the clay was 20.75 feet.

The consolidation property data supplied [6] was of the form

$$e = 29.43 \ (\sigma')^{-0.29} \hspace{3cm} (14)$$

$$k = (4 \times 10^{-6})(e)^{4.11} \hspace{3cm} (15)$$

where the units are pounds, feet and days.

Figure 13 presents a comparison of the measured and predicted void ratio profiles at the beginning and the end of the tank test. Figure 14 presents the predicted and measured average void ratios as a function of time. Also shown in Figure 14 is a calculation using conventional Terzaghi-Fröhlich theory. For this theory the average void ratio of 13.23 was assumed, the coefficient of permeability was taken to be 0.165 ft/day, and the coefficient of compressibility was taken to be 0.1046 ft^2/lb.

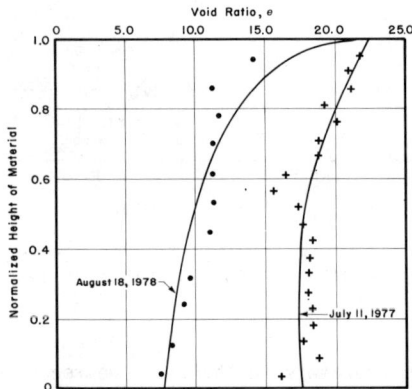

Fig. 13 - VOID RATIO PROFILE
TANK TEST
FLORIDA PHOSPHATIC CLAY

It is seen that here nonlinear finite strain theory is an excellent predictor of consolidation response.

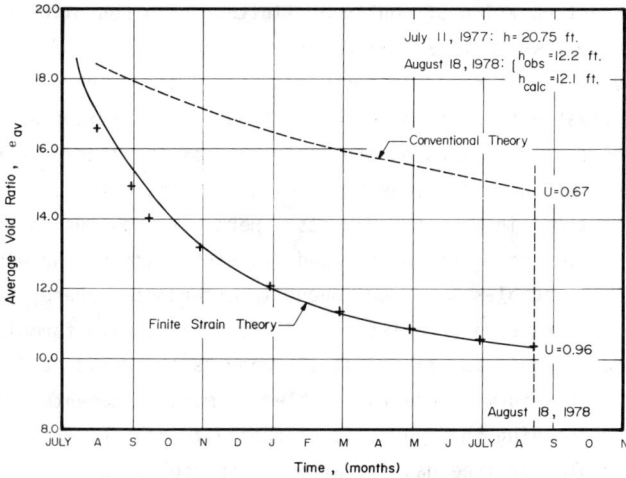

Fig 14 - PROGRESS OF CONSOLIDATION
TANK TEST
FLORIDA PHOSPHATIC CLAY

DISCUSSION

The theory and applications of nonlinear finite strain consolidation as presented above are based upon assumptions which are substantially less restrictive than Terzaghi-Fröhlich theory. In spite of this, the theory should be augmented or revised to enhance its relationship to reality.

The existing theory of nonlinear finite strain consolidation is limited to one-dimension. There is a clear need to extend this to two- and three-dimensional domains. It is not clear whether for engineering applications this extension should be as a coupled theory or as an uncoupled analysis. Since an uncoupled theory would be substantially more tractable from a mathematical and numerical viewpoint, its implementation probably carries first priority.

Some work along these lines has already been undertaken [7], but the study was restricted in some respects, for example the permeability was taken as constant.

The current equation of nonlinear finite strain consolidation is limited to a soil which is a mixture of incompressible particles, a deformable clay skeleton and an incompressible fluid. There are many real world situations in which these limiting considerations place disadvantageous restrictions on the usefulness of the theory. The most important of these concerns the presence of gas which is especially critical in marine sediments. Here, the gaseous phase may be partly in solution in the water, and its distribution otherwise may range from small bubbles to a continuous interconnected phase. In the latter case an extension of existing infinitesimal strain formulations [9] to encompass nonlinear finite strain appears to provide a fruitful beginning. In the former case the problem is more fundamental as information concerning the hydraulic conductivity of sediments containing bubbles of free gas is, at best, sparse and unreliable.

Multi-component mixtures of solids also provide an area of further investigation. Here we make specific reference to the consolidation of tar sand sludges, organic silts and sand-clay mixtures. In these cases one would need to consider two distinct deformable skeletons, each with its own deformation characteristics.

The soil particles for most natural inorganic sediments can safely be assumed to be incompressible within the range of pressure normally encountered. There are, however, classes of marine sediments (such as calcareous silts) where the particles themselves deform and, perhaps, fracture.

We have assumed that the void ratio is a unique single-valued function of the effective stress and that the permeability is a unique single-valued function of the void ratio. This implies monotonic, primary consolidation. In soft sediments where the difference between Terzaghi-Fröhlich and finite strain theory is likely to be important the assumption that the void ratio is an intrinsic function of the effective stress alone [12] is likely to be very wide of the mark. The void ratio would be expected to change with time even if the effective stress remained constant. This is usually refered to as secondary consolidation. Much more work needs to be done before this phenomenon can be taken properly into account in the theory but the relation between void ratio, effective stress and time will have the form of an imperfect differential

$$de = A(\sigma',t)d\sigma' + B(\sigma',t)dt \qquad (16)$$

since the void ratio will depend on the path traversed in (σ',t) space and

$$\frac{\partial A}{\partial t} - \frac{\partial B}{\partial \sigma'} \qquad (17)$$

is non-zero. The integrated form of this will lead to an expression of the character

$$e = E\{\sigma',t\} \qquad (18)$$

where E is a functional.

It is noted that equation (18) can take account of nonhomogeniety as well as secondary consolidation by inserting the Lagrangian coordinate within the functional. Just as conventional theory was extended to consider continuous variations of permeability and compressibility with depth [24] finite strain theory can be so augmented.

The applications presented in earlier sections were concerned exclusively with normally consolidated deposits. This limits the usefulness of the theory to cases of monotonic loading. We suggest that problems of erosion, surcharge removal and cyclic loading due to wave action can be accommodated by bookkeeping procedures within the numerical schemes used for monotonic loading.

The material function given by equation (18) and its companion for the permeability are not the most general constitutive relationships that occur in nature. They only refer to the consolidation generated by mechanical forces. To this one can add other motivations such as temperature, electrical and chemical gradients [19]. These can be of importance in certain mining operations, the disposal of nuclear wastes, dewatering by electro-osmosis and the planning and reclamation of chemical waste dumps. Particular care must be taken in the latter case to properly formulate the mass balance of the system. Equations (1) are predicated upon the assumption that the masses of the solid and fluid constituents of the mixture are separately conserved. This implies that all parts of the soil system are chemically inert. In dealing with problems of chemical wastes as a soil constituent this assumption must be carefully examined.

In this paper we have chosen to limit the discussion to quasi-static (non-dynamic) problems as this is the area in which the theory of consolidation is most generally applied. We recognize, however, that dynamic problems such as consolidation under time dependent loading and seabottom flow slides, have inertial components. This is certainly an area of potentially fruitful investigation.

CONCLUDING COMMENTS

In this paper we have shown applications where nonlinear finite strain consolidation theory provides rather better predictions than other companion theories. We have also indicated areas where we believe the theory can be improved and its application base broadened.

In the examples cited above two major features are clear. First, in the development of height-time relations for slowly deposited fills and geologic formations, finite strain theory is the only one which can be used directly. In this regard, where very soft soils or slurries are concerned, the development of an enhanced theory using the modified effective stress equation (5) would be mandatory.

The second feature noted is the dependence on a single, constant value of c_v in all the linear theories. The selection of an appropriate, all-encompassing value of c_v is little more than guesswork. A more appropriate approach is to use the void ratio-effective stress and the void ratio-permeability relationships. This, of course, would dictate the use of nonlinear theory.

ACKNOWLEDGEMENTS

We are pleased to acknowledge the financial support provided by the U.S. National Science Foundation to the U.S. authors. We are further pleased to acknowledge the stimulating and helpful comments by P. Croce (University of Rome), H.Y. Ko (University of Colorado), H.W. Olsen (U.S. Geological Survey and University of Colorado), and D. Znidarcic (Purdue University). Special acknowledgement is made to V. Sunara of the University of Colorado for her assistance in preparing the computations. We are further indebted to the personnel of Bromwell and Carrier, Inc., of Lakeland, Florida for providing us with the IMC tank test data.

REFERENCES

[1] Been, K., (1980), "Stress Strain Behaviour of a Cohesive Soil Deposited Under Water," Ph.D. Dissertation, University of Oxford, Oxford, United Kingdom.

[2] Been, K., and Sills, G.C., (1981), "Self-Weight Consolidation of Soft Soils: An Experimental and Theoretical Study," Geotechnique, 31, pp. 519-535.

[3] Biot, M.A., (1941), "General Theory of Three-Dimensional Consolidation," Journal of Applied Physics, 12, pp. 155-164.

[4] Biot, M.A., (1956), "General Solutions of the Equations of Elasticity and Consolidation for a Porous Material," Journal of Applied Mechanics, 23, pp. 91-96.

[5] Bryant, W.R., (1980), Private Communication.

[6] Carrier, W.D., III, (1980), Private Communication.

[7] Carter, J.P., Small, J.C., and Booker, J.R., (1977), "A Theory of Finite Elastic Consolidation," International Journal of Solids and Structures, 3, pp. 467-478.

[8] Croce, P., Pane, V., Znidarcic, D., Ko, H-Y., Olsen, H.W., and Schiffman, R.L., (1984), "Evaluation of Consolidation Theories by Centrifugal Modelling," Proceedings of the Conference on Applications of Centrifuge Modeling to Geotechnical Design, Manchester University, United Kingdom, pp. 380-401

[9] Fredlund, D.G., and Hasen, J.U., (1979), "One-Dimensional Consolidation Theory: Unsaturated Soils," Canadian Geotechnical Journal, 16, pp. 521-531.

[10] Gibson, R.E., (1958), "The Progress of Consolidation in a Clay Layer Increasing in Thickness with Time," Geotechnique, 8, pp. 171-182.

[11] Gibson, R.E., England, G.L. and Hussey, M.J.L., (1967), "The Theory of One-Dimensional Consolidation of Saturated Clays, I. Finite Non-Linear Consolidation of Thin Homogeneous Layers," Geotechnique, 17, pp. 261-273.

[12] Gibson, R.E., Schiffman, R.L., and Cargill, K.W., (1981), "The Theory of One-Dimensional Consolidation of Saturated Clays, II. Finite Non-Linear Consolidation of Thick Homogeneous Layers," Canadian Geotechnical Journal, 18, pp. 280-293.

[13] Kynch, G.J., (1952), "A Theory of Sedimentation," Transactions of the Faraday Society, 48, pp. 166-176.

[14] Lamb, H., (1932), Hydrodynamics, (Sixth Edition), Dover Publications, New York, New York.

[15] Lehner, F.K., (1984), "On the Consolidation of Thick Layers," Geotechnique, 34, pp. 259-262.

[16] Leung, P.K., Schiffman, R.L., Ko, H.Y., and Pane, V., (1984), "Centrifuge Modeling of Shallow Foundations on Soft Soil," Proceedings, Sixteenth Annual Offshore Technology Conference, Paper OTC 4808, 3, pp. 275-282.

[17] Michaels, A.S., and Bolger, J.C., (1962), "Settling Rates and Sediment Volumes of Flocculated Kaolin Suspensions," Industrial and Engineering Chemistry, 1, No. 1, pp. 24-33.

[18] Mikasa, M., (1963), The Consolidation of Soft Clay - A New Consolidation Theory and Its Application, Kajima Institution Publishing Co., Ltd., (In Japanese).

[19] Mitchell, J.K., (1976), Fundamentals of Soil Behavior, John Wiley & Sons, Inc., New York, New York.

[20] Mittal, H.K., and Morgenstern, N.R., (1976), "Seepage Control in Tailings Dams," Canadian Geotechnical Journal, 13, pp. 277-293.

[21] Pane, V., and Schiffman, R.L., (1981), "A Comparison Between Two Theories of Finite Strain Consolidation," Soils and Foundations, 21, No. 4, pp. 81-84.

[22] Pane, V., and Schiffman, R.L., (1985), "A Note on Sedimentation and Consolidation," Note submitted to Geotechnique.

[23] Schiffman, R.L., and Cargill, K.W., (1981), "Finite Strain Consolidation of Sedimenting Clay Deposits," Proceedings, Tenth International Conference on Soil Mechanics and Foundation Engineering, 1, pp. 239-242.

[24] Schiffman, R.L., and Gibson, R.E., (1964), "Consolidation of Nonhomogeneous Clay Layers," Journal of the Soil Mechanics and Foundations Division, ASCE, 90, No. SM5, Proceedings Paper 4043, pp. 1-30.

[25] Schiffman, R.L., Chen, A.T-F., and Jordan, J.C., (1969), "An Analysis of Consolidation Theories," Journal of the Soil Mechanics and Foundations Division, ASCE, 95, No. SM1, Proceedings Paper 6370, p. 285-312.

[26] Scully, R.W., Schiffman, R.L., Olsen, H.W., and Ko, H.Y., (1984), "Validation of Consolidation Properties of Phosphatic Clay at Very High Void Ratios," Proceedings, ASCE Symposium on Consolidation/Sedimentation, (This Volume).

[27] Terzaghi, K., (1923), "Die Berechnung der Durchlässigkeitsziffer des Tones aus dem Verlauf der Hydrodynamischen Spannungserscheinungen," Akademie der Wissenchaften in Wien, Sitzungsberichte, Mathematisch-naturwissenschaftliche Klasse, Part IIa, 132, No. 3/4, pp. 125-138.

[28] Terzaghi, K., and Fröhlich, O.K., (1936), Theorie der Setzung von Tonschichten; eine Einfuhrung in die Analytische Tonmechanik, F. Deuticke, Leipzig, Germany.

[29] Vick, S.G., (1983), Planning, Design, and Analysis of Tailings Dams, John Wiley & Sons, New York, New York.

PARTICLE INTERACTION AND STABILITY OF SUSPENDED SOLIDS

Raymond N. Yong[*] M.ASCE

ABSTRACT

The physics of particle interaction, which is responsible for the problem of the apparent tardy settling of suspended solids in tailings discharge slurries containing initially low concentrations of solids, is examined in this study. The degree to which the solids remain in a quasi-suspended state is identified as the dispersion stability of the suspension. This study first identifies the basic elements of the general situation, and discusses the characteristics of interactions of surface-active solids which are common to many types of slurries and fundamental to the development of suspension properties. The basic principles established from the physics of interaction of the suspended solids identify dispersion stability as a function of the composition of the solids and the chemistry of the suspending fluid. The equilibrium concentration of solids remaining in suspension at any one time can be calculated in terms of the balance of energy (internal and external) of the system. The concept of equilibrium suspension volumes is tested with comparisons between theoretically computed and measured values of solids concentrations.

INTRODUCTION

The phenomenon of settling of suspended solids (fines) is one which is common to (a) natural processes in soil sedimentation such as the initial stages for formation of sedimentary soils, and (b) management of tailings discharge from mineral resource industries. In both general types of situations, one of the major items of interest is the problem of prediction of the rate of settling of the suspended solids and the consolidation of the sediment layer.

The intent of this paper is to present a view of the phenomenon of interaction of surface-active solids in a fluid medium which produces the situation commonly identified as a solids-suspension. Because of the surface activities of the solids and their resultant interactions, the phenomenon of suspended solids occurs in many containment ponds associated with mineral resource industry discharges.

We begin by identifying the general situation of pond containment

---

[*]William Scott Professor of Civil Engineering and Applied Mechanics, and Director, Geotechnical Research Centre, McGill University, Montreal, Que. Canada.

and the establishment of a solids concentration profile in the pond. This is followed by a discussion of the characteristics of interactions of surface-active solids which are fundamental to the development of the suspension properties - with a view to establishing the thesis that the solids are dispersed in the suspending fluid at a particular equilibrium concentration, consistent with (a) the composition of solids, and (b) the balance of energy (internal and external) of the system. To demonstrate the thesis, several types of discharge slurries are examined in this study. The laboratory methods of compositional analysis and the pertinent compositional features of the slurries studied are also listed. Finally, using the concept of the equilibrium suspension volume described by each type of solid which comprises the types of solids present in the system, the theoretically computed solids concentrations for some of the slurries studied are compared with the actual measured solids concentrations.

THE PHYSICAL PROBLEM

The solids (fines) concentration by weight, in a solids-suspension tailings discharge from many mineral extraction process industries is generally very dilute - in the range of 1 to 3%. If the solids are non surface-active, and if they are silt-sized or larger, it is likely that they will settle in accord with the general predictions advanced by the simple Stokesian model. However, if the particles are clay-sized or less, and if they are surface-active in nature, simple Stokesian predictions will not accurately portray the settling behaviour of the solids.

As the solids settle, their concentration will increase to the point where proximal hindrances become significant. Thus, even if initial settling of the suspended solids can be predicted by the simple Stokesian model, subsequent settling of the solids will render the model invalid in application. The problem of the inability of the solids to settle in accord with gravitational mechanisms has been documented previously - e.g. [1,5,25] and will not be repeated here. Figure 1 which shows the essence of the problem, portrays the solids concentration profile in a settling pond. This profile is common to the settling performance of initially dilute suspensions from tailings discharge or natural processes after a period of several years.

Zone A which is sometimes defined as the top, water layer, is the supernatant water layer. This represents the water released in the immediate settling of the discharge tailings slurry plus the accumulation of the water derived from the sedimentation/consolidation processes occurring in the lower portions of the containment pond. Below Zone A, the solids concentration appears to increase to the point where a somewhat constant solids concentration is obtained. The transition solids concentration zone is identified as Zone B whilst the apparent "constant" solids concentration zone is labelled as Zone C. This zone is sometimes called the stagnant zone - in recognition of the fact that the rate of solids concentration increase is remarkably low in contrast to that observed in Zone B.

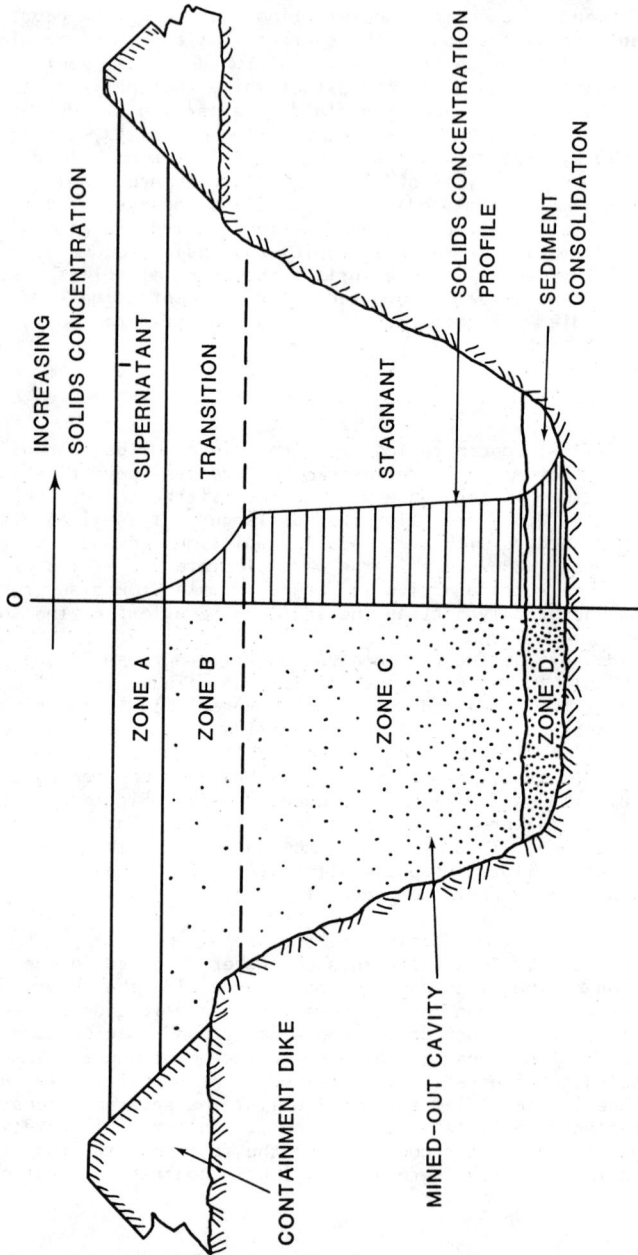

Fig. 1   Ideal representation of settling of suspended solids in a containment pond.

The transition between Zones C and D is not well-established.
Zone D represents the proper sediment layer where consolidation is known
to occur. However, there is considerable evidence available from field
measurements to show that even though the solids concentration in the
lower portion of Zone C might be low, (void ratios of about 5 or less),
some success in the application of large-strain consolidation methods of
analyses can be obtained in the prediction of settling rate. This is
discussed further in the next Section of this paper.

We should note however that the situation shown in Fig. 1 represents
the ideal containment pond in a stagnant situation, and that the pond
filling process is assumed to produce an initial uniform distribution of
solids. The real field situation however is not as easily described.
Figure 2 for example shows the development of a typical pond where the
tailings slurry is discharged from the "upper" end of the pond. This
idealized sketch represents the ongoing discharge occurring in several
operations presently under study. Relying on the characteristics of the
settling material in combination with natural solar drying processes
(where applicable), a kind of inclined layering effect is generally
achieved. The coarser particles or solids settle out in the near end
of the pond whilst the finer fractions will get "transported" to the
further extremes of the pond. In effect therefore, a segregative effect
is developed.

The three sections identified in Fig. 2 show an attempt in broadly
classifying the apparent segregative effect. Section I shows the delta
layering effect. Present experience in the ongoing two studies being
conducted in S.E. Asia and Jamaica shows that with proper sequencing of
tailings discharge, densification of the layers through partial drying
of the tailings discharge layers can be achieved. By moving the discharge pipe from one location to another to allow for this drying process to occur, a quasi-stable beach can be formed. Drilling and sampling through some of these beaches have shown that the solids distribution profile and especially the water content profile are not uniform.
This is because operational procedures generally do not permit the time
required for each tailings discharge layer to fully dry before accepting
the next load.

Section II in Fig. 2 shows a greater presence of coarser particles
in the suspension in contrast to Section III. The coarser particles
which sediment more readily in Section II and the other solids remaining
in suspension in Section II are obviously influenced by the input conditions at the pipe entry. The settling performance of the suspended
solids in Section III may be assumed to approach the ideal pond containment situation shown in Fig. 1.

Many terms have been used to describe the combination of suspended
solids and water, e.g. "slime", "slurry", "mud", etc. By and large,
these terms have generally been associated with certain industries, processes, or types of discharge. The fundamental phenomenon of suspended
fines or solids however remains the same. For the sake of simplicity in
communication in this presentation, the term "slurry waste" will be used
in discussions concerned with the general solids-suspension phenomenon
associated with tailings discharge. However, where specific terms have
been used in particular industries, e.g. red mud for bauxite slurry

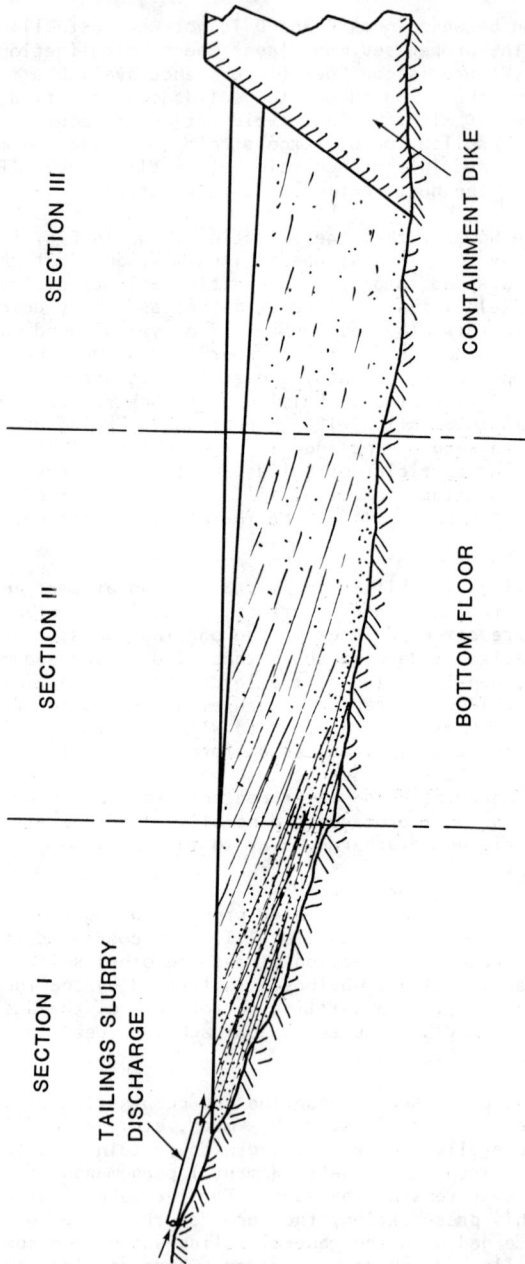

Fig. 2 Actual schematic representation of a typical containment pond showing slurry discharge from the input pipe at left. Note that the pipe can be moved periodically to other locations to build up a "competent" beach.

waste discharge, these will be used at the appropriate times.

INTERACTION CHARACTERISTICS IN DEVELOPMENT
OF DISPERSION STABILITY OF SLURRY WASTES

Since the composition and properties of a specific slurry waste is a direct function of (a) the nature and composition of the host material, (b) the extraction/process variables and techniques used, and (c) the end product requirement, it is apparent that considerable variations in exact compositional characteristics and properties of product slurries will be obtained. The preceding notwithstanding, the fundamental issues concerning the nature of the problem remain somewhat constant, i.e. nature and characteristics of the suspended solids (generally surface-active solids such as clay mineral particles), and chemistry of the suspending fluid. Some examples of typical slurry wastes, such as those considered in this study, include: (a) phosphatic clay slimes, (b) bauxite red mud, (c) tar sand sludge, (d) humic slimes, (e) tin mine slimes, and (f) clay slimes.

The phenomenon of slow self settling of suspended solids which is most likely due to the increased dispersion stability of the solids in suspension can be traced to at least three mechanisms:

(i)   Mechanism 1. Mutual net repulsion caused by the surface-active nature of the solids and the low salt concentrations, or predominance of (a) monovalent cations, or (b) potential determining anions.

(ii)   Mechanism 2. Adsorption or coating of amorphous material (small lyophilic colloid) on a large electronegative colloid. The affinity of amorphous material for water exceeds the net attraction of the van der Waals forces.

(iii)   Mechanism 3. Steric hindrance due to adsorption of organic molecules such as demonstrated in the tar-sand sludge.

In a suspension containing various kinds of constituents, all three mechanisms are expected to be present - with differing degrees of influence - in promoting the dispersion stability of the suspension. It is indeed likely that other mechanisms may exist which are complex variants of the above three mechanisms. These have yet to be fully identified and documented. All these mechanisms contribute not only to the slow self settling rate of the suspended (or dispersed) solids, but also to the suspension volume i.e. stagnant condition in Zone C.

The three mechanisms and complex variants of the three mechanisms are basic contributors to the dispersion stability of suspensions. The surface forces are mainly due to the electrical field developed from the charge balancing cations - generally identified as counter ions. The effect of surface electrical fields on the properties of adsorbed molecules needs to be understood. Approximate calculations for the surface electrical fields suggest that they are strong since a unit positive point charge away from a surface is about 1500 million volts per cm or 15 volts per Angstrom Unit.

In suspensions consisting of lyophobic colloids, such as slurry wastes, their stability (i.e. dispersion stability) depends on the characteristics of particle (solid) interaction - i.e. solid-to-solid interaction in the presence of a fluid medium containing various ionic species. Colloidal theory appears to be most useful in providing the basis for evaluation of interparticle action. The quantitative theory of stability of lyphobic colloids, identified as the DLVO theory (after Derjagin, Landau, Verwey and Overbeek), constructs its analytical model on the basis of (a) electrostatic repulsive forces due to interpenetration of the diffuse ionic layers, and (b) van der Waals attraction forces. Net repulsion occurs between particles (colloids) when double layer repulsion overwhelms van der Waals attraction. The presence of potential determining anions such as bicarbonates, carbonates, hydroxides, phosphates, etc. in the fluid phase, all contribute to the enhancement of net repulsion.

When attractive forces dominate, the system becomes unstable and coagulation occurs. At least two factors are important in the coagulation process: (a) Brownian motion of the constituent particles, and (b) particle interaction. In a system where the repulsive forces are vanishingly small and can be neglected, Brownian collision between particles will lead to agglomeration or aggregation of the particles - thus leading to the production of floc units. Not every collision however will result in aggregation. For two different systems, given the same collision frequency, the effectiveness of aggregation upon collision depends on the properties of the particle surfaces. Under such circumstances, this kind of phenomenon is identified as "slow coagulation".

When the rate of aggregation upon collision no longer depends on the properties of surfaces, and is conditioned only by the collision frequency, the process of "fast coagulation" has been attained. This state occurs when the properties of the surfaces of the constituents in the suspension are ineffective insofar as collision aggregation is concerned.

In solids-suspensions characterized by clay minerals, the zeta potential at any given clay:water ratio depends on the concentration of electrolyte in the suspension. The mono-molecular layer of the electrolyte on the clay surface is formed at a particular concentration. There does exist a strong relationship between the dispersion stability (of the suspension) and the zeta potential of the system, as shown previously by Yong and Sethi (1977) - see Table 1.

In considering the dispersion stability of suspensions consisting of solids and fluids, it is useful to note that the dispersions which are stable due to the presence of strong mutual repulsive forces can be agglomerated or flocculated in one of at least six different ways:

TABLE 1 - Relation between Dispersion Stability of Clays
and Zeta Potential

Stability characteristics	Average zeta potential mv	
1. Maximum agglomeration and precipitation	+3 to	0
2. Excellent agglomeration and precipitation	-1 to	-4
3. Fair agglomeration and precipitation	-5 to	-10
4. Threshold of agglomeration (tactoids or domains)	-11 to	-20
5. Plateau of slight stability (few domains)	-21 to	-30
6. Moderate stability (no domains)	-31 to	-40
7. Good stability, i.e. stable suspension	-41 to	-50
8. Very good stability	-51 to	-60
9. Excellent stability	-61 to	-80
10. Maximum stability	-81 to	-100

1.   Through lowering of the zeta potential of the system to zero with the use of a strong cationic electrolyte.

2.   Through the use of a strong cationic electrolyte in conjunction with an appropriate alkali. (Optimum pH is required).

3.   By adding a reagent which results in the formation of an insoluble matrix which engulfs and binds the water in the system.

4.   Agglomeration through the addition of sufficient cationic polyelectrolytes.

5.   Agglomeration with long chain or branched-chain anionic polyelectrolytes.

6.   Agglomeration with non-ionic long chain or branched-chain polymers.

The protective coating of amorphous material on the surfaces of primary minerals and larger-sized solids [24] may result in strong repulsion between particles, thus rendering stable dispersion conditions. In suspension systems derived with soil particulates, the amorphous material is seen to be composed of silica with or without sesquioxides. It has a strong affinity with water and its properties change on drying.

SETTLING/SEDIMENTATION PROCESS

The simplest procedure for description of the total settling/sedimentation and consolidation process is to trace the "life" or status of a typical representative solid, beginning with its initial state. Following introduction of the tailings slurry, the solids in the slurry will settle in a fashion more or less controlled by either gravitational forces or by interactive forces dictated by surface-active relationships. The concentration of solids at this time is not sufficient to account for proximal hindrances, and is identified in Fig. 1 as Zone A. It is indeed important to stress that the "line" separating any of the Zones shown in Fig. 1, e.g. Zone A from Zone B, is not a line but will be a transition Zone of variable thickness.

When initial settling of the representative solid has proceeded to the stage where neighbouring solids begin to interfere because of their highly active surfaces, a hindered settling performance characteristic becomes evident - shown as Zone B in Fig. 1. Beyond this stage, the settling of the representative solid becomes tediously slow and is apparently hindered not only by the physical interferences of neighbouring solids, but also by interactions controlled by surface active relationships. It is not clear where Zone B ends and where Zone C begins. The kinds of mechanisms operative in these two Zones are not totally different. The specific items which separate them are perhaps better thought of in terms of the more or less dominant solid-to-solid (i.e. interparticle) interactions established via physico-chemical forces.

Zone C can be said to represent the region where even though no physical contact between solids is established - void ratios of 4 to 6 - the physical evidence shows that some small value of excess hydrostatic pressure (pore pressure) can be measured [25]. A form of compression settling of the solids can be said to be occurring at this stage. When the representative solid undergoes further settling to the point where physical contact between adjacent solids is achieved, the consolidation process becomes the dominant mechanism (Zone D).

RECONCILIATION WITH ANALYTICAL MODELS

Table 2 shows the elements of the settling process in relation to representative analytical models. The comments made are not meant to be comprehensive; only the highlights are listed. In viewing the Table, it is obvious that at the present time, total analysis of the problem can only be achieved by breaking up the problem into the four Zones for separate analytical treatment using available classical models - or their variants thereof. This procedure might be expedient, but is by no means satisfactory. The first significant problem that comes to mind is the definition of initial conditions for each particular mode of analysis. Whilst this might not be especially critical in analysis, it is decidedly so if predictions are to be made! Where does one transit from Zone B to Zone C? What are the material output values forthcoming from Zone B - for use as input values for modelling of performance in Zone C?

Obviously, until a unified (continuous) theory can be successfully developed, which trespasses the various boundaries, one will be constrained to work within the limits of the present available theories. The problem obviously begins with the need to fully understand the nature of the interactions producing dispersion stability. As discussed in the previous Section, the surface active nature of the solids and the various interactions producing the three possible mechanisms for the suspended state can indeed provide a very complex system which is not easily portrayed by standard models (or variants thereof) which rely on gravitational principles. Table 3 for example, lists some relationships available for prediction of settling velocity of the solids as a function of their concentration [21] - thought to be applicable to the prediction of solids settling performance in Zones A and B. Surface active relationships however do not appear to have been considered directly - except via "appropriate" constants.

TABLE 2 - Settling and Theory Applications

	Stokesian	Hindered Theories	Consolidation Theories
Zone A			
No measurable pressures. (no excess hydrostatic pressures)	Limited to non surface-active particle suspensions.	Empirical models require extensive laboratory testing for derivation of correction coefficients.	Not applicable because of absence of relationships defining effective stress and material properties.
Zone B			
No measurable pore pressures	Not directly applicable.	Same restrictions as for Zone A	Same restrictions as for Zone A.
Zone C			
Detectable pore pressures	Not applicable	Not applicable	Large strain consolidation phenomenon. Requires accountability for non-linear relationships and self-weight.
Zone D			
Pore pressures and effective stresses.	Not applicable	Not applicable	Classical theories applicable if self-weight is accounted for.

The very common quandry facing analysts and modellers in the study of settling ponds or very soft sediments is the establishment of the point at which effective stress can be considered to be operative, i.e. when pore pressures are measurable in a solids suspension. Experience in matching predictions and actual field values suggest that the models relying on pore pressure development can be satisfactorily used at void ratios of between 4 and 5, and less. This can indeed be puzzling since computations will show that solid-to-solid contact (thought necessary for transfer of intergranular stress) is not established at these high void ratios. To overcome this apparent inconsistency, it is perhaps better to visualize the suspended solids system in terms of the dispersion stability of the suspension.

The dispersion stability of the solids suspension which follows from the fact that surface active solids interact in the fluid medium in the sense of lyophobic colloids, is indicative of the degree to which the solids remain in suspension. This is a direct outcome of the water-holding capability of the solids. At a zero osmotic pressure (i.e. zero

TABLE 3 - Relationship between Solids Concentration
and Settling Velocity - from [21]

Equation	Source
$v = v_o(1 - KC)^{4.65}$	Richardson and Zakie (1954)
$v = v_o(1 - KC)^2 \cdot 10^{-1.82KC}$	Steinour (1944)
$v = v_o 10^{-aKC}$	Thomas (1964)
$v = ac^{-b}$	Cole (1968)
$v = v_o[1 - 2.78(KC)^{2/3}]$	Bond (1960)
$v = v_o c^{-ac}$	Vesilind (1969)
$v = v_o[1 + \frac{3}{4} KC (1 - \frac{8}{KC - 3})]$	Brinkman (1948)
$v = v_o(1 - KC)^a$	Maude and Whitmore (1958)
$v = v_o(1 - aKC)[1 - b(KC)^{1/3}]$	Oliver (1961)
$v = \dfrac{v_o(3 - \frac{9}{2}(KC)^{1/3} + \frac{9}{2}(KC)^{5/3} - 3(KC)^2)}{3 + 2(KC)^{5/3}}$	Happel (1958)

where v = interface velocity of solids concentration, C;
        v_o = Stokes settling velocity for a single discrete particle;
        K = conversion factor, so that KC = volume fraction of solids
             in the slurry; and
    a, b = constants (unique to each equation)

midplane potential), the volume of water associated with each solid or
floc is at its maximum value. Figure 3 shows a mineral particle (solid)
surrounded by its equilibrium shell of water. The thickness of this
shell, or the volume of water, can be computed from theoretical considerations using the DLVO model [26]. The energy-separation distance calculations will give the most likely equilibrium interparticle spacing
and void ratio of the system. Figure 4, which has been idealized into a
parallel particle arrangement to portray the use of cationic and anionic
relationships, shows the development of the attraction and repulsion
forces for an ideal three parallel-particle system. Obviously, real
slurry systems will not have ideal parallel particle arrangements. As
can be seen from the relationships given in the Appendix, the DLVO model
permits one to consider various modes and configurations of particles
(flat plates and spheres) in its computations for equilibrium separation
distances. The relationships shown take into account (a) non-parallel
particle arrangements, (b) mixed minerals in the suspension, and (c)
salt concentrations in the suspending fluid. Calculations for long
range repulsive and attractive energies corresponding to face-to-face,
edge-to-edge and edge-to-face mode of particle or solid interaction in
the presence of various salt concentrations can be made.

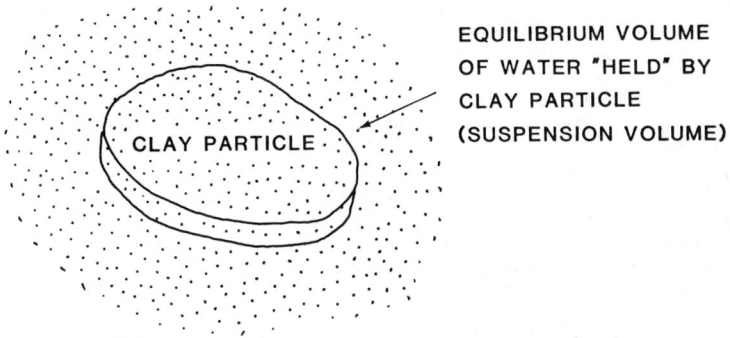

EQUILIBRIUM VOLUME
OF WATER "HELD" BY
CLAY PARTICLE
(SUSPENSION VOLUME)

CLAY PARTICLE

Fig. 3   Schematic representation of a water shell surrounding a clay
particle.

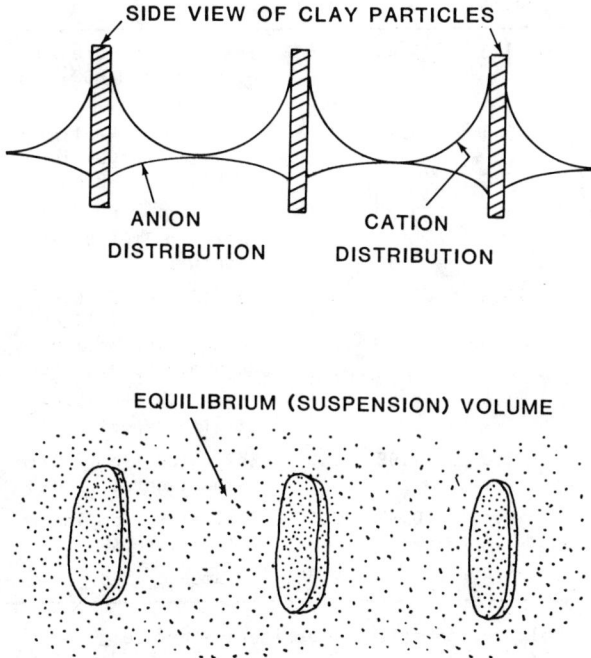

SIDE VIEW OF CLAY PARTICLES

ANION
DISTRIBUTION   CATION
DISTRIBUTION

EQUILIBRIUM (SUSPENSION) VOLUME

Fig. 4   Schematic diagram showing dispersion of particles and contained
water.   Note that particles are shown in parallel only to illustrate the distribution of ions above.   Actual conditions will
show non-parallel particles and flocs.

If each solid, or if each floc "holds" onto a specific volume of water as its "equilibrium" shell of water (defined as the suspension volume) - consistent with the balance of internal and external energies - the total stable dispersed state of the suspended solids at a particular depth in the containment pond, will show that the total volume of water retained in a representative unit volume would be equal to the sum of the suspension volumes of the individual component solids. In the experiments conducted by Yong and Sethi (1978) to determine the equilibrium (suspension) volume of water associated with specific minerals, the values obtained (Table 4) show good correspondence with those computed on the basis of the theoretical interaction model, - provided that proper accounting is given to the presence of the potential determining anions.

TABLE 4 - Suspension Volume for Various Minerals
and Method of their Determination
(adapted from [23])

Sample No.	Mineral	Suspension Volume (cc/g)	Void Ratio	Specific Gravity	Method of Determination
1	Kaolinite	1.3	3.44	2.65	Settling of pure kaolinite in $NaHCO_3$ for 18 months.
2	Illite	3.1	8.15	2.65	Settling of kaolinite: illite (40:60) in $NaHCO_3$ for 18 months.
3	Montmorillonite	21.5	57.0	2.65	Settling of Na-montmorillonite for 36 months.
4	Amorphous Fe_2O_3	20.5	82.0	4.00	Settling of Fe_2O_3+ kaolinite: illite in $NaHCO_3$ for 18 months, containing 4.5% Fe_2O_3.
5	Gibbsite	1.0	2.65	2.65	Estimated from kaolinite experiments.
6	Mica	3.0	7.95	2.65	Estimated from illite experiments.
7	Quartz	0.14	0.37	2.65	Estimated from Theory of Mixtures.
8	Others[*]	0.42	1.12	2.65	Assumed.

[*]Includes feldspar, calcite, ankerite, siderite and pyrite.

As seen from the extreme right hand column of Table 4, the methods for determination of the suspension volume for the various minerals required suspension/settling experiments in long settling columns. These columns were left in a controlled temperature-humidity chamber for periods of from 18 to 36 months. Continuous monitoring of the turbidity of the supernatant via the light transmittance technique [22] ensured that "equilibrium" suspension states were achieved prior to measurement of the solids content in the supernatant. The suspension volume, expressed in terms of cubic centimeters of fluid per gram of mineral solid, has been calculated from the amount of solids remaining in the supernatant. The suspension volumes measured do not require any preferred orientation of solids, and are thus used in comparing theoretically predicted equilibrium solids concentrations with actual measured values for the slurry samples studied.

LABORATORY ANALYSES OF SLURRY WASTES

To further elaborate on the thesis that surface-active solids participate through interparticle action in producing dispersion stability, and hence slow settling of the suspended solids, laboratory analyses of various slurry wastes were performed. Attention was paid to the physical, chemical and mineralogical composition of the slimes. The total types and numbers of tests conducted, by and large depended on the quantity of samples received. The following methods were used for analyses:

1. Mineralogy:
The oriented slide used for the X-ray diffraction test was prepared by depositing 4 ml of slime, diluted to about 1% solids concentration (by weight) onto a glass slide and subsequently air-dried. Horizontal particle orientation is generally achieved with this technique - thus allowing for enhancement of basal reflections and more ready mineral identification. Because the slurry sample contained the suspending fluid, no pre-treatment was given to the sample.

In other mineralogical studies on dried powdered samples, untreated and treated samples are used. Preparation of the samples for analysis required 100 mg of the powdered sample to be dispersed in 10 ml of distilled water, from which 4 ml of the dispersion was deposited on the glass slide. Subsequent to the first X-ray analysis, the slide was treated with glycerol/ethanol mixture and the test conducted again for detection of swelling minerals.

The Siemen's X-ray diffractometer with CuKα radiation was used for all the tests conducted. The percentage of minerals was generally estimated by comparing the peak heights of pure standard reference clay minerals with the peak heights of the minerals found in the test samples. This semi-quantitative technique is by no means accurate, but is used primarily to provide an indication of the approximate proportions of minerals present in the sample.

2. Pore (Suspension) Fluid Analysis:
The suspension fluid was obtained from the slime samples by vacuum filtration or through centrifugation and decantation of the clear supernatant. The fluid was subsequently analyzed for soluble ions and pH. The methods of analysis used includes:

Atomic Absorption Spectrophotometry (Emission): Na^+, K^+, Ca^{2+}

Atomic Absorption Spectrophotometry (Absorption): Mg^{2+}

Titration with H_2SO_4 to colorimetric end point: HCO_3^-, CO_3^{2-}, Cl^-

Turbidimetry: for SO_4^{2-}

Beckman pH meter: for pH

3. Organic Content:

Since only the total amount of organic matter (and not species) need to be determined in the initial phases of all the studies, and since precise methods are indeed too time consuming and complex, the simple approximate method of H_2O_2 washing was adopted. Two to three grams of dried and pulverized slime samples were weighed into a beaker, to which a small amount of distilled water was added to wet the sample. Following placement of the beaker on a hot plate at low heat, 30% H_2O_2 was periodically added until no further bubbling reaction was observed. The sample was then centrifuged and oven-dried and the resultant weight compared with the initial dry weight.

4. Cation Exchange Capacity:

To determine the cation exchange capacity of the suspended solids, 0.25 grams of the dried pulverized slurry sample was weighed into a centrifuge tube and 40 ml of silver nitrate-thiourea solution was added and shaken overnight. The sample was then centrifuged and the supernatant was tested for Ag^+. A Jarrel-Ash Atomsorb Atomic Absorption Flame Emission Spectrophotometer was then used for analysis. The CEC of a sample determined in this manner corresponds to the amount of Ag_3 absorbed on the clay surfaces.

5. Determination of Sand, Silt and Clay Fractions:

Prior to determination of the various size fractions in the slime, 15 meq/l sodium bicarbonate was added to the test sample and the combined solution shaken and placed in an ultrasonic bath. This was left overnight to equilibrate - to ensure dispersion of the suspended solids. Following the "curing" period, the dispersion was then wet-sieved and centrifuged for determination of sand ($>53\mu$), silt ($2 - 53\mu$), and clay ($<2\mu$).

LABORATORY ANALYSES OF SOME SLURRY WASTES

1. Soil-Organic Slime:

The source of the soil-organic slime samples examined in this study was the humates occurring as organic residues cementing sand grains in localized areas in the heavy minerals ore body in Northwest Florida. With the presence of the organic residues, the slimes obtained as the slurry waste product of wet milling processes associated with heavy mineral extraction in that region can be reasonably expected to contain soil-organic acids. In the samples obtained for examination the primary compositional constituents were humic acids, fulvic acids and soil minerals (quartz, feldspar, and chlorite). Since humates are generally composed of the two general classes of soil-organic acids - (a) humic, and (b) fulvic acids - it is therefore to be expected that the slimes obtained from the milling process would consist of these acids.

In preparing the samples used in the study for analysis of humic substances, the techniques used for separation of humic and fulvic acids required prewashing with 0.1 N HCl and water to remove the salts. Since (a) humic acids are soluble in dilute base but not soluble in alcohol

and acid, and since (b) fulvic acids on the other hand are water soluble
and are known to remain in solution after neutralization and after all
humic acid has precipitated out, the separation techniques used followed
procedures which first precipitated the humic acids.

The two types of samples received for analysis represented depth
sensitivity of the slimes - (a) near-surface samples where the solids
concentration was low because of the initial settling of the solids
following deposition of the slime, and (b) samples at depth where equilibrium solids concentration existed. The assignment of infra-red absorption bands in the humic and fulvic acids extracted from the test
samples is shown in Table 5.

TABLE 5 - Assignment of I.R. Absorption Bands in Humic
and Fulvic Acids

Frequency 1/cm	Assignment	Present in
3400	Hydrogen bonded OH	HA, FA*
1725	C = O of CO_2H, C = O stretch of ketonic carbonyl	HA
1630	Aromatic C = C or hydrogen bonded C = O or double bond conjugated with COO	FA
1610	Same as sample 1630 cm^{-1}	HA
1460	Aliphatic C-H	FA
1400	COO⁻, aliphatic C-H	HA
1200	C-O stretch or OH deformation of CO_2H	HA
1020	SiO of silicate impurities	HA

*HA and FA = humic and fulvic acids respectively.

For the samples taken at depth where equilibrium solids concentration was thought to exist, analyses conducted showed a solids concentration of about 19% by weight (and 81% water). The composition of the
solids showed, on the average, 6.7% minerals, 72% humic acid and 30.3%
fulvic acid. The minerals present included quartz, chlorite and field-
spar. In the samples obtained near the surface, the solids concentration
averaged about 2.0%. In the solids traction, the minerals concentration
was about 3.7%. The humic and fulvic acids were, on the average, about
68% and 28.3% respectively.

2. Phosphate Slimes:
    The problems and general properties of the Florida phosphate slimes
have received considerable attention, and have been well reported, e.g.
[5]. The results of studies on the composition of the slime indicate a
high degree of variability in the proportions of the constituent, obviously dependent on source material (host rock), processing, end product

requirements, spatial and temporal variational effects in the recovery of slime sample used for analysis. By and large, the following minerals have been recorded as present in most slimes:

Carbonate-flourapatite	Quartz	Montmorillonite
Attapulgite	Wavellite	Feldspar
Heavy minerals	Dolomite	Kaolinite
Crandallite	Illite	

In addition to the minerals listed above, the samples analyzed in this study showed the presence of amorphous materials. The dominant ions in the suspending fluid included:

Calcium	Magnesium	Sodium
Potassium	Bicarbonate	Sulphate

with pH variation from 7.0 to 8.0.

3.   Aggregate Slimes:
    The production of aggregate slimes arising from wash-extraction of coarse aggregates from weathered granite constitutes a major containment problem. In many regions, because of the scarcity of sources of coarse aggregates, surface layer soils derived from weathering of granite and other types of rock provide a ready supply of granular material - provided that these materials can be separated from the finer fractions (usually clay particles). Wash procedures are commonly used to extract the coarse fractions in gravitational separation processes.

    The aggregate slimes which were examined in this study came from S.E. Asia. Analysis of the physical composition of the slimes showed on the average:

Water	-	84.9% by weight
Organics	-	0.4%
Sand	-	0.1%
Silt	-	1.2%
Clay	-	13.5%

    Analysis of the minerals detected in the X-ray diffractograms using the approximate peak height matching technique described earlier, showed about 54% kaolinite, 35% montmorillonite, and 11% illite. These values are to be considered as approximate values.

    Analysis of the chemistry of the suspension fluid, obtained by centrifuging the slime, showed the presence of Na^+, K^+, Ca^{2+}, Mg^{2+}, and SO_4^{2-}. These were in very small quantities, with Na^+, at 0.5 meq/l being the highest. The suspending fluid pH = 7.5 and specific electrical conductivity was measured as 0.101 millimhos.

    Analysis of the exchangeable cations and cation exchange capacity of the mineral fractions of the suspended particles showed the presence of Na^+, K^+, Ca^{2+}, Mg^{2+}, the latter two being more dominant at 5.4 and 4.3 meq/100gm respectively. Na^+ and K^+ were registered as 2.5 and 1.5 meq/100gm respectively, bringing the total exchangeable cations to be 13.7 meq/100gm. The cation exchange capacity was determined as 31.1 meq/100gm.

As noted from the above, the exchangeable cations were found to be composed of mainly divalent cations. The total exchangeable cations of 13.7 meq/100gm and CEC of 31.1 meq/100gm indicate that only about one half of the exchangeable cations found in the suspended particles are made up of Na^+, K^+, Ca^{2+}, and Mg^{2+}. The remaining could be attributed to H^+ and Al^{3+} - since the pH of the slime was found to be acidic (pH = 4.7).

4.   Tin Mining Slimes:
The tin mining slimes examined in this study are located in areas of limestone formation in S.E. Asia. The tin ore bearing material derived from erosion of the parent rock was deposited in areas of highly weathered limestone formations. The resultant topography at the interface is rather complex, consisting of troughs, channels and pinnacles - typical of deeply weathered limestone. The slime material examined came from gravel-pump mining discharge of the wash material. Because of depositional techniques, sand segregation occurs at the immediate point of pump discharge and resultant finer material slimes are found at the distant points from pipe discharge. This situation is not unlike that represented in Fig. 2.

The solids concentration (weight basis) for the samples studied averaged around 52% - with solids specific gravity of 2.76. Analysis of the typical X-ray diffractograms obtained for the samples studied showed estimated mineral compositional contents as (about) 48% kaolinite, 2% gibbsite, 10% mica, 30% quartz, and 3% other types of solids.

Analysis of the pore water chemistry revealed the presence of Na^+, Ca^{2+}, K^+, and Mg^{2+}, with Ca^{2+}, being the dominant cation at 4.74 meq/l. The settled slime at the solids concentration of 52% does not provide any realistic strength for support of any type of load.

5.   Bauxite Red Mud:
The samples for examination of red mud obtained from the Bayer processing of bauxite for extraction of aluminium came from Jamaica. The suspended solids showed at least 80% of the solids (by weight) smaller than two microns. From an analysis of X-ray diffractograms of the samples, the main minerals appear to consist of hematite, goethite, and anatase, with some Bayer sodalite. Detailed studies conducted [27] using both scanning and transmission electron microscopy coupled with electron diffraction and energy dispersion analyses, show the system to be quite complex - especially in the resolution of whether the minerals are totally crystalline or whether they are indeed surrounded by a degraded amorphous shell. Because standard pure sample slides are not available for the minerals identified, no possible estimate of specific proportions of the minerals has been attempted.

The amorphous content determined using the method previously described by Yong et al. (1979) showed: Fe_2O_3 = about 50 to 60 mg/g of solid fraction, Al_2O_3 = 40 to 60 mg/g, with about less than 1 mg/g of SiO_2.

The pH of the samples varied from about 12.0 to 12.8, depending on the source of sample - primarily depth. By and large the solids concentration by weight also varied with respect to depth location, - from about 30 to 40%.

Analysis of the chemistry of the suspending fluid showed:

Cations	Meq/l	Anions	Meq/l
Na^+	222	HCO_3^-	41.5
K^+	0.58	Cl^-	0.25
Ca^{2+}	0.07	CO_3^{2-}	351
		SO_4^{2-}	11

6.   Clay Coating Slurry:

The clay coating slurries studied came from a clay production industry in the Southeastern part of the U.S. These clays constitute the base
material used in the paper coating industry, and more specifically in
paper machine wet end processes. Because of the requirements of the
paper industry, the slurries studied were fairly consistent insofar as
basic properties and compositional constitutents were concerned.

One of the primary problems encountered in the production of the
material for use in the paper industry is the "economic water content"
of the slurry. By this is meant the maximum amount of water that can be
extracted from the slurries before shipment to the paper industry so that
minimum bulk and weight can be attained. Excessive energy requirements
to remove water that is "held" to the clay material must be considered
in balance with production costs. Thus the samples received for study
represented the solids-suspension equilibrium volume state of interest.

Analysis of the chemistry of the suspending fluid showed the pre-
sence of CO_3^{2-}, HCO_3^-, Cl^-, SO_4^{2-}, Na^+, Mg^{2+}, and Ca^{2+}. Na^+, and SO_4^{2-}, were
the largest amount of ions present at 22.30 and 18.0 meq/l respectively.
The concentration of HCO_3^- and 5.0 meq/l and the remaining ions showed
concentrations less than 1 meq/l. The specific electrical conductivity
was 2.12 mmhos/cm and the pH was 8.69.

The minerals identified in the samples were illite, montmorillonite,
mixed layer material, chlorite, and quartz. The proportions of the minerals which were estimated again on the basis of peak ratios indicated:
(a) montmorillonite, 6.7%, (b) illite, 5.2%, (c) chlorite, 2.2%, (d)
mixed layer material, 2.1%, and (e) quartz, 83.8%. The solids concentration by weight was 35.5% (water = 64.5%), and the total mineral content in the solids was 95.8%. The remaining 4.2% of the solids was organics.

7.   Beneficiation Slurry:

The coal beneficiation slurry studied which came from Western Canada
was the product of procedures implemented to clean the clay coatings from
the ore deposits. The solids concentration of the slurry was 8.7% by
weight. The slurry remained in suspension for an interminable period
without apparent settling out of the solids. The estimated proportions
of the minerals present, determined from the X-ray analysis showed:

(a) montmorillonite, 42%, (b) illite, 21%, (c) feldspar, 14%, (d) kaolinite, 13%, and (e) chlorite, 10%.

Analysis of the suspending fluid showed the presence of CO_3^{2-}, HCO_3^-, Cl^-, SO_4^{2-}, Na^+, Ca^{2+}, K^+, Mg^{2+}. The dominant ions were SO_4^{2-} and Na^+, at 32.5 and 21.1 meq/l respectively. The specific electrical conductivity was 2.27 mmhos/cm whilst the pH was measured as 10.87.

8.   Tar Sands Sludge:
The analyses for the sludge obtained from the hot water processing of tar sands in the Western part of Canada for recovery of bitumen have been reported previously by Yong and Sethi (1978). The typical composition of the sludge material is shown in Table 6 whilst the basic characteristics and mineralogical analysis of some typical samples are shown in Table 7.

The material is seen to be composed of many types of minerals, with the major minerals identified as kaolinte, illite and some lesser fraction of montmorillonite. The presence of amorphous materials should also be noted.

SUSPENSION VOLUMES AND SOLIDS CONCENTRATIONS

Using the values shown in Table 4, the theoretical suspension volumes (of water) associated with the individual components can be summed to estimate the "as-is" solids equilibrium volumetric water content. These values have been used to compute the theoretical solids concentration and compared with the actual values of solids concentration de determined for the samples obtained from the ponds studied. To illustrate the computational procedure, the procedure used in computing the values for the tin mine sample from S.E. Asia is cited as follows:

Mineral	Suspension Volume cc/gram (from Table 4)	Specific Volume in Sample cc/100 gram
Kaolinite - 45%	1.3	58.5
Mica - 10%	3.0	30.0
Gibbsite - 2%	1.0	2.0
Quartz - 40%	0.14	5.6
Other - 3%	0.15	0.45
Total = 100% (or 100 gram of solids)		96.55 cc/100 gram of solids

Hence, weight of water = 96.55 gram
Total weight = (100 gram of solids) + (96.55 gram of water)
            = 196.55 gram.

Computed solids concentration = 100/196.55 = 50.8%
Measured solids concentration = 52%

The comparison between measured and computed solids concentration accords well.

TABLE 6 - Typical Composition of Tar Sands Sludge (from [23])

	Sludge		
	20' deep	40' deep	60' deep
Composition, wt. %			
Bitumen	3.3	3.9	4.5
Water	74.7	70.1	65.5
Mineral	22.0	26.0	30.0
Particle Size, wt. % Mineral			
Sand (> 44μ)	~0	~0	~0
Silt (2 - 44μ)	59	62	58.5
Clay (< 2μ)	41	38	41.5
Clay/Water	0.12	0.14	0.19

TABLE 7 - Basic Characteristics and Mineralogical Analysis of some Typical Samples of Tar Sands Sludge (from [23])

Depth (ft)	Particle Size Distribution			Sludge Composition			Percent Minerals in Bitumen-Free Sludge Solids											
	Sand % (> 44μ)	Silt % (2 - 44μ)	Clay % (< 2μ)	Mineral %	Water %	Bitumen %	Kaolinite	Illite	Mont-morillonite	Chlorite	Illite-Mont-morillonite Intergrade (70:30)	Quartz	Feldspar	Calcite	Siderite	Ankerite	Pyrite	Amorphous Fe_2O_3
20	1.9	50.1	48.0	24.6	73.2	2.2	54.4	8.4	1.1	0.7	1.8	24.8	1.1	0.9	2.6	0.0	0.0	3.9
45	13.5	51.6	34.9	33.0	62.5	4.5	42.8	8.4	1.1	0.6	1.1	38.5	1.1	0.8	2.5	0.0	0.0	3.5
60	44.6	31.3	24.1	40.8	56.4	2.8	27.2	5.8	0.9	0.7	1.0	59.1	0.9	0.7	1.4	0.0	0.1	2.2
70	47.5	34.5	18.0	47.5	51.1	1.4	25.1	4.2	0.5	1.1	0.6	63.8	0.8	0.6	1.2	0.0	0.0	2.1

Using this same procedure, the computed solids concentrations for the various types of samples examined, and the measured values are reported in Table 8. It must be stressed that both the theoretically computed and measured values must be considered as "average representative" values since: (a) the method used for determination of the proportion of various minerals from X-ray diffractograms, (b) sampling procedures, (c) quality of samples and how well they represent the actual conditions of the material in the ponds, are indeed pertinent and significant considerations. The preceding notwithstanding, it is noted that the theoretically computed values for solids concentrations do accord well with the measured values, - except for the aggregate slime - as seen in the extreme right hand column in Table 8.

TABLE 8 - Comparison of Theoretically Computed and Measured Solids Concentration (S.C.) for Samples Studied

Source	Theoretical S.C. %	Measured S.C. %	Theo. S.C. Meas. S.C. %
Phosphatic Slime	13.4	14	0.96
Aggregate Slime	10.5	14.3	0.71
Tin Mining Slime	50.8	52	0.98
Clay Coating Slurry	51.8	52.7	0.93
Beneficiation Slurry	9.1	8.7	1.05
Tar Sands Sludge	42.2	41.9	1.01
Bauxite Red Mud	Not computed	20-40	
Soil-Organic Slime	Not computed	2-19	

Note that since the equilibrium suspension volumes have not been determined for the humic and fulvic acids in the soil-organic slimes, and also for the various components in the red mud, no predicted values of solids concentrations have been obtained for these two types of waste slurries at this time. Much work remains to be done to assess (a) the apparent high dispersion stability of the soil-organic slimes, and (b) the theoretical and experimental values of suspension volumes for the humic and fulvic acids, and the red mud components.

In regard to the humic and fulvic acids which are generally referred to as humic substances, it is known that they are polycarboxylic acids with phenolic, alcoholic, and carbonyl groups, and aromatic rings. The structures for the humic and fulvic acids show properties for metal chelation and base exchange. They have a high affinity for protein and a capability for adsorption of many materials.

CONCLUDING REMARKS

From the preceding discussions, it is apparent that the nature of the suspension, at least for Zones A and B appears to be accountable in part by the water-holding capability of the solids - as a physical measure of the interaction of the solids. The initial settling process is clearly dominated by surface-active relationship, and sedimentation therefore is not readily described by simple gravitational relationships.

Theories which rely on Stokes' principles and solids concentration, which have been generally used to model the settling process in these two Zones [A and B] have not necessarily been successful because of the degree of surface activity (dispersion stability) of the solids. Corrections using a hindrance concept to augment these theories have also generally been unsuccessful if the dispersion stability of the suspension is high. The inability of the sedimentation models to account for the surface and physico-chemical interparticle actions has by and large been responsible for the less than accurate predictive capability of these models.

Insofar as the settling stages where effective stresses are opera-tive, models relying on the general principles of consolidation appear to be capable in predicting settling performance. The apparent lack of physical particle (solid) contact at void ratios of about 5, (for initial consideration of consolidation) where successful predictions have been made (e.g. [25]) does not necessarily mean the absence of "effective" stresses. The measurements made from field experiments show that for the settling ponds tested by Yong et al. (1984), measurable values of pore pressures were obtained. If "effective" stress is defined as the differ-ence between total stress and measured pore pressure, (disregarding the lack of computed physical contact from a void ratio value of 5 or more), application of the consolidation model appears as an acceptable procedure.

The mechanisms describing surface active relationships and inter-actions permit modelling for evaluation of the so-called equilibrium status of the suspended solids - i.e. the reasons for the amount of solids remaining in suspension. Modelling of the settling rate of the solids can take several forms - collision theory, relative fluxes for description of convection-diffusion phenomena, and consolidation theory. Since a continuous theory (i.e. unified model) does not as yet exist which would cover the entire spectrum of the settling phenomenon, it is necessary to consider appropriate submodels which would overlap in their predictions of the settling problem over its entire range. Two such theories are available at the present time: (a) the convection-diffusion model which encompasses solids settling performance in Zones B and C [28, 29], and (b) large-strain consolidation which encompasses the lower part of Zone C and all of Zone D, [7, 14, 25].

In making predictions of settling performance and comparisons be-tween predicted solids concentration profile and measured values, we should note that experimental procedures require that the suspensions studied in the laboratory for characterization of the void ratio and permeability or void ratio and effective stress relationships must realistically model the field conditions. By and large, long column settling tests need to be performed if the proper modelling of settling behaviour is to be obtained. Consolidation testing, using long columns, will not necessarily produce the kinds of information needed since the external stress situation can affect the self-settling characteristics of the suspended solids. This is particularly true if dilute solids-suspensions are to be examined.

Finally, it should again be noted that analytical modelling of settling of suspended solids in sedimentation processes for prediction of time-rate of settling of suspended solids, is complicated by several physical issues related to:

(a)   non-gravitational interactions - i.e. surface-active phenomena,
(b)   complications in laboratory measurements of properties of suspensions, and
(c)   defining initial conditions and effective stress-void ratio relationships.

It is important to recognize the limits of applicability of the total settling submodels in order that proper predictions of the total settling phenomenon can be made. Application of the concept of suspension
volumes in the development of dispersion stability of the solids-suspension, explains why the suspended solids appear reluctant to settle in
the time frame predicted by gravitational settling theories. It should
be noted that whilst the DLVO model, together with the relationships
presented in the Appendix, can be used to calculate the theoretical suspension volumes associated with the kinds of surface-active solids present in the system, the experimentally measured values of suspension
volume given in Table 4 are indeed more appropriate. This is because
the experimentally measured values implicitly account for the various
particle (solids) configurations in the suspension, and also the likely
presence of floc formations and interactions - without the need for
tedious calculations of interactions involving geometries and floc formation.

Much more work remains to be done in the study of dispersion stability - especially in regard to the magnitude and potential for floc formation in suspension. The results shown in Table 8 for the aggregate
slimes suggest very strongly that floc formation is indeed much more prevalent than anticipated - as witness the higher measured solids concentration in contrast to the lower computed values. Other work that needs
to be done concerns the presence of micro-organisms in contained slurries.
Development of gaseous products as a result of activities associated
with micro-organisms will undoubtedly affect both dispersion stability
and settling performance of the suspended solids.

ACKNOWLEDGEMENTS

This study constitutes part of the general study on Stability and
Settling of Suspended Fines, supported by a Grant-in-Aid of research
from the Natural Sciences and Engineering Research Council of Canada,
NSERC, under Grant No. A-882. The work concerned with S.E. Asia and
Jamaica is supported by the International Development and Research Centre,
Canada. The support from both Organizations is gratefully acknowledged.
The input provided by the various researchers in the Geotechnical
Research Centre, McGill University, is also acknowledged.

APPENDIX 1.

NET ENERGIES OF INTERACTION

Face-to-Face Interaction

Long range repulsive energies due to diffuse ion cloud interaction for constant surface charge density [18]:

$$V_R = \frac{64 \ n \ kT}{K} \ \gamma^2 \ \exp(-2 \ Kd) \tag{1a}$$

where

$$\gamma = \frac{e^{z/2} + 1}{e^{z/2} - 1} \tag{1b}$$

$$\sinh z = \sigma \left(\frac{\pi}{2n \ \varepsilon \ kT}\right)^{1/2} \tag{1c}$$

$$z = \frac{\nu \ e \ \Psi_o}{kT} \tag{1d}$$

d = interparticle half distance

K = Debye-Huckel reciprocal length

$$K = \left(\frac{8\pi \ ne^2 \nu^2}{\varepsilon kT}\right)^{1/2} \tag{2}$$

n = molarity x Avagadro's number
e = unit electronic charge
ν = effective cation valence
k = Boltzman constant
T = temperature (°K)
ε = dielectricity constant

Long range repulsive energies - for constant surface potential [19]:

$$V_R = \frac{4n \ kT \ z^2 \ \exp(-Kd)}{K[1+\exp(-2Kd)]} \tag{3}$$

Attractive long range interaction energies:

$$V_A = \frac{-A}{12\pi} \left[\frac{1}{(2d)^2} + \frac{1}{(2d+2\delta)^2} - \frac{1}{(2d+\delta)^2}\right] \tag{4}$$

A = Hamaker constant
δ = thickness of interacting particles

Edge-to-Face Interaction

The relationships developed for edge-to-face and edge-to-edge interactions make use of approximations in terms of spherical particle interactions.

Long range repulsive energies:

$$V_R = \varepsilon R_1 R_2 \left[\frac{(\psi_1^2 + \psi_2^2)}{4(R_1+R_2)} \left(\frac{2\psi_1\psi_2}{\psi_1^2+\psi_2^2} \cdot \ell n \frac{1+\exp(-2Kd)}{1-\exp(-2Kd)} + \ell n \; (1-\exp(-4Kd)) \right] \right. \quad (5)$$

ε = dielectric constant of the medium
R_1, R_2 = radii of interacting spheres
ψ_1, ψ_2 = surface charge potential of interacting spheres

Long range attractive energies [8]:

$$V_A = \frac{-A}{6} \left[\frac{2R_1 R_2}{x} + \frac{2R_1 R_2}{x+4R_1 R_2} + \ell n \left(\frac{x}{x+4R_1 R_2} \right) \right] \quad (6)$$

where $x = 4d^2 + 4d(R_1 + R_2)$

Edge-to-Edge Interaction

Long range repulsive energies - for constant surface potential [10]:

$$V_R = \frac{\varepsilon R \psi_o^2}{2} \ell n [1 + \exp(-2Kd)] \quad (7)$$

Long range attractive energies [8]:

$$V_A = \frac{-A}{6} \left[\frac{2R^2}{x} + \frac{2R^2}{x+4R^2} + \ell n \left(\frac{x}{x+4R^2} \right) \right] \quad (8)$$

where $R_1 = R_2 = R$

$x = 4d^2 + 8Rd$

PARTICLE INTERACTION FORCES

Long range attractive forces (van der Waals forces)

Face-to-Face Interaction

$$F_a = \frac{\pi^2}{240} \cdot \frac{hc}{(2d)^4} \cdot \frac{1}{(D_{30})^{1/2}} \cdot \frac{D_{10}-D_{30}}{D_{10}+D_{30}} \cdot \phi \left(\frac{D_{10}}{D_{30}} \right) \quad (9a)$$

where

D_{10} = static dielectric constant of clay particle
D_{30} = static dielectric constant of pore fluid
h = Planck's constant
c = velocity of light

Edge-to-Edge Interaction

$$F_a^{ee} = \frac{A}{12} \{16R^2 (d+R) [\frac{1}{(4d^2+8Rd)^2} + \frac{1}{4d^2+8Rd+4R^2} + \frac{2}{(4d^2+8Rd)^2+4R^2(4d^2+8Rd)}]\}$$

(9b)

R = the effective radius of each interacting particle edge.

Edge-to-Face Interaction

$$F_a^{ef} = \frac{A}{12} \{2R_1R_2 (8d + 4(R_1+R_2)) [\frac{1}{z(d)^2} \cdot \frac{1}{(z(d)+4R_1R_2} \cdot \frac{2}{z(d)^2+4R_1R_2z(d)}]$$

(9c)

$$z(d) = 4d^2 + 4d (R_1 + R_2)$$
R_1, R_2 = the effective radii of the interacting particle faces and edges.

Long range repulsive forces

Face-to-Face Interaction

The repulsive osmotic pressure P_{osm} [11]:

$$P_{osm} = 2n_o kT(\cosh y_c - 1)$$

(10a)

where for constant surface charge, y_c is given by

$$y_c = 2 n \frac{\pi}{0.32 \, \sigma \, \bar{c}_c \, d}$$

(10b)

and midplane electrical potential can be calculated by

$$K (d+x_o) = 2 \exp(\frac{-y_c}{2}) \int_{\phi=0}^{\phi=x/2} \frac{\phi d}{[1-\exp(-2y_c \sin^2\phi)]^{1/2}}$$

x_o = stern layer thickness

K = Debye-Huckel reciprocal length

For mixed mono-divalent ion system [3]:

$$P_{osm} = B [R(e^{y_c} - 1) + e^{2y_c} - 1 + (R + 2) e^{-y_c} - 1)]$$

(11)

$$B = k \, L_A \, T \, C_o /1000 \, (R+1)$$
$$R = c^+/c^{++}$$

Edge-to-Edge Interaction

$$F_R^{ee} = \frac{K \varepsilon R \Psi_o^2}{2} \cdot \frac{\exp(-2Kd)}{1+\exp(-2Kd)} \tag{12}$$

Edge-to-Face Interaction

$$F_R^{ef} = \frac{K \varepsilon R_1 R_2 (\Psi_1^2 + \Psi_2^2)}{2(R_1 + R_2)}[\frac{1}{1-\exp(-4Kd)} \{\frac{2\Psi_1\Psi_2}{\Psi_1^2+\Psi_2^2} \exp(-2Kd) -\exp(-4Kd)\}] \tag{13}$$

Interparticle Spacing

$$\overset{o}{(\text{Angstroms})} \frac{d}{} = \frac{10,000\ e}{G_s\ SA} \tag{14}$$

SA = specific surface area $(m^2\ gm^{-1})$

One-dimensional Theoretical Free Swell

$$\varepsilon_a = \frac{G_s \cdot \Delta_d\ (SA)}{100\ (1+e_o)} \tag{15}$$

G_s = specific gravity of soil

e_o = initial void ratio of soil

Δd = change in particle separation half distance.

Appendix 11.-REFERENCES

1. Been, K. and Sills, G. C. "Self-weight consolidation of soft soils:
    an experimental and theoretical study".Geotechnique, 31(4) 1981,
    pp. 519-535.
2. Bond, A. W. "Behaviour of suspension". Journal of Sanitary Engineer-
    ing Division, ASCE, Vol.86, No. SA3, 1960, pp. 57-85.
3. Bresler, E. "Interacting diffuse layers in mixed mono-divalent ionic
    systems". Soil Science Society of America, Proceedings, Vol.36, No.6,
    1972, pp. 891-897.
4. Brinkman, H. C. "On the permeability of media consisting of closely
    packed porous particles". Journal Applied Scientific Research, Vol.1,
    1948, pp. 81-84.
5. Bureau of Mines, "The Florida slimes problem: a review and biblio-
    graphy". U.S. Bureau of Mines, Information Circular No. 8668, 1975,
    41 p.
6. Cole, R. F. "Experimental evaluation of Kynch theory". Ph.D. Thesis,
    University of North Carolina, 1968, Chapell Hill.
7. Gibson, R. E., Schiffman, R. L. and Cargill, K. W. "The theory of
    one-dimensional consolidation of saturated clays. II. Finite non-
    linear consolidation of thick homogeneous layers". Canadian Geotech-
    nical Journal. Vol. 18. 1981, pp. 280-293.
8. Hamaker, H.C. "The London van der Waals attraction between spherical
    particles". Physica, Vol.4, 1937, pp. 1058-1072.
9. Happel, J. "Viscous flow in multi-particle system - slow motion of
    fluids relative to beds of spherical particles". Journal of American
    Institution of Chemical Engineers, Vol.4, No.2, 1958, pp. 197-201.
10. Hogg, R. Healy, T. W. and Fuerstenay, D. W. "Mutual coagulation of
    colloidal dispersions". Transactions Faraday Society, Vol. 62, 1966,
    pp. 1638-1651.
11. Langmuir, I. "Repulsive forces between charged surfaces in water and
    the cause of the Jones-Ray effect". Science, Vol. 88, 1938, pp. 430-
    432.
12. Maude, A. D. and Whitmore, R. L. "A generalised theory of sedimenta-
    tion". British Journal of Applied Physics, Vol.9, 1958, pp. 477-482.
13. Oliver, D. R. "The sedimentation of suspension of closely sized sphe-
    rical particles". Chemical Engineering Science, Vol. 15, 1961, pp.230-
    242.
14. Olson, R. E. and Ladd, C. C. "One-dimensional consolidation problems".
    ASCE Journal of the Geotechnical Engineering Division. 105 (GTI)
    Proceedings Paper 14330. 1979, pp. 11-30.
15. Richardson, J. F. and Zaki, W. N. "Sedimentation and fluidization,
    Part I". Transaction - Institution of Chemical Engineers, Vol. 32,
    1954, pp. 35-53.
16. Steinour, H. H. "Rate of sedimentation". Industrial and Engineering
    Chemistry, Vol. 35, No. 9, 1944, pp. 840-847.
17. Thomas, D. G. "Turbulent disruption of flocs in small particle size
    suspension". Journal of American Institution of Chemical Engineers,
    Vol. 10, No. 4, 1964, pp. 517-523.
18. Van Olphen, H. "An Introduction to Clay Colloid Chemistry". 2nd Ed.
    John Wiley and Sons, N.York, 1977, 308 p.
19. Verwey, E. J. W. and Overbeek, J. "Theory of the Stability of Lyopho-
    bic Colloids". Elsevier Press.
20. Vesilind, P. A. "Discussion of evaluation of activated sludge thick-
    ening theories - by R.I.Dick and B.B.Ewing". Journal of Sanitary
    Engineering Division, ASCE, Vol. 95, SA2, 1969, pp. 333.

21. Vesilind, P. A. "Treatment and Disposal of Waste Water Sludges".
    Ann Arbor Science Publishers. 319 p.
22. Yong, R. N. and Sethi, A. J. "Turbidity and Zeta Potential Measure-
    ments of Clay Dispersibility". ASTM Special Technical Publication
    623, 1977, pp. 419-431.
23. Yong, R. N. and Sethi, A. J. "Mineral particle interaction control
    of tar sand sludge stability". Journal of Canadian Petroleum Tech-
    nology, Vol.17, No.4, 1978, pp. 1-8.
24. Yong, R. N., Sethi, A. J. and LaRochelle, P. "Significance of amor-
    phous material relative to sensitivity in some Champlain Clays".
    Canadian Geotechnical Journal, Vol.16, No.3, 1979, pp. 511-520.
25. Yong, R. N., Siu, S. K. H. and Sheeran, D. E. "On the stability and
    settling of suspended solids in settling ponds. Part I. Piece-wise
    linear consolidation analysis of sediment layer". Canadian Geotech-
    nical Journal, Vol.20, No.4, 1983, pp. 817-826.
26. Yong, R.N., Sadana, M. L. and Gohl, W. B. "A particle interaction
    energy model for assessment of swelling of an expansive soil".
    Proceedings, Fifth International Conference on Expansive Soils,
    Australian Geomechanics Society. Adelaide, 1984. pp. 1-10.
27. Yong, R. N. and Wagh, A. S. "A study of Jamaican bauxite waste -
    Part II". First Annual Report for International Development Research
    Centre, #3-P-82-1016-03, Ottawa, 1984. pp. 1-95.
28. Yong, R. N. and Elmonayeri, D. S. "On the stability and settling of
    suspended solids in settling ponds. Part II. Diffusion analysis of
    initial settling of suspended solids". Canadian Geotechnical Journal,
    Vol. 21, No. 4, 1984a. In Press.
29. Yong, R. N. and Elmonayeri, D. S. "Convection-diffusion analysis
    of sedimentation in initially dilute solids-suspensions". ASCE
    Special Publication Volume SEDIMENTATION/CONSOLIDATION MODELS, 1984b.
    In Press.

General Report: Centrifuge Modelling of Time Dependent Deformations

by

F. C. Townsend, M, ASCE[1]

All the papers assigned to this reporter dealt with the use of centrifugal modelling techniques to validate "consolidation" or "time-deformation" numerical models. By separating these papers, the scope of this report becomes

(a) Centrifugal Modelling of Finite Strain Consolidation Behavior
and (b) Centrifugal Modelling of Finite Element Deformation Behavior

The objectives of this report were to review and comment on these submitted papers, and provide observations for future research.

Centrifugal Modeling of Finite Strain Consolidation

The papers by, (a) Bloomquist and Townsend and (b) Scully, Schiffman, Olsen, and Ko deal with the consolidation characteristics of two different Florida phosphatic waste clays. Both papers relate centrifugal model results to predictions based upon the Gibson, England, Hussey (1967) [GEH] finite strain theory. Bloomquist and Townsend show in their figures 3 and 4 that when a dilute clay slurry ($e \approx 88$) is pumped into a retention pond a settling/sedimentation/consolidation process begins, where a "consolidation" zone gradually increases at the expense of the "hindered settlement" zone until an equilibrium condition is achieved. The simultaneous presence of these two zones affects the time scaling exponent for centrifugal models, as the sedimentation time exponent is 1.0, while that for consolidation is 2.0. Results presented in Bloomquist and Townsend's figure 6 show that this time scaling exponent increases from 1.6 ($e_{ave}=16$) to 2.0 ($e_{ave}=11$). Correspondingly, Scully et al.'s figure 11 presents the time scaling exponent as ≈ 2.0 ($e_{ave}=15$ to 6.5) suggesting that little or no sedimentation occurred for their clay and models. The centrifugal models reported by Mikasa and Takada had average initial void ratios of ≈ 3.5, which is low enough to assure consolidation and presumably a time scaling exponent of 2.0 was used for their results.

The finite strain consolidation theories of Gibson, England, Hussey (1967) and Mikasa (1963) both incorporate self-weight. The former theory was compared by Bloomquist and Townsend and Scully, et al., while Mikasa and Takada compared the latter to centrifugal model tests. Pane and Schiffman (1981) have shown that both theories are quite similar;

---

1 Professor of Civil Engineering, University of Florida, Gainesville.

Editor's footnote: The papers by Mitchell, and Shen, which were reviewed in this Report were not available at time of printing, but have been re-submitted to the ASTM Symposium referred to in the INTRODUCTION, and should appear in the ASTM STP.

the only difference is that Mikasa (1963) assumes a layer instantaneously formed with a constant initial void ratio, while the GEH theory is unrestricted regarding initial conditions. Both theories rely upon an effective stress-void ratio ($\bar{\sigma}$-e) and a permeability-void ratio (k-e) relationship for input. Accordingly, the comparison of these theories with centrifugal models requires (a) the existence of effective stresses; i.e., "consolidation" not "sedimentation," and (b) the e-$\bar{\sigma}$ and e-k relationships.

The water content at zero effective stress has been designated as the "fluid limit" (Monte and Krizek, 1976), and was experimentally determined by Scully, et. al. for their clay as 840% (e=23). This value agreed well with the relationship reported by Ardaman and Associates (1983) for phosphatic waste clays; specifically, e_f = 7.59 + 0.186 LL or e_f = 22.4. Hence, Scully et al. centrifugal model tests had an e_0=15, and found a time scaling exponent of 2.0 (consolidation), while Bloomquist and Townsend's tests did not achieve a time scaling exponent of 2.0 until $e_0 \approx 11$. Yong (1983) reports that excess pore pressures can be measured at void ratios of 6 or less.

The void ratio-effective stress (e-$\bar{\sigma}$) relationship was determined by Scully et al. using settling tests, constant rate of deformation (CRD) consolidation tests and centrifugal model tests. Their results presented in their figure 8 show that consistent compressibility relationships were obtained for void ratios less than 7. For void ratios greater than approximately 9 (effective stress 0.5 kPa) the compressibility relationships are dependent upon the initial water content (void ratio). This observation is consistent with those of Imai (1981) and Been and Sills (1981) who independently observed a lack of a unique compressibility curve for effective stresses lower than approximately 0.1 kPa. Mikasa and Takada obtained their compressibility curve (Fig. 7) from water content profiles after complete consolidation. Bloomquist and Townsend's relationship was from slurry consolidation tests "tempered" with field observation (Lawver, 1982).

The void ratio-permeability relationship was determined by Scully et al. from CRD consolidation tests and flow pump permeability measurements on step loaded consolidation tests. These data presented in their figure 9 show that consistent results were obtained at void ratios less than 7 (effective stresses greater than about 1.0 kPa). Mikasa and Takada obtained their relationship (fig. 9) from Mikasa's equation $v = k(\gamma_0'/\gamma_w)$ for self-weight consolidation under single drainage centrifugal tests.

Bloomquist and Townsend, Scully, et al., and Mikasa and Takada present comparisons between centrifugal and computer models in their figures 10, 12, and 12, respectively. In all cases, the agreement is excellent with two of the three calculated results underpredicting those observed in the centrifuge (theory predicts faster settlement). In figure 8 of Bloomquist and Townsend, an excellent agreement between a controlled field test, centrifuge model, and computer predictions is presented; thereby demonstrating the viability of centrifugal modelling and numerical predictions.

From these papers, this reporter concludes that both centrifugal and finite strain consolidation models provide viable deformation predictions for soft clay deposits. However, users must be aware of the limitations and applicability conditions of these techniques. The time scaling exponent for consolidation in centrifugal modelling is 2.0; however, if "hindered settling" phenomena exists, the exponent is less than 2.0, and "modelling of models" is required to establish the appropriate scaling exponent. Finite strain consolidation is based upon effective stresses. However, what is the threshold void ratio for existence of effective stresses? In addition, void ratio-effective stress-permeability input relationships are non-linear and need to be determined.

Two papers, (a) Shen, Sohn, Mish, and Herrmann, and (b) Mitchell, Liang, and Tse applied centrifugal modelling techniques to validate finite element consolidation models. Shen et al. investigated the settlement and consolidation behavior of a storage tank founded on a soft kaolinite during filling and emptying cycles by centrifugation at 60 g's (see their figure 2). Surface settlement profiles and pore pressure changes at two locations were monitored and served as comparisons for finite element predictions using a version of Dafalias' bounding surface model. The bounding surface model used 18 parameters consisting of classical "critical state" parameters, material constants to describe the bounding surface, the hardening parameter, location of projection point and size of elastic zone. These parameters were determined from (a) standard consolidation tests and (b) triaxial compression and extension tests at several OCR values. Unfortunately, some of these parameters had to be estimated from similar soils and the centrifuge model boundary was assumed to be frictionless. Nevertheless, a good agreement was obtained between the observed and predicted pore pressure changes during loading and unloading of the model storage tank (see their figure 9). The surface deflection measurements compared less favorably as shown in their figure 10, possibly due to the necessary assumptions.

Mitchell et al. investigated the time-dependent pore pressure and deformation (with consideration for creep effects) behavior of a model embankment founded on a soft clay by stage centrifugation at 40 and 80 g's. The model presented in their figure 1 consists of a sand embankment and blanket top drainage layer founded on a doubly drained soft clay layer. Foundation pore pressures were measured with five miniature pressure transducers, while foundation deformations were determined from photographs of a grid of marker beads. The time dependent constitutive model used is an updated version of that by Kavazanjian and Mitchell (1977) and requires 16 soil parameters. Two approaches for generating the creep strain rate tensor were used; (a) the conventional volumetric creep equation for secondary compression ($C\alpha$ scaling) and (b) the Singh-Mitchell (1968) function (SM scaling). Comparisons between centrifugal model and predictions presented in their figures 3a through 3e demonstrated that the creep effect plays an insignificant role in pore pressure response and numerical models with or without creep qualitatively predicted measured pore pressures. Conversely, creep strains significantly affected the time-dependent deformation, with the $C\alpha$ scaling providing better predictions than the SM scaling as presented in their figures 5a through 5d. Again as in the case of pore pressure predictions the constitutive model only qualitatively predicted the total deformations primarily due to underestimation of the immediate deformations.

These two papers, Shen et al. and Mitchell et al., present the laudable combination of centrifuge and finite element models for constitutive model verification. Although predictions were qualitative, due to complexity in boundary conditions, further endeavors of this nature should be encouraged.

References

Ardaman & Associates (1983), "Evaluation of Phosphatic Clay Disposal and Reclamation Method, Vol. 4: Consolidation Behavior of Phosphatic Clays," Florida Institute of Phosphate Research, Bartow, Florida.

Been, K. and Sills, G. C. (1981), "Self-Weight Consolidation of Soft Soils: An Experimental and Theoretical Study," Geotechnique, Vol. 31, No. 4.

Bloomquist, D. G. and Townsend, F. C. (1984), "Centrifuge Modeling of Phosphatic Clay Consolidation" ASCE Symposium on Sedimentation/Consolidation, San Francisco.

Gibson, R. E., England, G. L. and Husey, M. H. L. (1967), "The Theory of One-Dimensional Consolidation of Saturated Clays, 1. Finite Non-Linear Consolidation of Thin Homogeneous Layers," Geotechnique Vol 17, p. 261.

Imai, G. (1981), "Experimental Studies on Sedimentation Mechanism and Sedimentation Formation of Clay Minerals," Soils and Foundations, Vol 21, No. 1, March.

Karazanjian, Jr., E. and Mitchell, J. K. (1977), "A General Stress-Strain-Time Formulation for Soils, Special Session #9, 9 ICSMFE, Tokyo.

Lawver, J. E. (1982) "IMC-Agrico-Mobil Slime Consolidation and Land Reclamation Study," Progress Report #6 IMC, Bartow, Florida.

Mikasa, M. (1963), "Consolidation Theory of Soft Clay - A new Consolidation Theory and its Application," Kajuna Schappankai.

Mikasa, M. and Takada, N. (1984), "Selfweight Consolidation of Very Soft Clay by Centrifuge" ASCE Symposium on Sedimentation/Consolidation, San Francisco.

Mitchell, J. K., Laing, R. Y. K., Tse, E. C. (1984), "Evaluation of a Constitutive Model for Soft Clay Using the Centrifuge" ASCE Symposium on Sedimentation/Consolidation, San Francisco.

Monte, J. L. and Krizek, R. J. (1976), "One-Dimensional Mathematical Model for Large-Strain Consolidation" Geotechnique, Vol. 26, p. 495.

Pane, V. and Schiffman, R. L. (1981), "A Comparison Between Two Theories of Finite Strain Consolidation" Soils and Foundations, Vol. 21, No. 4, December.

Scully, R. W., Schiffman, R. L., Olsen, H. W., and Ko, H. Y. (1984), "Validation of Consolidation Properties of Phosphatic Clay at Very High Ratios" ASCE Symposium on Sedimentation/Consolidation, San Francisco.

Shen, C. K., Sohn, J. and Mish, K., and Hermann, L. R. (1984), "Centrifuge Consolidation Study for Purpose of Plasticity Theory Validation," ASCE Symposium on Sedimentation/Consolidation, San Franciso.

Singh, A. and Mitchell, J. K. (1968), "General Stress-Strain-Time Function for Soils," ASCE Journal, SMFD, Vol. 94, No. SM1.

Yong, R. N., Siu, Sk. K. and Sheeran, D. E. (1983), "On the Stability and Settling of Suspended Solids in Setting Ponds, Part I. Piece-Wise Linear Consolidation Analysis of Sediment Layer," Canadian Geotechnical Journal, Vol. 20, No. 4, November.

Reporter Summary
Sedimentation/Consolidation Models: Theory

By Gustav A. Nystrøm,* M. ASCE

Introduction

Nine papers presented at this symposium deal with theories for modelling sedimentation or consolidation. This summary synthesizes the papers by touching on the wide range of topics. The nine papers making up this group are listed below.

Kavazanjian and Poepsel, Numerical analysis of embankment foundation deformations: two case histories.

Nystrøm, The use of soil mechanics capabilities in a general purpose finite element program.

Soulie and Silvestri, Prediction of consolidation performance on soft clays.

Tanal, Dredge spoil disposal predictions and performance: a case history.

Lechowicz, Szymanski and Wolski, Prediction of consolidation of soils with large secondary compression.

Yong and Elmonayeri, Convection-diffusion analysis of sedimentation in initially dilute solids-suspensions.

Adachi, Oka and Mimura, Interpretation of secondary consolidation by elasto-viscoplastic constitutive equations.

Tamura, Eigenvalue problem of consolidation and its application to problem of sand drain in anisotropic ground.

Takada and Mikasa, Consolidation of multi-layered clay.

Overview

The accompanying table provides an overview of this group of papers. First, it lists a dozen topics ranging from "secondary consolidation" to "bulking factors for dredging". Then, for each paper it identifies which topics are of primary importance and which are of secondary importance. The remainder of this summary will be based on the accompanying table.

Solids Suspensions

The term "solids suspensions" refers to the physical phenomenon whereby solid particles remain in suspension for very long times. Two papers deal with this topic. Yong's paper presents a theoretical formulation which predicts settling rates given

*Research Specialist, Exxon Production Research Co., Houston, Texas.

Importance of topics:

A = primary

B = secondary

	solids suspensions	linear-elastic primary consolidation	nonlinear-elastic primary consolidation	plastic primary consolidation	secondary consolidation	finite strain	multiple geometrical dimensions	new theory	computer capabilities	field validation	bulking factors for dredging
Kavazanjian		B	B	B	B		B		A	A	
Nystrom	B	B	B	B		B	B		A		
Soulie	B	B	B		A		B	A	B	A	
Tanal	B									B	A
Lechowicz	A		B	B	A			A		A	
Yong	A							A			
Adachi			B	B	A		B	A			
Tamura		A					A	A			
Takada			B		A	A		A			

measured diffusion coefficents. In cases where convective effects are small, the governing equation becomes a diffusion-like partial differential equation. Tanal's paper reports on the design and operation of a large dredge spoil disposal facility. Much of the discussion deals with observed bulking factors. They represent an increase in volume that most soils experience when dredged and redeposited.

Primary Consolidation

Primary consolidation refers to volume changes associated with pore water presure dissipation. Seven papers discuss primary consolidation calculations. Of these, four employ plasticity (which differentiates between loading and unloading) while three employ nonlinear elasticity.

The Kavazanjian, Nystrøm, and Adachi papers employ versions of the Cam-clay three dimensional plasticity relations. The Takada paper presents nonlinear elasticity results for multi-layered soil columns. And the Tamura paper presents an elegant eigenvalue solution for a linear elastic axi-symmetric problem.

Secondary Consolidation

Secondary consolidation refers to volume changes which are not associated with pore water dissipation. Thus, it involves continuing deformations while the effective stresses are constant.

Three papers introduce three ways to account for secondary consolidation effects. The Soulie paper uses data from the oedometer test. The Lechowicz paper makes use of rate process theory and laboratory data obtained from Rowe's cells. And the Adachi paper uses an extension of the Cam-clay plasticity model to treat rate effects.

Finite Strain

Most analyses assume that strains are infinitesimally small. Two of the papers in this group account for finite strain effects. The Nystrøm paper uses the continuum mechanics approach. The Takada paper follows the one-dimensional work of Schiffman and Gibson.

Multiple Geometric Dimensions

Four of the nine papers are limited to one-dimensional calculations. Four papers report on multi-dimensional calculations employing the finite element method. And Tamura's paper presents an eigenvalue solution for a drain problem having a simple geometry and constitutive relation.

New Theory and Computer Capabilities

Six of the papers introduce a new theory for treating sedimentation or consolidation. Three of the papers describe new computer capabilities to obtain finite element solutions.

Field Validation

Four of the papers compare theoretical predictions to large-scale field results. The papers by Kavazanjian, Soulie and Lechowicz deal with test embankments. The paper by Tanal deals with a dredge spoil disposal problem.

Closing

The nine papers in this group present a variety of theoretical approaches to treat sedimentation/consolidation problems. It is hoped that this brief summary will provide a framework for reading the papers themselves.

STUDY ON SECONDARY COMPRESSION OF CLAYS

Toshihisa Adachi*, Fusao Oka** and Mamoru Mimura***

The secondary compression of clays was investigated with an elasto-viscoplastic constitutive model. The model showed that a strain rate effect parameter m' is related to the coefficient of secondary compression $\alpha(=dv^P/\ln \cdot t)$ and the relation was verified experimentally by test results of four different clayey soils. The effect of a sample thickness on amount of primary consolidation was also discussed on the basis of the model.

Introdiction

Since clayey soils show remarkable time-dependent behavior such as creep, stress relaxation, the strain rate effect, but secondary compression as well, a constitutive model for the materials should describe these time-dependent mechanical behavior. Assuming normally consolidated clays as a strain-hardening plastic and rate-sensitive material with dilatancy characteristic, Adachi and Okano[4], Oka[11], and Adachi and Oka[3] proposed a constitutive model for normally consolidated clays based on the critical state energy theory[13] and Perzyna's theory of elasto-viscoplasticity[12]. A previous report[3] showed that the coefficient of secondary compression α can be predicted by a strain rate effect parameter m' which is determined from results of undrained strain rate controlled triaxial compression tests conducted at least with two different strain rates.

After showing the constitutive model and how to determine their material parameters, the relation between α and m' will be evaluated by using test results of four different undisturbed clayey soils. The practically important effect of sample thickness on the end of primary consolidation, i.e., secondary compression, will be also investigated on the basis of the constitutive model.

Elasto-Viscoplastic Constitutive Model and Determination of Parameters

Adachi and Oka[3] proposed the following constitutive model for normally consolidated clays by extending the critical state energy theory[13] to explain time-dependent behavior based on Perzyna's theory [12] and empirical evidence.

$$\dot{\varepsilon}_{ij} = \frac{1}{2G} \dot{s}_{ij} + \frac{\kappa}{3(1+e)} \frac{\dot{\sigma}'_m}{\sigma'_m} \delta_{ij} + \frac{1}{M^*\sigma'_m} \Phi(F) \frac{s_{ij}}{\sqrt{2J_2}}$$

*   Professor, Dept. of Transportation Engg., Kyoto University, Yoshida-Hon-machi, Sakyo-ku, Kyoto, Japan.
** Associate Professor, Dept. of Civil Engg., Gifu University, Yanagido, Gifu, Japan.
***Research Associate, Disaster Prevention Research Institute, Kyoto University, Goka-sho, Uji, Kyoto, Japan.

$$+ \frac{1}{3M*\sigma_m'} \Phi(F) [M* - \frac{\sqrt{2J_2}}{\sigma_m'}] \delta_{ij} \tag{1}$$

$$\Phi(F) = c_o \exp[m'\ln(\sigma_{my}'^{(d)}/\sigma_{my}'^{(s)})]$$

$$= c_o \exp[m'(\frac{\sqrt{2J_2}}{M*\sigma_m'} + \ln \frac{\sigma_m'}{\sigma_m'^{(s)}} - \frac{\sqrt{2J_2}^{(s)}}{M*\sigma_m'^{(s)}}] \tag{2}$$

in which ε_{ij} is the strain tensor, s_{ij} is the deviatoric stress tensor, σ_m' is the mean effective stress, $\sqrt{2J_2} = \sqrt{s_{ij}s_{ij}}$ is the second invariant of deviatoric stress, δ_{ij} is Kronecker's delta, G is the elastic shear modulus, κ is the swelling index, e is the void ratio, M* is the value of stress ratio $\sqrt{2J_2}/\sigma_m'$ in critical state, c_o and m' are the parameters representing time-dependent characteristic of materials, $\sigma_{my}'^{(s)}$ is the strain-hardening parameter, $\sigma_{my}'^{(d)}$ is the parameter for both the strain-hardening and the viscoplastic strain rate effect, and the values with the superscript (s) denote the values in static equilibrium state.

The constitutive model has seven material parameters, i.e., G, κ, e, M*, c_o, m' and $\sigma_{my}'^{(s)}$. Of these, the strain-hardening parameter $\sigma_{my}'^{(s)}$ can be determined by the consolidation index λ and the initial strain-hardening parameter $\sigma_{myi}'^{(s)}$ through the next relation.

$$v^p - v_i^p = \frac{\lambda-\kappa}{1+e} \ln[\sigma_{my}'^{(s)}/\sigma_{myi}'^{(s)}] \tag{3}$$

in which v_i^p and $\sigma_{myi}'^{(s)}$ denote the initial values of v^p and $\sigma_{my}'^{(s)}$. Therefore, eight material parameters have to be given to complete the constitutive model.

λ, κ and e can be obtained from consolidation and swelling test results. G and M* can be determined from triaxial compression test results. How to determine the remaining material parameters, c_o, m' and $\sigma_{myi}'^{(s)}$ from strain controlled undrained compression test results will be given under.

Since the volumetric change is negligibly small under an undrained condition, i.e., $\dot{\varepsilon}_{kk} = \dot{v} = 0$, the next relation is obtained from Eq.(1).

$$\dot{\varepsilon}_{kk} = \dot{v} = \dot{v}^e + \dot{v}^p = \frac{\kappa}{1+e} \frac{\dot{\sigma}_m'}{\sigma_m'} + \frac{1}{M*\sigma_m'} \Phi(F)[M* - \frac{\sqrt{2J_2}}{\sigma_m'}] = 0$$

or

$$\frac{\kappa}{1+e} \frac{d\sigma_m'}{\sigma_m'} + dv^p = 0 \tag{4}$$

Integrating Eq.(4) with the condition of $v^p = 0$ at $\sigma_m' = \sigma_{me}'$ gives

$$v^p = - \frac{\kappa}{1+e} \ln(\sigma_m'/\sigma_{me}') \tag{5}$$

in which v^p is the viscoplastic volumetric strain and σ_{me}' is the stress at the end of consolidation. Eq.(5) can be interpreted as follows- when the strain rates are different, the same viscoplastic volumetric strain v^p is possible to take place in the different deviatoric stress states on condition of the same mean effective stress σ_m' and the same initial consolidation pressure σ_{me}'. Fig.1 shows this schematically, namely the viscoplastic volumetric strain v^p is the same in both of

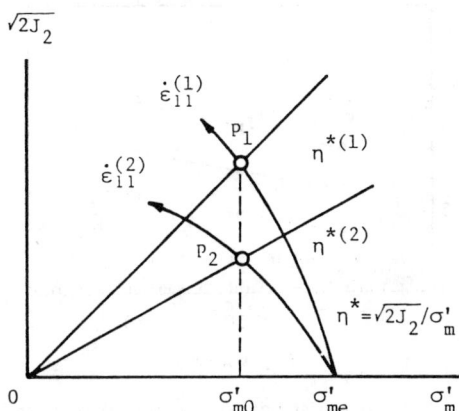

Figure 1. The same viscoplastic volumetric strain
(the same strain-hardening) states on
different effective stress paths

the stress states represented by p_1 and p_2 lying on two different effective stress paths corresponding to two different strain rates $\dot{\varepsilon}_{11}^{(1)}$ and $\dot{\varepsilon}_{11}^{(2)}$. Under an undrained condition, the total strain rate $\dot{\varepsilon}_{ij}$ are the same of deviatoric strain rate $\dot{e}_{ij}$, i.e.,

$$\dot{\varepsilon}_{ij} = \dot{e}_{ij} = \frac{s_{ij}}{2G} + \frac{1}{M*\sigma'_m} \Phi(F)\frac{s_{ij}}{\sqrt{2J_2}} \tag{6}$$

In the case of axisymmetric triaxial compression, the following relations are valid.

$$s_{11} = 2(\sigma_{11} - \sigma_{33})/3 = 2q/3, \quad \sqrt{2J_2} = \sqrt{2/3}(\sigma_{11} - \sigma_{33}) = \sqrt{2/3}q \tag{7}$$

in which $q = (\sigma_{11} - \sigma_{33})$. Using the relations in Eqs.(6) and (2), we obtain the next relations.

$$\dot{\varepsilon}_{11} = \frac{\dot{q}}{3G} + \frac{1}{M\sigma'_m} \Phi(F) \tag{7}$$

$$\Phi(F) = c_0 \exp[m'(\frac{q}{M\sigma'_m} + \ln \frac{\sigma'_m}{\sigma'_m(s)} - \frac{q(s)}{M\sigma'_m(s)})] \tag{8}$$

in which $M = \sqrt{3/2} M*$. Assuming $\dot{\varepsilon}_{11} = \dot{\varepsilon}_{11}^P$, that is the elastic strain rate is negligibly small, the following relation is given from Eqs.(7) and (8) by taking the difference between the strain rates $\dot{\varepsilon}_{11}^{(1)}$ and $\dot{\varepsilon}_{11}^{(2)}$ of the stress states p_1 and p_2 in Fig.1.

$$\ln \frac{\dot{\varepsilon}_{11}^{(1)}}{\dot{\varepsilon}_{11}^{(2)}} = \frac{m'}{M}[\frac{q^{(1)}}{\sigma'_m} - \frac{q^{(2)}}{\sigma'_m}] \tag{9}$$

in which the superscripts (1) and (2) denote the stress states of p_1 and p_2.

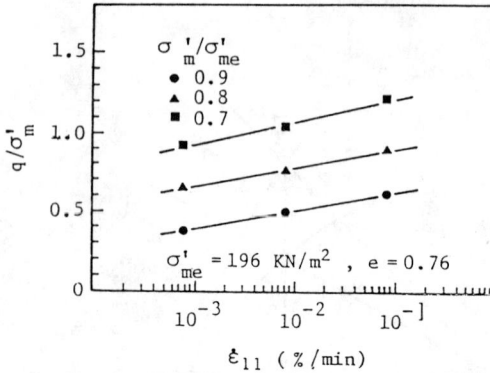

Figure 2. Relationship between stress ratio and
logarithm of strain rate (Kyoto Fujino-
mori-clay)

Fig.2 is an experimental result to evaluate Eq.(9) and shows that
a linear relationship between the logarithm of strain rate $\dot{\varepsilon}_{11}$ and the
stress ratio q/σ_m' is valid as an equi-viscoplastic volumetric strain v^P-
line(equivalency of σ_m'/σ_{me}'-value in the figure to the value of viscoplastic volumetric strain v^P is known in Eq.(5)). The strain rate
effect parameter m' can be determined from the slope of straight lines,
provided the M-value is prescribed.

To determine the remaining parameters c_0 and $\sigma_{myi}'^{(s)}$, only the viscoplastic deviatoric strain rate components $\dot{e}_{ij}^P$ are taken into account.

$$e_{ij}^P = \frac{c_0}{M*\sigma_m'} \exp[m'(\frac{\sqrt{2J_2}}{M*\sigma_m'} + \ln \frac{\sigma_m'}{\sigma_{my}'^{(s)}})] \frac{s_{ij}}{\sqrt{2J_2}} \qquad (10)$$

Paying attention to only newly occured viscoplastic volumetric strain
v^P, namely replacing $v^P - v_i^P$ in Eq.(3) by v^P, the next relation is
given.

$$v^P = \frac{\lambda-\kappa}{1+e} \ln \frac{\sigma_{my}'^{(s)}}{\sigma_{myi}'^{(s)}} \qquad (11)$$

Substituting this relation into Eq.(10) gives

$$e_{ij}^P = \frac{c_0}{M*\sigma_m'} \exp[m'(\frac{\sqrt{2J_2}}{M*\sigma_m'} + \ln \frac{\sigma_m'}{\sigma_{myi}'^{(s)}} - \frac{1+e}{\lambda-\kappa} v^P)] \frac{s_{ij}}{\sqrt{2J_2}}$$

$$= \frac{c_0}{M*\sigma_m'} \exp[- m'\ln \frac{\sigma_{myi}'^{(s)}}{\sigma_{me}'}] \exp[m'(\frac{\sqrt{2J_2}}{M*\sigma_m'} + \ln \frac{\sigma_m'}{\sigma_{me}'} - \frac{1+e}{\lambda-\kappa} v^P)] \frac{s_{ij}}{\sqrt{2J_2}}$$

$$= C \exp[m'(\frac{\sqrt{2J_2}}{M*\sigma_m'} + \ln \frac{\sigma_m'}{\sigma_{me}'} - \frac{1+e}{\lambda-\kappa} v^P)] \frac{s_{ij}}{\sqrt{2J_2}} \qquad (12)$$

in which C is defined by

$$C = \frac{c_o}{M*\sigma'_m} \exp[-m'\ln\frac{\sigma'^{(s)}_{myi}}{\sigma'_{me}}] \tag{13}$$

In the case of undrained axisymmetric triaxial compression, Eq.(12) becomes the next relation by assuming the elastic deviatoric strain rate to be negligibly small.

$$\dot{e}^P_{11} = \dot{\varepsilon}_{11} = \sqrt{2/3}\ C\ \exp[m'(\frac{q}{M\sigma'_m} + \ln\frac{\sigma'_m}{\sigma'_{me}} - \frac{1+e}{\lambda-\kappa}v^P)] \tag{14}$$

Introducing Eq.(5) in Eq.(14) gives

$$\dot{\varepsilon}_{11} = \sqrt{2/3}\ C\ \exp[m'(\frac{q}{M\sigma'_m} - \frac{\lambda(1+e)}{\kappa(\lambda-\kappa)}v^P)] \tag{15}$$

When the value of viscoplastic volumetric strain v^P is prescribed, both values of the strain rate component $\dot{\varepsilon}_{11}$ (term in the left hand side of Eq.(15)) and the stress ratio q/σ'_m (term in the right hand side of Eq.(5)) are determined by the relations experimentally obtained as shown in Fig.2. Substituting thus determined values of v^P, $\dot{\varepsilon}_{11}$ and q/σ'_m into Eq.(5), the parameter C is determined. Since both of the material parameters c_o and $\sigma'^{(s)}_{myi}$ can be expressed by the parameter C, it is sufficient to know the parameter C instead of obtaining individual values for c_o and $\sigma'^{(s)}_{myi}$.

Thus, all material parameters in the constitutive model can be determined from results of consolidation, swelling and strain-rate controlled undrained compression tests.

Interrelation Between Parameter m' And Coefficient Of Secondary Compression α

The secondary compression is one of the well known time-dependent behavior of clays. In the case of isotropic consolidation($s_{ij} = 0$), the viscoplastic volumetric strain rate $\dot{v}^P$ is given from Eqs.(1) and (2) as follows;

$$\dot{v}^P = \dot{\varepsilon}_{kk} = \frac{c_o}{\sigma'_m} \exp[m'\ln\frac{\sigma'_m}{\sigma'^{(s)}_m}] \tag{16}$$

and the next relation is also satisfied.

$$dv^P = \frac{\lambda-\kappa}{1+e}\frac{d\sigma'^{(s)}_m}{\sigma'^{(s)}_m} \tag{17}$$

With conditions of $\sigma'^{(s)}_m = \sigma'_m$ and $v^P = v^P_\infty$ at the end of secondary compression under a pressure σ'_m, Eq.(17) is integrated.

$$v^P_\infty - v^P = \frac{\lambda-\kappa}{1+e}\ln\frac{\sigma'_m}{\sigma'^{(s)}_m} \tag{18}$$

Substituting Eq.(18) into Eq.(16) gives

$$v^P = \frac{c_o}{\sigma'_m}\exp[\frac{1+e}{\lambda-\kappa}m'(v^P_\infty - v^P)]$$

$$= \frac{c_o}{\sigma'_m} \exp[\ \frac{1+e}{\lambda-\kappa} \ m'v^P_\infty] \ \exp[-\ \frac{1+e}{\lambda-\kappa} \ m'v^P]$$

$$= \frac{c_o}{\sigma'_m} \exp[\ \frac{1}{\alpha} \ v^P_\infty] \ \exp[-\ \frac{1}{\alpha} \ v^P]$$

$$= C_1 \exp[-\ \frac{1}{\alpha} \ v^P] \tag{19}$$

in which

$$C_1 = \frac{c_o}{\sigma'_m} \exp[\ \frac{1}{\alpha} \ v^P] \tag{20}$$

$$\alpha = \frac{\lambda-\kappa}{m'(1+e)} \tag{21}$$

Assuming C_1 to take a constant value during the process, Eq.(19) can be integrated with condition of $v^P = v^P_o$ at a time $t = t_o$.

$$v^P = \alpha \ \ln(t/t_o) + v^P_o \tag{22}$$

or

$$\alpha = (v^P - v^P_o)/(\ln t - \ln t_o) = \Delta v^P/\Delta \ln t \tag{23}$$

in which

$$v^P_o = \alpha \ \ln(\alpha/C_1 t_o) \tag{24}$$

Furthermore, in the case of isotropic consolidation, the initial viscoplastic volumetric strain rate $\dot{v}^P_o$ is obtained from Eqs.(13) and (16) as follows;

$$\dot{v}^P_o = C \ M^* \tag{25}$$

As seen in Eq.(23), α is obviously the coefficient of secondary compression and it can be predicted from the parameter m' through Eq.(21).

Experimental Evaluation Of Relation Between Parameters m' and α

In the previous section, it showed that the coefficient of secondary compression α is estimated by knowing the strain rate parameter m' from Eq.(21). To evaluate the relation experimentally, a series of consolidation, swelling and strain controlled undrained compression test was carried out for four different clays, i.e., Osaka alluvial clay, Osaka diluvial clay, Nishinomiya alluvium clay and Nishinomiya diluvial clay. The test results for each clay are given in from Fig.3 to Fig.6, and the material parameters determined from the results are listed in Table 1.

The void ratio e vs. the logarithm of mean effective stress p' relation is shown in (a) of each figure, and the material parameters λ and κ were determined from the e-log p' relation.

The relation between the stress ratio q/p' and the logarithm of axial strain rate is shown in (b), and the rate effect parameters m'

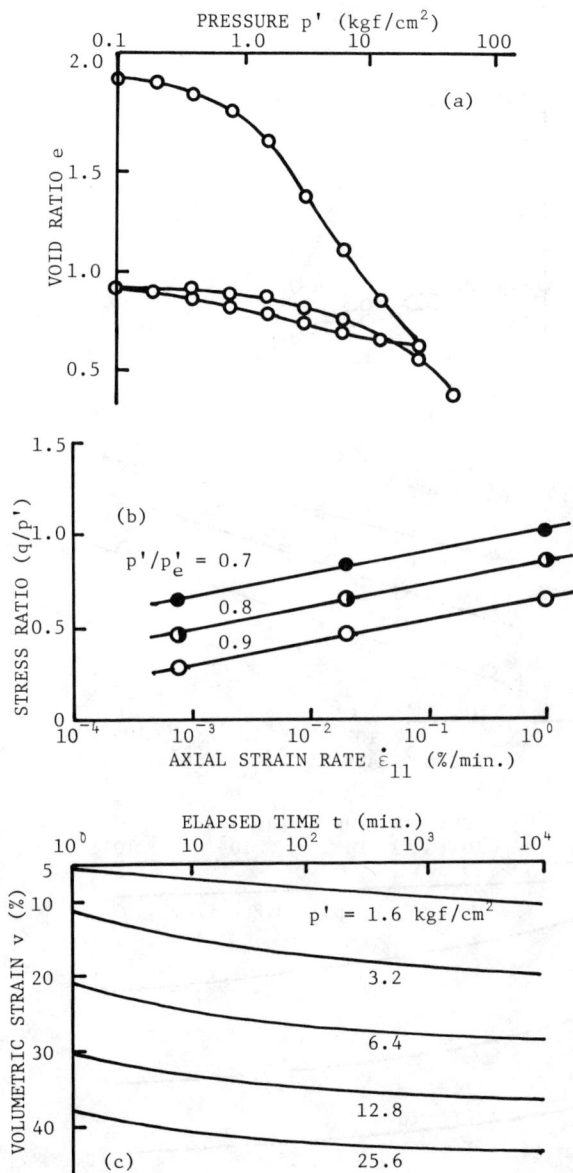

Figure 3. Experimental results of Osaka
          alluvial clay

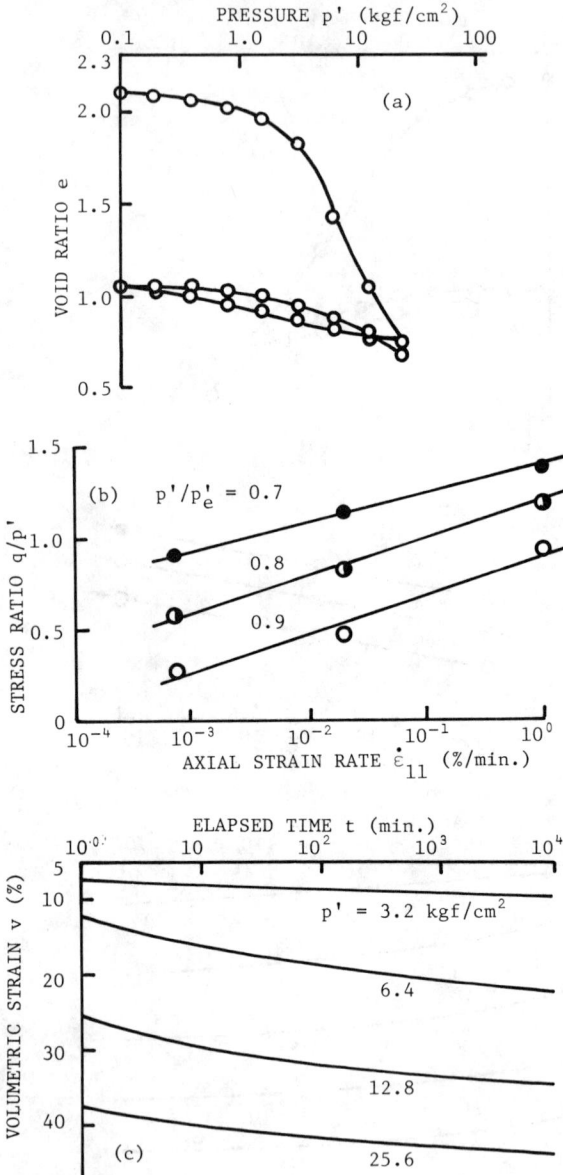

Figure 4. Experimental results of Osaka
             diluvial clay

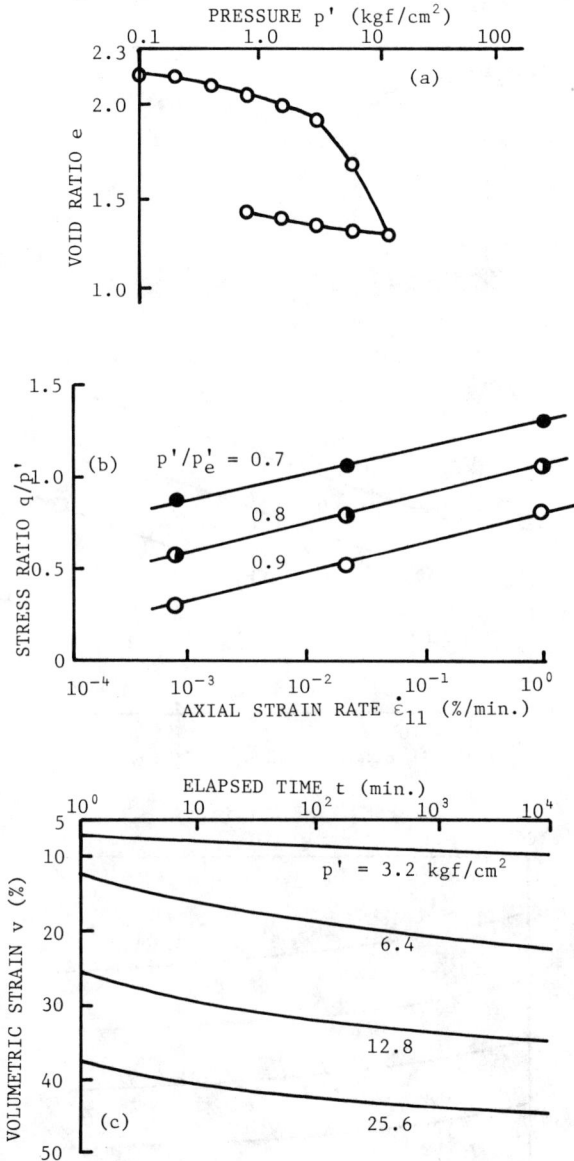

Figure 5. Experimental results of Nishinomia
alluvial clay

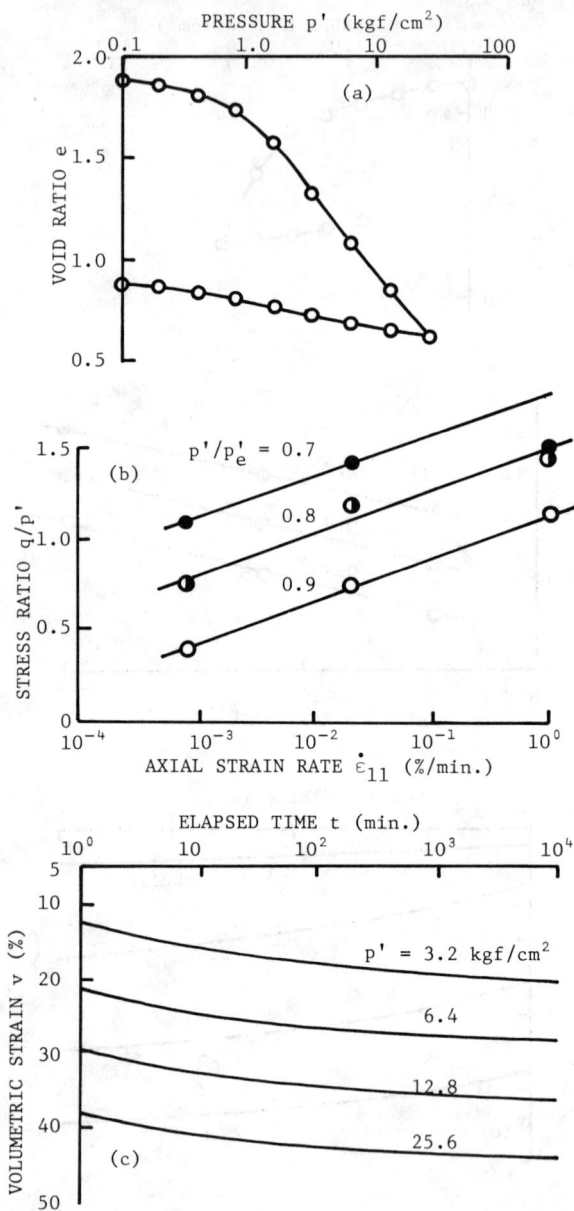

Figure 6. Experimental results of Nishinomiya
dilvial clay

Table 1. Material parameters of clays

	Osaka alluvium clay	osaka diluvial clay	Nishinomiya alluvium clay	Nishinomiya diluvial clay
λ	0.372	0.545	0.344	0.511
κ	0.054	0.058	0.047	0.026
M	1.28	1.32	1.50	1.50
m'	21.49	14.76	15.60	10.48
C	$4.5\ 10^{-8}$	$3.3\ 10^{-8}$	$1.03\ 10^{-7}$	$4.0\ 10^{-8}$
G	132.0	170.0	219.58	160.0
e_0	1.28	1.54	1.78	1.84
α	0.010	0.019	0.0075	0.010
$\alpha = \dfrac{\lambda-\kappa}{(1+e)m'}$	0.006	0.013	0.0068	0.016

and C were obtained from the straight lines given in (b) of each figure by taking the method described in the previous section. The lines in the figures are not rigorously parallel each other. However, m' was determined as an average value by assuming m' to take a constant value for a clay.

The measurement of pore water pressure is essential in this study, because the evaluation was made in terms of effective stress concept. Pore water pressure was measured at the center of bottom pedestal by means of a semiconductor type pressure transducer. The diameter of pressure sensing diaphram of the transducer is 5 mm and the volumetric change of diaphram due to pressure is so small as $1.8 \times 10^{-6} cm^3/kgf/cm^2$ even comparing with the compressibility of water. Pore water pressure has been correctly measured by the pressure transducer when the strain rate was less than 10 %/min.[5].

The volumetric strain and time curves of consolidation are shown in (c) and the coefficient of secondary compression α was determined by paying attention to the final straight portion of curves given in (c) of each figure. In addition to thus determined coefficient, the estimated value by Eq.(21) is also listed in Table 1.

Fig. 7 shows the correlation between both of the coefficients of secondary compression, one obtained directly from consolidation test and the other estimated by Eq.(21). The data are a little scattering, however at least they are in the same order. To collect further data for various clays is concluded to be necessary to verify the relation.

Effect Of Sample Thickness Of Consolidation Phenomena

To examine the effect of a sample thickness of clay on consolidation phenomena, the constitutive equations developed in the previous section were applied to a finite element consolidation analysis of clay strata of different heights. The material parameters used in the analysis are listed in Table 2. The coefficients of permeability k is assumed to be

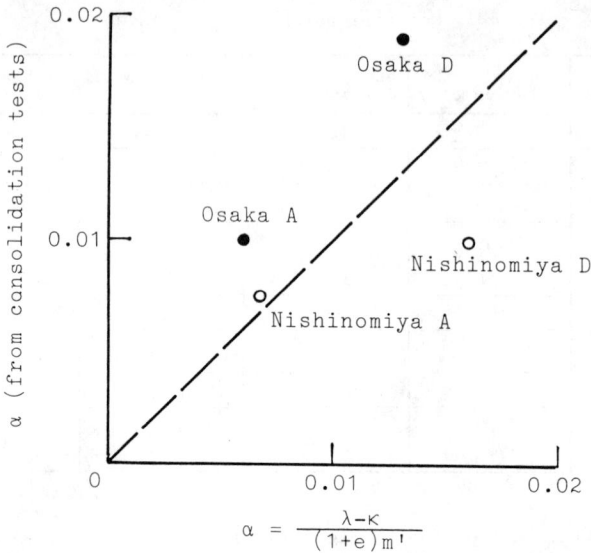

$$\alpha = \frac{\lambda - \kappa}{(1+e)m'}$$

Figure 7. Interrelation between α obtained from tests
and α estimated from m'

Table 2. Material parameters used in analysis

λ	κ	M*	m'	G_0	e_0
0.231	0.05	0.865	25.0	37.5 (kgf/cm²)	1.5

C_k*	k_0				
0.1	1.16×10^{-9}				

* The elastic shear modulus G is proportional to $\sqrt{\sigma'_m}$.
** Subscript (o) denotes the value in initial state

determined by

$$k = k_0 \exp[(e - e_0)/C_k^*] \tag{26}$$

in which k_0 and e_0 are the initial values of k and e, and C_k^* is a
material parameter.

Vertical settlement-time curves for different heights are shown in
Fig.8. The effect of the length of the drainage path has been well in-
vestigated([6], [7], [8], [1] and [14]).

Various hypotheses which explain the scale effect are illustrated
schematically in Fig.9. Curve A is supported by Ladd[9] and by Mesri

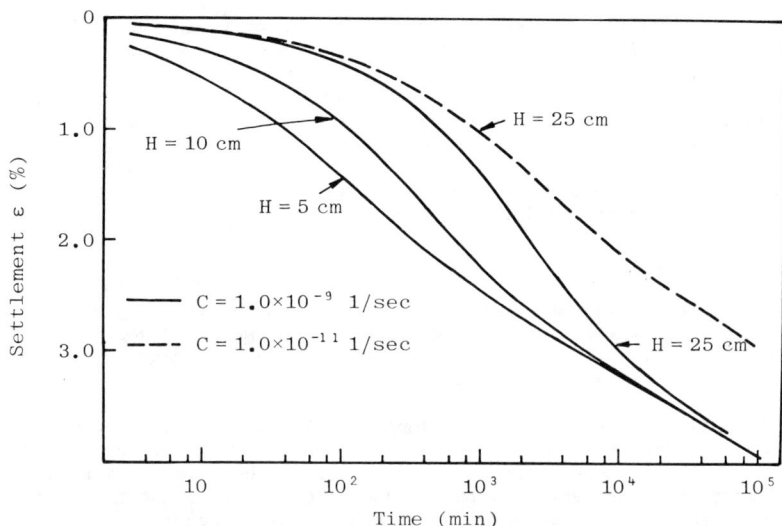

Figure 8. Strain versus time curves

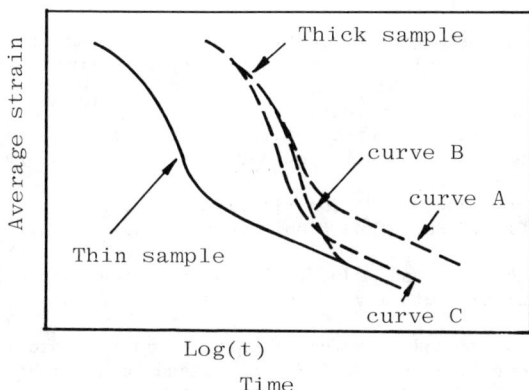

Figure 9. Schematic diagram of strain-time curves

and Rokhsar[10], curve B by Barden[6]. Curve C appears to correspond
to the experimental results of Aboshi[1]. The calculated results denoted
by the solid lines in Fig.8 are nearly equal to curve B. In this case,
the initial viscoplastic strain rates $\dot{v}_o^p$ are equal. In the experimental
study by Aboshi[1], [2], however, the initial strain on the four specimens of different heights are almost equal, but the initial volumetric
strain rate for the thick sample is lower than of the thin sample. As
has been pointed out by Sekiguchi and Toriihara[14], the value of the
initial volumetric strain rate $\dot{v}_o^p$ (= CM*) plays an important role in

consolidation analysis. Predictably, the values of $\dot{v}_o^p$ in thick clay deposits in the field may differ from those of the thin samples used in laboratory tests. If we hypothesize that $\dot{v}_o^p$ is proportional to H^{-2} (H is height of the specimen), the brocken line in Fig.8 is obtained. In this case, the tendency is for the strain-time curve to approach the value of curve A in Fig.9. The calculated results are very sensitive to the value of C as seen in Fig.8. Clearly it is then necessary to investigate the value of C for concluding the effect of a sample thickness of clay on consolidation phenomena.

Conclusions

1. The relation between the coefficient of secondary compression α and the strain rate effect parameter m' was investigated based on experimental results of four clays. The relation can estimate α-value from m' at least in the same order.

2. To investigate the effect of a sample thickness on consolidation phenomena, finite element consolidation analysis was carried out by using the proposed constitutive model. Since the calculated results were remarkably effected by the parameter C, it could not conclude whether the strain-time curve of a thicker sample becomes as curve A, curve B or curve C in Fig.9. Further investigation on the parameter C is required.

References

1. Aboshi, H.,"An Experimental Investigation on the Similitude of Consolidation of Clay Including Secondary Creep Settlement," Proc. 8th ICSMFE, Vol.4, No.3, 1973, p.88.
2. Aboshi, H.,"Personal Communication," 1983.
3. Adachi, T., and Oka, F.,"Constitutive Equations for Normally Consolidated Clay Based on Elasto-Viscoplasticity," Soils and Foundations, Vol.22, No.4, 1982, pp.57-70.
4. Adachi, T., and Okano, M.,"A Constitutive Equations for Normally Consolidated Clay," Soils and Foundations, Vol.14, No.4, 1974, pp.55 -73.
5. Akai, K., Adachi, T., and Ando, N.,"Existence of A Unique Stress-Strain-Time Relation of Clays," Soils and Foundations, Vol.15, No.1, 1975, pp.1-16.
6. Barden, L.,"Time Dependent Deformation of Normally Consolidated Clays and Peats," Journal of the Soil Mechanics and Foundations Division, ASCE, Vol.95, No.SM1, 1969, pp.1-31.
7. Berre, T., and Iversen, K.,"Oedometer Tests with Different Specimen Heights on a Clay Exhibiting Large Secondary Compression," Geotechnique, Vol.22, No.1, 1972, pp.53-70.
8. Garlager, J.F.,"The Consolidation of Soils Exhibiting Creep under Constant Effective Stress," Geotechnique, Vol.22, No.1, 1972, pp.71-78.
9. Ladd, C.C.,"Settlement Analysis for Cohesive Soils," Res. Report R71-2(272), Dept. of Civil Engg., Mass. Inst. of Tech., 1973, p.115.
10. Mesri, G., and Rokhsar, A.,"Theory of Consolidation for Clays," Journal of the Geotechnical Engineering Division, ASCE, Vol.100, No.GT8, 1974, pp.889-904.

11. Oka, F.,"Prediction of Time Dependent Behavior of Clay," Proc. 10th ICSMFE, Vol.1, 1981, pp.215-218.

12. Perzyna, P.,"The Constitutive Equations for Work Hardening and Rate Sensitive Plastic Materials," Proc. of Vibrational Problems, Vol.4, No.3, 1963, pp.281-290.

13. Roscoe, K.H., Schofield, A.N., and Thrairajah, A.,"Yielding of Clays in States Wetter than Critical," Geotechnique, Vol.13, No.3, 1963, pp.212-240.

14. Sekiguchi, H., and Toriihara, M.,"Theory of One-Dimensional Consolidation of Clays with Consideration of Their Rheological Properties," Soils and Foundations, Vol.16, No.1, 1976, pp.27-44.

NUMERICAL ANALYSIS OF TWO EMBANKMENT FOUNDATIONS

Edward Kavazanjian, Jr.[1] M.ASCE
and
Patrick H. Poepsel[2] AM.ASCE

ABSTRACT

Two case histories of the performance of a compressible embankment foundation are analyzed using the Cam-Clay constitutive model and a finite element consolidation program. The I-95 highway embankment test section and the East Atchafalaya Basin flood control levee test section were chosen for analysis because of the wealth of field and laboratory data available for these two cases. Performance criteria evaluated include the magnitude and rate of horizontal and vertical displacements and the generation and dissipation of excess pore pressures. Results of the analysis show generally good agreement between observed and predicted values after soil parameter calibration based upon initial construction deformations. The major discrepancy between observed and predicted results is an underprediction of lateral deformations in the softer, normally consolidated soils and a consequent underprediction of vertical centerline settlement. The discrepancy can be attributed primarily to the effects of undrained creep.

INTRODUCTION

Historically, analyses of the behavior of the saturated clay soils that comprise compressible embankment foundations have usually been limited to either fully drained or undrained analyses. Recent developments in soil mechanics have provided the tools to perform numerical consolidation analyses that "track" the behavior of compressible soils from the initial undrained response through primary consolidation to the final, fully drained state. In this study, the ability of one such numerical model to accurately predict soft clay deformation behavior and pore pressure response is demonstrated.

The performance of the foundations of two embankments built on soft, compressible soils is analyzed herein using a finite element consolidation code based on the Modified Cam-Clay soil model (19). Modified Cam-Clay theory uses an isotropic, elasto-plastic strain hardening model to describe the behavior of compressible clay soils.

---

1   Assistant Professor of Civil Engineering, Terman Engineering Center, Stanford, CA. 94305
2   Graduate Student, Research Assistant, Department of Civil Engineering, Terman Engineering Center, Stanford, CA. 94305

The two case histories selected for analysis, the I-95 test embankment used for the M.I.T. Foundation Deformation Symposium (17) and the East Atchafalaya Basin Flood Control Test Levees (21), provide an excellent basis for evaluation of the capabilities and limitations of the numerical model because both cases have well-documented records of field performance and laboratory soil tests. Only the highlights of the analyses are presented here. Complete details are presented by Poepsel (18).

Foundation response was predicted using the finite element code JFEST developed at Stanford University (7,10). JFEST combines the Modified Cam-Clay model with a coupled deformation-pore pressure consolidation scheme (2) to predict compressible soil behavior. Previous uses of this code were directed at modeling conventional and advanced shield tunneling techniques, including the consolidation deformations around the tunnel and subsequent surface movements.

Performance criteria evaluated in this study include horizontal and vertical displacements and the mobilization and dissipation of excess pore pressures. Numerical predications are compared to the field measurements of these parameters. Results show generally good agreement between observed and predicted values. The primary discrepancy between observed and predicted results appears to be an underestimation of lateral deformation and, consequently, of vertical centerline settlement. This discrepancy can be attributed to the influence of undrained creep upon the deformation behavior of the soft foundation soils.

Finite Element Program

The finite element program used in this study is called JFEST. JFEST, an acronym for 'Johnston-Finno Elasto-plastic Soil-Tunnel Interaction,' is a modification of Johnston's PEPCO (Program for Elasto-Plastic Consolidation). PEPCO and JFEST have been successfully used by the two developers at Stanford University to study consolidation around conventional and advanced shield tunnels in cohesive soils (7,10). The two case histories described herein provide the first opportunity for the applicability of the program to be evaluated for foundation consolidation problems. Details pertaining to the finite element algorithms and governing equations behind the code are given by Johnston (10).

Two of the three material types available in JFEST were used in the analyses described herein. An elasto-plastic Modified Cam-Clay model (22) was used to describe the behavior of the cohesive soils, while the non-linear psuedo-elastic Duncan and Chang hyperbolic model (6) was used to describe the cohesionless materials.

The modified Cam-Clay model uses an elliptical yield surface centered on the hydrostatic axis and an associative flow rule to describe plastic deformations. The parameters describing the yield surface are p'_c, the effective isotropic pre-consolidation pressure, and M, the slope of the critical state line connecting the peak point of the yield ellipse with the origin of a Mohr diagram plotted in terms of the octahedral shear stress, q, and the effective octahedral normal stress, p'. The magnitude of the plastic strains are based upon the slope of the isotropic virgin consolidation curve in a void ratio-ln stress space, λ. Elastic deformations are described by the shear modulus, G, and the slope of the rebound curve in a void ratio-ln. stress space, κ. The other parameter required by the model is the void ratio at an effective normal stress of 1.0 on the critical state line, e_{cs}.

The nonlinear hyperbolic stress–strain model required 8 parameters to describe drained soil behavior. These parameters are the Mohr–Coulomb strength parameters, the ratio of the shear stress at failure to the asymptote of the best fit hyperbola, modulus multipliers for initial loading and unloading/reloading, a modulus exponent, and initial and final values of Poisson's ratio. Details of the model are given by Duncan and Chang (6).

JFEST uses a coupled consolidation–deformation scheme for the Cam-Clay materials (2). Anisotropic permeability may be input to JFEST, if desired, by specifying the horizontal and vertical permeability separately. This feature is particularly advantageous when attempting to simulate the consolidation of natural soils beneath an embankment on soft clays since the horizontal permeability of natural clays can be often 5 to 10 times the vertical permeability.

Of the five element types available in JFEST, only two were used to model the subsurface conditions at the I-95 and Atchafalaya test sites. An eight-noded isoparametric quadrilateral element, termed Q8, was used in the drained analyses of cohesionless soils in this study. An eight-noded isoparametric quadrilateral element with pore pressure degrees of freedom at the corner nodes, termed Q8P4, was used in the consolidation analysis of the soft clays and peat.

THE I-95 TEST EMBANKMENT

Introduction

In November of 1974, a Foundation Deformation Symposium was held at the Massachusetts Institute of Technology in Cambridge, Massachusetts. The purpose of the Symposium was to provide an opportunity to examine current design techniques and methodologies for the prediction of the foundation response of an embankment constructed upon a soft clay. Ten separate teams and individuals submitted predictions of deformation, pore pressure, and stability performance characteristics prior to the emplacement of additional fill upon an existing highway embankment located north of Boston. The Massachusetts Department of Public Works and the Federal Highway Administration sponsored the research program and the additional constuction to the embankment.

The predictions of the foundation performance reported herein are based on the information supplied to the predictors at the symposium. The results of laboratory tests on the embankment foundation material are described in the Proceedings of the Foundation Deformation Symposium (17). Interested readers are referred to D'Appolonia, Lambe and Poulos (3) and Lambe, et al. (13) for detailed information on the embankment site. The embankment cross-section together with the locations and types of instrumentation installed at the site are shown in Figure 1.

Over five years of consolidation took place between the end of construction of the original embankment in May of 1969 and the resumption of construction in October of 1974. None of the participants in the MIT symposium could model this consolidation behavior. Predictions were generally made using undrained analyses with parameters calibrated based

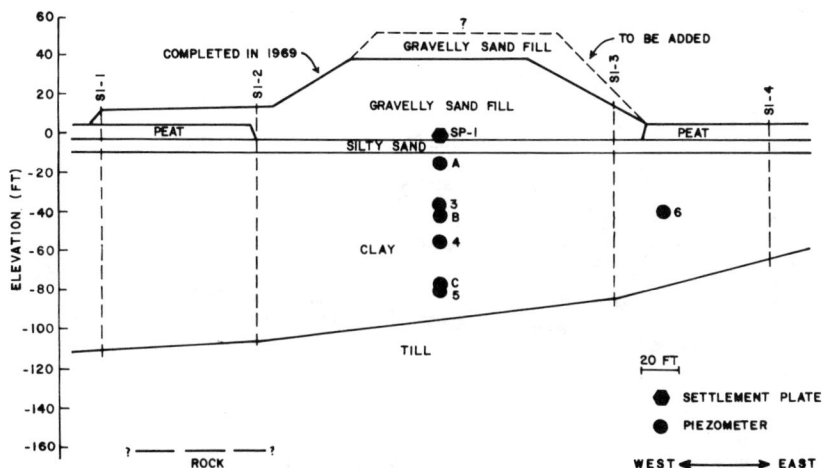

Figure 1.   I-95 Embankment Cross Section.

upon deformations during initial embankment construction and then adjusted to account for subsequent consolidation effects. In the case study described herein, a complete analysis of the foundation response, including the initial undrained construction deformations, consolidation deformations, undrained deformations during the second stage of embankment construction, and pore pressure response is performed.

Soil Parameters of the I-95 Test Site

Complete details of the development of the soil profile at the I-95 test section are described by Poepsel (18). A total of nine materials were used to model the soil conditions at the I-95 site for the finite element analysis. The low plasticity (CL) Boston Blue clay (BBC) stratum was modeled as six separate materials due to the overconsolidation ratio decreasing with depth. The OCR value varies from 5.0 at the top of the clay to 1.0 near the base. The peat and Boston Blue clay were described using the Modified Cam-Clay model, while the silty sand and embankment fill were modeled with the drained hyperbolic stress-strain relationship.

The soil profile used in the finite element analysis is shown in Figure 2. The values used for the BBC Cam-Clay parameters are the same as those used by Wroth in his predictions for the symposium (17). The same values for Cam-Clay parameters kappa (κ = .06), lambda (λ = .147), critical state void ratio (e_{cs} = 1.74), and the slope of the critical state line (M = 1.05) were used for all six BBC sublayers. The M-value was based on an effective friction angle of 26.5°. A saturated unit weight (γ_t)of 115 pcf (18.1 KN/m^3) was used for all sublayers.

Perhaps the most difficult soil parameters to ascertain were the horizontal and vertical permeabilities. Since prior analyses only

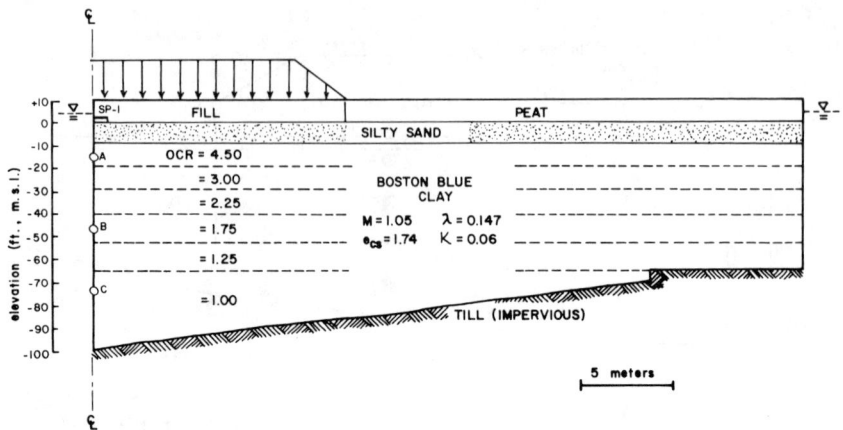

Figure 2.   I-95 Embankment Soil Profile.

considered undrained loading, permeability values were not required. Only a range of values for the vertical coefficient of consolidation were available from the symposium proceedings. The vertical coefficient of consolidation was used to backcalculate the vertical permeability (k_y). In this study, normally consolidated clays were assigned a coefficient of consolidation of 0.14 ft^2/day (1.3 x 10^{-2} m^2/day), and a value of 0.42 ft^2/day (3.9 x 10^{-2} m^2/day) was assigned to the overconsolidated clays. These values were in good agreement with other available consolidation results (11).

The vertical permeability values of the BBC obtained from the coefficient of consolidation were found to decrease with depth, varying from 7 x 10^{-4} ft/day (2.1 x 10^{-4} m/day) at the top of the layer to 2.3 x 10^{-4} ft/day (7.0 x 10^{-5} m/day) at the bottom. These values were compared to Lambe and Whitman's (14) recommendations for the permeability of Boston Blue clay and were found to be in good agreement. Since no values of the horizontal coefficient of consolidation were available from the symposium, the horizontal permeability ($k_x = k_z$) was initially estimated as 6.0 times the vertical value and was then adjusted to match the field pore pressures during construction, as described subsequently.

Overconsolidated shear moduli were extrapolated from soil property correlations presented by Mitchell (16) which show the influence of overconsolidation ratio and plasticity on shear modulus.

The Cam-Clay parameters for the peat were estimated using Mayne's (15) compilation of Cam-Clay properties for a variety of clays and silts. Strength and deformation parameters were based on typical values from this compilation for very soft, organic soils with similar water contents. Because of its loose nature, a critical state void ratio of 4.00 and a unit weight of 75 pcf (11.8 KN/m^2) were assigned to the peat. Since the peat located beneath the embankment was excavated and replaced with fill, the material properties for the peat beyond the embankment toe were not considered very critical to the results of the analysis.

The Duncan and Chang drained hyperbolic stress-strain properties for the embankment fill were determined from the drained triaxial results supplied to the predictors. Since no stress-strain data was available for the silty sand, the hyperbolic parameters were estimated from typical values for cohesionless soils documented by Duncan, et al. (5).

The Finite Element Mesh and Boundary Conditions

The plane strain finite element mesh consisted of 176 elements and 587 nodes. The mesh describes only the compressible foundation soils and not the constructed embankment. The load applied by embankment construction above the ground surface was modeled as a progression of surface loads. The mesh represented only the eastern portion of the subsoil and embankment profile, assuming conditions of symmetry at the centerline of the original embankment. General failure of the embankment and foundation soils occurred in this easterly direction because of the non-symmetric loading sketched in Figure 1. The assumption of symmetry at the original embankment centerline was made to simplify the finite element mesh representation. In addition to the non-symmetrical loading, the underlying glacial till actually continues to slope downward on the western side of the centerline. However, these non-symmetrical conditions were neglected in the analysis performed herein in the interest of economy.

The BBC and peat were modeled using Q8P4 elements with a pore pressure degree of freedom at each of the corner nodes. For the silty sand and embankment fill, Q8 elements were used since fully drained behavior is assumed for these materials. Because of the high stiffness and low permeability of the glacial till, the bottom boundary of the mesh was assumed rigid and impervious.

Description of Embankment Construction

The construction sequence for the embankment can be separated into seven distinct phases, consisting of five loading and two consolidation periods. Table 1 summarizes the construction sequence by presenting the order, duration, and applied load of each stage of construction. The duration and height of fill placed during each phase were determined based upon the proceedings of the symposium.

Calibration of Finite Element Model

Calibration of the finite element model to the soil system took place during the initial three phases of construction. Vertical displacements were matched to the field measurements by reducing the value of the shear modulus in the Boston Blue clay during the first eight time increments. From this point on, no further changes in the shear modulus values were made. The maximum adjustment of the shear modulus was only a 26% reduction of those values determined from the results of triaxial compression tests or empirical correlations. Table 2 presents the initial and the final adjusted shear modulus values for the Boston Blue clay.

The values of the coefficient of permeability were also calibrated during the first eight time increments. The vertical values

Table 1   I-95 Test Embankment Construction Sequence

Phase	Elapsed Time (days)	Total Time (days)	Number of Increments	Description of Constructional and Consolidation State
I	75.	75.	3	Excavation of peat and replacement with fill net pressure = 450 psf
II	75.	150.	3	Initial construction 4' in 75 days, pressure = 488 psf
III	100.	250.	4	Initial construction 21' in 100 days, pressure = 2562 psf
IV	120.	370.	4	Consolidation during winter; time = 120 days
V	25.	395.		Initial construction completed; 4' in 25 days, pressure = 488 psf
VI	1800.	2195.	16	Long term consolidation; time = 1800 days
VIIA	9.	2204.	2	Test fill placed; 6' in 9 days, pressure = 650 psf
VIIB	12.	2216.	4	Loaded till failure; 13' in 12 days, pressure = 400 psf

Table 2   Boston Blue Clay Shear Moduli

Clay Sublayer	Initial G psf (MPa)		Adjusted G psf (MPa)	
1	15500	(742)	17500	(838)
2	23300	(1116)	21700	(1039)
3	32600	(1561)	24250	(1161)
4	33000	(1580)	25000	(1197)
5	30300	(1451)	22350	(1070)
6	29500[1]	(1412)	29500	(1412)

1.   Based upon G/S_u = 74 from Ref. 17. (All other initial values based on empirical correlations from Ref. 16).

as determined from one-dimensional consolidation theory were not changed. The horizontal permeability values were adjusted to match the measured pore pressures. Adjustments at this point were minimal. The initial ratio of horizontal to vertical permeability used in the analysis for the Boston Blue clay was 6, and the final value was 5.

Results of the Analysis

Settlement Plate 1 (SP-1) was installed at the embankment center-line following the excavation of the peat layer, and is seated on the silty sand overlying the Boston Blue clay. The predicted response of SP-1 obtained from a displacement node at the same location in the finite element mesh is compared to the field measurements in Figure 3. Also shown in this figure is the corresponding embankment elevation versus time.

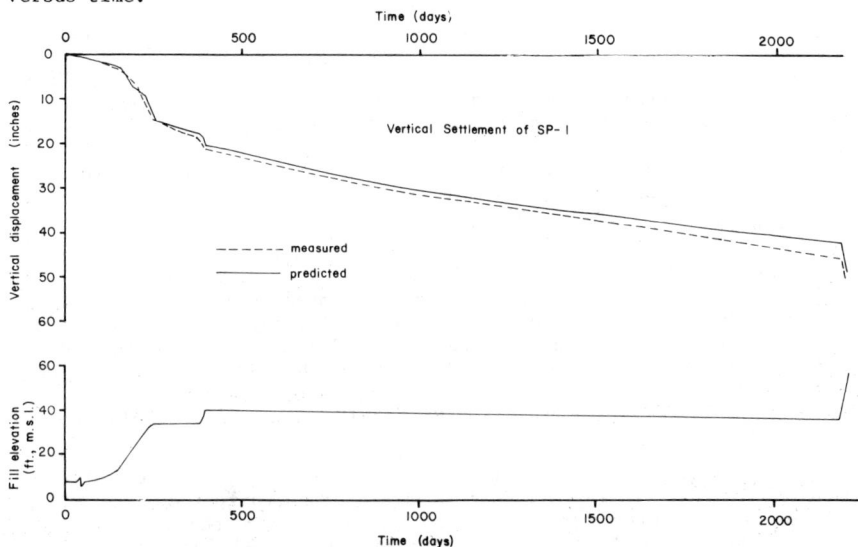

Figure 3.   Settlement of SP-1 and Fill Height, I-95 Embankment.

The JFEST results appear in excellent agreement with the measured response of SP-1 during the initial loading and short-term consolidation stages (Stages I through V), with a maximum deviation of only 4%. The predicted settlements during the long-term consolidation period were very consistent with the measured results for the first 1000 days following the completion of the original embankment. From this point in time, the two curves began to diverge. At the end of the consolidation period, just prior to the symposium, the predicted vertical settlement was underestimated by 10%, and the difference between measured and predicted response was almost five inches.

The placement of the six feet of additional fill for the symposium caused SP-1 to displace only 0.66 inches (1.68 cm.), which was considerably less than predicted by JFEST, or by anybody at the symposium. The predicted settlement from JFEST during this stage of loading was 1.89 inches (4.8 cm.). Wroth and his colleagues, who also based their predictions on the Cam-Clay soil model, estimated an additional settlement of 2.60 inches (6.6 cm.). A similar overprediction of vertical settlement of SP-1 was apparent during the placement of the remainder of the test fill.

Despite the slight inaccuracies mentioned above, JFEST predictions of vertical settlements are considered to be in very good agreement with field measurements.

The horizontal displacements occurring in the easterly direction at the I-95 test site were measured with slope inclinometers SI-3 and SI-4 at centerline offsets of 92 and 181 ft., respectively. All inclinometers were installed at the test site during the early stages of embankment construction.

The predicted and measured response for three readings of inclinometer SI-3 prior to the Symposium is shown in Figure 4. These horizontal displacement results show good agreement with the measured displacements in the overconsolidated BBC occurring between elevations 0 and −40 ft. (12.2 m.), but moderately underpredict the horizontal movement of the lightly overconsolidated BBC between elevations −40 and −80 ft (−12.2 and −24.4 m.). Deviations between predicted and measured results appeared to increase during the long-term consolidation phase.

Figure 4 also displays the additional horizontal movement of SI-3 after the placement of the initial six feet of the symposium test fill. Note that a smaller displacement scale is used for this plot than for the other three. In contrast to earlier predictions of horizontal movement, JFEST results overpredict this response, in some places by more than 50% of the measured values. Also plotted in this figure is an inclinometer reading taken three days after the six feet of fill was placed, but prior to any additional fill placement. These displacements show better agreement with the JFEST predictions, which should remain essentially unchanged over the three day period.

The measured and predicted displacements of SI-4, presented in Figure 5, show a similar trend in predicted results to those for SI-3. However, the magnitude of the discrepancy between the observed and predicted results was not as large as observed in SI-3, which is located closer to the embankment than SI-4.

The predicted excess pressure head and the readings from functional piezometers along the embankment centerline are presented in Figure 6 for the entire construction sequence. The incremental pore pressures computed during the initial embankment construction stages (Stages II, III, V) show excellent agreement with those measured in the Boston Blue clay. The amount of pore pressure registered in Piezometer B founded at elevation −42 ft. (−12.8 m.) in lightly overconsolidated clays and in Piezometer C at elevation −77 ft. (−23.5 m.) in the normally consolidated clays were practically identical to the JFEST predictions. Piezometers B and C did exhibit a slight discrepancy in the pattern of pore pressure behavior. While JFEST predicts dissipation during the short term consolidation stage (Stage IV), the measured pore pressures of Piezometer C tended to slightly increase. Piezometer B exhibited a much slower rate of dissipation than predicted in the long term consolidation phase, until it stopped functioning.

The measured and predicted pore pressures of Piezometer A yielded anamolous results. One possible sources of this discrepancy is that the surrounding clay was much more overconsolidated than is assumed in the analysis. Since Piezometer A was founded at elevation −15 ft. (−4.6 m.), which is five feet above the node used for comparison, and since the OCR

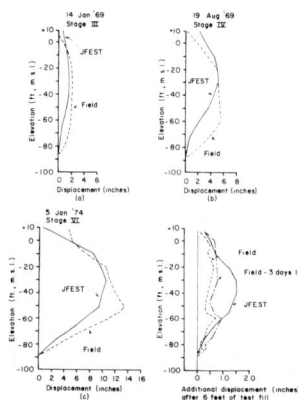

FIGURE 4   INCLINOMETER SI-3, I-95 TEST SECTION

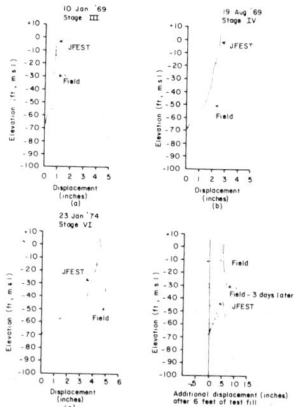

FIGURE 5   INCLINOMETER SI-4, I-95 TEST SECTION

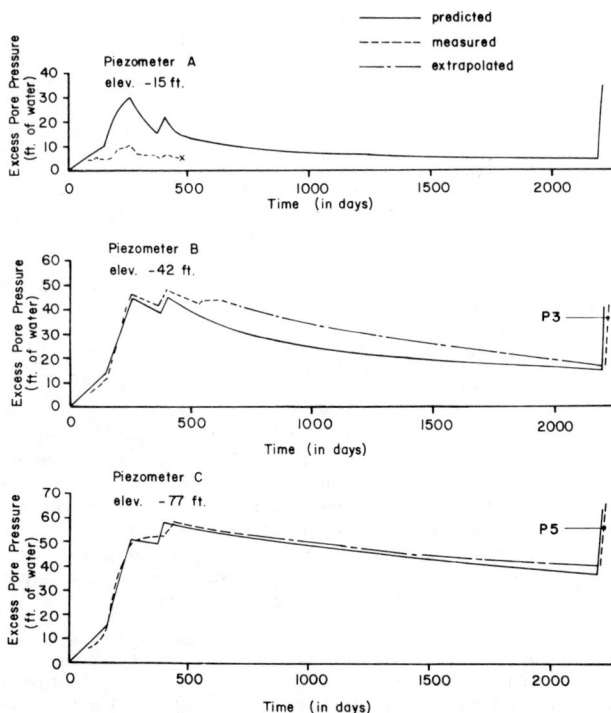

FIGURE 6   CENTERLINE PIEZOMETERS, I-95 TEST SECTION

increases upward through the Boston Blue clay profile, the piezometer was most likely founded in clay with an OCR greater than the 4.50 value used in the computer analysis. Thus, the OCR is probably much too low to accurately predict the response of Piezometer A, resulting in an over-prediction of pore pressures by JFEST.

Unfortunately the three centerline piezometers A, B, and C ceased to function shortly after initial embankment construction. Piezometers B and C were, however, replaced just prior to the symposium with Piezo-meters P3 and P5 at the same approximate locations. Shortly after their installation, the new piezometers indicated that excess pore pressures still existed within the Boston Blue clay at this time. These new readings are compared to the JFEST predictions in Figure 6. Comparison shows that the estimated and predicted excess pore pressures dissipated at approximately the same rate. The measured pore pressures retained about 2 feet of pressure head more than the predicted values at the end of the long consolidation interval. The excess pore pressures predicted during the placement of the test fill were in good agreement with the measured response of the centerline piezometers, as shown in Figure 6.

Discussion of Results

In general, JFEST effectively simulated the foundation response during the initial embankment construction and consolidation. All pre-dicted displacements and excess pore pressures were very consistent with the field measurements during this period (Stages II through V). During the long-term consolidation stage, the predicted and measured displacements initially coincided, but diverged slightly by the end of consolidation prior to the symposium. Horizontal displacements were moderately underpredicted at this time, whereas vertical settlements were only slightly underpredicted.

During placement of the test fill, all soil movements were over-estimated by JFEST. The loads that produced failure at the test site did not cause failure in any element within The Boston Blue clay in the finite element analysis. A zone of local yielding was observed in the silty sand beyond the embankment toe.

THE ATCHAFALAYA FLOOD CONTROL LEVEE
Introduction

The high potential for damaging floodwaters in the lower Mississippi Valley and Atchafalaya Basin has resulted in the construc-tion of an extensive series of flood control levees in these areas. The original levees built along the Atchafalaya River during the 1930's have experienced significant crest settlements and progressive siltation that have greatly reduced their effectiveness. Because of these difficulties, additional placement of fill upon these levees was proposed by the U.S. Army Corps of Engineers in 1950 to raise the crest elevation in order to increase the flow capacity of the flood control system.

To evaluate a stage construction design strategy, the U.S. Department of the Army and Corps of Engineers constructed two adjacent test sections in 1964. The test sections, called Test Sections II and III, were each approximately 1500 feet (460 m.) in length. Both test sections were built over existing levees at a location thought to be one of the more critical areas with the worst foundation conditions within the Atchafalaya Basin. The predictions described in this study are made for the performance of Test Section II only.

Detailed investigations of the soil conditions at the site of Test Section II (9,12,21) show the profile to consist of 50 to 400 ft. (15-120 m.) of highly plastic fat clay (CH) underlain by sands and gravels. The thick layer of clay present at the test site can be characterized as a backswamp deposit, laid down as uniform layers in the shallow ponds and lakes left by floodwaters. This soft clay deposit exists in a very loose state and possess very little strength. Intermixed between the layers of clay are very thin lenses of silt that were deposited during short periods when the river flows increased slightly. These lenses are often undistinguishable in boring logs.

Figure 7 shows the locations and types of instrumentation installed at the test site. The instrumentation program consisted of 26 closed-system piezometers, five surface settlement plates, seven centerline deep settlement plugs, and six inclinometers.

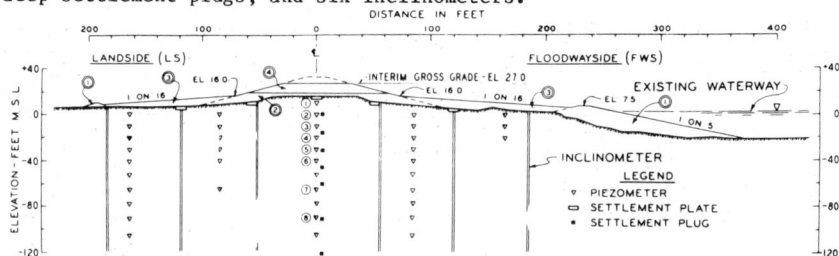

Figure 7. Atchafalaya Test Embankment Cross Section.

Soil Parameters of the Atchafalaya Test Site

Four separate phases of subsurface testing focused on studying the engineering properties of the thick clay stratum. The first wave of testing was part of the initial site investigation conducted by the U.S. Army Corps of Engineers prior to the construction of the test sections (21). In the second phase, Duncan (4) performed a series of unconsolidated-undrained (UU) triaxial tests as a followup to the initial research. The final two phases were the MIT Testing Programs. These results provided by Ladd, et al. (12) and Fuleihan and Ladd (9) comprised the primary source of information for quantifying the strength and deformation parameters used in the computer analysis.

The foundation and compacted levee soils were modeled as Cam-Clay materials. The subsurface profile was divided into sublayers according to the trend and variability of soil properties. The spatial locations of each soil type are presented in Figure 8. Table 3 summarizes the characteristic properties assigned to each material. A total of 9 materials were used to describe the subsurface conditions at Test Section II.

The stress history results of Foott and Ladd (8) provided the basis for ascertaining the degree of overconsolidation within the Atchafalaya clay at the test site. From the results of consolidation tests performed

Table 3   Cam-Clay Material Properties-Atchafalaya Test Section II

Material Number	γ_t (pcf)	OCR	$\dfrac{S_u}{p'_o}$	G (psf)	$k_x = k_z$ (ft/day)	k_y (ft/day)	e_{cs}	e_o	M	κ	λ
1	104	1.00	.24	125*	.00675	.00080	5.50	2.12	.88	.079	.458
2	100	1.00	.24	100*	.01500	.00175	5.50	2.45	.88	.079	.475
3	104	1.22	.29	17875	.00675	.00075	5.50	1.98	.88	.079	.458
4	87	1.29	.29	10500	.08750	.00525	5.50	2.91	.88	.079	.496
5	100	1.42	.32	14250	.01420	.00135	5.50	2.37	.88	.079	.455
6	87	2.35	.48	11500	.03250	.00326	5.50	2.87	.88	.079	.496
7	87	3.19	.62	12000	.01250	.00331	5.50	2.65	.88	.079	.496
8	102	7.50	1.10	10750	.01050	.00275	5.50	3.11	.88	.072	.384
9	90	1.00	.22	75*	.15030	.00425	5.50	2.75	.88	.079	.577

* G/S_u for OCR = 1.0

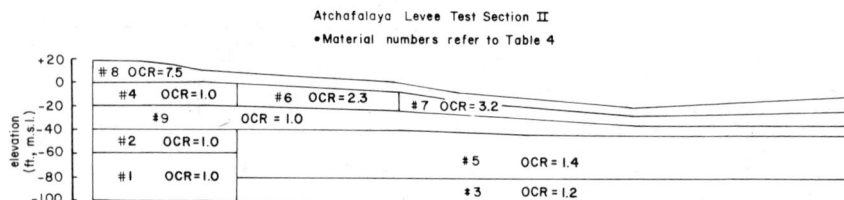

Figure 8.   Atchafalaya Embankment Soil Profile.

on clays beneath the original levee centerline, it appears that the clays exist here for the most part in a normally consolidated state. Investigations of maximum past pressure in the foundation were also conducted at centerline offsets of 105 and 180 feet (32 and 55 m.). Bands of slightly overconsolidated material were revealed at both of these locations. The amount of overconsolidation in these layers steadily increased with distance from the centerline, indicating that the decrease in OCR in the vicinity of the centerline is due to the original embankment loads. The coefficient of earth pressure at rest was ascertained as a function of stress history by Brooker and Ireland (1) for the Atchafalaya clays.

Fuleihan and Ladd (9) used a statistical approach to quantify the virgin and rebound compression ratios at several locations within the profile for the range of stresses that were expected for the foundation soils beneath the original levees. To determine the compression indices λ and κ required for this analysis, the value of the initial void ratio, e_o, was required. It was computed based upon natural water content, a specific gravity of 2.70, and complete saturation of the clay.

The isotropic virgin compression index, λ, showed a considerable variation of values among the normally consolidated soils. Mean values tended to slightly decrease with increasing depth and increasing density. An exception is the material located between elevations -20 and -40 ft. (-6 and -12 m.) (material #9), which appeared to be softer and more compressible that the other normally consolidated soils, with virgin compression indices up to 25% higher. Furthermore, pore pressure readings in the vicinity of this layer give an indication that there is considerable lateral drainage, possibly as the result of a silt or sand lens.

The slope of the failure line, M, was based on an average effective internal friction angle of 24° obtained from several triaxial and direct simple shear tests. The undrained shear strength for each material was expressed as a fraction of the vertical effective stress, S_u/p_o'.

The critical state void ratio is directly related to the values of κ, λ, and e_o. The initial stress generation routine in JFEST was employed to determine the appropriate value of e_{cs}. An iteration procedure of adjusting the input value of e_{cs} was used until reasonable values of initial void ratio were obtained in the initial stress generation output. A value of 5.50 for e_{cs} gave satisfactory results.

The shear modulus, G, was estimated from available stress-strain information on the Atchafalaya clays and the correlations presented by Mitchell (16) where necessary. For normally consolidation soils, modulus values were based upon the ratio of the shear modulus to the undrained strength, G/S_u.

The value of the vertical permeability was computed from vertical coefficient of consolidation values using one-dimensional consolidation theory and the results of Fuleihan and Ladd's tests on the normally and overconsolidated Atchafalaya clays. For the normally consolidated clay, a vertical coefficient of consolidation of 0.105 ft^2/day (9.8 x 10^{-3} m^2/day) was used, whereas overconsolidated clays were assigned a value of 0.20 ft^2/day (1.9 x 10^{-2} m^2/day).

Vertical permeability values were found to decrease with depth, ranging from 0.00525 ft/day (1.6 x 10^{-3} m/day) near the surface of the clay later to 0.0008 ft/day (2.4 x 10^{-4} m/day) at the base. Due to the subtle presence of silt lenses within the Atchafalaya clays, a relatively large ratio of horizontal to vertical permeability was anticipated. Based on the calibration with the field measurements, a ratio of approximately eight was used for most of the clay layers. A permeability ratio of 35 was used to account for the additional lateral drainage anticipated in the layer located between elevations - 20 and - 40 ft (- 6.1 and - 12.2 m).

The soil densities of the test section levees used to quantify the magnitude of applied loads for each construction stage differed with method of fill placement. The stabilizing berms were pumped into place as a slurry of soil and water and settled to an assumed density of 67 pcf (10.5 KN/m^3). The clay mechanically compacted at the centerline has a greater density of 98 pcf (15.4 KN/m^3).

The Finite Element Mesh and Boundary Conditions

The finite element mesh composed of 132 elements and 453 nodes, described only the cohesive foundation soils and represents the floodway side of the levee profile. As in the I-95 case study, the construction of the new levee embankments is simulated by a progression of surface loads.

The thick clay deposit is modeled using Q8P4 elements. The response of the underlying sands and gravels were not modeled in this study. Instead, this layer is represented as a rigid, free-draining boundary. Vertical deformations on the right hand boundary were unrestrained. In this manner, the assumption of "free field" conditions at this boundary can be evaluated from the finite element results.

Test Section II Construction

The computer simulation of the test section construction is presented in Table 4. Test Section II had a very complex construction sequence during its initial stages. There was considerable uncertainty in the magnitude, timing and location of the loads simulating the construction of the floodwayside berms.

Table 4   Atchafalaya Test Section II Construction Sequence

Phase	Elapsed Time (days)	Total Time (days)	Number of Increments	Description of Construction/ Consolidation State
I	66	66	2	2' centerline lift; pressure = 196 psf. 40% of major berm completed; pressure = 300 psf
II	24	90	1	60% of major berm; pressure = 450 psf
III	17	107	1	48% of connecting berm; pressure = 190 psf
IV	48	155	2	Consolidation; time = 48 days
V	9	174	1	52% of connecting berm; pressure = 210 psf
VI	42	216	4	10' centerline lift; pressure = 980 psf
VII	240	456	6	Post construction consolidation; time = 240 days

Calibration of Finite Element Model

Shear modulus and compression indices evaluated from the results of laboratory tests were calibrated based upon the measured field response of Stage VI of the construction process (final centerline lift). Only minor adjustments were required in the values of the compression indices, however shear modulus values were reduced significantly (by

over 50% in one case). A large portion of this reduction in the elastic
modulus is believed to be attributable to undrained creep. Table 5
presents the initial and reduced values of shear modulus. Permeability
values were also calibrated based upon initial field measurements, as
previously discussed.

Table 5   Atchafalaya Clay Shear Moduli

Material Number	Initial G psf (MPa)		Adjusted G psf (MPa)	
1	31000	(1484)	24000	(1150)
2	30000	(1436)	21500	(1029)
3	23625[1]	(1131)	19875	(952)
4	17000	(814)	10500	(503)
5	22800[1]	(1092)	14250	(682)
6	20000	(958)	11500	(551)
7	21300[1]	(1020)	12000	(575)
8	24500	(1173)	10750	(515)
9	20250	(970)	13000	(622)

1. Based upon empirical correlations from Ref. 19. (All other initial
values based upon $\overline{CK_oU}$ direct simple shear test results from Ref.
14).

Results of the Analysis

The final centerline lift of embankment construction was a fairly
straightforward operation. The measured deformation and pore pressure
response during this stage and the consolidation that followed provided
the primary basis for the comparison of predicted results.

Surface settlement plates and deep settlement plugs were installed
throughout the clay foundation at the test site to measure the vertical
displacements during and after construction of the test levees. Surface
settlement plates were seated on the heavily overconsolidated clays
comprising the original levees. The deep settlement plugs were installed
in the normally consolidated clays along the centerline at various
depths (Figure 7).

The measured vertical settlements of the surface plates and deep
plugs and the JFEST predictions of their response are presented in
Figures 9 and 10 for the entire six month construction period and for
eight subsequent months of consolidation.

The measured settlements during the floodwayside berm construction
observed in settlement plates 1, 2, and 3 were somewhat inconsistent
with those predicted by JFEST. Surface settlement plate 3 founded at the
centerline displayed a smaller amount of vertical settlement than JFEST
had predicted during this stage, whereas the settlement measured in

FIGURE 10   DEEP SETTLEMENT PLUGS 3-7, ATCHAFALAYA EMBANKMENT

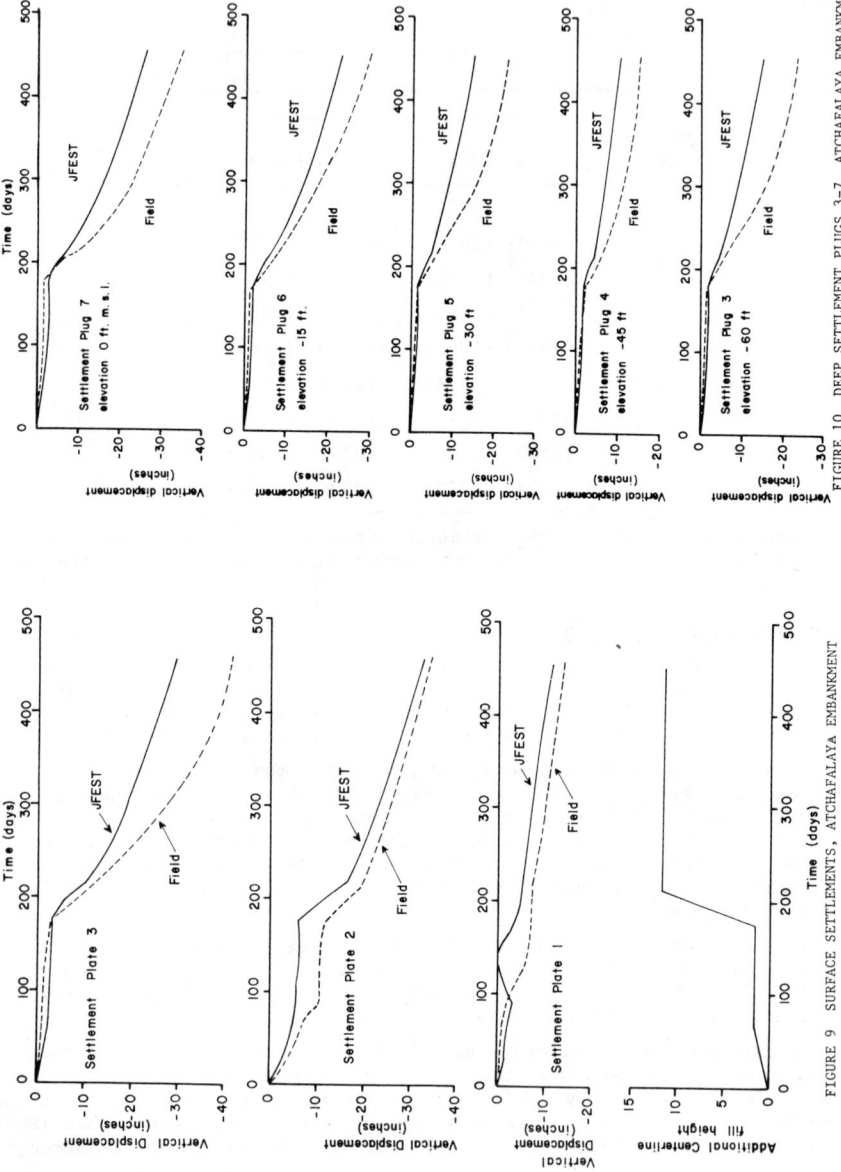

FIGURE 9   SURFACE SETTLEMENTS, ATCHAFALAYA EMBANKMENT

plate 2 at a distance of 50 feet (15 m.) from the centerline was considerably underpredicted. Further from the centerline at an offset of 120 feet (37 m.), predicted results were again somewhat distorted. Plate 1 displayed a gradual downward settlement during Stages III-V, while JFEST predicted some degree of heaving.

During Stages I-V, the settlement plugs did record downward vertical displacements in accordance with the predicted results. The vertical settlements measured in the deep plugs as the major centerline lift (Stage VI) was placed were approximately equal in magnitude to those predicted. At centerline settlement plate 3, however, deformations were overpredicted during this loading by 20%. This overprediction indicates that the value of the recompression index used for the original levee material was most likely too high. During the 240 days of consolidation that followed, measured centerline settlements increased at a faster rate than predicted by JFEST. At the end of the consolidation period, differences between measured and predicted settlements were at a maximum of 35%, observed in Plate 3 and Plug 7. Inspection of the centerline settlements reveals that a significant portion of the total settlement is occurring in the very compressible clay layer located between elevations -20 and -40 ft. (- 6 and - 12 m.), as previously anticipated.

The measured and predicted displacement of the inclinometers at the end of levee construction (May 1965) and 240 days thereafter (December 1965) are presented in Figure 11. As in the predicted vertical settlements, the predicted horizontal displacements for both inclinometers were qualitatively good, but underestimated the observed displacements, especially in the clays above elevation -20 ft. (- 6 m.), where predicted displacements were up to 50% less than the measured values. Excessive lateral movements in the soft layer between elevations -20 and -40 ft. (- 6 and - 12 m.) were observed in both the inclinometer data and the JFEST predictions. The previously mentioned underprediction of vertical settlement can be directly related to the underprediction of lateral displacements, as increased lateral spreading will result in increased centerline settlements.

A comparison of the measured and predicted deformations for the three inclinometers show that the tendency to underpredict lateral movement decreases with increasing distance from the centerline. The measured lateral soil movements for the inclinometer at the 55 ft. (17 m.) offset are significantly underpredicted by JFEST, while further from the centerline and nearer the floodwayside berms, the measured and predicted displacements begin to converge.

The horizontal movement of the original levee material (centerline elevation 0 to +18 ft. (+ 5.5 m.) was not modeled accurately. Predicted displacements in this layer were very small compared to the field measurements. Note that the Cam-Clay model does not really apply to heavily overconsolidated material such as this.

Figure 12 presents the field measurements of centerline piezometers 2,6 and 8 and the JFEST predictions of their response. There is considerable disagreement in the pore pressure predictions and measured values at the centerline during the floodway side berm construction. During the major centerline loading, all piezometers registered a lower pore pressure response than is predicted by JFEST. Following the

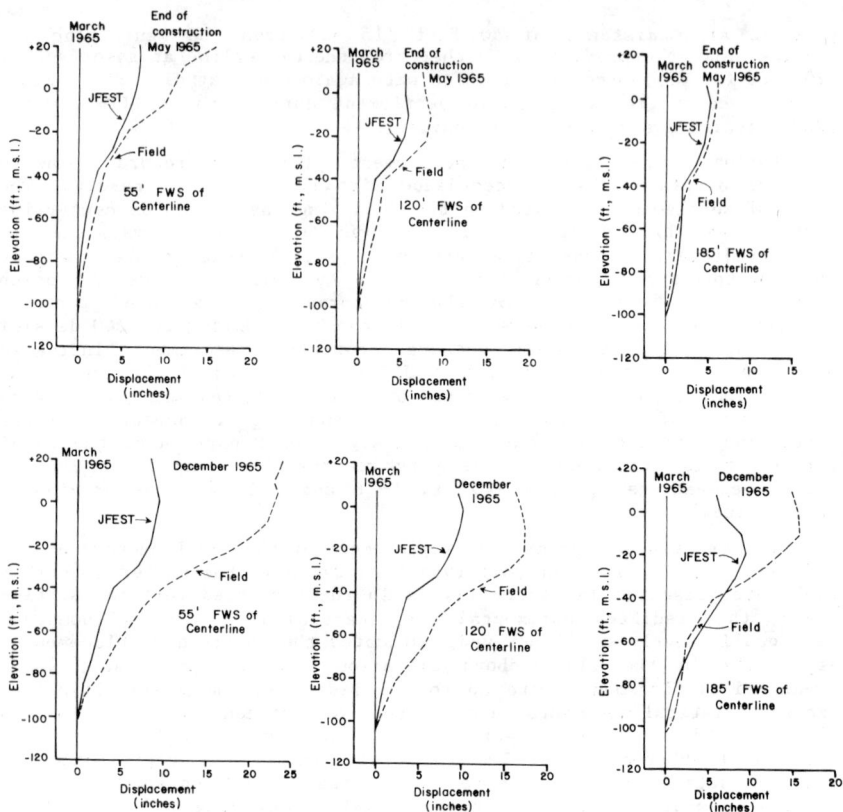

FIGURE 11 INCLINOMETER DISPLACEMENTS, ATCHAFALAYA EMBANKMENT

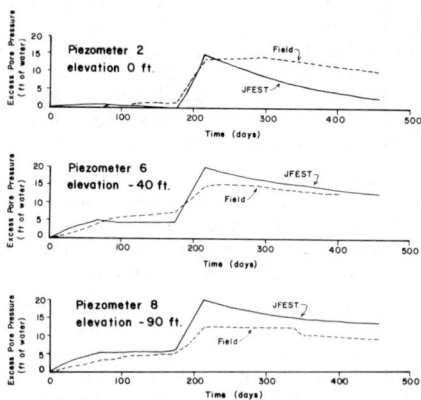

FIGURE 12 CENTERLINE PIEZOMETERS, ATCHAFALAYA EMBANKMENT

completion of the centerline levee, the piezometer readings taken over the next 240 days indicate that the foundation clays are retaining most of the generated excess pore pressures. JFEST, on the other hand, predicts that the excess pore pressures will dissipate during this consolidation phase. The readings from Piezometer 2 situated at elevation 0 ft. actually showed a slight increase in pore pressure during consolidation, while the other piezometers showed little or no change. Pore pressures at this piezometer achieved their maximum value about 40 days after the end of construction.

Discussion of Results

The early stages of the levee construction produced inconsistent results, so attention was directed at the major centerline lift and the foundation consolidation that followed. During this loading phase, both vertical and horizontal displacements were considerably underestimated, whereas the pore pressures due to this loading were overpredicted. As the foundation consolidated following the fill addition, measured pore pressures dissipated only slightly. JFEST, on the other hand, predicted significant dissipation to occur. Settlements at the centerline during consolidation were underpredicted. These observations indicate that the discrepancy between observed and predict centerline settlements was due to an underprediction of lateral spreading and not an underprediction of consolidation deformations. All of these discrepancies can be attributed to undrained creep. Undrained creep will significantly increase lateral deformations, and creep-generated pore pressures will retard the observed rate of pore pressure dissipation and could conceivably cause an increase in excess pore presure with time under no additional load as observed at piezometer 2..

SUMMARY AND CONCLUSIONS

Summary

Case histories of compressible embankment foundation performance at the I-95 highway embankment test section and the East Atchafalaya Basin flood control levee test section are analyzed using the Cam-Clay constitutive model and a finite element consolidation program. After calibrating soil parameters based upon field measurements during initial construction, predictions of deformations and pore pressures at both sites show varying degrees of agreement with measured values.

For the I-95 case, only minor reductions in values of initial shear moduli derived from laboratory test data were made during property calibration. These reduced moduli moderately underpredicted lateral deformations in the bottom half of the Boston Blue clay layer, where the clay is only lightly overconsolidated or normally consolidated. Initial "undrained" pore pressure response is only slightly underpredicted in the lightly overconsolidated and normally consolidated clay, and the predicted rate of pore pressure dissipation in these layers during consolidation is in good agreement with field observataions. In the heavily overconsolidated clay, where the Cam-Clay model is not really valid, predicted pore pressures are seriously in error.

For the Atchafalaya case major reductions in the undrained shear modulus obtained from laboratory test data were required to match initial construction deformations. Lateral deformations in the soft, creep susceptible Atchafalaya clays were seriously underpredicted. Vertical settlement was also underpredicted, but this can be directly attributed to the underestimation of lateral deformations and not to the consolidation model. Initial undrained pore pressure response is only slightly overpredicted, but the rate of excess pore pressure dissipation is seriously overpredicted.

Conclusions

The major discrepancies observed for these two case studies are:

1. An over-prediction of shear moduli from laboratory test data.

2. A slight overprediction of undrained pore pressure response.

3. An underprediction of lateral deformations in the softer, normally consolidated soils and a consequent underprediction of centerline settlements.

4. An overprediction of the rate of pore pressure dissipation for the Atchafalaya clays.

Most of these discrepancies can be directly attributed to undrained creep. Tavenas, Mieussens, and Bourges (20) have previously noted the underprediction of lateral deformations and vertical settlements of embankments on normally consolidated soft clays by conventional analyses and attributed this discrepancy to unaccounted for undrained creep deformations. This would explain why predicted lateral deformations in the stiffer overconsolidated upper half of the Boston Blue clay stratum were in good agreement with measured values, while predicted deformations in the normally consolidated or lightly overconsolidated lower layer were too small.

The overprediction of I-95 embankment settlement during additional construction for the MIT symposium may be attributable to an increase in soil stiffness subsequent to undrained creep.

The occurrence of undrained creep will also account for the need to reduce laboratory derived shear moduli to match field behavior, the required reduction being greater in the creep-susceptible Atchafalaya clays than in the Boston Blue clay. Similarly, creep generated pore pressures are probably the reason why pore pressures in the Atchafalaya clay dissipated slower than predicted during consolidation, and even increased at one location.

Thus, for clay soils with little potential for undrained creep, the soil model described herein should provide an accurate and reliable estimate of the field behavior of a compressible foundation using parameters derived from laboratory test data. For creep susceptible soils, undrained lateral deformations and vertical displacements will be seriously in error if moduli based on laboratory tests are used. Even if soil moduli are backfigured from initial construction field measurements for a creep susceptible soil, long term predictions of pore pressure dissipation and lateral displacements will still be in error. A creep-inclusive soil model is required to accurately predict the consolidation behavior of soft, creep-susceptible clay soils for other than one-dimensional situations. Such a model is currently under development at Stanford University as part of a collaborative Stanford-University of California, Berkeley research project on the time-dependent behavior of cohesive soils.

Acknowledgement

The work described in this paper is part of an NSF funded collaborative Stanford-University of California, Berkeley research project on the time-dependent behavior of cohesive soil, Contract No. NSF-CEE 8204320.

References

1.   Brooker, E.W., and Ireland, H.O., "Earth Pressure at Rest Related to Stress History," Canadian Geotechnical Journal, Vol. 2, No. 1, 1965, pp. 1-15.

2.   Carter, J.P., Booker, J.R., and Small, J.C., "The Analysis of Finite Elasto-Plastic Consolidation," Numerical and Analytical Methods in Geomechanics, Vol. 3, 1979.

3.   D'Appolonia, D.J., Lambe, T.W., and Poulos, H.G., "Evaluation of Pore Pressures Beneath an Embankment," Journal of the Soil Mechanics and Foundations Div., ASCE, Vol. 97, No. SM6, June 1971, pp. 881-987.

4.   Duncan, J.M., "Strength and Stress-Strain Behavior of Atchafalaya Levee Foundation Soils," University of California, Berkeley, College of Engineering, Research Report TE 70-1, 1970.

5.   Duncan, J.M., Byrne, P., Wong, K.S., and Marby, P., "Strength, Stress-strain and Bulk Modulus Parameters for Finite Element Analyses of Stresses and Movements in Soil Masses," Report No. UCB/GT/80-01, University of California, Berkeley, August 1980.

6.   Duncan, J.M. and Chang, C.Y., "Nonlinear Analysis of Stress and Strain in Soils," Journal of the Soil Mechanics and Foundations Div., ASCE, Vol. 96, No. SM5, September 1970, pp. 1629-1653.

7.   Finno, R.J., "Response of Cohesive Soil to Advanced Shield Tunneling, Ph.D. Dissertation, Stanford University, Stanford, CA, 1983.

8.   Foott, R. and Ladd, C.C., "The Behavior of Atchafalaya Test Embankments During Construction," M.I.T. Dept. of Civil Engineering, Research Report R73-27, Soils Publication 322, May 1973.

9.   Fuleihan, N.F. and Ladd, C.C., "Design and Performance of Atchafalaya Flood Control Levees," M.I.T. Dept. of Civil Engineering, Research Report R76-24, Soils Publication 543, 2 Vols., May 1976.

10. Johnston, P.R., "Finite Element Consolidation Analyses of Tunnel Behavior in Clay," Ph.D. Dissertation, Stanford University, Stanford, CA, 1981.

11. Ladd, C.C., Bovee, R.B., Edgers, L., and Rixner, J.J., "Consolidated-Undrained Plane Strain Shear Tests on Boston Blue Clay," M.I.T. Dept. of Civil Engineering, Research Report R71-13, Soils Publication 273, March 1971.

12. Ladd, C.C., Williams, C.E., Connell, D.E., and Edgers, L., "Engineering Properties of South Foundation Clays at Two South Louisiana Sites," M.I.T. Dept. of Civil Engineering, Research Report R72-26, Soils Publication 304, December 1972.

13. Lambe, T.W., D'Appolonia, D.J., Karlsrud, K., and Kirby, R.C., "The Performance of the Foundation Under a High Embankment," Journal of the Boston Society of Engineers, ASCE, Vol. 59, No. 2, April 1972, pp. 71-94.

14. Lambe, T.W. and Whitman, R.V., Soil Mechanics, John Wiley and Sons, New York, 1968, 553 pp.

15. Mayne, P.W., "Cam-Clay Predictions of Undrained Strength," Journal of the Geotechnical Engineering Div., ASCE, Vol. 106, No. GT11, November 1980, pp. 1219-1242.

16. Mitchell, J.K., "Soil Property Correlations," prepared for Woodward-Clyde Consultants' Symposium on Recent Developments in the Understanding and Characterization of Soil Properties, July 30-31, 1977.

17. Massachusetts Institute of Technology Constructed Facilities Division, "Proceedings of the Foundation Deformation Prediction Symposium," U.S. Federal Highways Administration Report No. FHWA-RD-75-516, 2 vols., July 1975.

18. Poepsel, P.H., "Finite Element Analyses of Embankment Construction and Foundation Consolidation: Two Case Histories," thesis presented in partial fullfillment of the requirements for the degree of Engineer, Dept. of Civil Engineering, Stanford University, 1984.

19. Roscoe, K.H. and Burland, J.B., "On the Generalized Stress-Strain Behavior of 'Wet' Clay," Engineering Plasticity, Cambridge University Press, 1968, pp. 535-609.

20. Tavenas, F., Mieussens, C. and Bourges, F., "Lateral Displacements in Clay Foundations under Embankments," Canadian Geotechnical Journal, 16, 1979, pp. 532-550.

21. U.S. Army Corps of Engineers "Interim Report on Field Tests of Levee Construction, Test Sections I, II, and III, EABPL, Atchafalaya Basin Floodway, La.," Dept. of the Army, New Orleans District, La., 1968.

CONSOLIDATION-STRENGTH ANALYSIS FOR SOFT SOILS

Zbigniew Lechowicz[1], Alojzy Szymanski[2], Wojciech Wolski[3]

ABSTRACT: A one-dimensional consolidation analysis, including prediction of consequent shear strength increase, for soft soils which exhibit large secondary compression is proposed. Consolidation and strength characteristics of two kinds of organic soils have been determined from laboratory tests carried out in the Rowe cell and standard triaxial cell. The observations carried out at a test embankment site have shown good agreement between the predicted and measured settlement and shear strength values.

INTRODUCTION

In soft soils large secondary compression plays a significant part in the consolidation process (1,6,7). Consolidation analyses of these soils, requiring the derivation of a stress-strain-time relationship, are usually based on rheological models representing the behaviour of the soil skeleton. The rate process theory provides a way of describing the viscous behaviour of the soil skeleton (6,16). Many authors (2,3,5,11) point to the fact that organic soils exhibit more immediate elastic deformation and larger secondary compression as compared to soft mineral soils. Investigations have also disclosed the nonlinear variation of consolidation characteristics and the simultaneous occurrence of all phases of the consolidation process (4,5,8). The test results indicate a significant increase, with consolidation, of initially low shear strengths (10,12,13). These findings indicate that a proper stability analysis of foundations should include the coupled effects of consolidation and consequent shear strength increase.

A new method of one-dimensional consolidation analysis with the prediction of shear strength increase during consolidation is presented. This method is based on rheological models with the assumption of non-linear changes of the governing parameters, with void ratio, during the consolidation process.

[1] D.Sc., Dept. of Geotechnics, Warsaw Agricultural University.

[2] D.Sc., Dept. of Geotechnics, Warsaw Agricultural University.

[3] Prof., Head of Dept. of Geotechnics, Warsaw Agricultural University, 02-766 Warsaw, Nowoursynowska 166, Poland.

CONSOLIDATION ANALYSIS

The soil skeleton of highly decomposed organic soils predominantly consists of colloidal organic particles. Free interparticle spaces are filled with water and gas bubbles. In the initial moment of loading, a sudden deformation of gas bubbles takes place as a result of the increase in pore pressure. The excess pore pressure causes drainage of pore water. The resultant increase in effective stress is carried by the particle contacts. During the consolidation process, the stronger bonds, with thin hygroscopic water film between particles, show elastic character of deformation. Weaker bonds, with thicker water film between particles after initial elastic deformation, show an elastic-viscous character of deformation. This viscous deformation of the skeleton of organic soils plays a significant part in total compression.

The Stress-Strain-Time Relationship

The behaviour of the organic soil skeleton under consolidation stress can be represented by the rheological model given in Fig. 1 (14). In this model the spring constant a_1 represents the effect of the

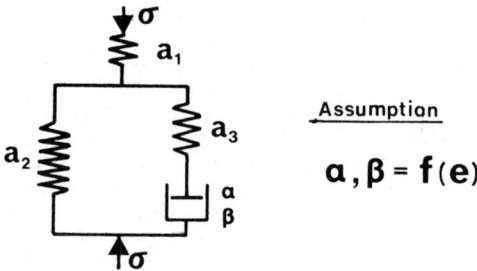

Fig. 1   Rheological model of stress-strain-time relationship of organic soils skeleton

initial deformation caused by compression of gas bubbles in pore water. The spring constants a_2 and a_3, respectively, represent elastic response of stronger and weaker contacts. The viscous response of weaker contacts is characterized by a dashpot according to the rate process theory (6). The stress-strain-time relationship derived from the rheological model (15) can be expressed as

$$\frac{\partial U}{\partial t} = \frac{1}{\varepsilon_f} \{ \frac{1}{a_1} \frac{\partial \sigma}{\partial t} + \frac{1}{a_2+a_3} \frac{\partial \sigma}{\partial t} + \frac{a_3}{a_2+a_3} \beta(e) \sinh [\alpha(e) .$$

$$. (\frac{a_1+a_2}{a_1}) \sigma - U\sigma_f)] \} \tag{1}$$

in which U = degree of consolidation; t = time, ε_f = final strain;

a_1, a_2, a_3 = spring constants; σ = effective stress; $\alpha(e)$, $\beta(e)$ = viscosity parameters as functions of void ratio e; σ_f = final stress.

Continuity Equation

Consolidation analysis assumes the one-dimensional flow of water. In the consolidation process of the organic soils considered the change in water volume is equal to the change in porosity less the immediate compressibility of gas bubbles in pore water. The continuity equation can be given as (15)

$$\frac{1}{g\,\rho_w\,\varepsilon_f\,H_i^2}\,\frac{\partial}{\partial x_i}\,[\frac{k(e)}{1-\varepsilon_f U}\,\frac{\partial\sigma}{\partial x_i}\,] = \frac{\partial U}{\partial t}\,(1 + \frac{1}{a_1}\,\frac{\partial\sigma}{\partial t}) \tag{2}$$

in which g = the acceleration of gravity; ρ_w = density of water; H_i = initial thickness of consolidated layer; z_i = initial distance from the impermeable boundary, $x_i = z_i/H_i$; k(e) = coefficient of permeability as a function of void ratio.

Laboratory Tests

Two kinds of organic soils were used in this investigation: an amorphous peat and a calcareous-organic soil called "gyttja". Gyttja commonly occurs as lake bottom sediments and consists primarily of calcium carbonate as well as organic and clay particles. The investigations have been performed on peat and gyttja at an experimental site located in Notec River Valley in northern Poland (8). Test specimens were obtained from large samples 30 cm in diameter, which were taken from a depth of 1.5 m for peat and 3.0 m for gyttja. The index properties and general composition of the organic soils are listed in Table 1. The organic content was determined by the weight loss on ignition at a temperature of 550°C. Degree of humification was estimated visually with a microscope. Consolidation tests have been performed in Rowe cells with a specimen size of 15 cm diameter and 5 cm height. The test program comprised of six series of long-term consolidation tests for peat and five similar series for gyttja.

TABLE 1 Properties of Organic Soils

	Peat	Gyttja
Water content (%)	430	135
Cone liquid limit (%)	442	121
Plastic limit (%)	290	65
Organic content (%)	88	45
Degree of humification (%)		60
$CaCO_3$ content (%)		51
Sensitivity by field vane	~7	~8

All the individual samples were initially saturated and reconsolidated
to the in situ stress conditions and subsequently loaded to various
maximum stresses. In each series, the individual samples of peat and
gyttja, respectively, were loaded to maximum stresses of 10, 15, 20, 40,
80, 160 kPa, and 10, 20, 40, 80, 160 kPa. A total of 42 tests were
performed on the peat samples and 40 on the gyttja. In all consolidation tests, the samples were first tested in undrained and subsequently
in drained, conditions. Vertical deformations, volume of water released
and pore pressures were measured at regular intervals. The experimental
data were used for the determination of the coefficients of permeability
and parameters for the rheological model. The method used to calculate
the parameters a_1, a_2, a_3, α and β is shown in Appendix I. The results,
presented in Fig. 2, indicate significant variation of the consolida-
tion parameters k, α and β with void ratio.

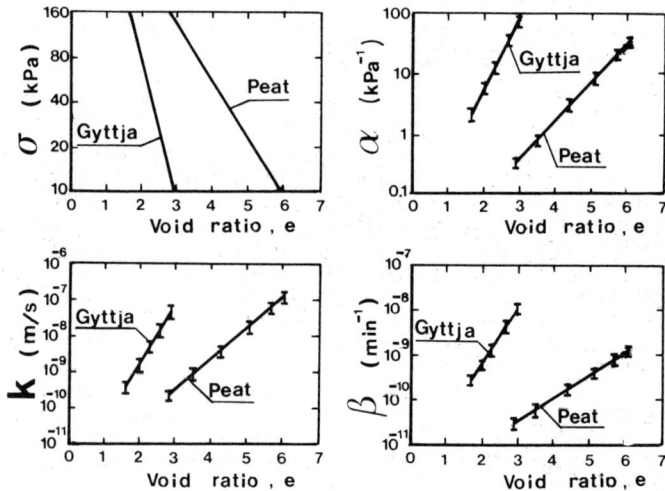

Fig. 2   Consolidation characteristics

Consolidation Equations

On the basis of experimental results, the change in consolidation parameters during consolidation can be expressed as

$$\alpha = \alpha_i \ (\alpha_f/\alpha_i)^U \qquad (3)$$

$$\beta = \beta_i \ (\beta_f/\beta_i)^U \qquad (4)$$

$$k = k_i (k_f/k_i)^U \qquad (5)$$

in which the subscript i and f refer to initial and final conditions, respectively. Using the empirical Eqs. 3 and 4, the final form of the stress-strain-time relationship is obtained as

$$\frac{\partial U}{\partial t} = \frac{1}{\varepsilon_f} \{(S_1+S_2) \frac{\partial \sigma}{\partial t} + S_3 B^U \sinh [\alpha_i A^U (S_4\sigma - U\sigma_f)]\} \qquad (6)$$

in which $S_1 = 1/a_1$; $S_2 = 1/(a_2+a_3)$; $S_3 = a_3\beta_i/(a_2+a_3)$; $B = \beta_f/\beta_i$; $A = \alpha_f/\alpha_i$; $S_4 = (a_1+a_2)/a_1$.

Taking into consideration the empirical Eq. 5, the final form of the continuity equation is obtained as

$$R \frac{\partial}{\partial x_i} [\frac{K^U}{1-\varepsilon_f U} \frac{\partial \sigma}{\partial x_i}] = \frac{\partial U}{\partial t} (1 + S_1 \frac{\partial \sigma}{\partial t}) \qquad (7)$$

in which $R = k_i/g \, \rho_w \, \varepsilon_f \, H_i^2$; $K = k_f/k_i$.

Eqs. 6 and 7 together are sufficient to predict the consolidation process of organic soils.

SHEAR STRENGTH ANALYSIS

Due to the complex character of the deformation of organic soils, it is necessary to take into account the visco-plastic deformations in the description of the stress-strain relationships during the shearing process. The behaviour of organic soils subjected to shear can be represented by the rheological model given in Fig. 3 (10). The model consists of an elasto-viscous element, which represents the weaker contacts of particles, and an elasto-plastic element which represents behaviour of stronger contacts.

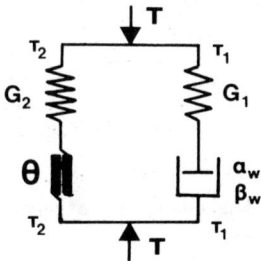

Fig. 3   Rheological model of organic soils distortion

Shear Strength

The rheological model shown in Fig. 3 was used for the mathematical description of shear strength of organic soils.

The equation, describing shear strength τ_F of organic soils, can be expressed as

$$\frac{d\tau_F}{d\gamma} = G_1 + G_2 - \frac{G_1\beta_w}{\delta}\sinh\left[\alpha_w\left(\tau_F-G_2\gamma_{cr}\right)\right] \tag{8}$$

in which γ = shear strain; G_1 = modulus of elastic shear deformation of weaker contacts; G_2 = modulus of elastic shear deformation of stronger contacts; α_w, β_w = viscosity parameters; δ = value of constant rate of shear strain, γ_{cr} = critical shear strain. The derivation of the above equation is presented in Appendix II.

Eq. 8 can predict the undrained shear strength of samples tested at constant deformation rates.

Laboratory Tests

In this investigation the same peat and gyttja were used as in the consolidation tests. The test program consisted of two types of tests: undrained triaxial, creep and shear, tests. All samples were initially saturated and reconsolidated to the in situ stress conditions. The program of creep tests included seven series for peat and five series for gyttja. The samples of peat and gyttja, respectively, in all but the first series, were consolidated to 10, 20, 35, 40, 50, 80 kPa and 15, 20, 30, 40 kPa. After consolidation in all cases the deviator stress was applied in one increment. Vertical and lateral deformations of soil samples were measured at regular intervals. A total of 42 test were performed on the peat samples and 35 on the gyttja. Undrained triaxial shear tests were carried out under constant rate of axial deformation with the cell pressure held constant. Vertical and lateral deformations, as well as pore pressure and axial load, were measured at selected increments of axial strain. The experimental data were used for determination of parameters required in the rheological model. The method used in the evaluation of parameters G_1, G_2, α_w, β_w and γ_{cr} is shown in Appendix I. The results, presented in Fig. 4, indicate significant variation of parameters G_1, G_2, α_w and β_w with void ratio.

Increase in Shear Strength

On the basis of experimental results, the change in strength parameters during consolidation can be expressed as

$$G_1 = G_{1i}\left(G_{1f}/G_{1i}\right)^U \tag{9}$$

$$G_2 = G_{2i}\left(G_{2f}/G_{2i}\right)^U \tag{10}$$

$$\alpha_w = \alpha_{wi}\left(\alpha_{wf}/\alpha_{wi}\right)^U \tag{11}$$

$$\beta_w = \beta_{wi}\left(\beta_{wf}/\beta_{wi}\right)^U \tag{12}$$

Substituting the above equations into Eq. 8, the following equation to predict the change in shear strength during consolidation is obtained

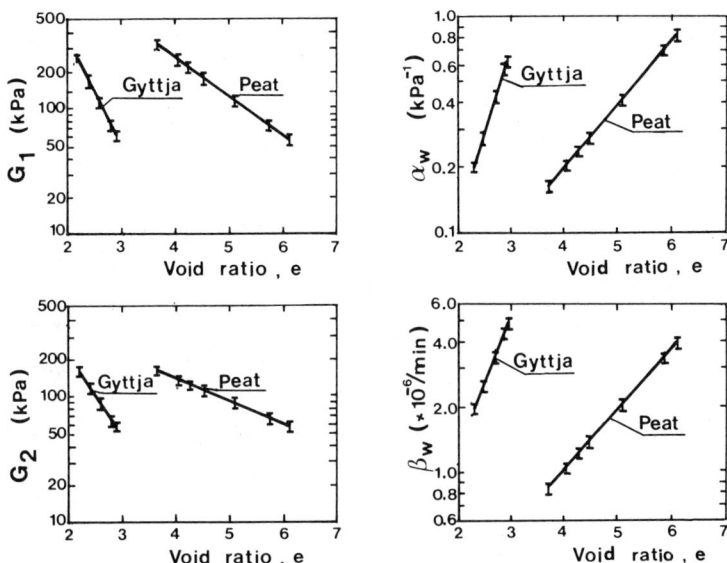

Fig. 4   Strength characteristics

$$\frac{d\tau_F}{d\gamma} = (G_{1i}E_1^U + G_{2i}E_2^U) - \frac{G_{1i}E_1^U\beta_{wi}B_w^U}{\delta} \sinh\left[\alpha_{wi}A_w^U(\tau_F - G_{2i}E_2^U\gamma_{cr})\right] \quad (13)$$

in which $E_1 = G_{1f}/G_{1i}$; $E_2 = G_{2f}/G_{2i}$; $A_w = \alpha_{wf}/\alpha_{wi}$; $B_w = \beta_{wf}/\beta_{wi}$.

FIELD TESTS

Field tests have been performed in the experimental site located in the Notec River Valley, where levees of more than 40 km length have been constructed on predominantly organic subsoil. Verification of the proposed method has been made on four test embankments (10,14). In this paper field investigations for a test embankment constructed on peat-gyttja subsoil are presented. Geotechnical characteristics of organic subsoil are shown in Fig. 5. The vertical displacements of subsoil were measured at seven locations with screw settlement gauges located at different depths under the embankment. Vane shear strengths were measured before construction and 1, 3, 5, 12, and 40 weeks after construction of the embankment. At each time, a sufficient number of vane tests were performed to describe the lateral variation of shear strength profiles. In addition, the mean vane shear strength, measured in the middle of each of the two organic layers,

SEDIMENTATION/CONSOLIDATIONS MODELS

were determined.

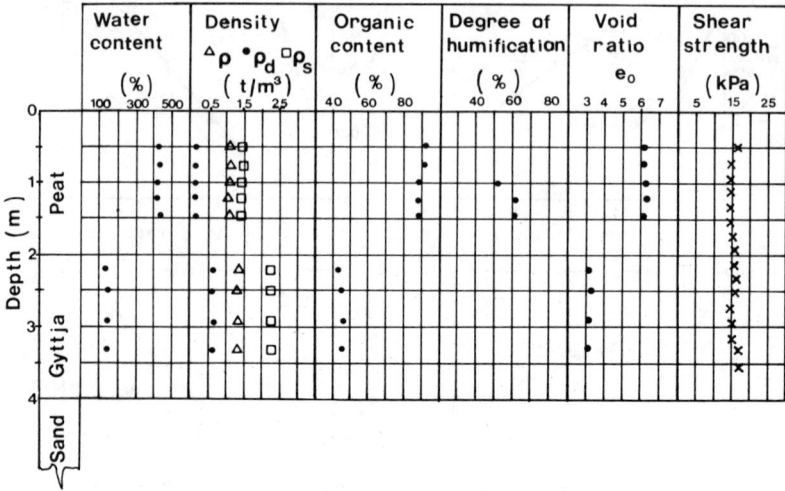

Water content (%) 100 300 500	Density $\Delta \rho$ $\bullet \rho_d$ $\square \rho_s$ (t/m³) 0,5 1,5 2,5	Organic content (%) 40 60 80	Degree of humification (%) 40 60 80	Void ratio e_0 3 4 5 6 7	Shear strength (kPa) 5 15 25

Fig. 5 Geotechnical properties of the test embankment
subsoil

For comparison purposes only, values of settlement and shear
strength measured under the center of the embankment are presented.
The measured field vane shear strength was corrected according to
correction factors which were the subject of a separate study (9,10).
The prediction of settlement of peat and gyttja has been made on the
basis of Eqs. 6 and 7. The comparison of observed settlements with
the predicted values is given in Fig. 6. The comparison of the
corrected vane shear strength of peat and gyttja with predicted values
calculated from Eq. 13 is given in Fig. 7. The rate of deformation
during vane test was substituted for parameter δ in the calculation of
shear strength by Eq. 13. The results of the comparison show good
agreement between measured and calculated values of settlement and
shear strength.

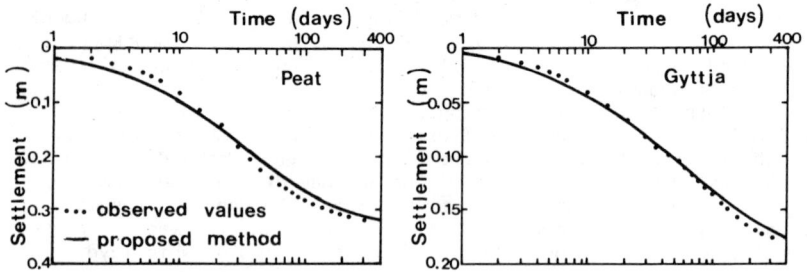

Fig. 6 Measured and predicted settlements

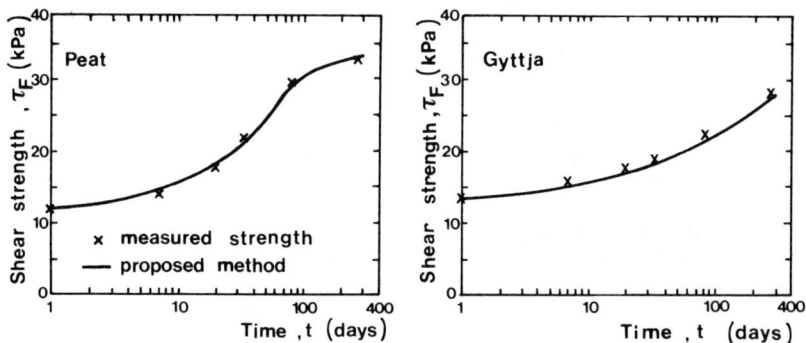

Fig. 7 Measured and predicted shear strength

CONCLUSIONS

A method of the consolidation-strength analysis of organic soils which takes into account both the consolidation process and consequent shear strength increase has been presented. This nature of analysis is necessary for the geotechnical design of safe loading schemes for organic subsoils characterized by very low virgin shear strengths and significant strength increases during consolidation. A one-dimensional consolidation analysis with nonlinear variation of the permeability and viscosity parameters with void ratio, is proposed. The presented consolidation theory describes immediate compression of gas bubbles, primary consolidation and secondary compression, assuming that all the phases of deformation take place simultaneously. Consolidation characteristics for amorphous peat and calcareous-organic soil (gyttja) have been determined from laboratory tests carried out in Rowe cells. The description of shear strength changes during consolidation has been based on a non-linear rheological model of the behaviour of organic soils during shear deformation. Strength characteristics have been estimated from tests carried out on these organic soils in triaxial cells. The equations presented make it possible to predict the consolidation and shear strength increase in loaded organic subsoil. Good agreement has been obtained between the values predicted and those observed at the test embankment site.

APPENDIX I - DETERMINATION OF PARAMETERS

Parameters for the Consolidation Model

The required parameters for the proposed consolidation model can be determined from oedometer test results, including the measurements of pore pressure and strain over long periods of time. The parameter a_1 is calculated on the basis of final values of pore pressure u_{1f} and strain ε_{1f} measured in undrained conditions. However, parameter a_2 is determined from final values of stress σ_f and strain ε_{2f} in drained conditions. The method of the evaluation of parameters a_1 and a_2 is shown in Fig. 8(a). In order to calculate parameters a_3, α and β, test results should be plotted as degree of consolidation versus time function (Fig. 8(b)). Then parameters C and D must be chosen from the theoretical relationship between degree of consolidation and time which best fits the actual test results. From the values of parameter a_2 and degree of consolidation U_1, parameter a_3 is calculated. The parameters α and β are obtained on the basis of C, a_2, a_3 and D, a_2, a_3, α parameters, respectively.

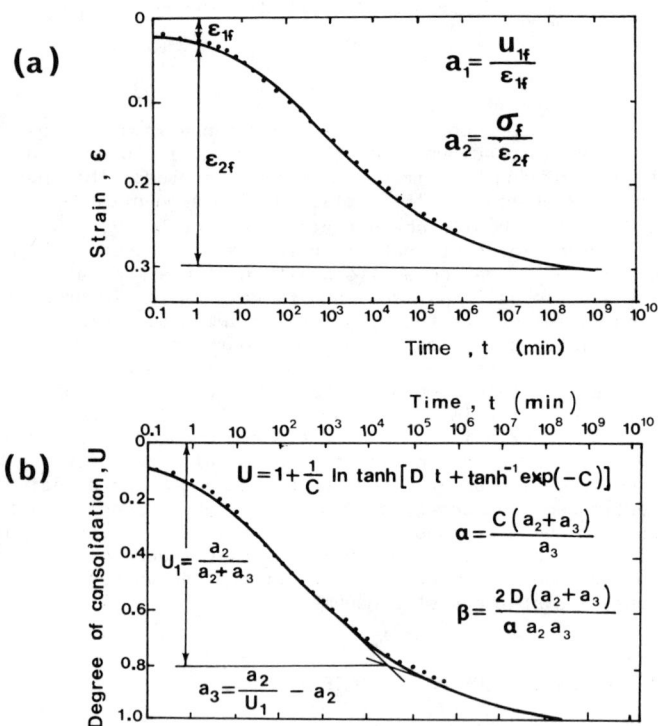

Fig. 8 Determination of consolidation parameters:
(a) a_1 and a_2; (b) a_3, α and β

Parameters for the Strength Model

The parameters G_1, G_2, α_w and β_w can be determined from the undrained triaxial creep test, however, parameter γ_{cr} is determined from the undrained triaxial shear test. To calculate parameters G_1 and G_2, the vertical strain-time relationship and vertical strain-rate vertical strain relationship are required. Parameter G_2 is determined on the basis of the amount of deviator stress $(\sigma_v - \sigma_h)$ and final vertical strain ε_{vf} which is obtained by extrapolation of the $\varepsilon_v - \dot{\varepsilon}_v$ relationship to $\dot{\varepsilon}_v = 0$ (Fig. 9(b)). To determine the parameter G_1,

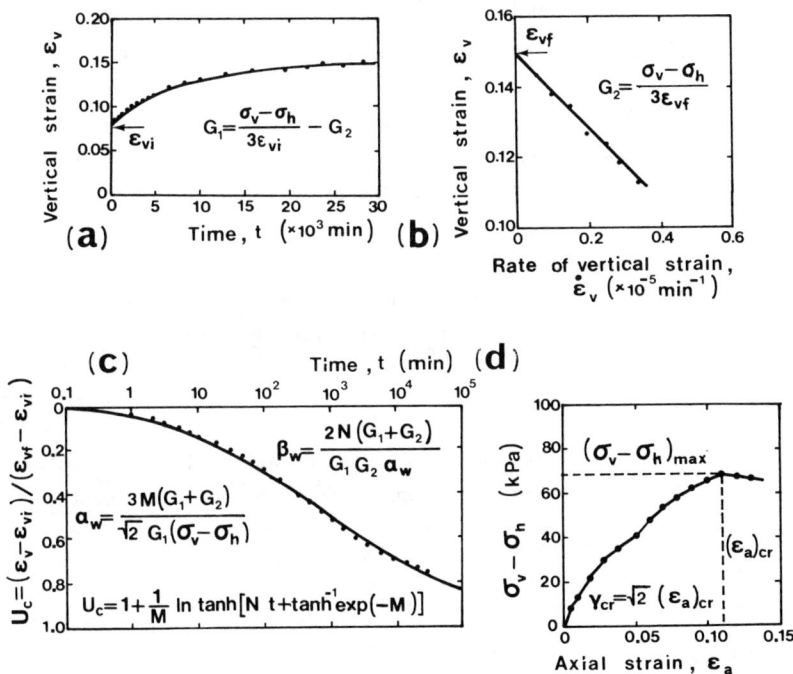

Fig. 9 Determination of strength parameters; (a) G_1;
(b) G_2; (c) α_w and β_w; (d) γ_{cr}

the initial vertical strain ε_{vi} from the ε_v-t relationship is obtained by extrapolation to t = 0. The equation yielding parameter G_1 is given in Fig. 9(a). In order to obtain parameters α_w and β_w, the degree of creep strain-time relationship is needed (Fig. 9(c)). These parameters are calculated using previously evaluated parameters and parameters M and N of the theoretical curve U_c-t that gives the best fit with measured values of the degree of creep strain with time. Method of the evaluation of parameter γ_{cr} from deviator stress-axial strain relationship obtained from the triaxial shear test is shown in Fig. 9(d).

APPENDIX II - RHEOLOGICAL MODEL FOR SHEAR STRENGTH

The model shown in Fig. 3 represents the relationship between shear stress τ and shear strain γ which develops when soil is subjected to shear. The stress τ in rheological model is equal to

$$\tau = \tau_1 + \tau_2 \tag{14}$$

in which τ_1 = shear stress in elasto-viscous element; τ_2 = shear stress in elasto-plastic element. The strain γ of the model is

$$\gamma = \gamma_{ep} = \gamma_{ev} \tag{15}$$

in which γ_{ep} = strain of elasto-plastic element; γ_{ev} = strain of elasto-viscous element. Elastic shear response of weaker contacts is represented by spring G_1, therefore

$$\tau_1 = G_1 \ (\gamma - \gamma_1) \tag{16}$$

in which G_1 = modulus of elastic shear deformation of weaker contacts; γ_1 = viscous shear strain. The viscous behaviour of those contacts is realized by the dashpot element, which is described according to the rate process theory (6) as

$$\dot{\gamma}_1 = \beta_w \sinh \ (\alpha_w \tau_1) \tag{17}$$

in which $\dot{\gamma}_1$ = rate of viscous shear strain; α_w, β_w = viscosity parameters. Elastic shear response of stronger contacts before the plastic limit has been reached is represented by spring G_2, therefore

$$\tau_2 = G_2 \gamma \tag{18}$$

in which G_2 = modulus of elastic shear deformation of stronger contacts. Using the above equations, the following equation is obtained

$$\frac{1}{G_1} \dot{\tau} = \frac{G_1 + G_2}{G_1} \dot{\gamma} - \beta_w [\alpha_w(\tau - \tau_2)] \tag{19}$$

Since shear stress τ depends on shear strain γ, we have the following equation, when τ_2 is lower than plastic limit

$$\frac{1}{G_1} \frac{d\tau}{d\gamma} \dot{\gamma} = \frac{G_1 + G_2}{G_1} \dot{\gamma} - \beta_w \sinh [\alpha_w(\tau - \tau_2)] \tag{20}$$

The proposed model is used for the mathematical description of shear strength of organic soils when soil is subjected to shear under constant rate of strain during shear deformation, as shown by

$$\dot{\gamma} = \delta = \text{const.} \tag{21}$$

in which δ = value of constant rate of shear strain. For the above conditions the following equation of model state can be written

$$\frac{d\tau}{d\gamma} = G_1 + G_2 - \frac{G_1 \beta_w}{\delta} \sinh [\alpha_w (\tau - \tau_2)] \tag{22}$$

Shear strength τ_F is defined as the limit shear stress τ, which is developed when

$$\gamma = \gamma_{cr} \tag{23}$$

in which γ_{cr} = critical shear strain. At this moment in the plastic element, the plastic limit Θ is reached, therefore

$$\tau_2 = \Theta \tag{24}$$

hence

$$\Theta = G_2 \gamma_{cr} \tag{25}$$

The equation describing shear strength of organic soils, derived on the basis of the above model, can be expressed as

$$\frac{d\tau_F}{d\gamma} = G_1 + G_2 - \frac{G_1 \beta_w}{\delta} \sinh [\alpha_w (\tau_F - G_2 \gamma_{cr})] \tag{26}$$

APPENDIX III - REFERENCES

1. Barden, L., "Primary and secondary consolidation of clay and peat", Geotechnique, Vol. 18, No. 1, 1968, pp. 1-24.
2. Barden, L., "Time dependent deformation of normally consolidated clays and peats", Journal of the Soil Mechanics and Foundations Division, ASCE, Vol. 95, No. SM1, 1969, pp. 1-31.
3. Berry, P.L., "Application of consolidation theory for peat to the design of a reclamation scheme by preloading", Quarterly Journal of Engineering Geology, Vol. 16, 1983, pp. 103-112.

4. Berry, P.L., Poskitt, T.J., "The consolidation of peat", Geotechnique, Vol. 22, No. 1, 1972, pp. 27-52.
5. Berry, P.L., Vickers, B., "Consolidation of fibrous peat", Journal of the Geotechnical Engineering Division, ASCE, Vol. 101, No. GT8, 1975, pp. 741-753.
6. Christensen, R.W., Wu, T.H., "Analysis of clay deformation as a rate process", Journal of the Soil Mechanics and Foundations Division, ASCE, Vol. 90, No. SM6, 1964, pp. 125-153
7. Dhowian, A.W., Edil, T.B., "Consolidation behaviour of peats", Geotechnical Testing Journal, GTJODJ, Vol. 3, No. 3, 1980, pp. 105-114.
8. Furstenberg, A., Lechowicz, Z.,Szymanski, A., Wolski, W., "Effectiveness of vertical drains in organic soils", Proceedings 8th European Conference on Soil Mechanics and Foundation Engineering, Vol. 2, Helsinki, Finland, 1983, pp. 611-616.
9. Golebiewska, A., "Vane testing in peat", Proceedings, 7th Danube-European Conference on Soil Mechanics and Foundation Engineering, Vol. 1, Kishinev, U.S.S.R., 1983, pp. 49-53.
10. Lechowicz, Z., "The increase in shear strength of organic soils during consolidation", Doctoral Thesis, Warsaw Agricultural University, Warsaw, 1982, (in polish).
11. Raymond, G.P., Wood, E.A., Hollingshead, G.W., "Consolidation of undisturbed fine fibrous peat", Proceedings, 4th International Peat Congress, Vol. 2, Otaniemi, Finland, 1972, pp. 209-219.
12. Samson, L., LaRochelle, P., "Design and performance of an expressway constructed over peat by preloading", Canadian Geotechnical Journal, National Research Council, Ottawa, Ontario, Vol. 9, No. 3, 1972, pp. 447-466.
13. Sasaki, H., "A case history of national road construction over peat desposits in Hokkaido, Japan", Proceedings, 34th Canadian Geotechnical Conference, Fredericton, New Brunswick, Canada, 1981, pp. 1-13
14. Szymanski, A., "Deformation characteristics of selected kinds of organic soils", Doctoral Thesis, Warsaw Agricultural University, Warsaw, 1982 (in polish)
15. Szymanski, A., Furstenberg, A., Lechowicz, Z., Wolski, W., "Consolidation of organic soils", Proceedings, 7th Danube-European Conference on Soil Mechanics and Foundation Engineering, Vol. 1, Kishinev, U.S.S.R., 1983, pp. 273-278
16. Wu, T.H., Resendiz, D., Neukirchner, R.J., "Analysis of consolidation by rate process theory", Journal of the Soil Mechanics and Foundations Division, ASCE, Vol. 92, No. SM6, 1966, pp. 229-248

Selfweight Consolidation of Very Soft Clay by Centrifuge

Masato Mikasa* and Naotoshi Takada**

1. Introduction

This paper reports a series of selfweight consolidation tests of a very soft clay by centrifuge and its analysis by Mikasa's consolidation theory.

Mikasa derived a generalized one-dimensional consolidation equation in 1960 (7), (9) to analyze the field consolidation test of very soft dredged clay in Osaka South Port (8), (9). This field test was conducted as the research work on the improvement of soft clay fill anticipated in the reclamation project of Osaka City (17). The new theory could successfully explain the behavior of a soft clay fill 2 m thick that was consolidated by combined load of selfweight and dewatering in the underlaid sand mats (8), (9), (17).

The equation was derived just in time at an early stage of the field consolidation test and produced a time-settlement curve that showed a remarkable fitting with one month's settlement record. The settlement of the fill traced the predicted curve correctly for four months till it settled in as short a time as one fourth of the Terzaghi theory's prediction. This success was in what Lamb classified as type-A prediction, and the validity of the new equation might be said to have been proved already by this field test -- its first application.

However it is only one case, and the selfweight consolidation is a problem to be investigated more extensively. Now, the new equation suggested that the centrifuge will provide an effective and powerfull tool for this problem, making it possible to simulate the big prototype behavior in a small model shortening the test period substantially. We have since developed Mark 1 to 5 centrifuges in our laboratory, and conducted several series of selfweight consolidation model tests by them for the past two decades (12), (13), (14), (15), (16). We observed the selfweight consolidation behavior of soft clays in a variety of prototype conditions both from basic (we dare not say "theoretical" here) and practical standpoints.

This paper presents four typical results in a test series conducted in 1976 - 1978 as a representative of our works on this problem. In this test series, our main target was the case of coastal reclamation project, and the initial consistency of the "soft clay" was so chosen as to make a certain soil structure that would not allow any particle segregation. The reason for this is twofold. Firstly, any consolidation equation, including Mikasa's, is not applicable for a soil-water suspension under the process of sedimentation that has no definite soil structure and, consequently, no effective stress (this may be called a pre-soil state). Secondly, from practical view-

*Professor, Civil Engineering Department, Osaka City University, Sugimoto Sumiyoshi-ku Osaka Japan
**Associate Professor, Civil Engineering Department, Osaka City University, Sugimoto Sumiyoshi-ku Osaka Japan

point, it is not necessary to study minutely the process of sedimentation, including Imai's "zone settling" (4) that may rather be identified as "soil", because such sorts of settlement will proceed and cease very rapidly before any engineering measures are to be applied in the field. Some discussion on this point will be given at the end of this paper.

2. General Consolidation Equation and Similarity Law in Centrifuge

The general equation of one-dimensional consolidation of saturated clay published by Mikasa in 1963 (9) is given as

$$\frac{\partial \zeta}{\partial t} = \zeta^2 \left[c_v \frac{\partial^2 \zeta}{\partial z_0^2} + \frac{dc_v}{d\zeta}\left(\frac{\partial \zeta}{\partial z_0}\right)^2 - \frac{d}{d\zeta}(c_v m_v \gamma')\frac{\partial \zeta}{\partial z_0} \right] \qquad (1)$$

where c_v is the coefficient of consolidation, m_v is the coefficient of volume compressibility (=$d\varepsilon/dp$, where ε is the compression strain and p is the effective pressure), t is time, ζ is the consolidation ratio (=f_0/f, f denotes the volume ratio which is equal to (1+e)), γ' is the submerged unit weight, z_0 is the original coordinate or the coordinate in the original state* in which the clay is assumed to have a certain uniform volume ratio f_0 and submerged unit weight γ' throughout the depth. z_0 is measured positively in downward direction.

This equation is free from the assumptions employed in the Terzaghi consolidation equation that 1) the coefficient of permeability k, 2) m_v, 3) c_v, 4) thickness of the clay layer and 5) the consolidation pressure are all constant during consolidation period, and 6) the selfweight of the clay does not affect the consolidation process.

The assumptions used in deriving Eq.(1) are 1) clay is homogeneous, 2) clay is saturated, 3) one-dimensional consolidation, 4) soil particls and water are incompressible, 5) Darcy's law is applicable, 6) f-log p and f-log k relations are not time-dependent, 7) soil water mixture should constitute a structure that carries an effective stress, though it may be feeble ; particle segregation is not allowed. This is a prerequisite to treat a soil-water mixture as a "soil", and usually is not stated explicitly as an assumption.

By using the expressions

$$\phi(\zeta) = c_v / c_{vo} \qquad (2)$$

$$T_v = c_{vo} t / (H_0/2)^2 \qquad (3)$$

$$Z = z_0 / H_0 \qquad (4)$$

Eq.(1) is transformed as

$$\frac{\partial \zeta}{\partial T_v} = \frac{\zeta^2}{4} \left[\phi(\zeta) \frac{\partial^2 \zeta}{\partial Z^2} + \frac{d\phi}{d\zeta}\left(\frac{\partial \zeta}{\partial Z}\right)^2 - \frac{d}{d\zeta}(\phi(\zeta) m_v \gamma') H_0 \frac{\partial \zeta}{\partial Z} \right] \qquad (5)$$

* Original state is not equivalent to the initial state from which the calculation starts ; the former is an imaginary state assumed for convenience of calculation of finite strain consolidation, while the latter is a given condition in which the soil need not be uniform. In ordinary cases the value of f_0 is taken equal to or larger than the maximum f value in the initial (generally non-uniform) state of the layer. For an initially uniform clay as in the present case the value of f_0 is usually taken equal to the initial volume ratio of the specimen, and the "original state" coincides with the "initial state".

where c_{vo} is the original c_v value, T_v is the nondimensional time, Z is the nondimensional depth, and H_o is the original thickness of the clay. Eq.(5) shows that in this case the H^2-similarity rule of time-consolidation relation is not applied in the gravitational force field because of the existence of H_o in the third term in the bracket. This term represents the effect of selfweight of clay and gives the consolidation process a significant effect when the clay is compressible (m_v is large) and/or its thickness H_o is large.

Now, if a model clay specimen having thickness H_o/n is subjected to a centrifugal acceleration of ng, where g is the acceleration of gravity, its submerged unit weight becomes $n\gamma'$. Then by replacing γ' and H_o in Eq.(5) with $n\gamma'$ and H_o/n, respectively, the equation remains the same. This indicates that the consolidations of the prototype and the centrifuged model are to proceed similarly as a function of T_v, and H^2-similarity rule is valid again between these two as follows :

$$\frac{t_m}{t_p} = \frac{H_m^2}{H_p^2} = \frac{1}{n^2} \qquad (6)$$

$$\frac{S_m}{S_p} = \frac{1}{n} \qquad (7)$$

where S denotes settlement and suffixes m and p indicate the model and the prototype, respectively. Examples to illustrate the validity of this similarity rule is shown in literature (15).

3. Method of Numerical Calculation

3.1 Finite Difference Form of Governing Equation

Governing equation, Eq.(1), is transformed into a finite difference form, as

$$\Delta \zeta_{z_0} = \frac{\Delta t\, c_{vo}}{(\Delta z_0)^2} \left[\phi(\zeta)(\zeta_{z_0+\Delta z_0} - 2\zeta_{z_0} + \zeta_{z_0-\Delta z_0}) + \frac{1}{4}\frac{d\phi(\zeta_{z_0})}{d\zeta}(\zeta_{z_0+\Delta z_0} \right.$$

$$\left. - \zeta_{z_0-\Delta z_0})^2 - \frac{\Delta z_0\, d}{2\, d\zeta}(\phi(\zeta_{z_0})\, m_v\, \gamma')(\zeta_{z_0+\Delta z_0} - \zeta_{z_0-\Delta z_0})\right] \qquad (8)$$

where $\Delta \zeta_{z_0}$ is an increment of ζ corresponding to the time increment Δt at the depth z_0 in the original coordinate. Referring to Fig.1, consolidation ratio at the time $t+\Delta t$ is then

$$\zeta_{z_0, t+\Delta t} = \zeta_{z_0, t} + \Delta \zeta_{z_0} \qquad (9)$$

Assuming linear f-log p and f-log c_v relations as shown in Fig.2, $\phi(\zeta)$, $d\phi(\zeta)/d\zeta$ and $d(\phi(\zeta)m_v\gamma')/d\zeta$ in Eq.(8) are expressed as the function of ζ as follows :

$$\phi(\zeta) = \exp\left[\frac{f_o}{0.4343\, Cc_v}\left(1 - \frac{1}{\zeta}\right)\right] \qquad (10)$$

$$d\phi(\zeta)/d\zeta = \frac{f_o}{0.4343\, Cc_v}\frac{1}{\zeta^2}\phi(\zeta) \qquad (11)$$

$$d(\phi(\zeta)\, m_v\, \gamma')/d\zeta = \frac{d\phi(\zeta)}{d\zeta}\frac{0.4343\, Cc}{f_o\, p}\zeta^2\, \gamma_o'$$

$$- \left(1 - \frac{0.8686\, Cc}{f_o}\zeta\right)\frac{\gamma_o'}{p}\phi(\zeta) \qquad (12)$$

where Ccv is the index of c_v change as defined in Fig.2.

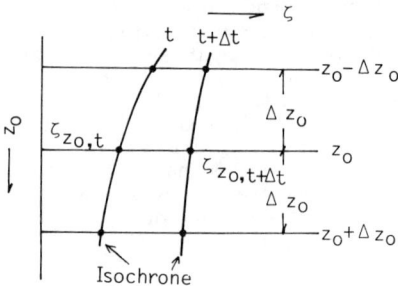

Fig.1   Schematic illustration
        of isochrones

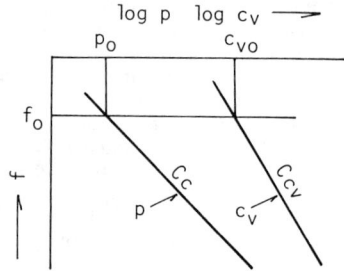

Fig.2   f-log p and f-log c_v
        relations

3.2 Initial and Boundary Conditions

In the present test series, the initial state of the clay layer was set homogeneous with a volume ratio f_o and an effective stress p_o determined from the f_o value on the f-log p relation of the clay. This p_o is also defined as the consolidation yield stress (pseud-preconsolidation stress) p_y of the clay.

In the case of selfweight consolidation, there is a region from the surface to a certain depth z_{oy}, where the effective overburden pressure is less than p_y and the volume ratio is kept as f_o under its selfweight as illustrated in Fig.3. This depth z_{oy} is given as

$$z_{oy} = p_y / \gamma_o' \qquad (13)$$

Thus the upper boundary condition of the clay is given as $f = f_o$ at $z = z_{oy}$. Although z_{oy} is negligibly small for a very soft clay, this boundary condition is a theoretical requirement.

The lower boundary condition when the bottom surface is pervious is that the effective stress is equal to the overburden effective pressure, which is constant during consolidation in the present test condition. The boundary condition at the impermeable base, where no seepage force exists, is given as the condition that the effective pressure gradient is equal to the submerged unit weight of the clay as follows :

$$\partial p/\partial z = \gamma'$$

using the relations $dz = dz_o / \zeta$ and $\gamma' = \gamma_o' \zeta$ we get

$$\partial p/\partial z_o = \gamma_o' \qquad (14)$$

Eq.(14) indicates that the gradient of an isochrones of effective stress against the original coordinate z_o at the impermeable boundary are the same as γ_o', a constant, during the consolidation as illustrated in Fig.4. This relation is conveniently used in numerical calculation.

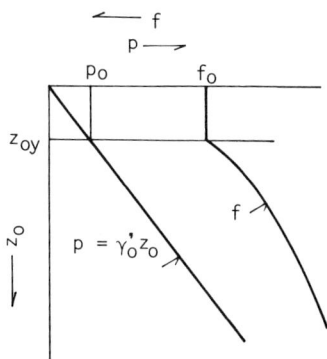

Fig.3 Over consolidation range

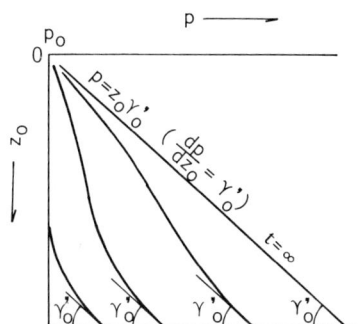

Fig.4 Boundary condition at
       impermeable base

4. Centrifugal Test

4.1 Testing Apparatus

Fig.5 shows the Mark 4 centrifuge developed in our laboratory (in operation 1975 - 1983). It is, as all of our centrifuges are, a swing bucket type without any device to fix the specimen container in flight. Therefore, the direction of the resultant body force always comforms to the vertical axis of the specimen. Its maximum acceleration at the nominal radius of 1.53 m is 200 g. For selfweight consolidation test, two steel vessels shown in Fig.6(a) are attached to both ends of the rotor, and two transparent acrylite specimen cylinders (Fig.6(b)) are set in them. One cylinder with impermeable base is for single drainage test, and the other with a pervious ceramic disk at the bottom is for double drainage test.

4.2 Material and Test Procedure

Test material is an alluvial marine clay taken from the reclaimed land of Osaka South Port with specific gravity G_s = 2.67, liquid limit w_L = 98.9 % and plasticity index I_p = 69.0 %. It was fully remoulded with sea water, sieved by 0.3 mm mesh and adjusted to the water content of 120 or 150 percent. After deaired in a vacuum chamber, it was poured into the specimen cylinder to a thickness of 10 cm, above which sea water was filled to prevent desiccation of the specimen. In the case of double drainage, the water levels at the upper and lower surface of the specimen were kept equal by means of vertical drain pipe connecting them (Fig.6(b)). After the specimen cylinders were set in steel vessels, the rotor was accelerated to give the specimen centrifugal acceleration up to 100 g in 30 to 40 seconds, and was kept in steady rotation. The origin of time was taken at the midpoint of accelerating time. The settlement of the clay surface, which was always very distinct to observe, was measured visually at proper intervals with the help of a synchronized stroboscope. For each test condition two or three specimens were prepared to obtain a set of isochrones of volume ratio : at 100 percent degree of consolidation and at some lesser degrees. When a specimen reached the pre-determined degree of consolidation, the test was terminated and the distribution of water content along the depth

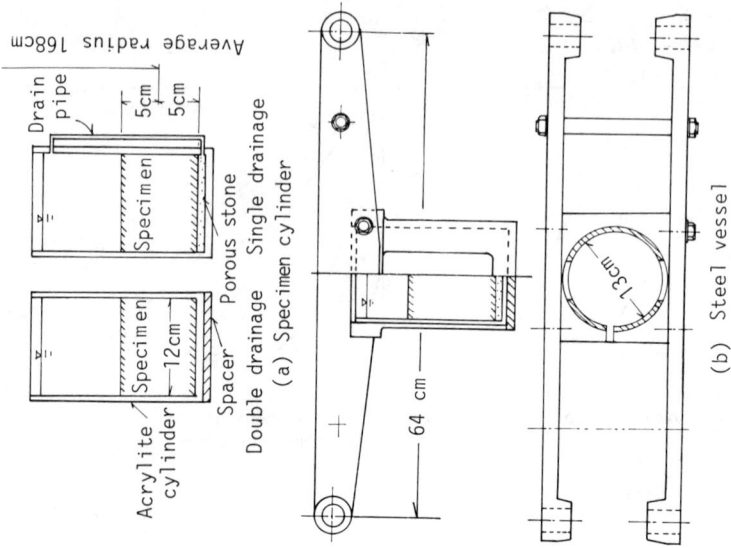

Fig.6 Steel vessel and specimen cylinders

(a) Specimen cylinder

(b) Steel vessel

Average radius 168cm

Drain pipe

5cm 5cm

Specimen

Porous stone

Single drainage

Specimen

Spacer

Acrylite cylinder

12cm

Double drainage

64 cm

13cm

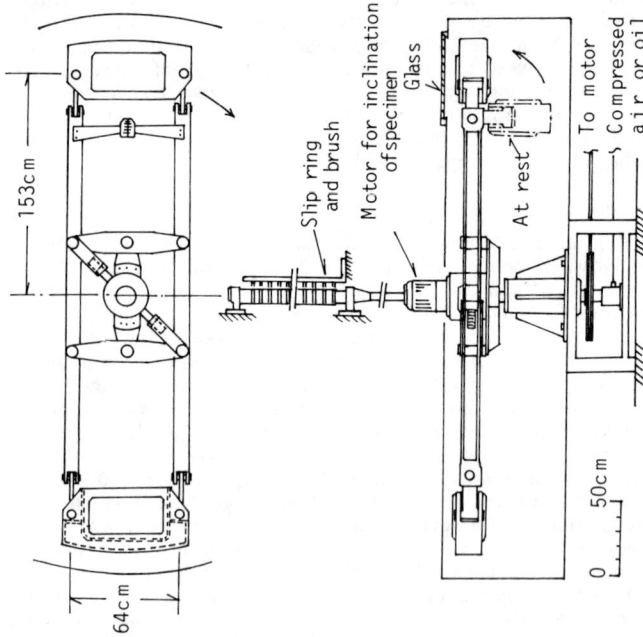

Fig.5 Centrifuge (Mark 4)

153cm

64cm

Slip ring and brush

Motor for inclination of specimen

Glass

At rest

To motor

Compressed air or oil

0   50cm

was quickly measured.

Throughout the test, the friction between the specimen and the acry-lite cylinder is considered negligibly small. More important is the problem of thin water path along the boundary between the specimen and the cyl-inder that is often seen in sedimentation test of soil water mixture that has much higher water content making the quantitative evaluation of the test difficult. Since the ratio of (specimen height/specimen diameter) was not so large and the water content was not too high, the effect of such water pass, if it had existed at all, was considered very small, and the con-dition of one-dimensional consolidation is considered to have been secured.

4.3 Consolidation Characteristics

Since the conventional oedometer test is not suitable to soft slurry-clay in low pressure range, consolidation parameters were determined from a series of selfweight consolidation test in centrifuge as follows.

The f-log p relation was obtained from the relationship between the volume ratio f and the overburden effective pressure p at the steady state after selfweight consolidation, where the value of p was obtained by sum-ming up the submerged unit weight γ', and both f and γ' were calculated from measured water content and specific gravity $G_s = 2.67$, assuming full saturation. Fig.7 shows several f-log p curves with different initial volume ratios. The scale of water content is also shown in the right side of the figure*. The f-log p relation with larger initial volume ratio occupies the upper position holding larger f value for whole stress range. The linearity of these curves show that the particle segregation did not take place ; in the case of clay slurries with much higher water content such as 500 % or more, particle segregation usually takes place and the f-log p line thus obtained show a marked drop at the rightest end that corresponds to the lowest part of sedimented soil that contains a lot of coarser particles (16). The deviation of the leftest measured points from the linear f-log p rela-tion may have some positive implications. But since the measurement of water content at the uppermost thin element of selfweight consolidated specimen was rather difficult, this deviation was ignored in the following analysis.

The coefficient of permeability k was obtained from the initial settle-ment rate in selfweight consolidation under single drainage condition. According to Mikasa's consolidation theory the upper part of a soft uniform clay layer under single drainage condition settles initially at a constant rate so that its submerged unit weight γ_0' balances the upward seepage force j. Then

$$\gamma_0' = j = i \gamma_w$$

Combining this relation and Darcy's law, v = ki, the velocity of upward seepage flow relative to the soil skeleton is expressed as

$$v = k \frac{\gamma_0'}{\gamma_w} \tag{15}$$

---

* Since the specific gravity of the clay was $G_s = 2.67 = 8/3$, the water con-tent w and the void ratio e relate as $e = G_s w = (8/3)w$, or $e/w = 8/3$. Thus $\Delta f = \Delta e = 0.8$ corresponds to $\Delta w = 30$ percent. Since this G_s value is considered to be a standard of inorganic soils, this relation is conven-iently used for calculating e or f of a saturated soil from w or vica versa.

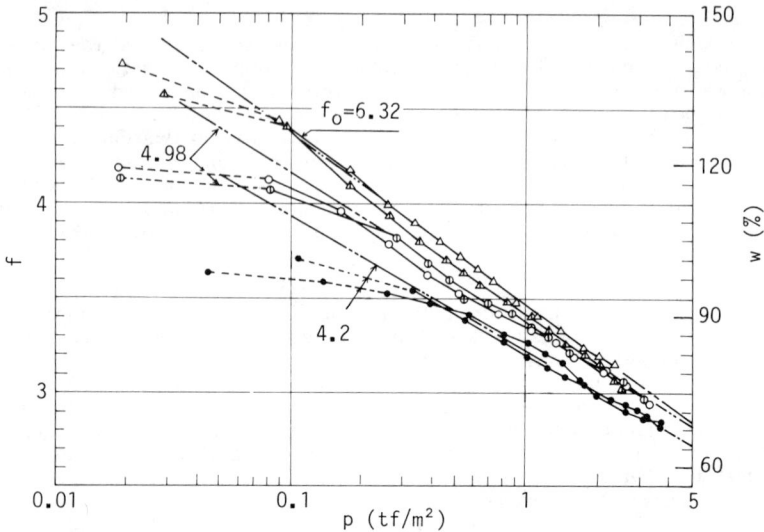

Fig.7   f-log p relations

This velocity of pore water flow is the same, but opposite in direction, as the settlement rate of upper part of the clay layer (including the surface of the layer) in the stationary water.   The values k and γ_0' being both the functions of volume ratio, this settlement rate depends only on the initial volume ratio of the clay, and is constant for some time range in the early stage of consolidation.   Fig.8 shows two time-settlement curves of self-weight consolidation under single drainage condition, which are the same curves as in Figs.11 and 12 but in an arithmetic time scale.   They show linear time-settlement relations up to about 50 percent degree of consolidation*.

A series of selfweight consolidation test by centrifuge with different initial volume ratios produced the relationship between k and f shown in Fig.9, which has a continuity with the data by oedometer test of the clay with lower initial f values, the value k in the latter being calculated in the way proposed by Mikasa (11), (17).

The consolidation parameters to be used in the analysis of the model test were determined as follows.   The value k was assumed to be a unique function of f as shown in Fig.9, and m_v was calculated from the linear f-log p relation assumed as in Fig.7 by using the relation

$$m_v = 0.4343 C_c/(f \cdot p) \tag{16}$$

where Cc is the compression index.   The value of c_v as shown in Fig.10 was calculated for each volume ratio using the corresponding values of k and m_v.   The thin lines show the change of c_v value for several different initial volume ratios, and the thick line is the locus of their initial values. Both lines have linearity on semi-log plot.

Table 1 shows the consolidation parameters of the clay with two diffe-

* This settling process is not equivalent to the "hindered settling" of soil suspension without any effective stress that is identified as pre-soil state.

rent initial volume ratios, f_o = 4.2 and f_o = 4.98. To represent the change in c_v value during consolidation, the ratio δ is introduced as

$$\delta = c_{vf}/c_{vo} \qquad (17)$$

where c_{vo} is the initial, and c_{vf} is the final c_v value at the lower boundary where the minimum volume ratio f_f will be seen in selfweight consolidation. The index Ccv defined in Fig.2 is related to δ as

$$Cc_v = \frac{\log c_{vf} - \log c_{vo}}{f_o - f_f} = \frac{\log \delta}{\Delta f} \qquad (18)$$

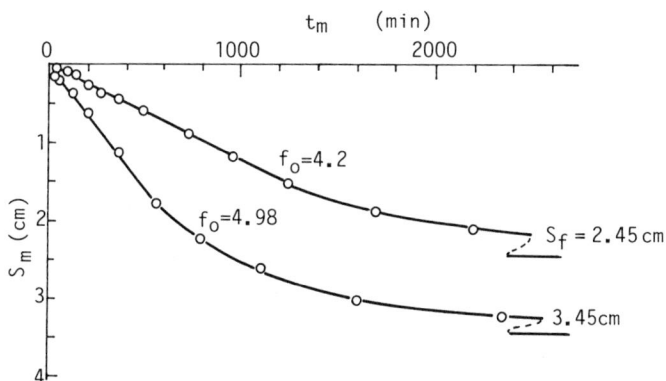

Fig.8   Settlement to arithmetic time

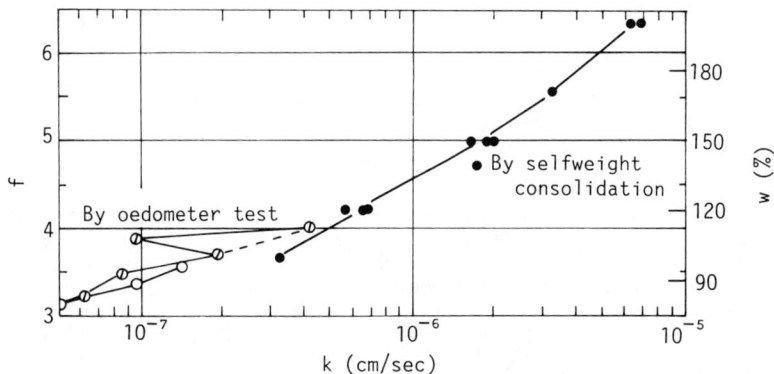

Fig.9   f-log k relation

The values of δ to be used in the analysis were roughly estimated with the initial and final c_v values shown in Fig.10 : $\delta = 8$ for the initial volume ratio $f_o = 4.2$ and $\delta = 16$ for $f_o = 4.98$. Smaller δ values were also tried for comparison.

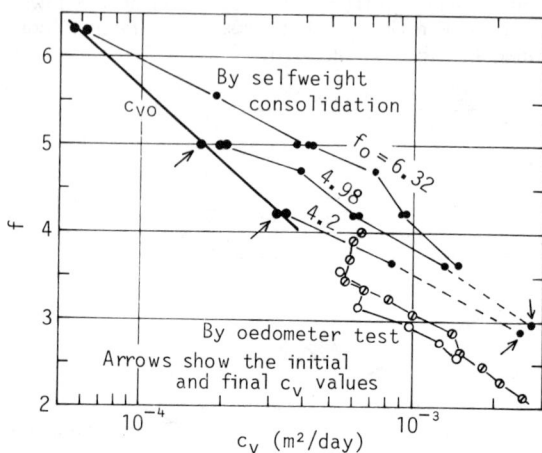

Fig.10   f$-$log c_v relation

Table 1   Consolidation characteristics

Initial condition			f-log p relation	c_{vo}
w_i	f_i	w_{Ri}*		
120%	4.2	1.31	$f=3.22-0.71 \log p$	$3.2 \times 10^{-4} \text{m}^2/\text{day}$
149	4.98	1.75	$f=3.31-0.77 \log p$	1.7×10^{-4}

* Relative water content $w_R=(w-w_P)/(w_L-w_P)$ $(=I_L)$
**p expressed in tf/m^2

5. On Numerical Calculation

The numerical calculation of difference equation was performed taking the space mesh as $\Delta z_o = H_o/80$, and the time mesh as $\Delta T_v (=c_{vo} \Delta t/(H_o/2)^2)$ = $1/(12800 \delta)$. These values were determined after some preliminary trial calculations to secure sufficient stability of calculation. Note here that δ is an important factor to determine the time mesh. When the space and time mesh are not properly selected, an unstable oscillation of isochrone will occur, particularly in the uppermost part of the clay layer where ζ shows sharp change in its value and gradient in a small increment of z_o. This is peculiar to the selfweight consolidation of very soft clay, and needs special attention in the finite difference calculation of such cases.

To obtain good approximation, particularly in the early stage of consolidation, we adopted following procedure : the value ζ in Eqs.(10) to (12) and ζ^2 in Eq.(8) is taken as the average value $(\zeta_{ZO} + \Delta\zeta_{ZO} /2)$ within the time span of Δt, and effective stress p in Eq.(12) is obtained from this value of ζ. Actual calculation was done by succesive approximation as follows : the first approximation of increment of consolidation ratio $\Delta\zeta_{ZO,1}$ is obtained by taking ζ in the functions as ζ_{ZO}. Then second approximation $\Delta\zeta_{ZO,2}$ is obtained by taking ζ as $(\zeta_{ZO} + \Delta\zeta_{ZO,1} /2)$, and so forth.

Actually in our calculation very small time mesh Δt and space mesh Δz_O were chosen to secure the stability of numerical calculation, and consequently the successive approximation converged very rapidly.

6. Comparison of Experiment and Calculation

Time-consolidation relations of centrifuged models together with those calculated are shown in Figs.11 (f_O= 4.2) and 12 (f_O= 4.98) as U_s-log t curves on logarithmic time scale both of the model and prototype, where U_s denotes the percent degree of consolidation in terms of settlement. Now, it may be said a matter of course that the initial part of the test curves of single drainage condition show good agreement with those calculated, because k used in the calculation was originally determined from observed initial settlement rates in a series of centrifuged model test including the tests shown in the figures. The overall agreement between the experimental and calculated time-settlement curves and isochrones, therefore, is what we have concern about. For the single drainage case, the curves calculated by taking δ = 16 for f_O= 4.98 and δ = 8 for f_O= 4.2 show good agreement with experimental ones. The latter, however, might show a better fitting by decreasing c_{vo} and increasing δ, both a little. For the double drainage cases, the observed consolidation rates in both cases were a little lower than those calculated, showing apparent smaller c_{vo} value than in single drainage cases, the reason of which is not clarified yet.

Figs.13 to 16 show measured and calculated isochrones of volume ratio plotted against the original coordinate z_o. The figures reveal the mechnism of the selfweight consolidation quite different from the ordinary consolidation by the applied load that starts from the pervious boundary. The soft clay structure settles in the stationary water, and consolidates from the bottom even in the case of impermeable lower boundary. In the case of pervious lower boundary, a consolidation of ordinary mechanism is added to the settling mentioned above. Eq.(1), of course, takes into account both of the two mechanisms of consolidation. In this context, we had better consider the movement of soil structure in the stationary water than the pore water flow through the soil structure, not only for the soil suspension (hindered settling) but also for a soft clay that has a definite structure. This way of thinking may help to understand the mechanism of selfweight consolidation.

In all cases, however, the upper part of the layer or the part in low effective stress shows a marked decrease in f value up to 100 percent degree of consolidation in comparison with the calculation. This corresponds to the deviation of f-log p curves from the linear ones assumed in the analysis (Fig.7). The whole deviation of the model behavior from that calculated, however, seems difficult to be explained only from the deviation in f-log p relations. No definite conclusion is obtained yet, but it might partly be attributed to some peculier property of the clay structure itself, e.g. an electro-chemical inter-particle forces in its very soft state, which is not considered in Mikasa's consolidation theory.

Calculated time-settlement curves in Figs.11 and 12 were obtained

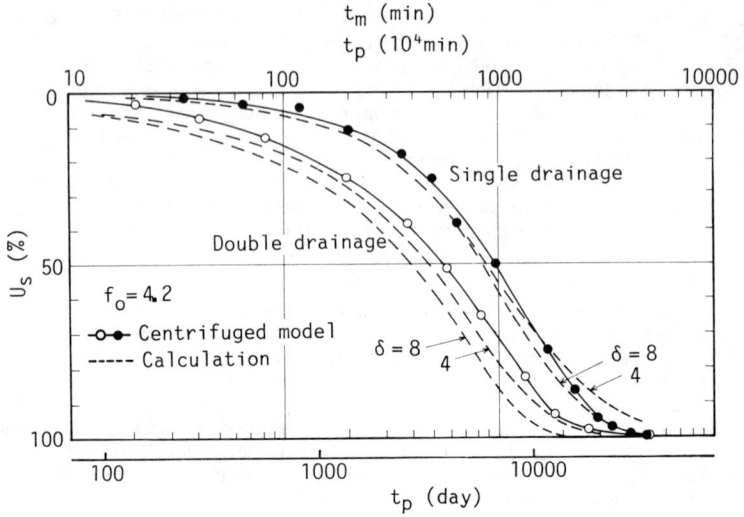

Fig.11   Time-consolidation curves (f_o= 4.2)

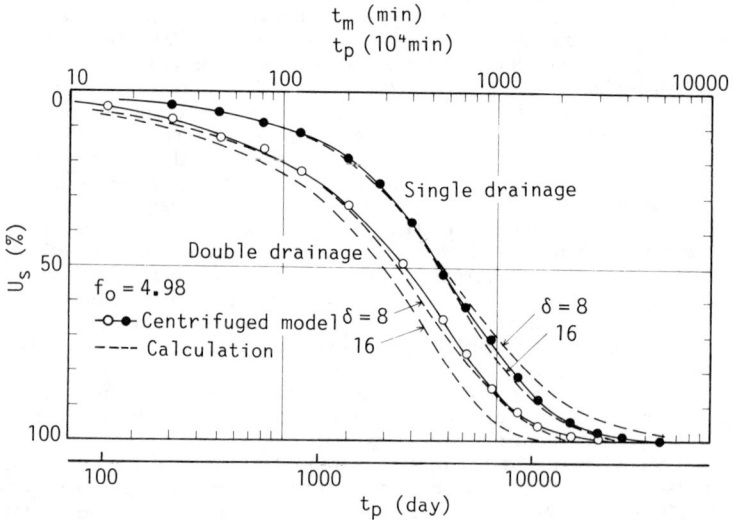

Fig.12   Time-consolidation curves (f_o=4.98)

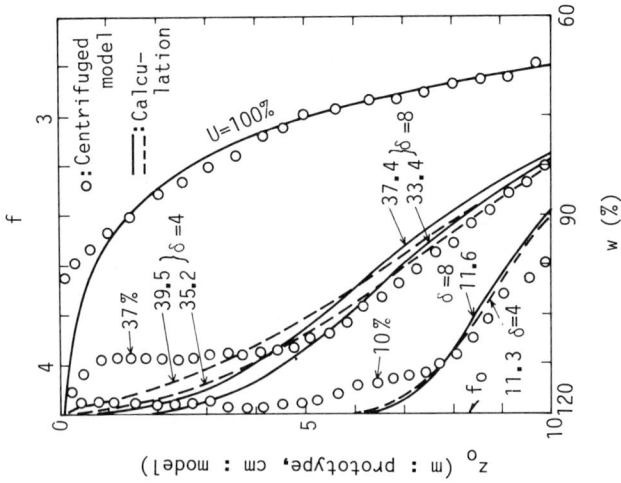

Fig. 14 Isochrones of volume ratio (single drainage f_o =4.2)

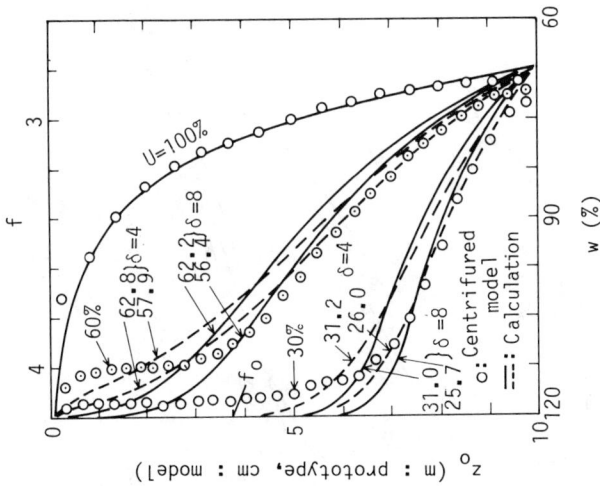

Fig. 13 Isochrones of volume ratio (double drainage f_o =4.2)

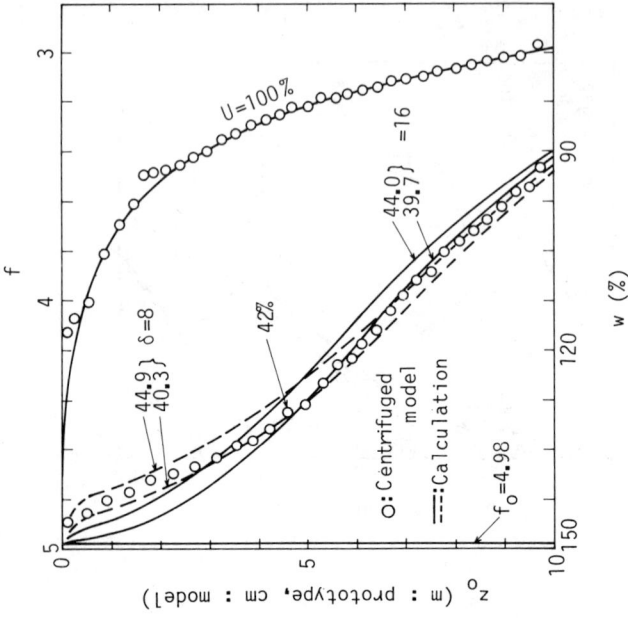

Fig.16   Isochrones of volume ratio
(single drainage f_o=4.98)

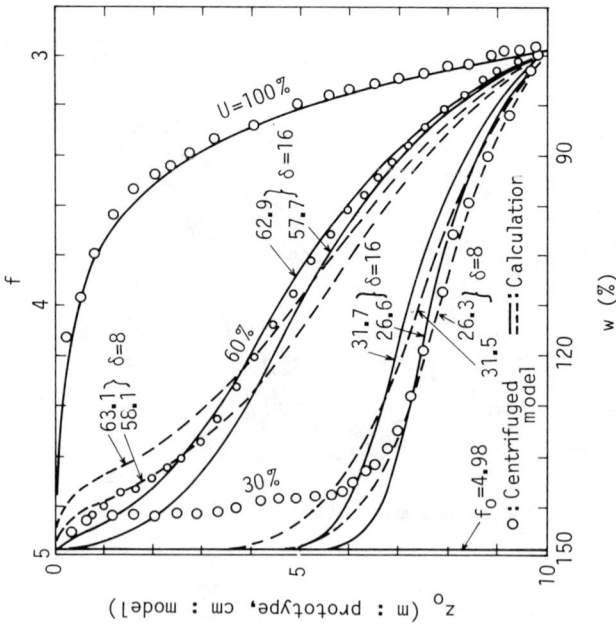

Fig.15   Isochrones of volume ratio
(double drainage f_o=4.98)

from Figs.13 to 16 by integrating the area between f_o-z_o as follows :

$$s = \frac{1}{f_o} \int_0^{H_o} \Delta f \, dz_o$$

because $\Delta f/f_o$ gives the nominal (arithmetic) strain $\bar{\epsilon}$ (=decrease in depth/ initial depth) of each element.

7. Conclusive and Supplementary Remarks

In the test series presented here, centrifuged models and numerical models by Mikasa's consolidation theory showed good agreement in time-settlement curves and isochrones of selfweight consolidation of very soft clay. Though some minor discrepancies still remain unsolved, major aspects of the soil behavior in the actual stress level were clarified sufficiently, and we can use centrifuge or Mikasa's theory in dealing with this problem for practical application.

Now, a few comments will be added here to make our standpoint clear in reference to the recent achievements by other authors that apparently concern with the content of this paper.

As theoretical achievements, Gibson et al (1967, 2), Gibson and Schiffman (1981, 3), Monte and Krizek (1976, 19) published their papers that stand almost on the same propositions as Mikasa's (1963, 9) in somewhat different forms. Pane and Schiffman (1981, 20) compared the theory by Gibson et al (2) with Mikasa's and showed that they are interchangeable. But they criticized the latter in that it could be applied only to the case in which a soft clay had a homogeneous state (a certain volume ratio throughout the depth) as its initial condition. This is quite a misunderstanding. Already in 1960 (8),(9), a field consolidation test with a non-homogeneous initial soil condition was successfully analyzed by Mikasa's equation. Another example of its application to a complicated ground condition is reported in another paper in this symposium (18). Thus the theoretical achievements on nonlinear, finite strain and selfweight consolidation by the other authors were not referred to in this report ; though they might be useful in some other occasions, we need not them in our present research.

As to the research work on the sedimentation or settling of very soft soil suspension, the notion of "hindered settling" should be discussed. This was first described by Kynch (5), and was studied extensively by McRoberts and Nixon (6) to clarify the mechanism by which a thick dispersion of grains and water settles and consolidates into a soil that has effective stress. In these papers the hindered settling is defined for the state of suspension of soil grains in water without any effective stress. Imai (4) applied this notion of sedimentation for clay, and discussed his observations on the very soft clay behavior. Been and Sills (1) also investigated this problem and tried some theoretical approach on some simplified assumptions.

Now, it should be noted that Mikasa described the difference between selfweight consolidation and sedimentation already in 1963 (9). According to his definition, sedimentation occurs in the soil suspension without any effective stress. Since the soil suspension does not have a certain structure, particle segregation will take place. In that case the phenomenon is beyond the scope of "consolidation theory" because the soil is not formed yet. But when the soil grains flocculate and begin to make a soil structure keeping their mutual spatial relationship, there should emerge an effective stress between the grains, though it may be feeble. If the soil structure becomes stable enough to have a certain stress-strain relationship, f-log p relation may be obtained by some suitable testing method. We call this state a "very soft clay", and its settlement can be analyzed by Mikasa's consolidation theory quantitatively (actually there is an upper limit of water content for a soil from feasibility of calculation).

Miscellaneous qualitative findings on sedimentation by Imai are very interesting from the scientific viewpoint. In a practical field problem, however, the settling of soil with such very high water content as to cause "zone settling" will proceed very rapidly in a rate of 2 mm/min = 2.88 m/day or more (4), and need not be analyzed in most practical cases at all. According to Mikasa's experiences in the field, the water content at the surface of a newly placed very soft dredged marine clay seldom exceeded twice of its liquid limit in a couple of weeks after the last filling, during which the "sedimentation" and "zone settling" should have taken place (9), (10). The problem for engineers is how the soil of such consistency will settle thereafter. The initial state of the soft clay chosen in the present research work of selfweight consolidation was based on these experiences.

Conclusively, abundant information about sedimentation of soil suspension, though they may provide a useful knowledge on the initial state of soil after it settled, mainly relates to another phase of soil history (rather a pre-history) from what this paper dealt with.

Appendex I.   Summary of Mikasa's General One-dimensional Consolidation Theory

A general theory for 1) one-dimendional consolidation of 2) saturated clay with 3) homogeneous consolidation characteristics, published by Mikasa in 1963, is briefly summarized in the following.

The assumptions employed in this theory, besides above stated 1), 2) and 3) that define the scope of the problem, are the following four : 4) soil grains and water are incompressible, 5) Darcy's law is valid, 6) the compressibility and permeability of soil are the function of volume ratio and not time-dependent and 7) the soil suspension without any effective stress is excluded.

Using assumption 1), 2), 4), the following equations are derived from the continuity of pore water flow.

Non-steady
(during consolidation)
$$\frac{\partial \varepsilon}{\partial t} = \frac{\partial v}{\partial z} \qquad (1)$$

steady
(after consolidation)
$$v = v_0 \qquad (2)$$

where $d\varepsilon$ (= $-df/f$) is the increment of compressive strain (f = 1+e ; volume ratio) and v is the superficial velocity of the pore water flow. z and v are measured positively in downward direction. Assumption 5) gives

$$v = k \cdot i \qquad (3)$$

where k is the coefficient of permeability and i is the hydraulic gradient.
The seepage force of pore water acting on the clay structure is

$$j = i \cdot \gamma_w \qquad (4)$$

Thus the total body force acting on the clay structure downwards is

$$\frac{\partial p'}{\partial z} = j + \gamma' \tag{5}$$

where γ' is the submerged unit weight of clay, and p' is the effective stress. Assumption 6) allows us to define the volume compressibilty m_v by

$$m_v = \frac{d\varepsilon}{dp'} \tag{6}$$

Combining Eqs.(3), (4), (5) and (6), we obtain

$$v = c_v \left(\frac{\partial \varepsilon}{\partial z} - m_v \gamma'\right) \tag{7}$$

where c_v (= $k/m_v \gamma'$) is the coefficient of consolidation.

Inserting Eq.(7) into Eqs (1) and (2), and taking c_v as a function of ε, we obtaine the following consolidation equations.

Non-steady $\quad \dfrac{\partial \varepsilon}{\partial t} = c_v \dfrac{\partial^2 \varepsilon}{\partial z^2} + \dfrac{dc_v}{d\varepsilon}\left(\dfrac{\partial \varepsilon}{\partial z}\right)^2 - \dfrac{d}{d\varepsilon}(c_v \, m_v \gamma')\dfrac{\partial \varepsilon}{\partial z}$ $\qquad$ (8)

Steady $\quad \dfrac{d\varepsilon}{dz} = m_v \gamma' + \dfrac{v_0}{c_v}$ $\qquad$ (9)

Now in the case of highly compressible clays, z coordinate of each clay element changes its value during consolidation process, and consequently Eqs.(8) and (9) cannot duly be integrated for a finite strain. To overcome this difficulty, a new notion of original coordinate z_0 (z coordinate in the original state in which the clay is supposed to have a homogeneous state with a certain volume ratio f_0 throughout the layer) is introduced together with the following three quantities concerning compressive strain.

Consolidation ratio $\quad \zeta = f_0 / f$ $\qquad$ (10)

Natural strain (logarithmic strain) $\quad \varepsilon = \displaystyle\int_{f_0}^{f} - df / f = \log_e(f_0 / f) = \log_e \zeta$ $\qquad$ (11)

Nominal strain (arithmetic strain) $\quad \bar{\varepsilon} = \displaystyle\int_{f_0}^{f} - df / f_0 = (f_0 - f)/f_0 = 1 - 1/\zeta$ $\qquad$ (12)

The relationship of the three variables ε, $\bar{\varepsilon}$ and ζ is illustrated in Fig.A-1 together with the examples of volume ratio change for the cases of $f_0 =$ 5, 4, 3 and 2.

Using these quantities Eqs.(8) and (9) are transformed into the following equations integrable for the case of finite strain.

Non-steady $\quad \dfrac{\partial \zeta}{\partial t} = \zeta^2 \left[c_v \dfrac{\partial^2 \zeta}{\partial z_0^2} + \dfrac{dc_v}{d\zeta}\left(\dfrac{\partial \zeta}{\partial z_0}\right)^2 - \dfrac{d}{d\zeta}(c_v \, m_v \gamma')\dfrac{\partial \zeta}{\partial z_0}\right]$ $\qquad$ (13)

Steady $\quad \dfrac{d\zeta}{dz_0} = m_v \gamma' + \dfrac{v_0}{c_v}$ $\qquad$ (14)

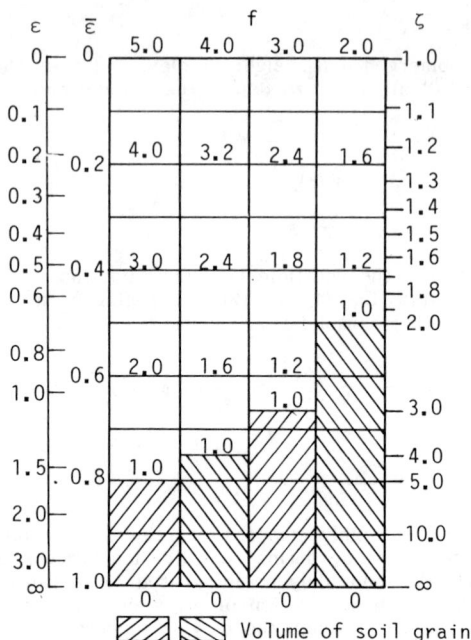

Fig.A-1 Relation between ε, $\bar{\varepsilon}$, ζ and f

These equations have a very wide generality being free from the following six assumptions which are necessary for the Terzaghi theory : 8) finite strain and 9) selfweight do not affect the consolidation ; 10) c_v, 11) consolidation pressure, 12) k and 13) m_v remain constant during consolidation.

Assumptions 8), 9) and 10) may effectively be used to simplify Eqs.(13) and (14), when the effects of such factors are considered to be small.   A series of equations obtained in this way were applied for many cases and showed good agreement with the experimental results (8), (9), (12), (13), (14), (15), (18). The effects of finite strain and selfweight of clay were both found to increase considerably the consolidation rate of compressible clays.

If assumptions 8), 9) and 10) are applied together, consolidation equations are reduced to the following simple forms

Non-steady $$\frac{\partial \varepsilon}{\partial t} = c_v \frac{\partial^2 \varepsilon}{\partial z^2}$$ (15)

Steady $$\frac{d\varepsilon}{dz} = \frac{v_o}{c_v}$$ (16)

Since Eq.(15) is of the same form as the Terzaghi equation, it would not be worthwhile transforming the former into the latter by applying three additional assumptions 11), 12) and 13), and restricting unduly the applicability of the consolidation theory. There is abundant evidence, in the numerous consolidation test data on compressive strain, that supports Eq.(15), and that has long been considered to support the Terzaghi consolidation theory.

Appendex II. References

1. Been, K. and Sills, G. C., "Selfweight Consolidation of Soft Soils : An Experimental and Theoretical Study," Geotechnique, vol.31, No.4, pp.514—535, 1981.
2. Gibson, R. E., England, G. L. and Hussey, M. J. L., "Theory of One--dimensional Consolidation of Saturated Clays," Geotechnique, vol.17, pp.261-273, 1967.
3. Gibson, R. E. and Schiffman, R. L., "The Theory of One-dimensional Consolidation of Saturated Clays, II Finite Nonlinear Consolidation of Thick Homogeneous Layers," Canadian Geotechnical Journal, vol.18, pp.280-293, 1981.
4. Imai, G., "Settling Behavior of Soil Suspension," Soils and Foundations, vol.20, No,2, pp.61-77, 1980.
5. Kynch, G. J. "A Theory of Sedimentation," Trans. Farady Soc., No.48, pp.166-176, 1952.
6. McRoberts, E. C. and Nixon, J. F., "A Theory of Soil Sedimentation," Canadian Geotechnical Journal, vol.13, pp.294-310, 1976.
7. Mikasa, M., "The Consolidation Theory of Soft Clay," Proc. of Autumn Annual Meeting of JSSMFE (in Japanese), 1960.
8. Mikasa, M., Tsutsumi, M., Akune, S. and Kubo, S., "Field Test of Improvement Method of Soft Clay by Sand Mat," Tsuchi-to-Kiso, JSSMFE, Special Issue, No.4, (in Japanese), 1961.
9. Mikasa, M., "The Consolidation of Soft Clay -- A New Consolidation Theory and its Application," Kajima Shuppan-Kai (in Japanese), 1963.
10. Mikasa, M. and Maeda, K., "Settlement and Soil Volume in a Reclaimed Land," Proc. of 19th Annual Convention of JSCE, III-56, (in Japanese) 1964.
11. Mikasa, M., "Determination of Consolidation Parameters from Oedometer Test," Proc. of Annual Convention of JSCE, III-7, (in Japanese), 1964.
12. Mikasa, M., Takada, N. and Kishimoto, Y., "Selfweight Consolidation Test in Centrifuge (1st Report)," Proc. of 20th Annual Convention of JSCE, III-25 (in Japanese), 1965.
13. Mikasa, M., Takada, N. and Kishimoto, Y., "Selfweght Consolidation Test in Centrifuge (2nd Report)," Proc. of Annual Convention of JSCE Kansai Branch, III-8 (in Japanese), 1965.
14. Mikasa, M. and Takada, N., "Selfweight Consolidation Test in Centrifuge (3rd Report)," Proc. of 21th Annual Convention of JSCE, III-46 (in Japanese), 1966.
15. Mikasa, M. and Takada, N., "Significance of Centrifugal Model Test in Soil Mechanics," Proc. of 8th ISSMFE, vol. , pp.273-278, 1973.
16. Mikasa, M., Takada, N. and Li, K., "Consolidation Characteristics of Very Soft Clay," Proc. of 11th Annual Meeting of JSSMFE, pp.185-186 (in Japanese), 1976.
17. Mikasa, M. and Ohnishi, H., "Soil Improvement by Dewatering in Osaka South Port, Geotechnical Aspects of Coastal Reclamation Project in Japan," Proc. of 9th ISSMFE, Case History Volume, pp.639-664, 1981.
18. Mikasa, M. and Takada, N., "Investigation of settlement of Kobe Port

Island," Proc. of Symposiun on Prediction and Validation of Consolidation, 1984.

19. Monte, J. L. and Kriezek, R. J., "One Dimensional Mathematical Model for Large Strain Consolidation," Geotechnique, vol.26, No.3, pp.495-510, 1976.

20. Pane, V. and Schiffman, R. L., "A Comparison Between Two Theories of Finite Strain Consolidation," Soils and Foundations, vol.21, No.4, pp.82-84, 1981.

21. Takada, N., Imai, G. and Kiyama, M., "Settlement of Dredged Soft Clay -- In a Series of Consolidation Problem," Tsuchi-to-Kiso, JSSMFE, pp.101-109 (in Japanese), 1979.

The Use of Soil Mechanics Capabilities in a
General Purpose Finite Element Program

By Gustav A. Nystrom[1], M. ASCE

ABSTRACT

Over the last few years a number of soil mechanics capabilities
have been added to a general purpose nonlinear finite element program.
The soils capabilities include a critical state soil mechanics constitutive relation, effective stress coupling between the solid and fluid
phases, a finite strain theory, and a permeability which varies
throughout the numerical solution. This paper describes the soils
capabilities, points out unique features of implementation in a
general purpose code, and describes some numerical considerations.

The paper makes use of illustrative solutions of one-dimensional,
plane strain, and axisymmetric problems. The one-dimensional
consolidation problem shows the effects of finite strain and a
solution dependent permeability which can result in a less permeable
"crust." Other problems show unstable strain softening behavior and
stable hardening at very large strains. Finally, a more complicated
axisymmetric problem outlines how the capabilities can be used to
evaluate the installation, consolidation, and axial loading of a deep
pile driven into normally consolidated clay.

INTRODUCTION

The finite element method has seen widespread application to the
solution of geotechnical problems. This is due to the simplicity with
which the method treats irregular boundaries, the adaptability of the
method to various nonlinear material models (e.g. clay, sand, rock,
etc.) and the proliferation of computer programs available for such
analyses.

Most soils applications have employed special-purpose codes
developed specifically to handle a limited range of problems. For
example, there are special-purpose codes which handle only
one-dimensional consolidation problems, and others which handle only
one-dimensional wave propagation problems.

An alternative to special-purpose codes is a general-purpose
finite element code. A general-purpose code is one which has a wide

[1]Research Specialist, Exxon Production Research Company, Houston, Texas.

range of computational capabilities and is designed for widespread usage. A general-purpose code will likely solve static, dynamic, and heat-dissipation problems employing a great variety of constitutive relations (elastic, plastic, viscous, etc.).

Many considerations are involved in the choice between using a special- or general-purpose program. Some considerations are: reliability, level of verification, ability of the program to treat variations of the problem at hand, ease of use, quality of documentation, level of support provided by the code developers, availability of pre- and post-processors, size of computer available, ease of making modifications, cost of acquiring the program, numerical efficiency, and robustness. Many special-purpose codes are excellent research tools in the hands of their developers. On the other hand, general-purpose codes are aimed at production-oriented usage by a large number of different users.

Based on these considerations, and anticipating a broad range of applications, it was decided to implement a number of soils capabilities in the general purpose computer program ABAQUS (Reference 7). This code is acknowledged as the "state of the art" in nonlinear general purpose computer programs (Reference 5), and has very strong algorithms for treating geometric, material, and boundary nonlinearities.

Over the last several years, the soils capabilities added to ABAQUS include constitutive models appropriate for modeling clays and sands, a finite strain theory, an algorithm for the efficient solution of coupled fluid-solid equations associated with effective stress analyses, and a permeability which depends on the current state of compaction. These capabilities are intended for use with plane-strain, axisymmetric, and three-dimensional elements.

The first part of this paper briefly describes each of the soils capabilities. Then the second part makes use of example solutions to illustrate the various capabilities and point out associated numerical considerations. Most of the illustrative problems are academic demonstrations of individual soils capabilities. The last problem is a real engineering problem sketched out to show how the various capabilities can be combined.

INTRODUCTION TO SOILS CAPABILITIES

This section provides an introduction to the soils capabilities that have been incorporated in the general purpose program. More detailed and all-inclusive documentation is contained in References 7 and 16.

Coupled Effective Stress Analysis - One major difference between soils and conventional engineering materials is that soils may contain three phases (solid grain, pore liquid, and pore gases). Since each phase behaves very differently, modelling the mechanical behavior of the soil composite is highly complex.

If the soil is fully saturated with a pore fluid, the effective stress principle can provide quite reasonable results. It states that the total stress at a point is made up of the sum of pore pressure in the fluid phase and effective stress carried by the solid phase. The pore fluid dissipation has been found to obey a D'Arcy's dissipation relation, while the effective stress controls the deformation of the solid. Such an effective stress formulation was first investigated by Terzaghi (Reference 17) for one-dimensional conditions. Subsequently, Biot (References 1 and 2) extended Terzaghi's theory to three dimensional conditions.

The first soil capability consists of the capacity to solve the coupled solid-fluid problem efficiently. The solution of fluid and solid degrees of freedom is carried out simultaneously. Usually, the time step increment is selected automatically as the solution progresses. The maximum change in pore pressure is compared with a user-specified tolerance to select the next step size. Automatic time step selection is an invaluable feature, since for an efficient solution the initial time step should be orders of magnitude smaller than the final time step.

It should be noted that for consolidation problems, element size and time step are related. If one wants to solve the "skin" problem which takes place at very small times right at the soil surface, then the mesh and time step must be selected accordingly. On the other hand, if one wants to solve the problem involving a significant volume of soil, a different mesh and time step should be selected. This issue is discussed in detail in Reference 20, where a simple criterion is suggested for selecting the minimum step size Δt in terms of the typical element dimension at the boundary Δh. That criterion is

$$\Delta t > \Delta h^2 \, \gamma_w \, / \, (6 \, E \, k)$$

where E is the Young's modulus, k is the permeability, and γ_w the unit weight of the fluid.

For a more detailed discussion of effective stress analysis and the associated coupled equations, please refer to a soil mechanics text. Illustrative Problems One and Four will discuss effective stress solutions of consolidation problems.

Variable Permeability - Permeability is the measure of a soil's resistance to fluid flow. Many analyses assume that permeability is a constant. But the soil's resistance to fluid flow is known to depend strongly on the level of compaction (Reference 8).

For this implementation, the user is allowed to define how the permeability varies with calculated void ratio (a measure of compaction). Illustrative Problem One will show how this effect can create an impermeable crust which significantly affects the consolidation time.

Modified-Cam-Clay Constitutive Relations - One of the most popular constitutive relations for modelling the mechanical behavior of clay is called modified-Cam-clay. It is one of the family of critical state plasticity models developed by Roscoe and his Cambridge colleagues (References 14 and 15). It captures many of the observed features of clay using a fairly simple constitutive relation. Unlike some critical state models, its yield surface has no discontinuous corners.

The modified-Cam-clay yield surface may be stated as

$$(p-a)^2 + (\frac{q}{M})^2 = a^2$$

where

p = mean normal stress invariant = $-\frac{1}{3}(\sigma_{xx} + \sigma_{yy} + \sigma_{zz})$

q = Mises stress invariant = $[\frac{1}{2}(\sigma_{xx} - \sigma_{yy})^2 + \frac{1}{2}(\sigma_{yy} - \sigma_{zz})^2$

$\quad + \frac{1}{2}(\sigma_{zz} - \sigma_{xx})^2 + 3(\sigma_{xy}^2 + \sigma_{yz}^2 + \sigma_{zx}^2)]^{1/2}$

M = input soil parameter
and
a = yield surface semi-radius.

One key feature of modified-Cam-clay is that plastic strain increments are normal to the yield surface. In other words, the flow rule is associated. A second key feature is that the size of the yield surface is controlled by plastic volume changes.

Figure 1 illustrates the yield surface ellipse in p-q space. If the state of stress at a point lies inside of the yield surface, the strain increment is elastic. If the state of stress lies on the yield surface, elastic and plastic strain increments are summed. Under some yield stress conditions, the plastic volume decreases, and the yield surface expands. For other yield stress conditions, the plastic volume increases, and the yield surface shrinks. Finally, for the intermediate critical state conditions, shear strain can continue indefinitely at constant stress conditions. The three types of plastic deformation are illustrated on Figure 1. They correspond to strain-hardening, strain-softening, and perfect plasticity.

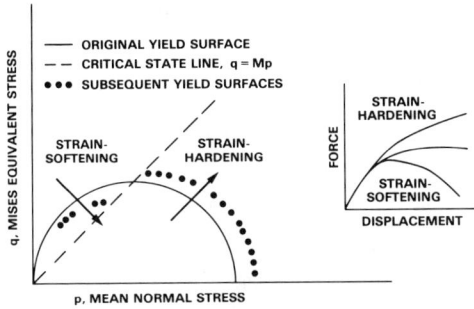

FIG. 1. MODIFIED-CAM-CLAY YIELD SURFACE AND STRAIN HARDENING

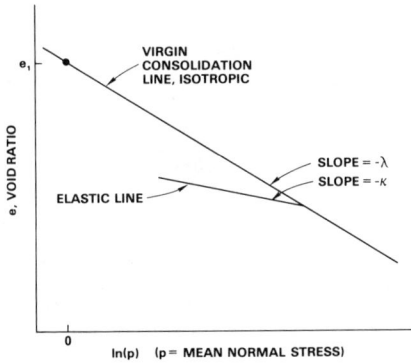

FIG. 2. MODIFIED-CAM-CLAY PRESSURE – VOLUME RELATIONS

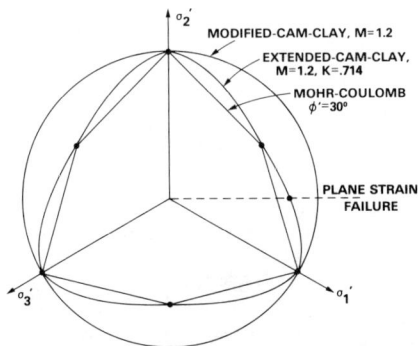

FIG. 3. COMPARISON OF THREE DIFFERENT FAILURE SURFACES IN THE π PLANE

Figure 2 illustrates the pressure-volume relationship for modified-Cam-clay. For isotropic stress conditions, one line represents virgin compression behavior, while the second line represents elastic behavior.

To illustrate the simplicity of modified-Cam-clay, the list below summarizes the required effective-stress soil parameters.

obtainable from a consolidation test:

e_1 = void ratio at the intersection of the virgin
    consolidation line and $p = 1$
k = permeability
κ = elastic logarithmic bulk modulus
λ = plastic logarithmic bulk modulus

obtainable from a triaxial test:
M = ratio of q to p at the critical state
a second elastic parameter

For additional information about modified-Cam-clay relations, please see References 14, 15, and 21. Problems Two and Three will illustrate simple solutions.

Extended-Cam-Clay Constitutive Relations - The extended-Cam-clay relations represent an improvement over the modified-Cam-clay constitutive relations. One major fault of modified-Cam-clay is that, contrary to laboratory observation, the undrained shear strength predicted for a triaxial extension test is the same as that predicted for a triaxial compression test. The extended-Cam-clay relations correct the situation by introducing the effect of a third stress invariant in a simple way. The Cam-clay extension was developed by J. D. Murff based on the concepts presented in References 22 and 19.

The extended-Cam-clay yield surface is given by

$$(p-a)^2 + \left(\frac{q}{Mg} \right)^2 = a^2$$

where
$$g = \frac{2K}{(1+K) + (1-K) (r/q)^3} ;$$

K is an input parameter which governs the shape of the yield surface in the π plane (for which mean stress, p is held constant)

$$= \frac{\text{triaxial extension shear strength}}{\text{triaxial compression shear strength}} ;$$

and r = third stress invariant
= [4.5 determinant (stress deviator matrix)]$^{1/3}$.

The function g introduces the third stress invariant, r, in order to modify the yield surface somewhat. To match Mohr-Coulomb predictions at triaxial compression and triaxial extension, one sets

$$M = \frac{6 \sin \phi}{3 - \sin \phi} \quad \text{and} \quad K = \frac{3 - \sin \phi}{3 + \sin \phi}$$

Figure 3 shows the modified-Cam-clay, extended-Cam-clay, and generalized-Mohr-Coulomb yield surfaces in the deviatoric (π) plane. (The deviatoric plane plots principal stresses on a plane of constant mean normal stress, p). Figure 3 shows the extended-Cam-clay yield surface to be as smooth as that for modified-Cam-clay. Yet it comes much closer to the generalized-Mohr-Coulomb yield surface.

Setting the input parameter K to 1.00, extended-Cam-clay becomes identical to modified-Cam-clay. If K is less than 7/9, it ceases to be convex. Yet, numerical solutions with such non-convex surfaces have revealed no unusual numerical problems.

References 7 and 16 contain additional information about extended-Cam-clay. Problem Four will make use of this constitutive relation.

Basic Sand Capabilities - Some basic sand capabilities have also been implemented. They are mentioned here only for the sake of completeness.

The basic sand capabilities are a generalization of the Drucker-Prager elastic-perfectly-plastic relations. One generalization allows the user to specify an associated flow rule, no plastic volume change, or any condition in between. A second generalization is that effects of the third stress invariant can be accounted for as in the extended-Cam-clay relations.

Finite Strain Theory - Most geotechnical analyses employ an infinitesimal strain theory. For such a theory, there is a linear relation between strain and displacement. For linear-elastic materials, load-displacement results are linear. The basic assumption for this theory is that the initial geometry is indistinguishable from the deformed geometry.

However, for many geotechnical problems, the displacements and strains are definitely not infinitesimal. Examples include pile installation and many consolidation problems. In these cases, the nonlinear kinematic effects may be important.

There have been several investigations of the finite deformation of soil (References 3, 4, 6, and 13). This implementation is based on References 3, 4, and 10.

One key question is how the infinitesimal strain rate relation between stress and strain is applied to the finite strain theory. For this application, the soil constitutive relations are carried over directly. But the infinitesimal strain theory's "stress" and "strain" are replaced by the finite strain theory's "Kirchoff stress" and "logarithmic strain" respectively.

Problem One will illustrate the effect of finite strain on a one-dimensional problem. Problem Three will show that modified-Cam-clay material behaves well up to very large strains. Finally, Problem Four will sketch out the use of finite strain theory for a pile installation simulation.

ILLUSTRATIVE PROBLEMS

Problem One: Terzaghi Consolidation - The one dimensional Terzaghi consolidation problem (Reference 18) illustrates salient features of coupled effective stress - pore fluid flow problems. A fully saturated column of soil is loaded instantaneously and the surcharge is held constant. Initially, all of the surcharge is carried as excess pore pressure in the fluid. As drainage proceeds, the pore pressure dissipates and the effective stress on the soil skeleton increases. This increasing effective stress results in a reduction of the soil column's height.

The plane strain model analyzed is sketched in Figure 4. The material is treated as linear elastic, and fluid drainage is allowed along the top surface. This illustrative model does not reflect any actual soil conditions. Three analyses were run: one using small strain theory, one using finite strain theory and constant permeability, and one using finite strain theory and a permeability which varies linearly with void ratio as shown on Figure 4.

Figure 5 shows the consolidation histories for all three cases. Different steady state compaction results are predicted by the small strain and finite strain analyses due to the different strain measures used. The steady state results predicted by the two finite strain analyses agree, but consolidation is markedly slower in the case where permeability varies with void ratio. This reflects the physical effect of a low permeability crust delaying dissipation of excess pore pressure in soil below the crust. Such crust effects have received quite a bit of attention for simple geometries. For example, see Reference 6.

It should be noted in Figure 5 that the solution time scale spans many orders of magnitude. This fact makes automatic time stepping essential for efficient solution of consolidation problems. The technique used in ABAQUS of basing the time step for integration of the diffusion equation on the maximum calculated change in pore pressure worked well in this case. The small strain solution used a total of 22 time increments to complete the analysis, with time increments ranging from 0.001 minutes to 2.0 minutes. It seems clear that cost-efficient, production consolidation analyses require such an automatic time incrementation scheme.

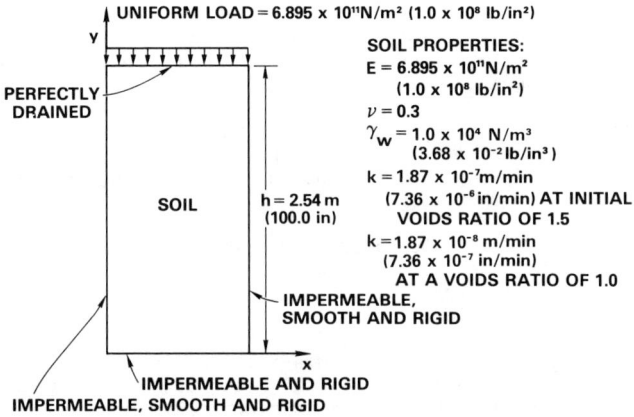

FIG. 4. TERZAGHI CONSOLIDATION: PROBLEM DEFINITION

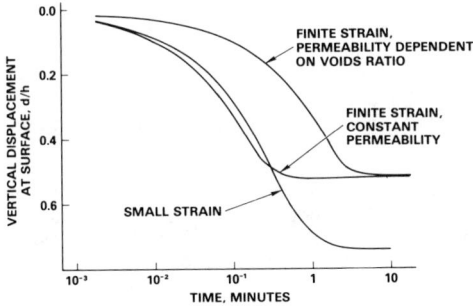

FIG. 5. TERZAGHI CONSOLIDATION: SOLUTIONS

Problem Two: Drained Triaxial Test - The analysis of a drained
triaxial test provides a good illustration of the modified-Cam-clay
implementation.   For this axisymmetric problem, the lateral stress is
held constant.   The axial displacement is brought to failure in
compression and then in extension.   Small strain theory is assumed.
Figure 6 shows the model dimensions and material properties.

Modified-Cam-clay exhibits strain hardening behavior in
compression and strain softening behavior in extension.   This was
illustrated in Figure 1.

Figure 7 shows the calculated stress path.   We note that in p-q
space a drained triaxial test normally has a slope of +3 or -3.   That
figure shows that during compression, the yield surface expands, while
during extension it shrinks.

Figure 8 shows the calculated load-displacement curve.   Starting
at the initial, isotropic stress conditions, the solution is elastic.
Then it shows strain hardening.   Upon reversal of the specified axial
displacement, the solution is elastic until the expanded yield surface
is reached.   Past this point, plastic strains are associated with
strain softening, and the calculated axial load is reduced.   Finally,
the stress state ends near the critical state condition, and the soil
continues to deform at constant stress conditions.

Figures 7 and 8 indicate the cumulative number of increments to
obtain the solution.   They show that very small increments were
required to obtain a meaningful solution during the strain-softening
phase.   This illustrates the fact that an automatic increment-size
selection feature is essential for obtaining efficient and reliable
soil mechanics solutions.

The solution during the strain-softening stage is particularly
interesting.   At this stage, all of the soil is unstable in the sense
that the "stiffness" is negative.   For such unstable materials, the
deformation tends to become localized.   The nature of the displacement
solution can be controlled by introducing a non-uniformity or flaw
(Reference 12).

This solution was obtained by specifying the platen motion.   But
it should be noted that the code does have the capability to treat
unstable response under load control (a modified Riks algorithm), and
this capability may be used in conjunction with soils analyses which
exhibit instabilities.

Problem Three: Simple Shear at Very Large Strain - In the terminology
of plasticity theory, the modified-Cam-clay relations involve a
combination of kinematic and isotropic strain-hardening.   In other
words, the size and center of the yield surface both change with
straining.   For some implementations of finite strain theory with
kinematic plasticity, an unusual result has been found: for a

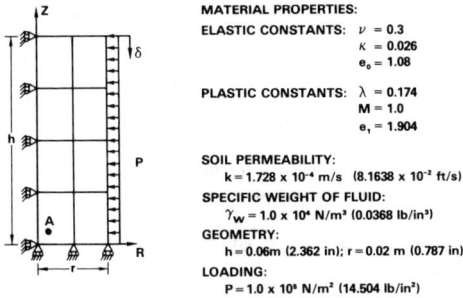

MATERIAL PROPERTIES:

ELASTIC CONSTANTS: $\nu = 0.3$
$\kappa = 0.026$
$e_o = 1.08$

PLASTIC CONSTANTS: $\lambda = 0.174$
$M = 1.0$
$e_1 = 1.904$

SOIL PERMEABILITY:
$k = 1.728 \times 10^{-4}$ m/s (8.1638×10^{-2} ft/s)

SPECIFIC WEIGHT OF FLUID:
$\gamma_W = 1.0 \times 10^4$ N/m³ (0.0368 lb/in³)

GEOMETRY:
h = 0.06m (2.362 in); r = 0.02 m (0.787 in)

LOADING:
$P = 1.0 \times 10^5$ N/m² (14.504 lb/in²)

FIG. 6. TRIAXIAL TEST: GEOMETRY AND SOIL PROPERTIES

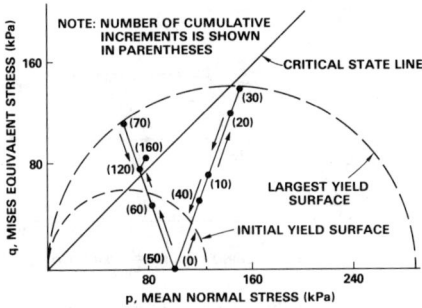

FIG. 7. TRIAXIAL TEST: STRESS PATH

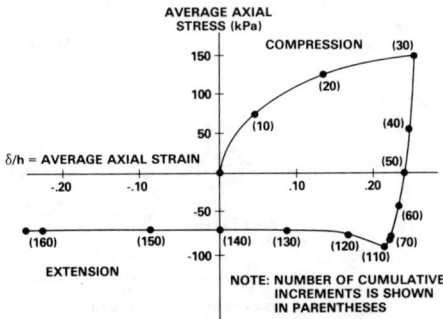

FIG. 8. TRIAXIAL TEST: LOAD-DISPLACEMENT CURVE

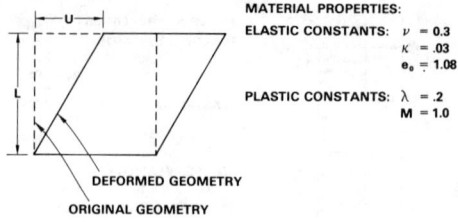

FIG. 9. SIMPLE SHEAR: GEOMETRY AND SOIL PROPERTIES

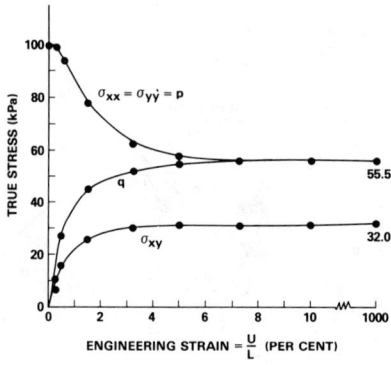

FIG. 10. SIMPLE SHEAR: CALCULATED STRESSES

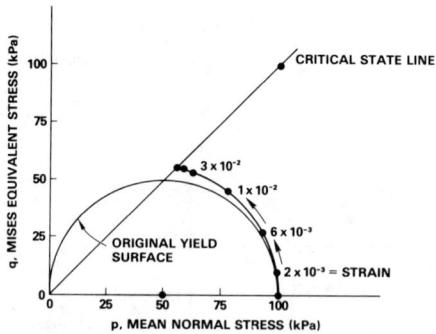

FIG. 11. SIMPLE SHEAR: STRESS PATH

monotonically increasing simple shear strain, the calculated shear stress oscillates for large strains (Reference 9).

A simple shear solution was carried out to very large strains in order to verify that the stresses can behave reasonably. For this simple demonstration problem, nine isoparametric plane-strain elements were used. Boundary displacements were specified for this finite strain solution. Figure 9 identifies the problem geometry and soil parameters.

Figures 10 and 11 show the calculated homogeneous stresses. Figure 10 plots true (Cauchy) stress components versus engineering strain, while Figure 11 plots the stress path in p-q space. The solution is seen to be very well behaved and there is no evidence of stress oscillation.

For the simple shear problem, there is no volume change. Since modified-Cam-clay hardening is governed by changes in plastic volume, the amount of hardening is small. (The small elastic volume changes offset small plastic volume changes). It may be that the small amount of hardening rules out the possibility of stress oscillations at large strain.

A second consequence of the lack of volume change is that there are no changes in pore water pressure. Thus, the existence of fluid in the pores has no effect on the solution.

Numerically, an interesting aspect of this solution is that a fairly large number of increments (about 100) was required to obtain an accurate solution. The algorithm which is usually used to automatically select the increment size failed to select small enough increments. The algorithm control is based on nodal equilibrium, which is usually a good measure of solution accuracy. Apparently, for this very simplified problem equilibrium is not a good measure of solution accuracy. To control the accuracy of the stress-strain integration a fairly small increment was required.

Problem Four: Axisymmetric Pile - This example illustrates the use of finite-element soils capabilities in an engineering application of high interest: the analysis of a deep pile. Here the description is qualitative; complete details of the problem and its solution are presented in Reference 11.

The purpose of the analysis was to develop a better understanding of the behavior of a long pile embedded in clay and subjected to various loads. The finite element mesh is shown in Figure 12. 166 second order, isoparametric elements were used in the model together with 25 linear springs to model the soil flexibility far from the rigid pile. Extended-Cam-clay material and the finite strain theory were used.

SEDIMENTATION/CONSOLIDATIONS MODELS

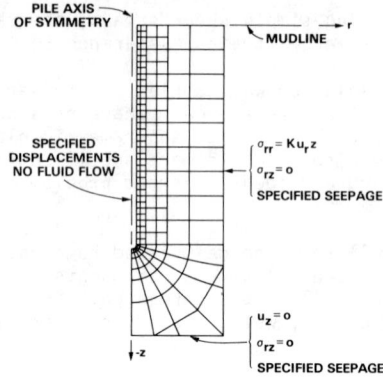

FIG. 12. AXISYMMETRIC PILE: FINITE ELEMENT MESH

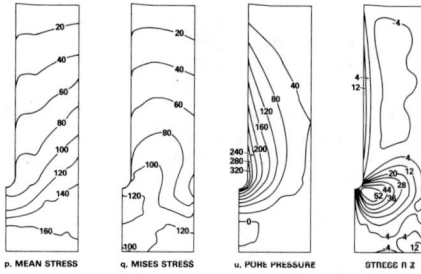

FIG. 13. AXISYMMETRIC PILE: STRESS CONTOURS RIGHT AFTER CAVITY EXPANSION

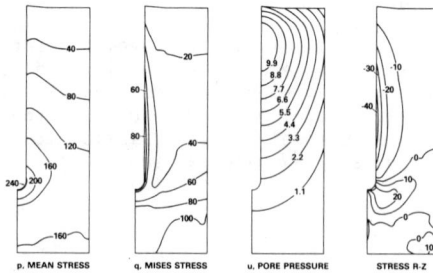

FIG. 14. AXISYMMETRIC PILE: STRESS CONTOURS AFTER 3 YEARS OF CONSOLIDATION

The analysis proceeded by defining the in-situ conditions and
then simulating pile installation by expanding a pre-existing pile
shaped cavity. This expansion involves 60% strains, which clearly
call for finite strain calculations. Following installation, pile
set-up was simulated by allowing excess pore pressures generated
during pile installation to dissipate. Finally, the axial pile was
loaded by translating the pile in both directions and at various
rates.

The complete results of this analysis are documented in Reference
11. Here some qualitative results are shown. Figure 13 presents
contours of stress intensity around the pile right after cavity
expansion and Figure 14 shows the same quantities three years after
installation.

CLOSURE

A number of soil mechanics capabilities have been added to a
general purpose finite element program. This paper has provided an
introductory discussion of the capabilities and illustrated them by
showing four illustrative problems.

ACKNOWLEDGEMENTS

The computer program ABAQUS is the property of Hibbitt, Karlsson
and Sorensen, Inc. The ABAQUS soils capabilities were developed as a
joint effort between Hibbitt, Karlsson and Sorensen, Inc. and Exxon
Production Research Co. This successful development effort has
benefitted from the efforts of many, but particularly E. P. Sorensen,
J. D. Murff, S. S. Wilder, H. D. Hibbitt, and A. B. Potvin.

Appendix I. REFERENCES

1. Biot, M. A., "General Theory of Three-Dimensional Consolidation,"
    Journal of Applied Physics, vol. 12, 1941, pp 155-164.

2. Biot, M. A., "Consolidation Settlement Under a Rectangular Load
    Distribution," Journal of Applied Physics, vol. 12, 1941,
    pp 426-430.

3. Carter, J. P., Booker, J. R., and Davis, E. H., "Finite
    Deformation of an Elastic-Plastic Soil," International Journal
    for Numerical and Analytical Methods in Geomechanics, Vol. 1,
    1977, pp 25-43.

4. Carter, J. P., Small, J. C., and Booker, J. R., "A Theory of
    Finite Elastic Consolidation," International Journal of Solids
    and Structures, Vol. 13, 1977, pp 467-478.

5. Fong, H. H., "A Comparison of Eight General Purpose Finite Element Computer Programs," Structural Mechanics Software Series, ed. by W. D. Pilkey, University Press of Virginia, Charlottsville, 1983.

6. Gibson, R. E., Schiffman, R. L., and Cargill, K. W., "The Theory of One-Dimensional Consolidation of Saturated Clays. II: Finite Nonlinear Consolidation of Thick Homogeneous Layers," Canadian Geotechnical Journal, Vol. 18, 1981, pp 280-293.

7. Hibbitt, Karlsson and Sorensen, Inc., ABAQUS Documentation: User's Manual, Theory Manual, Example Problems Manual, and Systems Manual, Providence, Rhode Island, 1983.

8. Lambe, T. W., and Whitman, R. V., Soil Mechanics, John Wiley and Sons, New York, 1969.

9. Lee, E. H., Mallett, R. L., and Wertheimer, T. B., "Stress Analysis for Anisotropic Hardening in Finite-Deformation Plasticity," Rensselaer Polytechnic Institute Metal Forming Report No. 4/82, November 1982.

10. McMeeking, R. M., and Rice, J. R., "Finite Element Formulations for Problems of Large Elastic - Plastic Deformation," International Journal of Solids and Structures, 1975, pp 601-616.

11. Nystrom, G. A., "Finite-Strain Axial Analysis of Piles in Clay," Proceedings of ASCE Symposium on Recent Numerical Procedures for Pile Foundations, San Francisco, October 1984.

12. Prevost, J. H., and Hughes, T. J. R., "Finite Element Solution of Boundary Value Problems in Soil Mechanics," International Symposium on Soils under Cyclic and Transient Loading, Swansea, January 1980.

13. Prevost, J. H., "Mechanics of Continuous Porous Media," Journal of Engineering Science, vol. 18, 1980, pp 787-800.

14. Roscoe, K. H., and Burland, J. B., "Stress-Strain Behavior of 'Wet' Clay," Engineering Plasticity, ed. by J. Heyman and F. A. Lechie, Cambridge University Press, 1968.

15. Schofield, A., and Wroth, C. P., Critical State Soil Mechanics, McGraw Hill, New York, 1968.

16. Sorensen, E. P., and Nystrom, G. A., "Implementation of Soil Mechanics Capabilities in a General Purpose Finite Element Program," paper under preparation.

17. Terzaghi, K., Theoretical Soil Mechanics, John Wiley and Sons, New York, 1943.

18. Terzaghi, K., and Peck, R. B., Soil Mechanics in Engineering Practice, second edition, John Wiley and Sons, New York, Article 25, 1948.

19. Van Eeckelen, H. A. M., "Isotropic Yield Surfaces in Three Dimensions for Use in Soil Mechanics," International Journal for Numerical and Analytical Methods in Geomechanics, vol. 4, 1980, pp 89-101.

20. Vermeer, P. A., and Verruijt, A., "An Accuracy Condition for Consolidation by Finite Elements," International Journal for Numerical and Analytical Methods in Geomechanics, vol. 5, 1981, pp 1-14.

21. Zienkiewicz, O. C. and Naylor, D. J., "The Adaptation of Critical State Soil Mechanics Theory for Use in Finite Elements," Stress-Strain Behavior of Soils, ed. by R. H. G. Parry, Cambridge, 1971.

22. Zienkiewicz, O. C., and Pande, G. N., "Some Useful Forms of Isotropic Yield Surfaces in Soil and Rock Mechanics," Finite Elements in Geomechanics, ed. by G. Gudehus, John Wiley and Sons, 1977.

Appendix II - NOTATION

a yield surface semi-radius

e_o initial void ratio

k permeability

K third stress invariant soil parameter

M critical state ratio q/p at critical state

p mean normal stress; first stress invariant

q Mises shear stress intensity; second stress invariant

r third stress invariant

κ elastic logarithmic bulk modulus

λ plastic logarithmic bulk modulus

γ_w unit weight of water

ϕ effective stress friction angle

VALIDATION OF CONSOLIDATION PROPERTIES OF PHOSPHATIC CLAY AT VERY HIGH VOID RATIOS

Richard W. Scully[1], Robert L. Schiffman[2], Harold W. Olsen[3], and Hon-Yim Ko[2]

ABSTRACT

The void ratio-effective stress and void ratio permeability relationships for a "phosphate slime" mine tailing material were measured by several methods. The measurements were used in nonlinear finite strain consolidation theory to predict the progress of settlement of a centrifugal prototype.

The void ratio-effective stress and void ratio-permeability relationships were determined from self-weight settling tests, constant rate of deformation consolidation tests, and step loading tests with flow pump permeability measurements. Results from different types of tests were consistent at effective stresses greater than 1.0 kPa (void ratios less than 7.0). At lower stresses, the void ratio-effective stress relationship varied with the initial water content of the slurry, and exhibited curvature that suggested an apparent preconsolidation behavior.

Nonlinear finite strain consolidation theory, employing the measured void ratio-effective stress and void ratio-permeability relationships, was used to predict the consolidation settlement behavior of a centrifugal prototype. The measured properties were for a similar initial condition as the prototype, which was a soil with a uniform void ratio of 15. The prediction was compared with

---

[1] IT Corporation, Englewood, Colorado
[2] Department of Civil Engineering, University of Colorado at Boulder
[3] U.S. Geological Survey, Denver, Colorado

measurements of the settlement behavior of four centrifugal models at different acceleration levels. The settlements of the centrifugal models were evaluated, and from them the time scale factor was determined to be generally consistent with consolidation settlement. This factor is inversely proportional to the square of the acceleration ratio. The accuracy of the measured compressibility relationship was confirmed by satisfactory agreement with the void-ratio-effective stress relationship derived from the void ratio profiles of the centrifugal models. The predicted and measured settlements were in only approximate agreement, with discrepancies attributed primarily to experimental errors in the centrifugal modeling and probable overestimation of permeabilities by the constant rate of deformation consolidation tests.

INTRODUCTION

It has been shown by centrifugal modeling that nonlinear finite strain consolidation theory is an accurate predictor of prototype response for kaolinite clays at void ratios as high as 2.8 (Croce et al., 1984) and 3.6 [11]. The consolidation property relations, which were used in the numerical analysis of theory, were determined by step loading consolidation tests with flow pump permeability measurements [5,11] and the compressibility relation also was deduced from the void ratio profiles of centrifugal models [5]. The flow pump permeability measurements [13,15] were performed by inducing very low hydraulic gradients across the samples so as to minimize seepage-induced consolidation.

This paper investigates, for a very soft phosphatic clay at void ratios much higher than 4, the validation by centrifugal modeling of the consolidation properties determined by several methods. The tested hypothesis is that, if the consolidation properties of the very soft clay are accurately measured, then the numerical analysis [14] of nonlinear finite strain consolidation theory will successfully predict [7] the settlement of a valid and accurate centrifugal model

that settles solely by the process of self-weight consolidation.
Settlement of the centrifugal model by the consolidation process alone
is confirmed if the time-scale-factor exponent, determined by a series
of modeling of models experiments, is equal to the theoretical value,
2.0. At the same time, if the predicted and the measured centrifugal
model results are consistent, the consolidation properties are
confirmed as accurate, within the limitations set by experimental
errors and analytical approximations. In other words, if both
suppositions are answered positively, it will have been demonstrated
that the settlement of the very soft clay is due to consolidation and
that finite strain theory describes the consolidation process for the
clay in this very soft state.

SOIL CHARACTERIZATION

The material used in this study was a Florida phosphatic clay.
Phosphatic clays are a waste product of the mining of sedimentary
phosphate ore. The clays are discharged into tailing ponds at a
solids content of a few percent. After an initial phase of
sedimentation, the clays are in a very soft state and they consolidate
very slowly under their own weight.

The material, provided by Bromwell Engineering [3], was collected
in late 1982 from the N-1 area of the CF mine in Hardee County,
Florida. The material was shipped as collected, at a void ratio of
about 9 (solids content about 23%). The material is a clay that
consists primarily of phosphate minerals (40%) and clay minerals
(44%). The clay minerals are predominantly smectite (19%) and
palygorskite (13%) (Bromwell Engineering, 1983). The pore fluid is a
calcium-magnesium-bicarbonate water and has a neutral pH. The clay
mineral exchange sites are overwhelmingly occupied by bivalent cations
[17]. The physical index properties are:

            Liquid Limit: 161
            Plastic Limit: 44
            Activity: 1.67
            Specific Gravity: 2.74

CONSOLIDATION PROPERTY MEASUREMENTS

The consolidation property relations of interest were the relations of void ratio to effective stress and permeability (respectively, the compressibility relation and permeability relation). The five kinds of tests performed were: settling tests, constant rate of deformation consolidation (CRD) tests, step loading tests, flow pump permeability tests, and centrifuge tests. Settling tests were performed to determine the compressibility relation at very low effective stresses and the highest possible void ratio of the material as a soil. This void ratio is equivalent to the fluid limit (water content of the soil at zero effective stress) [12]. CRD tests were used to determine both the compressibility and permeability relations over the range of effective stresses from 0.01 to 20 kPa. Much reliance has been placed on the CRD test because it is applicable over a wide range of effective stresses and it can be performed in a relatively short time (1-1/2 days including setup). Therefore, it was important to confirm that the properties determined by the CRD test were accurate. Step loading consolidation tests with flow pump permeability measurements were used to measure the compressibility and permeability relations at effective stresses greater than 0.3 kPa; the results served to validate the CRD results at these stresses.

A compressibility relation was derived from the void ratio profiles, at full consolidation, of the self-weight consolidated centrifugal models. The compressibility relations determined by other methods were compared to this one that applies in the centrifugal prototype.

In all of the tests the samples were prepared as slurries by first blending with an egg beater, followed by addition of soil supernatant water, and then vigorous shaking. Finally, the slurries of void ratio 20 or less, which were soils, were deaired under a vacuum to help ensure full saturation.

Results of Self-Weight Settling Tests

In the settling tests, slurried soil was allowed to settle by sedimentation and self-weight consolidation. The water content profile was determined at the completion of primary consolidation, as judged by the progress of settlement [10]. The void ratio profile was derived from the water content profile assuming full saturation.

The settling tests were either singly or doubly drained. For doubly drained tests the settling tube was a plastic cylinder with bottom drainage through a filter paper and a porous metal disk connected by tubing to the cylinder top. The whole volume of the cylinder and tubing was filled by soil supernatant and the soil specimen. The singly drained tests were performed in 1000-ml glass cylinders with drainage to the top only.

Sampling for the water content determinations was performed by suction of the soil through a 6-mm I.D. glass tube into a collecting flask. The soil was removed in lifts, starting from the top.

Particle size segregation was neglected because the initial void ratios were near or below the fluid limit (the void ratio of the soil at zero effective stress) and the columns were short [2,8].

The compressibility relation was determined from the void ratio profile. The effective stress was taken as the total bouyant weight of the overlying soil per unit area, because it was assumed that the excess pore water pressures had dissipated completely. Friction between soil and sidewalls was neglected. It has been observed that, for muds of void ratios from 4 to 10, the magnitude of soil-sidewall friction was small enough to be neglected for approximate estimates of stresses [2].

The tests began with slurries at initial void ratios of 30, 20 and 15. The initial and boundary conditions and the durations of the tests are summarized in Table 1. The void ratio profiles of the fully consolidated specimens are shown in Figure 1 and the derived compressibility relations are shown in Figure 2.

Table 1
SUMMARY OF SETTLING TESTS

Initial Uniform Void Ratio	Drainage	Sample Diameter (cm)	Initial Height (cm)	Final Height (cm)	Test Duration (days)	No. of Test Replications
30	single	6.00	36.0	18-19.4	8-12	3
20	double	6.35	10.0	7.0	4	1
15	double	6.35	8.2	6.9	6	1

The fluid limit [12], which is equivalent to the water content at zero effective stress [2], is the water content of a soil newly formed from a suspension. The fluid limit is the highest water content at which the mineral-water system behaves as a soil. It varies somewhat

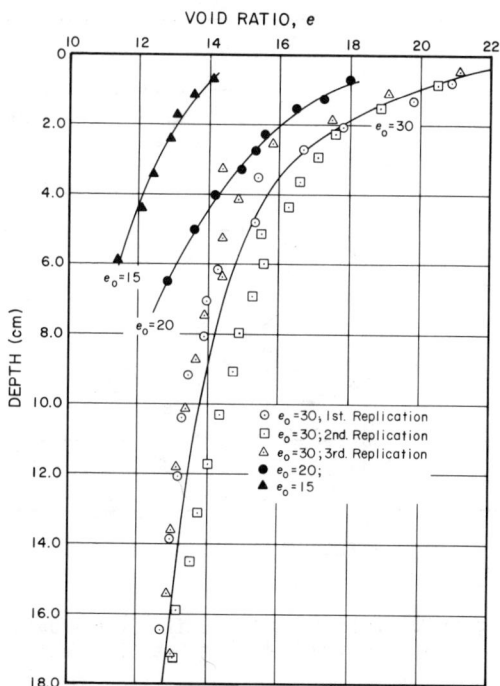

Figure 1. Void Ratio Profiles of Fully Consolidated Samples from Self-Weight Settling Tests

depending on the initial concentration of the suspension from which it formed, with a lower initial water content causing a lower fluid limit [2,8].

In properly conducted CRD tests the tested slurry must be a soil; that is, the water content of the slurry sample must be less than the fluid limit. Otherwise, the CRD analysis, which assumes that the principles of soil mechanics are applicable, would be invalid. Therefore it was necessary to determine the lowermost fluid limit of the phosphatic clay so that the slurries used in the CRD tests could be prepared at water contents at which the slurries would behave as soils.

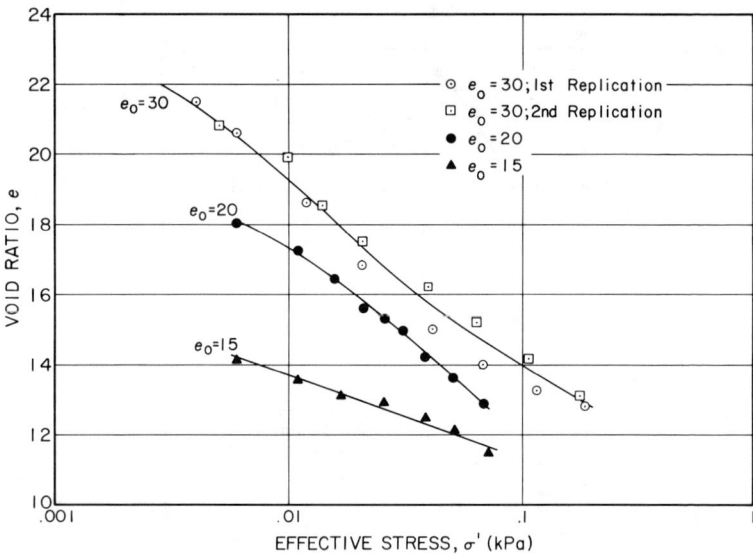

Figure 2.   Void Ratio-Effective Stress Data
from Self-Weight Settling Tests

The lowermost fluid limit was determined from the water contents of the uppermost soil layers formed in the settling tests with an initial void ratio of 30. By extrapolating the void ratio profile to

the soil surface, while treating the values of the uppermost water
content as too high because they were slightly contaminated by the
supernatant, it was estimated that the zero effective stress void
ratio was 23 or 24, which is equivalent to a water content (fluid
limit) of 840 to 875%.

The final settled void ratio, e_f, from 30-day settling tests (a
different method than the one used here), was determined previously
for a variety of Florida phosphatic clays (Ardaman and Associates,
1983). It was observed that e_f could be correlated with the liquid
limit, w_ℓ, according to the equation:

$$e_f = -7.59 + 0.186\, w_\ell \qquad (1)$$

For our clay the equation gives an e_f of 22.4, which agrees closely
with our determination of the lowermost fluid limit.

Comparing the compressibility relations from tests with different
initial void ratios, as shown in Figure 2, it is observed that the
initial void ratio has a strong influence on the compressibility.
Each sample with a different initial void ratio apparently had a
unique virgin compression curve. Because the tests had comparable
durations, it is unlikely that secondary compression or time-dependent
chemical alterations might have caused the differences in the
relations [17].

These compressibility relations derived from settling tests
support Imai's finding [8] that for some natural clays (not including
kaolin) there is no unique compressibility curve because the
compressibility depends on the initial water content, especially at
low effective stresses. Imai's data show the effect to be observable
at effective stresses less than approximately 0.1 kPa. At these
effective stresses Been and Sills [2] also observed that
compressibility curves were highly variable.

The reproducibility of results from settling tests was
demonstrated by the void ratio profiles and the compressibility
relations determined for the series of three identical settling tests

with an initial void ratio of 30. For a first approximation of one of the errors in the settling test method of determining compressibility, the error introduced by neglecting the uppermost layer of a sample, with a 5-mm thickness and a void ratio of 20, is equivalent to 0.004 kPa of effective stress.

Results of Constant Rate of Deformation Consolidation (CRD) Tests

The experimental apparatus used was the University of Colorado Mark II Device [20]. The samples were backpressured to ensure full saturation. The limit of resolution in stress measurements was about 0.007 kPa.

Analyses of the test results were performed with the TEST1 computer program [19]. The program output includes data for a single void ratio-effective stress relation and two void ratio-permeability relations. One permeability relation is calculated from the measured stress conditions at the undrained boundary of the sample and the other is calculated from the g-function. The g-function is defined as [18]

$$g(e) = - \frac{k(e)}{\gamma_w (1+e)} \frac{d\sigma'}{de} \qquad (2)$$

where e is the void ratio, σ' is the effective stress, γ_w is the unit weight of water, and k is the coefficient of permeability. The g-function includes the consolidation properties of the material and is a function of the void ratio; it is similar to the coefficient of consolidation, c_v.

CRD tests were performed on slurried soils with initial void ratios of 9.5, 10.7, 13 and 20. Three replicated tests were performed for the case with an initial void ratio of 13. The two tests with an initial void ratio of 20 were run at different velocities; one (20-Slow) was run so slowly that the excess pore pressure at the undrained boundary of the sample was negligible during most of the test and only the compressibility relation was determined for this test.

An apparent preconsolidation effect is evident in every compressibility relation from the CRD tests as shown in Figure 3. The compressibility relations from tests with an initial void ratio of 13 are consistent with the settling tests with an initial void ratio of 15; the apparent preconsolidation effect in these CRD tests is most

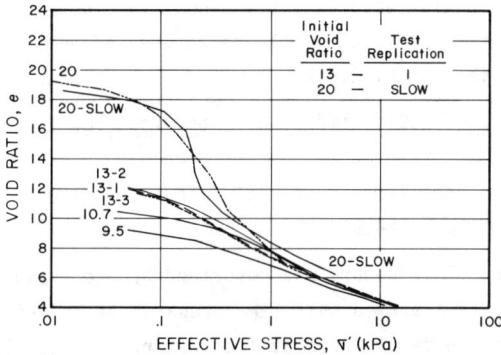

Figure 3. Void Ratio-Effective Stress Data from CRD Tests

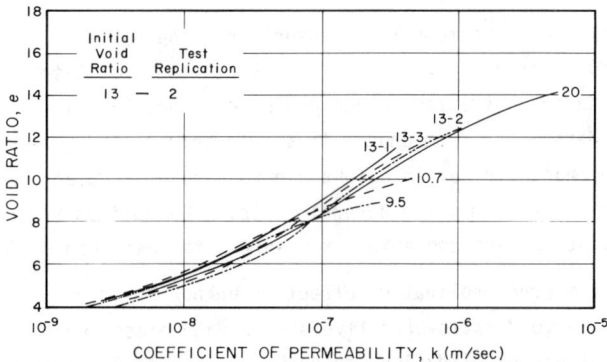

Figure 4. Void Ratio-Permeability Data from CRD Tests

probably the result of the initial void ratio. The cause of the apparent preconsolidation effect in the tests with an initial void ratio of 20 is unknown. The results of these tests do not coincide with the results of the settling test with the same initial void ratio. It probably is not an artifact of the test analysis because in

the slow test with an initial void ratio of 20, the effective stresses
are equal to the total stress measurements and the void ratio is
calculated directly from volume measurements without using the
computer analysis.

Permeability determinations were consistent among the tests
begun at various initial void ratios as shown in Figure 4, except for
the results near the beginning of the tests when effective stress
levels were less than about 0.5 kPa.

Results of Step Loading Consolidation Tests with Flow Pump Permeability Measurements

The apparatus consisted of several Anteus consolidometers
connected to a flow pump [4,9].

The experimental procedure was described by Croce [4]. In our
tests the samples of slurried soil were prepared at an initial void
ratio of 8.3. The permeant was distilled water. The samples and the
permeant were backpressured to ensure full saturation. Loads were
applied by elevating a reservoir.

During flow pump permeability measurements the hydraulic
gradients induced across the samples were kept less than 2 to minimize
seepage-induced consolidation. Seepage-induced consolidation--as much
as it can be determined from change in sample height [15]--was
minimal. The change in sample height due to seepage-induced
consolidation was recorded for several tests. The maximum reductions
in sample height and average void ratio were both less than 0.4%.

An apparent preconsolidation effect of unknown origin is
suggested by the void ratio-effective stress data presented in Figure
5. An apparent preconsolidation stress of about 1 to 2 kPa is evident
at a void ratio of about 7, even though the samples were slurries and
the initial load of 0.3 kPa was solely the result of the overlying
porous stone. The virgin compression portion of the compressibility
relation closely parallels the CRD test results.

The void ratio-permeability data are presented in Figure 6. At
void ratios less than 7, the permeabilities are similar to the CRD

results. At void ratios greater than 7 the slope of the permeability
relation tends to flatten in a manner that is analogous to the
apparent preconsolidation effect observed in the compressibility data.

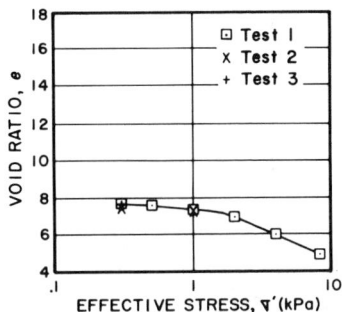

Figure 5. Void Ratio-Effective
Stress Data from Step Loading
Consolidation Tests

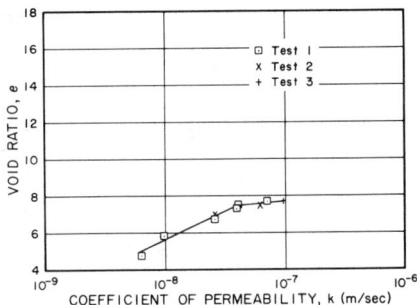

Figure 6. Void Ratio-Permeability Data from Flow Pump
Permeability Tests

Results from Centrifuge Tests of Self-Weight Consolidation

The compressibility relation for the centrifugal prototype was
determined from the water content profiles of the fully consolidated
centrifugal models immediately after centrifugal testing. No rebound
was observed. Soil-sidewall friction was neglected. Full saturation
was assumed.

The void ratio-effective stress data from the four
centrifugal models, shown in Figure 7, are consistent for effective
stresses greater than 1 kPa. The scatter of results at lower
effective stresses is attributable to experimental errors in sampling
very thin layers of the soft soils at the top of the models for water
content determinations.

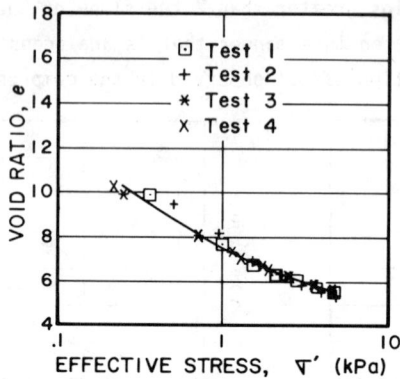

Figure 7.   Void Ratio-Effective Stress Data from
            Self-Weight Consolidation Centrifuge Tests

Summary of Consolidation Properties Determinations

The compressibility and permeability relations from the various
tests are summarized respectively in Figures 8 and 9.   The data from
replicated tests are represented by a single curve.

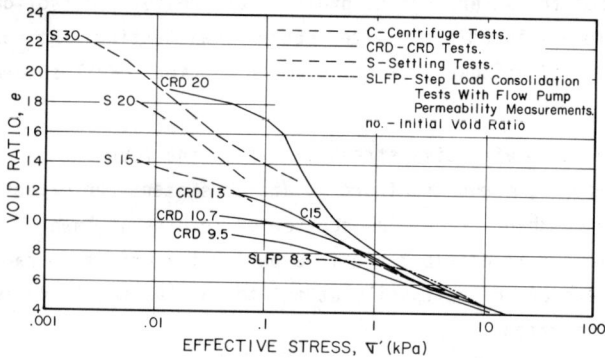

Figure 8.   Summary of Compressibility Relations
            Determined by Various Methods

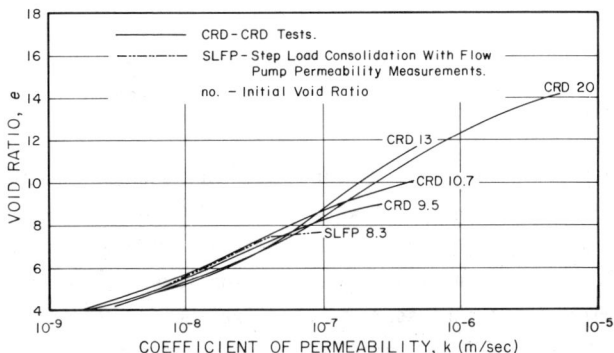

Figure 9. Summary of Permeability Relations
Determined by Various Methods

The various compressibility relations determined by the CRD, step
loading, and centrifuge tests are consistent at void ratios less than
7.

Much of the variation between compressibility relation
determinations at void ratios above very approximately 9 (effective
stress of about 0.5 kPa) can be attributed to the influence of the
initial water content. The close agreement of the compressibility
curves from the settling test with an initial void ratio of 15 and the
CRD tests with an initial void ratio of 13 is evidence that the
apparent preconsolidation effect in these CRD results is a real effect
resulting from the initial water content. Conversely for tests with
an initial void ratio of 20, the CRD results exhibit an apparent
preconsolidation effect that is unconfirmed by the settling test at
the same initial void ratio; the apparent preconsolidation effect in
these CRD results is of unknown origin.

The permeability relations determined by step loading
consolidation tests with flow pump permeability measurements and by
CRD tests of various initial void ratios were consistent, except for
some preconconsolidation-like effects. It appears that these effects
in the permeability relations are related to the apparent
preconsolidation effects in the compressibility relations.

CENTRIFUGAL MODELING

The apparatus and experimental procedure were described previously [4,5]. The consolidometer was the same used by Croce for self-weight consolidation centrifuge tests.

Because the samples were very soft, even the shortest samples would consolidate under their own weight. In the modeling of models experiments a single prototype with an initially uniform void ratio was modeled at several centrifugal acceleration levels. To have identical samples it was necessary to minimize the time between the placement of the sample into the consolidometer and the start of the test.

The consolidometer was backpressured to ensure full saturation.

Experimental Verification of the Scaling Relations

The scaling relations for centrifugal modeling of soil consolidation are:

(length) $$n = \frac{\ell_p}{\ell_m}$$

(time) $$\tau = \frac{t_p}{t_m} = n^\nu = n^2 \tag{4}$$

where ν is the time-scale-factor exponent

(stress) $$\alpha = \frac{\sigma_p}{\sigma_m} = 1 \tag{5}$$

(strain) $$\beta = \frac{\epsilon_p}{\epsilon_m} = 1 \tag{6}$$

$$n = \frac{\text{centripetal acceleration}}{\text{earth's gravitational acceleration}} \tag{7}$$

where the subscripts "m" and "p" stand for model and prototype, respectively. These scaling relations are appropriate for processes that are governed only by self-weight and seepage forces. These relations have been confirmed for the centrifugal modeling of the consolidation of soft kaolinite clay [4,5].

The centrifuge tests were conducted as a series of four modeling of models experiments to verify that the centrifugal scaling relations for soil consolidation applied. The scaling factor for time was of particular concern. When settlement by both sedimentation and consolidation occurs in a centrifuge test, the exponent in the time scale factor for the settlement of the whole model is between the values of 1 for sedimentation only and 2 for consolidation only [6].

For the centrifugal modeling of the consolidation of slurried soil an initial soil void ratio of 15 was chosen. Settling tests had shown that the lower fluid limit of this soil corresponded to a void ratio of approximately 23. The initial void ratio had to be less than 23 so that the slurry model would settle only by the process of consolidation. With an initial void ratio of 15 the water content was well below the fluid limit and therefore the slurry was a soil and not a suspension. Also, the full range of permeability measurements, at void ratios up to 14, could be validated.

All four centrifuge experiments, summarized in Table 2, modeled a single, fictitious prototype that was initially 5 meters high and doubly drained (at top and bottom). A different acceleration level was used in each experiment. Height measurements were recorded during the test by an in-flight camera. Consistency among the ultimate settlement strains for the whole model (Table 2) confirmed that the strains of models and prototype were equal.

Table 2

SUMMARY OF DOUBLY DRAINED, SELF-WEIGHT CONSOLIDATION CENTRIFUGE TESTS

Initial Void Ratio Distribution = 15 (uniform)
Sample Diameter = 6.35 cm

Test No.	Centrifugal Acceleration Level, n (g)	Initial Model Height h_m (cm)	Final Model Height $h_{m\infty}$ (cm)	Initial Prototype Height h_p (cm)	Final Prototype Height $h_{p\infty}$ (cm)	Ultimate Settlement Strain ε (%)
1	90.9	5.5	2.6	500	236	53
2	76.6	6.5	3.0	500	230	54
3	64.1	7.8	3.7	500	237	53
4	56.8	8.8	4.0	500	227	55

The prototype void ratio profiles derived from the models at full consolidation and, therefore, also the void ratio-effective stress data (Figure 7) were consistent. These data confirmed that the prototype and model stresses, like the strains, were equal.

To ascertain that the time-scale-factor exponent equaled the theoretical value of 2 for consolidation settlement, a comparison was made of the times taken by the models to reach various degrees of settlement (percentage of ultimate settlement). The model times were obtained from a plot of degree of settlement versus model time as shown in Figure 10. If there is similarity of models and prototype, then equation (4) must hold at any stage of settlement and so the equation can be rewritten as:

$$t_{U_p} = t_{U_{m_1}} (n_1)^\nu = t_{U_{m_2}} (n_2)^\nu \qquad (8)$$

where U is the degree of settlement. The time-scale-factor exponent, ν, can be determined for any degree of settlement by linear regression analysis of equation (8) rearranged as:

$$\nu \log(n) + \log(t_{U_m}) = \text{constant} \qquad (9)$$

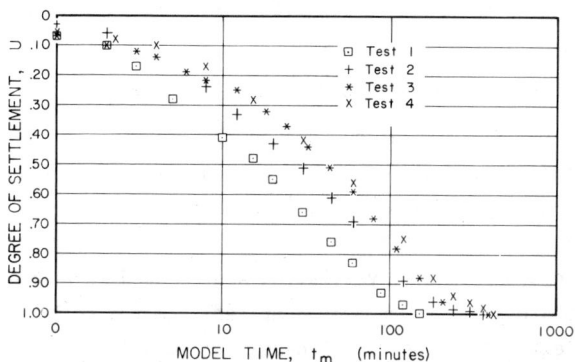

Figure 10. Progress of Settlement of Centrifugal Models with Time

The variation of the time-scale-factor exponent is shown in Figure 11.

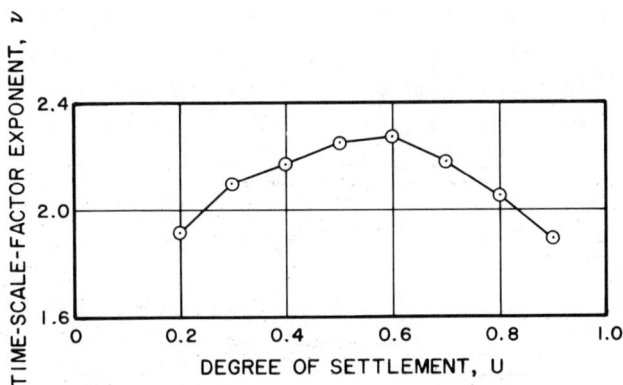

Figure 11.   Variation of the Time-Scale-Factor
Exponent with Degree of Settlement

The time-scale-factor exponent was generally greater than 2,
indicating that sedimentation probably did not occur in the tests.
Therefore, it was assumed that settlement was solely the result of
consolidation and a time-scale-factor exponent of 2.0 was used for the
analysis of the modeled prototype.

Comparison of Analytical Predictions with Prototype Behavior

The numerical analysis of nonlinear finite strain consolidation theory was performed with the FSCON4 computer program [16].

The compressibility relation used in the analysis is shown in Figure 12. It coincided at lower void ratios with the results of the CRD tests with an initial void ratio of 13 and at higher void ratios with the results of the settling test with an initial void ratio of 15. The chosen compressibility relation was in good agreement (Figure 8) with the compressibility relation derived from the fully consolidated centrifugal models. Because the compressibility relation was well defined and thought to be accurate, the effect of variations in the compressibility relation on the progress of settlement were neglected. In fact, a comparison of predicted settlements by the analysis using compressibility relations determined by settling tests with initial void ratios of 15 and 20, has demonstrated that the influence of these initial void ratios (that is, water contents) has no significant effect on the prediction of settlement progress for the prototype of interest here [17].

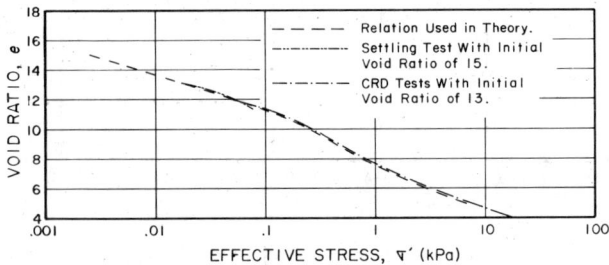

Figure 12. Compressibility Relation Used in the Comparison (Fig. 14) of Analytical Prediction with Centrifugal Prototype

The permeability relation was varied to determine, by comparison of the resulting settlement predictions with the centrifugal prototype, the part of the range of permeability measurements most likely to be correct. Moreover, at high void ratios the permeability was measured only by the CRD test and it was desirable to corroborate the data by

another method. The two permeability relations chosen for the
comparison are shown in Figure 13. One relation was representative of
the average permeabilities determined (the average of the calculations
from the g-function and at the undrained boundary) from the CRD tests
[18]. The other relation was representative of only the g-function
permeability calculations. At a given void ratio, the permeability
value calculated from the g-function is almost always less than the
value calculated at the undrained boundary.

The two theoretical predictions of settlement, which employ
the different permeability relations, are shown together for
comparison with the prototype behavior of the centrifugal models in
Figure 14. The permeability relation derived solely from the
g-function calculation gives the prediction that best agrees with the
centrifugal prototype. It is noteworthy that this relation is also in
better agreement with the flow pump permeability measurements.

There is fairly good agreement of the theory with the centrifugal
prototype although the theory predicts faster settlement. The slower
settlement of the centrifugal prototype may be attributable to
friction between the soil models and the consolidometer sidewall. The
agreement is of the same order as in some similar tests performed on
kaolinite clays [4,5,11]. It has been found, by measuring permeability at hydraulic gradients smaller than those used in the flow pump

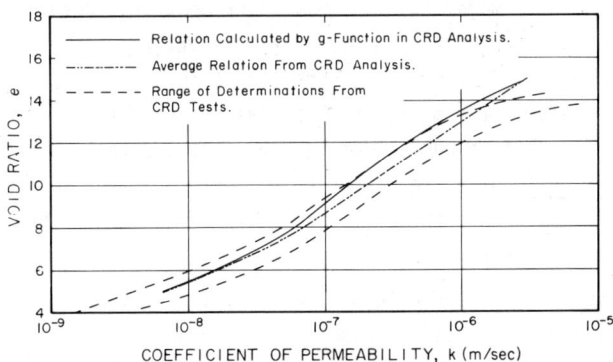

Figure 13. Two Permeability Relations Used in the Comparison (Fig. 14)
of Analytical Prediction with Centrifugal Prototype

tests here, that lower permeability values were observed that gave a much better agreement of theory with centrifugal prototype [5].

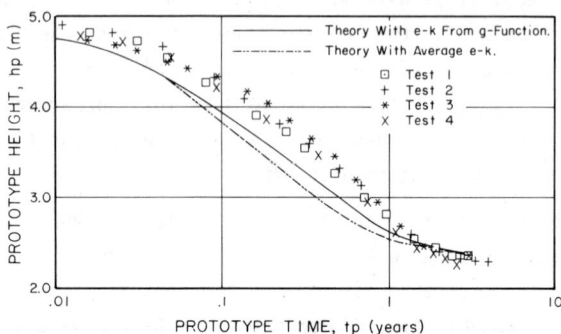

Figure 14.   Comparison of Analytical Predictions, Based on
            Different Permeability Relations (see Figure 13),
            Versus Centrifugal Models of the Prototype

CONCLUSIONS

The approximate agreement between centrifugal model settlements and settlement predictions by nonlinear finite strain consolidation theory suggested that, for a slurried phosphatic clay with an initial void ratio of 15, the settlement behavior could be accounted for on the basis of consolidation theory. Further support for this conclusion came from a modeling of models experiment that determined the time-scale-factor exponent, mostly 2 or greater, was explainable only by the consolidation process. Therefore, this slurry was a soil and existing techniques for measuring material consolidation properties, for predicting consolidation behavior, and for centrifugal modeling, all of which have been shown to be applicable for soft kaolinites [5, 11], can be applied to such very soft or slurried soils.

Independently of the centrifugal modeling, measurements of compressibility and permeability properties by several techniques showed that these properties can be reliably determined for this slurried soil over a large range of void ratios. At void ratios less than 7 (effective stresses greater than about 1.0 kPa), the results of CRD tests and step loading consolidation tests with flow pump permeability measurements were consistent. Permeability determinations by CRD tests were also consistent at higher void ratios, except for some anomalous curvature in the void ratio-permeability relations from the beginning of the tests.

At void ratios higher than 7, there was considerable variation (on a log scale of effective stresses) in the observed compressibility relations. The relations exhibited a curvature that is characteristic of a preconsolidation effect. For the most part, these differences among the various observed compressibility relations are attributable to the influence of the different initial water contents.

Centrifuge testing served to validate the compressibility measurements. A compressibility relation was derived from the consolidated centrifugal models and it confirmed that the results of CRD tests and a settling test on slurries with similar initial void ratios were accurate at effective stresses of 0.25 kPa and greater (void ratios of 10 and less). The results of these three types of tests on similar slurries formed a single compressibility relation for the slurry at the given initial void ratio of 15.

Since the compressibility relation for the centrifuge tests had been well defined, then the comparison between the settlement predictions by nonlinear finite strain consolidation theory and the measured settlements of the centrifugal models served to evaluate whether the CRD permeability determinations at high void ratios were reasonably accurate. However, measured settlements were somewhat slower than predicted settlements. Further work is needed to examine whether for soils at very high void ratios, the permeabilities determined by CRD tests may be too high.

ACKNOWLEDGEMENTS

This work was part of a project on nonlinear finite strain consolidation theory sponsored by the National Science Foundation.

The authors appreciate the assistance of the staff of Bromwell and
Carrier, Inc., Lakeland, Florida for sampling the phosphatic clay used
in this study and for making their data available.

REFERENCES

[1] Ardaman and Associates (1983). "Evaluation of phosphatic clay
    disposal and reclamation methods, Volume 4: Consolidation
    behavior of phosphatic clays," sponsored by Florida Institute of
    Phosphate Research, Bartow, Florida.

[2] Been, K. and Sills, G.C. (1981). "Self-weight consolidation of
    soft soils: an experimental and theoretical study," Geotechnique,
    31, 519-535.

[3] Bromwell Engineering, Inc. (1983). Private communication,
    Lakeland,Florida.

[4] Croce, P. (1982). "Evaluation of consolidation theories by
    centrifugal model tests," M.S. Thesis, Department of Civil
    Engineering, University of Colorado at Boulder.

[5] Croce, P., Pane, V., Znidarcic, D., Ko, H.-Y., Olsen, H.W. and
    Schiffman, R.L (1984). "The validation of non-linear finite
    strain consolidation theory by centrifuge modelling",
    Applications of Centrifuge Modelling to Geotechnical Design,
    University of Manchester, Manchester, United Kingdom.

[6] Davidson, J.L. and Bloomquist, D. (1980). "Centrifuge modeling
    of the consolidation/sedimentation process in phosphatic clays,"
    University of Florida-Gainesville Engineering and Industrial
    Experiment Station, Contract Report 245*W65.

[7] Gibson, R.E., England, G.L. and Hussey, M.J.L. (1967). "The
    theory of one-dimensional consolidation of saturated clays, I.
    Finite nonlinear consolidation of thin homogeneous layers,"
    Geotechnique, 17, 261-273.

[8] Imai, G. (1981). "Experimental studies on sedimentation
    mechanism and sediment formation of clay materials," Soils and
    Foundations, 21, (7), 7-20.

[9] Ketcham, S.A. and Znidarcic, D. (1981). "Consolidation/
    permeability equipment, procedures and results for bottom
    sediment in gas charged areas of Norton Sound," Department of
    Civil Engineering, University of Colorado at Boulder, Final
    Report for U.S. Geological Survey, Contract 14-08-0001-18760.

[10] Lambe, W.T. and Whitman, R.V. (1969). Soil Mechanics, John Wiley and Sons, New York, p. 411.

[11] Leung, P.K., Schiffman, R.L. and Ko, H.-Y. (1984). "Centrifuge modeling of shallow foundations on soft soil," Offshore Technology Conference, in press.

[12] Monte, J.L. and Krizek, R.J. (1976). "One-dimensional mathematical model for large-strain consolidation," Geotechnique, 26, 495-510.

[13] Olsen, H.W. (1966). "Darcy's law in saturated kaolinite," Water Resources Research, 2, 287-295.

[14] Pane, V. (1981). "One-dimensional finite strain consolidation," M.S. Thesis, Department of Civil Engineering, University of Colorado at Boulder.

[15] Pane, V., Croce, P., Znidarcic, D., Ko, H.-Y., Olsen, H.W. and Schiffman, R.L. (1983). "Effects of consolidation on permeability measurements for soft clay," Geotechnique, 33, 67-72.

[16] Pane, V. and Schiffman, R.L. (1980). "FSCON4 - One-dimensional finite strain consolidation of thick layers," Department of Civil Engineering, University of Colorado at Boulder, Geotechnical Engineering Report.

[17] Scully, R.W. (1984). "Determination of consolidation properties of phosphatic clay at very high void ratios," M.S. Thesis, Department of Civil Engineering, University of Colorado at Boulder.

[18] Znidarcic, D. (1982). "Laboratory determination of consolidation properties of cohesive soil," Ph.D. Dissertation, Department of Civil Engineering, University of Colorado at Boulder.

[19] Znidarcic, D. and Schiffman, R.L. (1982). "TEST1, Version 1, Level A, Constant rate of deformation test analysis," Department of Civil Engineering, University of Colorado at Boulder.

[20] Znidarcic, D. and Schiffman, R.L. (1983). "Constant rate of deformation testing," Department of Civil Engineering, University of Colorado at Boulder, Geotechnical Engineering Report.

CONSOLIDATION PERFORMANCE OF SOFT CLAYS:
PART 1.   MODEL

M. Soulié[1] and V. Silvestri[2], M.ASCE

ABSTRACT

The object of this paper is to present the prediction of the consolidation performance of soft clays. The paper is divided into two
parts: in Part 1, a simplified version of the soil model used is presented; in Part 2, field predictions are presented and an improved version of the model is described.

The simplified version of the soil model used in the Finite Element (FE) scheme is a linear anisotropic elastic model. This model was
retained for the analyses because (1) it permits to represent adequately
the soil behaviour during consolidation, and (2) it is quite simple in
its formulation as it needs only three parameters to represent anisotropy.

Part 1 defines also the parameters needed for the model and the
laboratory tests that are required for their determination.

INTRODUCTION

The prediction of the in-situ consolidation performance of soft
natural clays by means of the Finite Element Method (FEM) has been the
subject of several investigations in the past ten years. It is generally admitted that the first papers dealing with this subject were presented at the ASCE Specialty Conference on the Performance of Earth and
Earth-Supported Structures, held at Purdue University in June 1972 (3,
13,27,46). Some papers of practical interest have also appeared in the
Proceedings of the W.E.S. Symposium on the Applications of the Finite
Element Method in Geotechnical Engineering, held at Vicksburg in May
1972 (6,24). Several Type A (16) predictions of considerable practical
interest appeared in the Proceedings of the Foundation Deformation Symposium held at M.I.T. in November 1974 (14,22). Other related papers
have appeared either in various geotechnical journals and books or in
the proceedings of recent geotechnical conferences (1,7,8,9,20,21,28,29,
31,32,35,42,43).

Table 1, which summarizes the principal characteristics of the
FEM soil models most commonly used, indicates that these models have become, in general, quite involved. In the most recent approaches, all
the investigators mentioned in Table 1 are simulating the constitutive

---

[1]   Professor of Civil Engineering, Ecole Polytechnique, Montreal,
        Canada
[2]   Associate Professor of Civil Engineering, Ecole Polytechnique,
        Montreal, Canada

182

TABLE I

PRINCIPAL MODELS AND CASE HISTORIES
(Excluding M.I.T. (22))

Reference	Site	Object	Model
Bozozuk and Leonards (3)	Gloucester (Canada)	Stresses and deformations	Linear isotropic elastic
Wroth and Simpson (46)	King's Lynn (England)	Short-term and long-term deformations	Cam-Clay
Raymond (27,28)	Kars and New Liskeard (Canada)	Short-term and long-term deformations	Non-linear isotropic elastic
Palmerton (24)	Atchafalaya (U.S.A.)	Long-term deformations	Non-linear creep
Domaschuk and Valliapan (7)	Winnipeg (Canada)	Short-term deformations	Non-linear elastic
Foott and Ladd (9)	Atchafalaya (U.S.A.)	Short-term deformations	Bi-linear elastic, hyperbolic and plastic
Shibata et al. (31)	Mizushima (Japan)	Long-term deformations	Hybrid: Consolidation and dilatancy
Stille et al. (35)	Kalix (Sweden)	Deformations near rupture	Hyperbolic
Thoms et al. (43)	I90 (U.S.A.)	Settlements	Non-linear with creep
Soulié (34)	Cubzac B (France)	Deformations versus time	Bi-linear anisotropic elastic
Tham (42)	Stotel (Germany)	Deformations versus time	Isotropic, visco-elastic with consolidation
Adachi et al. (1)	St-Alban (Canada)	Short-term and long-term deformations	Elasto, visco-plastic
Magnan et al. (20,21)	Cubzac A,B (France)	Deformations versus time	Modified Cam-Clay

behaviour of soft clays in terms of effective stresses.

Ecole Polytechnique has been developing for the past ten years a finite element system, called SOL, which comprises several programs devoted to the numerical treatment of geotechnical problems. The present paper briefly describes the numerical method used in SOL to predict the consolidation response of soft natural clays. The soil model used is a simplified version of both the Cam-Clay model (4,30) and the YLIGHT model (37,38,39). In addition, the paper presents a Type B prediction of the consolidation performance of embankment B of the Cubzac-les-Ponts experimental site, built in 1975 by the Laboratoires des Ponts et Chaussées, near the city of Bordeaux in France. It is also shown that the principal characteristics of the model have been derived on the basis of experimental observations obtained from both the M.I.T. (22) and St-Alban (18,19) test fills.

FINITE ELEMENT PROGRAM

The main feature of the program SOL is the use of a non-conforming triangular element, as shown in Fig. 1. The mathematical aspects

● DISPLACEMENT NODES
X PORE PRESSURE NODES

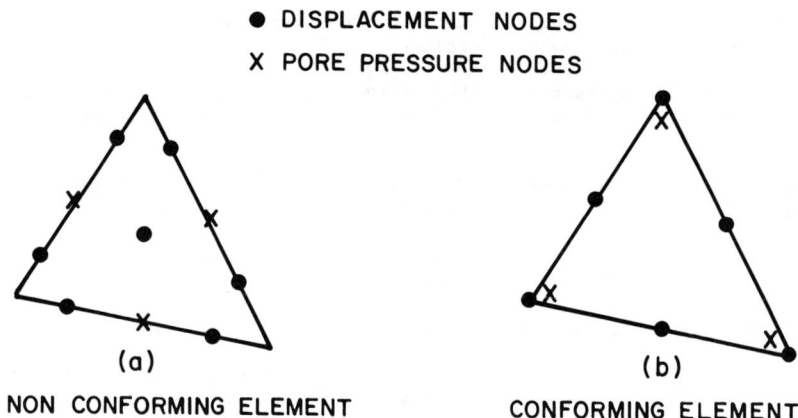

(a)
NON CONFORMING ELEMENT

(b)
CONFORMING ELEMENT

Figure 1.   Non Conforming Element Used in Program SOL

of this element have been described by Fortin and Soulié (10). The non-conforming element has quadratic interpolation for the displacement field and linear interpolation for the pore pressure field within each triangular element (Fig. 1). The main characteristics and advantages of the non-conforming element over conventional elements such as the conforming one are the following:
i)   The displacements are continuous from one element to the adjacent one only at the nodes;

ii)   Under undrained conditions and if the material properties of ad-
        jacent elements are identical, the pore pressures are found to
        be continuous at the corresponding pore pressure nodes, without
        imposing such a condition. During consolidation the continuity
        of the pore pressures follows as a direct consequence of the
        continuity equation of flow and volume change;

iii)   The equilibrium equations in terms of total stresses are veri-
        fied at every point inside each element; and

iv)   The total stress tensor is continuous from one element to the
        adjacent one at the mid-point of the common side, if the mate-
        rial properties are identical. In this case, the continuity of
        the effective stress tensor is a direct consequence of the con-
        tinuity of the pore water pressures. If the material properties
        of adjacent elements are different, continuity of the total nor-
        mal and tangential component of the stress vector is found to
        exist.

SOIL MODEL

Requirements of a Practical Model

Even though the soil models appearing in Table 1 are numerous,
a close examination shows that they can be grouped into two major classes, that is, isotropic and anisotropic models (linear, nonlinear, elastic, elastic-plastic and time dependent). It should be noted that when
one uses an isotropic elastic model for the consolidation response, it
can be shown (5,6) that the lateral movements predicted decrease with
time and may even become negative or directed towards the center-line of
the fill. Such a behaviour is in direct contradiction with field observations (22,38,39).

As the mathematical complexity involved in the application of
the various models shown in Table 1 does not constitute a problem for
the FEM technique, the practical implementation of some of them is lost
because of their requirement of a large number of parameters necessary
to describe the soil behaviour. Indeed, it is generally very difficult
if not impossible to obtain all the required parameters for field studies. When considering the prediction of the field consolidation response of soft clays, it is believed that a useful model should have the following characteristics:

i)   It should described at least qualitatively the observed behaviour;

ii)   It should be simple, that is, it should require a limited number
        of parameters necessary to describe its response; and

iii)   It should quantitatively reproduce the observed behaviour, the
        accuracy of the prediction being linked to other factors like
        sample quality, heterogeneity, etc...

In addition, it may be necessary to use not only one but two
models: one for the description of the soil behaviour during construction, the other for the response during consolidation.

Model Characteristics

The model used in this study has been developed on the basis of
the requirements mentioned above, that is, simplicity and quality of the

SEDIMENTATION/CONSOLIDATIONS MODELS

prediction. Its basic characteristics are:

1) Volumetric Yield Curve

A volumetric yield curve or Limit State Curve (LSC) separates the overconsolidated from the normally consolidated state, as shown in Fig. 2. The LSC corresponds to the straight lines σ'_{vc} and $\sigma'_{hc} = K_0 \, \sigma'_{vc}$,

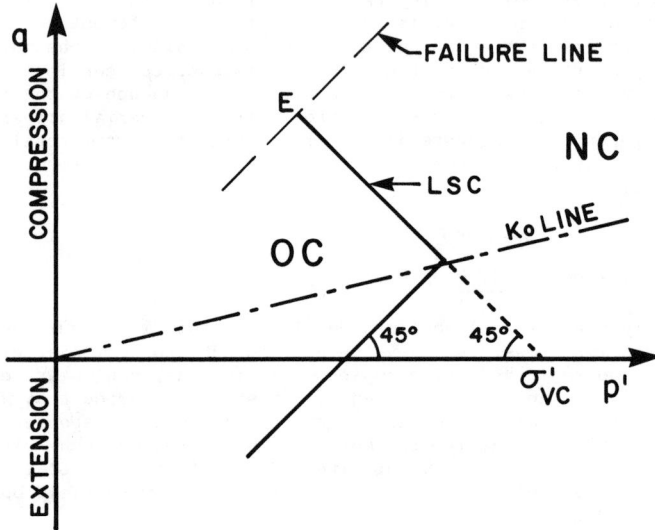

·Figure 2.   Soil Model Representation

where σ'_{vc} is the effective preconsolidation pressure and K_0 is the in-situ coefficient of earth pressure at rest.   The LSC is thus a simpli-fied representation of the volumetric yield curve of soft natural clays (17).

2) Linear Anisotropic Elastic Response

The soft clay is assumed to behave as a linear cross-anisotropic elastic material in both the overconsolidated and normally consolidated states.   Elastic materials having this of anisotropy may be characteri-zed by the following set of independent effective parameters (26):
E_v = Young's modulus in the vertical direction;
E_h = Young's modulus in the horizontal plane;
ν_v = Poisson's ratio for the strain in the horizontal direction due to a vertical stress;
ν_h = Poisson's ratio for the strain in the horizontal direction due to a horizontal stress; and
G_v = Shear modulus in the vertical plane.

In the overconsolidated state, these parameters have constant va-lues as the soil structure is intact.   However, in the normally consolidated state, the values of the elastic parameters vary throughout the deformation process and the soil behaves as a nonlinear elastic-plastic material. In this case, a piecewise linear approximation may be used and the values of the parameters are selected according to the stress level. It should be noted that, for the consolidation prediction presented in Part 2 of the paper, a single linear approximation was found to be adequate to represent the observed response of the clay in the normally consolidated state. The model is, thus, a bilinear anisotropic elastic model.

The restriction that the energy of incremental deformation has to be always positive (26) gives:

$$E_h, \ E_v \ \text{and} \ G_v > 0 \qquad\qquad\qquad (1a)$$

$$-1 < \nu_h < 1 \qquad\qquad\qquad (1b)$$

$$1 - \nu_h - 2n \ \nu_v^2 > 0 \qquad\qquad\qquad (1c)$$

where $n = E_h/E_v$ is called the modulus ratio. Fig. 3 shows these conditions for several values of the modulus ratio n and for positive values of ν_h and ν_v.

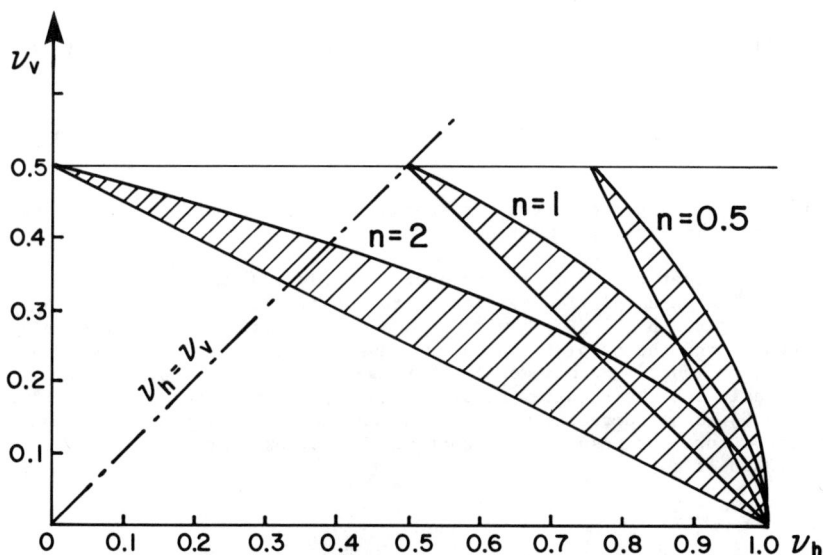

Figure 3.   Zones of Poisson's Ratio Values Corresponding to an Anisotropic Deformation Under an Isotropic Stress

As the independent elastic parameters required are quite numerous, i.e., $(E_v, G_v, n, \nu_v, \nu_h)$, and difficult to measure, a parametric study has been performed (33) in order to determine whether it could be possible to obtain a simplified model by reducing the number of elastic parameter, on the basis of experimental observations. The specific aims of the study were:

i)   To be able to fit any direction of effective stress path in undrained condition, for both overconsolidated and normally consolidated responses; and

ii) To be able, following an isotropic increase of effective stresses, corresponding approximately to the conditions prevailing under an embankment during consolidation, to obtain an anisotropic increase of deformation and a corresponding decrease in volume.

It may be shown (33) that the first condition mentioned above corresponds to the following relationship in the triaxial test:

$$m = \frac{\Delta \sigma_1'}{\Delta \sigma_3'} = \frac{2(n \nu_v - 1 + \nu_h)}{n(1 - 2\nu_v)} \tag{2}$$

where m is the slope of principal stress increments. It should be noted that m is linked to the gradient of the stress path of Wroth (45). The second condition, which corresponds to the necessity of having outward increments of horizontal displacements during consolidation (20,21,34, 38,39), may be expressed as:

$$1 - 2 \nu_v > 0 \tag{3a}$$

and

$$1 - \nu_v - n \nu_h < 0 \quad \text{or} \quad m > 0 \tag{3b}$$

In the (ν_v, ν_h) plane, these two inequalities correspond to the hatched zones shown in Fig. 3.

A further result of the parametric study (33) mentioned above has been that a simplified version of the model could be arrived at by considering E_v, $n = E_h/E_v$, $\nu = \nu_v = \nu_h$ and by imposing a relationship between G_v and the triplet (E_v, n, ν) thus, resulting effectively in a 3-parameter model. Such a model, expressed in terms of (E_v, n, ν), still satisfies the requirement of the theory of anisotropic elasticity. Because G_v is not directly measured in conventional tests, a review of the literature has resulted in adopting the following relationship (2, 11,12,33,36):

$$G_v = 0.5 \ E_v \tag{4}$$

Determination of the Parameters

1.   Normally Consolidated Clay

Let point A in the (p',q) plane represent the effective stress state of a typical soil element at the end of the construction stage in the field, as shown in Fig. 4. During the consolidation process, the

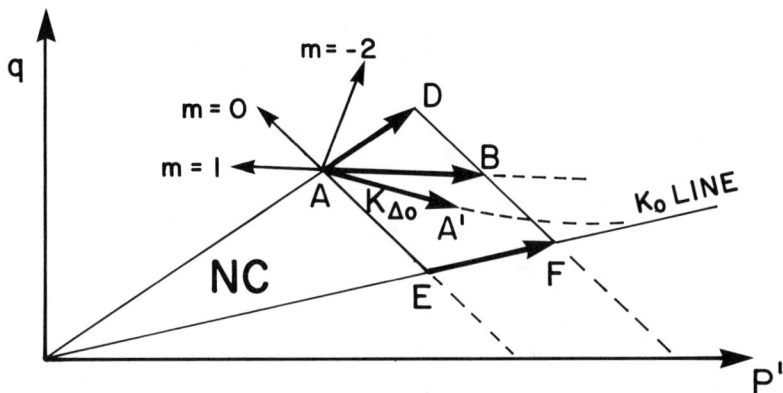

Figure 4. Modelling the Normally Consolidated State

effective stress path will be roughly along AB whose slope is given by $K = \Delta\sigma'_3/\Delta\sigma'_1 = 1$. In order to model the behaviour of the soil in the vicinity of this stress path, it will be assumed that for the proportional loading AD, the resulting decrease in volume will be the same as that obtained in moving from E to F corresponding to the stress path in an oedometer test, as in the case of the Cam-Clay model (9,25,30,44). Such a condition will result in the following condition (33):

$$\gamma_s = \frac{E_v}{E_{oed}} = \frac{\lambda}{n}\left[2(1-\nu) - n\,\nu\right] + 1 - 2\nu \qquad (5)$$

where λ is the effective stress ratio, σ'_h/σ'_v, at point A and E_{oed}, which is the constrained deformation modulus, is given by (25,45):

$$E_{oed} = \frac{1 + e_o}{0.435\,C_c}\,\bar{\sigma}'_v \qquad (6)$$

where e_o = void ratio,
C_c = compression index, and
$\bar{\sigma}'_v$ = average value of the vertical effective stress during consolidation.

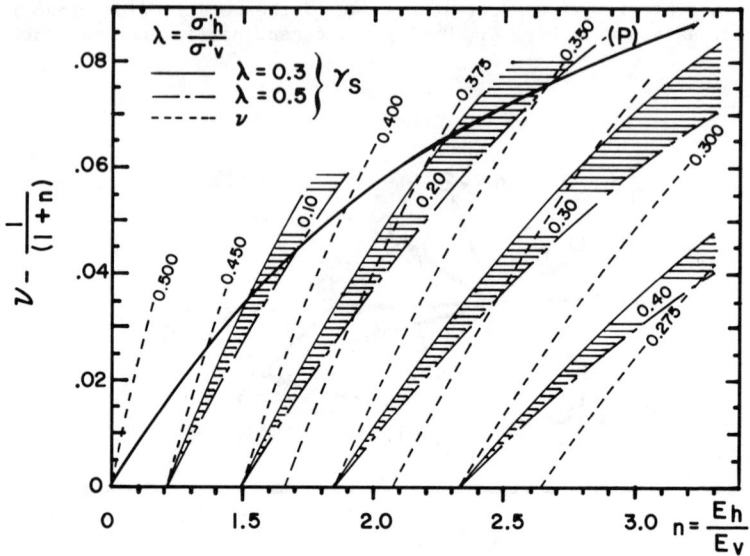

Figure 5.   Relation Between Young's Modulus E_v and Constrained Modulus E_{oed}

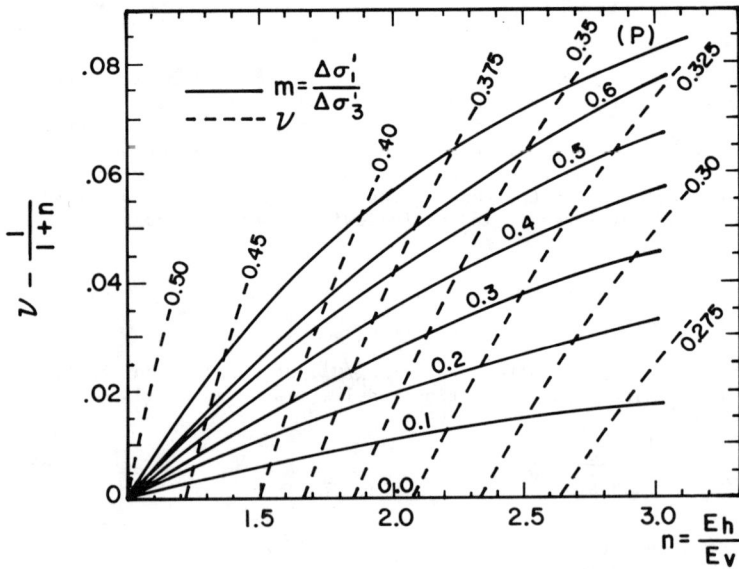

Figure 6.   Effective Stress Path Direction, $m = \Delta\sigma_1'/\Delta\sigma_3'$, in an Undrained Triaxial Test

Eq. 5 has been put in graphical form in Fig. 5 for λ equal to 0.3 and 0.5, respectively. Therefore, for the consolidation response in the field, the value of Young's modulus, E_v, may be obtained by performing a consolidation test and by using:

$$E_v = \gamma_s \, E_{oed} = 2.3\gamma_s \, \bar{\sigma}_v' \, \left(\frac{C_c}{1+e_o}\right)^{-1} \tag{7}$$

Depending upon the choice of the values of n and ν, the effective stress path in an one-dimensional consolidation test starting at point A, will be along AA' in Fig. 4 corresponding to a slope given by:

$$K_{\Delta o} = \frac{\Delta\sigma_3'}{\Delta\sigma_1'} = \frac{n\nu}{1-\nu} > 1 \tag{8}$$

Such a condition corresponds to the necessity of having outward increments of deformation, as mentioned in the previous section. The strain increment ratio is given by:

$$\frac{\Delta\varepsilon_h}{\Delta\varepsilon_v} = \frac{1 - \nu - n\nu}{n(1 - 2\nu)} \tag{9}$$

where $\Delta\varepsilon_h$ and $\Delta\varepsilon_v$ represent the horizontal and vertical strain increments, respectively. It has been found (33,34) that the values of n and ν that satisfy the above requirements are:

$$0.3 < \nu < 0.4 \tag{10a}$$

and

$$1.5 < n < 2.0 \tag{10b}$$

In addition, to a given pair (n,ν), there corresponds an undrained effective stress path whose slope is given by m in Eq. 2. Fig. 6 presents this equation in graphical form.

By using the restrictions on the values of ν and n mentioned above, typical values for m are found to between zero and one, during the in-situ consolidation process.

2.   Overconsolidated Clay

In its natural state in the ground an element of highly overconsolidated clay is acted upon by vertical and horizontal stresses σ_v' and σ_h', respectively, as represented by point A in Fig. 7. The volumetric yield locus for the soil element is represented by CDE where E lies on the failure line. If the total stresses are increased without drainage, as represented by the line AB, the effective stresses will follow the path AB' where B' lies on the yield locus. Since the stresses are axially symmetric the initial elastic portion AB' has a slope $m = \Delta\sigma_v'/\Delta\sigma_h' = \Delta\sigma_1'/\Delta\sigma_3' = -2$ (4). Inserting this value of m in Eq. 2, one gets n = 1 for the simplified model. A simple linear isotropic model is thus capable to predict adequately the soil response due to initial loading. In this case, Young's modulus may be obtained either from a triaxial or a consolidation test performed on an overconsolidated clay sample. It should be noted also that, had the general 5-parameter model been retained for the analysis, the value on n would have been (1,51):

$$n = \frac{1 - \nu_h}{1 - \nu_v} \qquad\qquad (11)$$

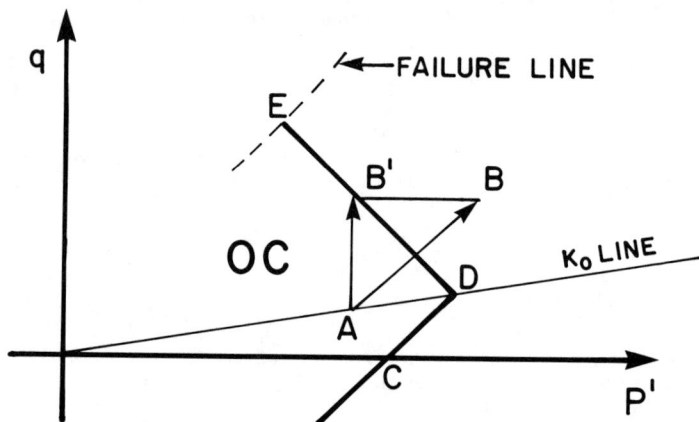

Figure 7. Overconsolidated Clay Behaviour

3. Coefficients of Permeability and Consolidation

The coefficients of permeability in the vertical and horizontal directions, k_v and k_h, to be used in the analysis of the rate of consolidation, may be determined either, directly, from permeameter tests or, indirectly, from consolidation tests (40,41). Concerning the value of k_h, even though field tests would, in theory, give better results, considerable difficulty is experienced in interpreting the field data and, as a consequence, laboratory tests are preferred (23). In addition, one should perform consolidation tests and calculate the values of C_v, by taking into account their dependence on the stress level and the fact that the laboratory values are almost always lower than those inferred from field observations (15,18,19,38).

CONCLUSIONS

A finite element program, SOL, that utilizes a bi-linear anisotropic elastic model, is presented in this paper. The parameters necessary for the implementation of the model and the laboratory tests required for their determination are also described.

ACKNOWLEDGEMENTS

The authors gratefully acknowledge the financial assistance provided by the Ministry of Education of Quebec and the National Research Council of Canada.

APPENDIX 1 - REFERENCES

1. Adachi, T., Oka, F., and Tange, Y., "Finite Element Analysis of Two Dimensional Consolidation Using an Elasto-Viscoplastic Constitutive Equation", *Proceedings of the Fourth International Conference on Numerical Methods in Geomechanics*, Edmonton, Alberta, Canada, Vol. 1, 1982, pp. 287-296.

2. Barden, L., "Stresses and Displacements in a Cross-Anisotropic Soil", *Géotechnique*, Vol. 13, No. 3, 1963, pp. 198-210.

3. Bozozuk, M., and Leonards, G.A., "The Gloucester Test Fill", *Proceedings of the Specialty Conference on Earth and Earth-Supported Structures*, ASCE, Purdue University, Lafayette, Indiana, U.S.A. Vol. 1, Part I, June 1972, pp. 299-318.

4. Burland, J.B., "A method of Estimating the Pore Pressures and Displacements Beneath Embankments on Soft Natural Clay Deposits", *Proceedings of the Roscoe Memorial Symposium, Stress-Strain Behaviour of Soils*, R.H.G. Parry Ed., G.T. Foulis and Co. Ltd., England, 1972, pp. 505-536.

5. Christian, J.T., "Undrained Stress Distribution by Numerical Methods", *Journal of the Soil Mechanics and Foundation Engineering Division*, ASCE, Vol. 94, No. SM6, Nov. 1968, pp. 1333-1345.

6. Christian, J.T. and Watt, B.J., "Undrained Visco-Elastic Analysis of Soil Deformations", *Proceedings of the Symposium on Applications of the Finite Element Method in Geotechnical Engineering*, U.S. Army Engineer Waterways Experiment Station, Vicksburg, Mississipi, U.S.A., May 1972, pp. 533-580.

7. Domaschuck, L., and Valliappan, P., "Nonlinear settlement Analysis by Finite Element", *Journal of the Geotechnical Engineering Division*, ASCE, Vol. 101, No. GT7, July 1975, pp. 601-614.

8. Duncan, J.M. and Poulos, H.G., "Modern Techniques for the Analysis of Engineering Problems in Soft Clay", *Soft Clay Engineering*, E.W. Brand and R.P. Brenner Eds., Elsevier North-Holland Inc., New York, 1981, pp. 367-406.

9. Foot, R., and Ladd, C.C., "Behaviour of Atchafalaya Levees during Construction", *Géotechnique*, Vol. 27, No. 2, 1977, pp. 137-160.

10. Fortin, M., and Soulié, M., "A Non-Conforming Piecewise Quadratic Finite Element on Triangles", *International Journal for Numerical Methods in Engineering*, Vol. 19, 1983, pp. 505-520.

11. Garnier, J., "Tassement et Contraintes. Influence de la Rigidité de la fondation et de l'anisotropie du massif", Ph.D. Thesis, University of Grenoble, France, 1973, 200 p.

12. Koning, H., "Stress Distribution in a Homogeneous, Anisotropic Elastic Semi-Infinite Solid", *Proceedings of the Fourth International Conference of Soil Mechanics and Foundation Engineering*, London, Vol. 1, 1957, pp. 335-338.

13. Ladd, C.C., "Test Embankment on Sensitive Clay", *Proceedings of the Specialty Conference on Performance of Earth and Earth-Supported Structures*, ASCE, Purdue University, Lafayette, Indiana, U.S.A., Vol. 1, Part 1, June 1972, pp. 101-128.

14.   Ladd, C.C., "Predicted Performance of an Embankment on Boston
       Blue Clay", *Proceedings of the Foundation Deformation Prediction
       Symposium*, Massachusetts Institute of Technology, Cambridge,
       Nov. 1974, Department of Transportation, Federal Highway Adminis-
       tration, Vol. 2, 1975, pp. B-1-B-111.
15.   Ladd, C.C., Foott, R., Ishihara, K., Schlosser, F., and Poulos,
       H.G., "Stress-Deformation and Strength Characteristics", *Procee-
       dings of the Ninth International Conference on Soil Mechanics
       and Foundation Engineering*, Tokyo, Vol. 2, 1977, pp. 421-494.
16.   Lambe, T.W., "Predictions in Soil Engineering", Thirteen Rankine
       Lecture, *Géotechnique*, Vol. 23, No. 2, 1973, pp. 149-202.
17.   Larson, R., "Drained Behaviour of Swedish Clays", *Swedish Geo-
       technical Institute*, Report No. 12, 1981, 157 p.
18.   Leroueil, S., Tavenas, F., Trak, B., LaRochelle, P., and Roy, M.,
       "Construction Pore Pressures in Clay Foundations under Embank-
       ments. Part I: the Saint-Alban Test Fills", *Canadian Geotechni-
       cal Journal*, Vol. 15, No. 1, Feb. 1978, pp. 54-65.
19.   Leroueil, S., Tavenas, F., Mieussens, C., and Peignaud, M.,
       "Construction Pore Pressures in Clay Foundations under Embank-
       ments. Part II: Generalized Behaviour", *Canadian Geotechnical
       Journal*, Vol. 15, No. 1, Feb. 1978, pp. 66-82.
20.   Magnan, J.P., Humbert, P., Belkiziz, A., and Mouratidis, A.,
       "Finite Element Analysis of Soil Consolidation with Special Re-
       ference to the Case of Strain Hardening Elasto-Plastic Stress-
       Strain Models", *Proceedings of the Fourth International Confe-
       rence on Numerical Methods in Geomechanics*, Edmonton, Alberta,
       Canada, Vol. 1, 1982, pp. 327-336.
21.   Magnan, J.P., Humbert, P. and Mouratidis, A., "Finite Element
       Analysis of Soils Deformations with Time under an Experimental
       Embankment at Failure", *Proceedings of the International Sympo-
       sium on Numerical Models in Geomechanics*, Zurich, 1972, pp. 601-
       609.
22.   Massachusetts Institute of Technology, *"Proceedings of the Foun-
       dation Deformation Prediction Symposium"*, Nov. 1974, Vols. 1 and
       2, U.S. Department of Transportation, Federal Highway Administra-
       tion, 1975.
23.   Mieussens, C., and Ducasse, P., "Mesure en Place des Coefficients
       de Perméabilité et des Coefficients de Consolidation horizontaux
       et verticaux", *Canadian Geotechnical Journal*, Vol. 14, No. 1,
       Feb. 1977, pp. 76-90.
24.   Palmerton, J.B., "Creep Analysis of Atchafalaya Levee Foundation",
       *Proceedings of the Symposium on Applications of the Finite Ele-
       ment Method in Geotechnical Engineering*, U.S. Army Engineer Wa-
       terways Experiment Station, Vicksburg, Mississipi, U.S.A., May
       1972, pp. 843-862.
25.   Parry, R.H.G., and Wroth, C.P., "Shear Stress-Strain Properties
       of Soft Clay", *Soft Clay Engineering*, E.W. Brand and R.P.
       Brenner Eds., Elsevier/North-Holland Inc., New York, 1981, pp.
       311-366.
26.   Pickering, D.J., "Anisotropic Parameters for Soils", *Géotechni-
       que*, Vol. 20, No. 3, 1970, pp. 271-276.
27.   Raymond, G.P. "The Kars (Ontario) Embankment Foundation", *Pro-
       ceedings of the Specialty Conference on the Performance of Earth
       and Earth-Supported Structures*, ASCE, Purdue University, Lafa-

yette, Indiana, U.S.A., Vol. I, Part 1, June 1972, pp. 319-340.
28. Raymond, G.P., "Prediction of Undrained Deformation and Pore Pressures in Weak Clay under Two Embankments", *Géotechnique*, Vol. 22, No. 3, 1972, pp. 381-401.
29. Sagaseta, C., and Arroyo, R., "Limit Analysis of Embankments on Soft Clay", *Proceedings of the International Symposium on Numerical Models in Geomechanics*, Zurich, 1972, pp. 618-625.
30. Schofield, A.N., and Wroth, C.P., *"Critical State Soil Mechanics*, McGraw-Hill, London, 1968, 310 p.
31. Shibata, T., Tominaga, M. and Matsuoka, H., "FE Analysis of Soil Movements Below a Test Embankment", *Proceedings of the Second International Conference on Numerical Methods in Geomechanics*, Blacksbur, Virginia, U.S.A., Vol. 2, June 1976, pp. 599-610.
32. Simon, R.M., Christian, J.I., and Ladd, C.C., "Analysis of Undrained Behavior of Loads on Clay", *Proceedings of the Specialty Conference on Analysis and Design in Geotechnical Engineering*, ASCE, University of Austin, Texas, U.S.A., Vol. 1, June 1974, pp. 51-84.
33. Siwe, A.S., *"Influence de l'Anisotropie Mécanique du sol pendant la Consolidation"*, M.Sc.A. Thesis, Ecole Polytechnique, Montreal, 1977, 193 p.
34. Soulié, M., *"Les Eléments Finis: Outils de Prédiction"*, New Study Methods of Fills on Soft Soils, Laval University (Quebec), Ecole Nationale des Ponts et Chaussées (Paris), Dec. 1978, 50 p.
35. Stille, H., Fredriksson, A., and Broms, B.B., "Analysis of a Test Embankment Considering the Anisotropy of Soil", *Proceedings of the Second International Conference on Numerical Methods in Geomechanics*, Blacksburg Virginia, U.S.A., Vol. 2, June 1976, pp. 611-622.
36. Tan, T.K., "Consolidation and Secondary Time Effect of Homogeneous Anisotropic Saturated Clay Strata", *Proceedings of the Fifth International Conference on Soil Mechanics and Foundations Engineering*, Paris, Vol. 1, 1961, pp. 367-373.
37. Tavenas, F., and Leroueil, S., "Effects of Stresses and Time on Yielding of Clays", *Proceedings of the Ninth International Conference on Soil Mechanics and Foundation Engineering*, Tokyo, Vol. 1, 1977, pp. 319-326.
38. Tavenas, F., and Leroueil, S., "The Behaviour of Embankments on Clay Foundations", *Canadian Geotechnical Journal*, Vol. 17, No. 2, May 1980, pp. 236-260.
39. Tavenas, F., Mieussens, C., and Bourges, F., "Lateral Displacements in Clay Foundations Under Embankments", *Canadian Geotechnical Journal*, Vol. 16, No. 3, Aug. 1979, pp. 532-550.
40. Tavenas, F., Leblond, P., Jean, P., and Leroueil, S., "The Permeability of Natural Soft Clays. Part I: Methods of Laboratory Measurement", *Canadian Geotechnical Journal*, Vol. 20, No. 4, Nov. 1983, pp. 629-644.
41. Tavenas, F., Jean, P., Leblond, P., and Leroueil, S., "The Permeability of Natural Soft Clays. Part II: Permeability Characteristics", *Canadian Geotechnical Journal*, Vol. 20, No. 3, Nov. 1983, pp. 645-660.
42. Tham, B.R., "Numerical Analyses of Embankments over Soft Subsoils", *Proceedings of the Third International Conference on Numerical Methods in Geomechanics*, Aachen, West Germany, Vol. 2, pp. 723-731.

43.   Thoms, R.L., Pecquet, R.A., and Arnran, A., "Numerical Analyses of Embankments over Soft Soils", *Proceedings of the Second International Conference on Numerical Methods in Geomechanics*, Blacksburg, Virginia, U.S.A., Vol. 2, 1976, pp. 623-637.

44.   Wroth, C.P., "Some Aspects of the Elastic Behaviour of Over-Consolidated Clay", *Proceedings of the Roscoe Memorial Symposium*, Stress-Strain Behaviour of Soils, R.H.G. Parry Ed., G.T. Poulis and Co. Ltd, England, 1972, pp. 347-361.

45.   Wroth, C.P., "The Predicted Performance on Soft Clay under a Trial Embankment Loading Based on the Cam-Clay Model", *Finite Elements in Geomechanics*, G. Gudehus Ed., John Wiley and Sons, Inc., London, 1977, pp. 191-208.

46.   Wroth, C.P., and Simpson, B., "An Induced Failure at a Trial Embankment: Part II - Finite Element Computations", *Proceedings of the Specialty Conference on Earth and Earth Supported Structures*, ASCE, Purdue University, Lafayette, Indiana, U.S.A., Vol. 1, Part 1, June 1972, pp. 65-80.

APPENDIX II - NOTATION

C_c	= compression index
C_v	= coefficient of consolidation
E_{oed}	= constrained modulus
E_h	= horizontal modulus
E_v	= vertical modulus
G_v	= shear modulus
$K_{\Delta o}$	= slope of effective stress path
K_o	= coefficient of earth pressure at rest
e_o	= initial void ratio
k_h	= horizontal coefficient of permeability
k_v	= vertical coefficient of permeability
m	= slope of effective stress path
n	= modulus ratio
p'	= $(\sigma_1' + \sigma_3')/2$
q	= $(\sigma_1' - \sigma_3')/2$
Δ	= increment
ε_h	= horizontal deformation
ε_v	= vertical deformation
γ_s	= E_v/E_{oed}
λ	= stress ratio
ν	= Poisson's ratio
ν_h	= Poisson's ratio for the strain in the horizontal direction due to a horizontal stress
ν_v	= Poisson's ratio for the strain in the horizontal direction due to a vertical stress
σ_h'	= effective horizontal stress
σ_{vc}'	= effective vertical preconsolidation pressure
σ_v'	= effective vertical stress
σ_1'	= effective major principal stress
σ_3'	= effective minor principal stress

CONSOLIDATION PERFORMANCE OF SOFT CLAYS:
PART 2. PREDICTIONS

M. Soulié[1] and V. Silvestri[2], M.ASCE

ABSTRACT

The simplified version of the model, described in Part 1 of the Paper, was firstly used to predict the short-term performance of the St-Alban test fill. Secondly, the long-term predictive capability of the model was tested and implemented by means of seven-year long observations of the I95 test fill built near Boston. On the basis of these observations, a first version of the anisotropic model was developed without the concept of undrained creep. This model which was found to predict quite correctly both vertical and horizontal displacements, yielded pore water pressures that were lagging somewhat behind the vertical settlements. In order to make the evolution of pore water pressure and settlements coincide with the observed values, a new version of the model was implemented which takes into account undrained creep.

This new version of the model was used to predict (Type B prediction) the long-term behaviour during consolidation of the Cubzac Les Ponts test fill and the soft clay stratum on which it is founded.

The present paper describes the prediction of the consolidation performance at the Cubzac site and compares it with the observed behaviour.

INTRODUCTION

Geotechnical engineers are often faced with the problem of predicting the consolidation performance of soft natural clays. The correct analysis of this problem requires a good knowledge of the soil behaviour during and after construction of the fills.

In Part 1, the simplified version of the soil model used in the Finite Element (FE) scheme, was described. In Part 2, this model is used to predict the consolidation behaviour of the Cubzac Les Ponts test fills. Because of the time-lag associated with the dissipation of the pore water pressures, a new version of the model that takes into account such a phenomenon, is used. It is shown that this improved version of the model has been developed on the basis of the consolidation performance of the M.I.T. test fill.

---

[1]   Professor of Civil Engineering, Ecole Polytechnique, Montreal, Canada
[2]   Associate Professor of Civil Engineering, Ecole Polytechnique, Montreal, Canada

197

PREDICTION DURING CONSTRUCTION

In order to illustrate the application of the proposed model for the prediction of displacements and pore water pressures during construction, an analysis of the performance of embankment D of the St-Alban experimental test fills is presented.

Detailed informations of this site, test fills and performance, have been reported elsewhere (7). For a better understanding of the analysis, a summary of the soil profile, geometry of the test fill and construction sequence, is presented in Fig. 1. Relevant geotechnical parameters are shown in Table 1. The parameters used in the model have

Table 1.   Geotechnical parameters: Saint-Alban

Depth	e_o	σ'_{vo}	σ'_{vc}	C_c	C_v	k(N.C.)
(m)		kPa	kPa		cm^2/s	cm/s
0-1.2 m	1.8	40	70	0.5	5×10^{-4}	4.8×10^{-8}
1.2-3	1.8	50	41	2.2	10^{-4}	3.4×10^{-8}
3-5	1.8	60	61	1.9	2×10^{-4}	4.9×10^{-8}
5-8	1.8	70	66	1.1	2×10^{-4}	2.4×10^{-8}
8-10	1.8	82	107	1.1	2×10^{-4}	2.0×10^{-8}
10-12	1.8	97	133	0.2	2×10^{-4}	1.7×10^{-8}

been obtained from undrained triaxial tests (E_V, n=1) and consolidation tests (C_V) for the soil in the overconsolidated state; for the normally consolidated clay, consolidation tests were used to obtain the pertinent parameters (C_V, C_c and n=2). In both cases, $\nu = 0.375$ and $G_V = 0.5\ E_V$. These parameters appear in Table 2. The simplified Limit State Curve

Table 2.   Parameters used in the analysis: Saint-Alban

Depth	ν	O.C. State E_V	n	G_V	N.C. State E_V	n	G_V
(m)		kPa		kPa	kPa		kPa
0-1.2	0.375	870	1	435	105	2	52
1.2-3	0.375	23000	1	11500	30	2	15
3-5	0.375	35500	1	17750	42	2	21
5-8	0.375	41500	1	20750	84	2	42
8-10	0.375	82500	1	41250	98	2	49
10-12	0.375	100000	1	50000	1160	2	580

(LSC) has been obtained from the vertical preconsolidation pressure profile and with a value of K_o in the normally consolidation stage equal to 0.5. The initial effective stress state was estimated on the basis of $K_o = 0.7$ (7).

The predicted values of the vertical displacements and pore water pressures at the depths of -2.5 and -5.0 m, and of the pore water distribution underneath the axis of the fill, are compared with the measured values in Fig. 2. The diagrams shown in these figures indicate

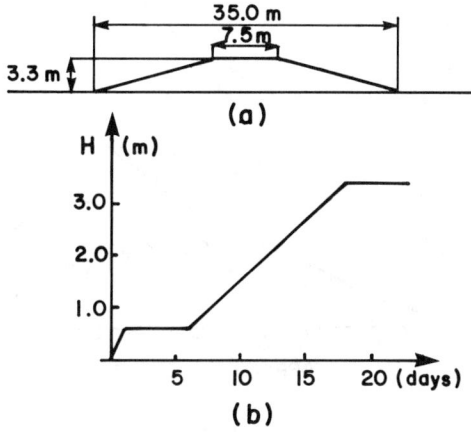

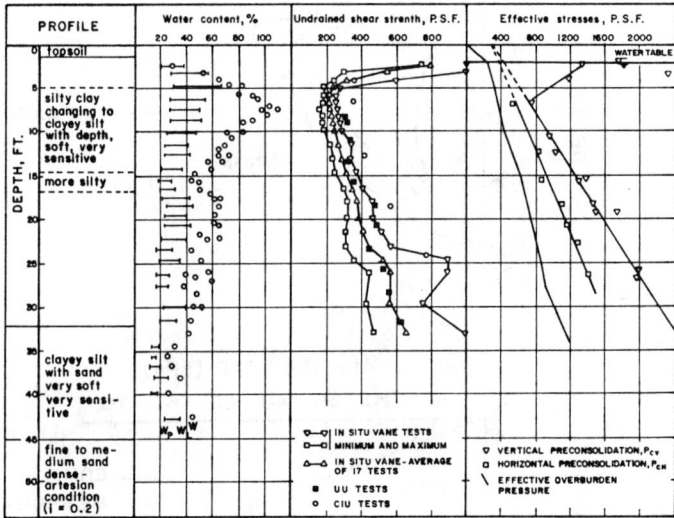

Figure 1.   Saint-Alban Test Fill D. a) Geometry; b) Construction
             Sequence; c) Soil Properties

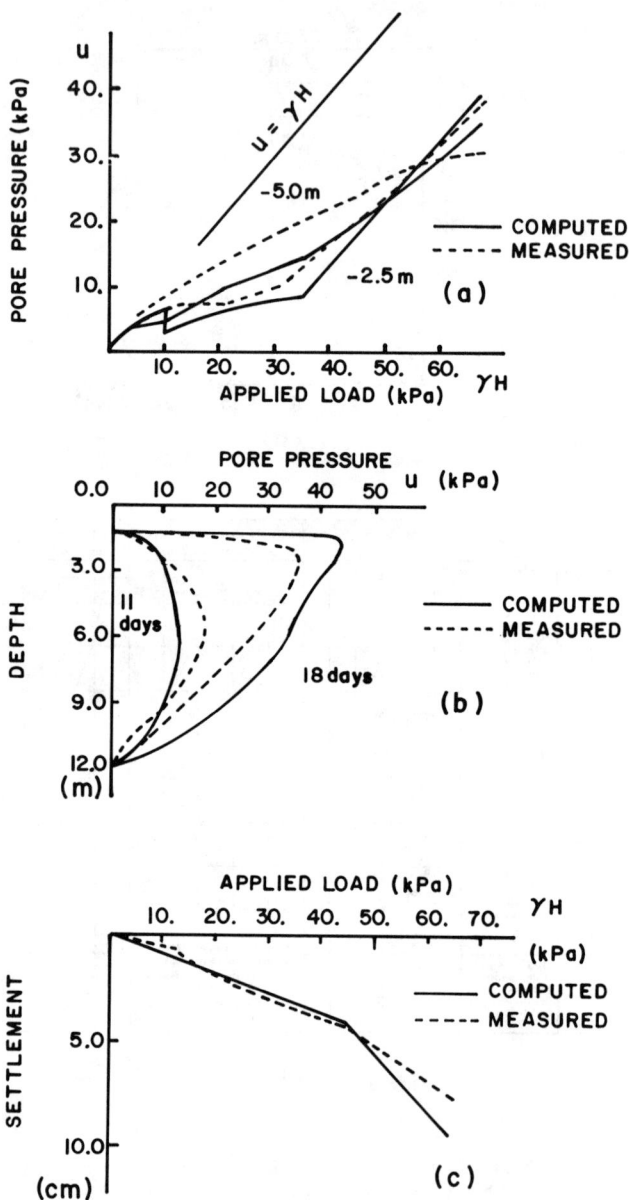

Figure 2.　Saint-Alban Test Fill D. a) Pore Pressure Response;
b) Pore Pressure Distribution; c) Settlement

that, in general, there is good agreement between the predicted and measured values. As it is noticed, there exists a break point in the curves showing settlement and pore water pressure generation versus height
of the fill for the soil at the depth of -2.5 m (Fig. 2a). Such a
change in behaviour is due to the fact that the effective stress path
is reaching the Limit State Curve and that the original soil properties
are modified, as shown in Fig. 3a (See, also, 7). In fact, the analysis of both the effective and total stress paths of the point A', corresponding to the soil at the depth of -2.5 m, shows that once the
stress state reaches the limit state at B', any subsequent increase of
the loading, increases the total vertical stress from B to C with practically the same increase for the horizontal stress. Moreover, because
of the much lower value of C_V for the clay in the normally consolidated
state, the rate of pore water pressure dissipation is very low, with the
result that points B' and C' are practically identical. Such a phenomenon explains the observed break point in the curve of pore water pressure versus height of fill. In addition, it should be noted that for the
clay in the overconsolidated state, the effective stress path $A'B'$ corresponds to a partial drainage due to the high value of C_V. The corresponding undrained effective stress path in this case would be $A'A''$.

    At the point E (Fig. 3b), corresponding to the soil at the depth
of -5.0 m, the effective stress path remains entirely in the overconsolidated zone. It is interesting to note that because of the sudden increase in the pore water pressure in the neighborhood of point E, i.e.,
at point A, due to the fact that the soil has reached the normally consolidated state, the effective stress path will bend upward ($F'G'$).

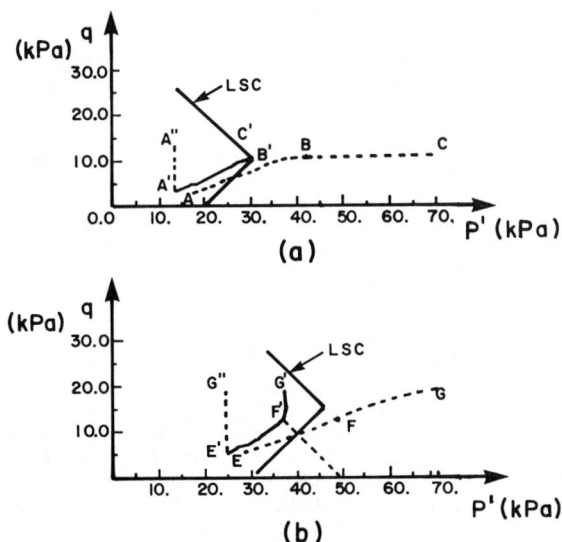

Figure 3.   Effective Stress Paths Under Centerline. a) Depth of
             -2.5m; b) Depth of -5.0 m

This is due to the fact that the high value of C_v for the soil in the overconsolidated state which surrounds the normally consolidated clay, will aid in the spreading and increase of the pore water pressure. The practical significance of such a phenomenon is that point E is now closer to failure.

In summary, the model can be used for the prediction of both the settlements and pore water pressures during the construction stage. Nevertheless, the analysis shows that if partial drainage occurs during construction, due to the high values of C_v for the soil in the overconsolidated state, it should be taken into account, as the results for the prediction of pore pressures are very sensitive to the choice of the values of C_v and E_v, as well as to the correct determination of the Limit State Curve.

CONSOLIDATION PREDICTION

In order to illustrate the application of the model for the prediction of settlements, horizontal displacements and pore water pressure dissipation, the analysis of two field cases will be presented.

The first one which is shown in Fig. 4 corresponds to the seven year long consolidation history of the I95 embankment which was used for the Foundation Deformation Symposium held at M.I.T. in 1974 (10). In this case, a type C1 prediction, or more exactly a fitting procedure, has been performed, using for the pertinent parameters the values which were obtained in laboratory tests (10).

The results which are presented in Fig. 5 show that for a variation of the coefficient of permeability by a factor of five, a good fitting is obtained for the progress of settlement. However, even by using the best value for the coefficient of permeability, there still exists such a large discrepancy between the predicted and measured values of the pore water pressure, that the adequacy of the first version of the model has to be questioned (Fig. 5b). In this case, the dissipation of the excess pore water pressure is highly overestimated.

The second field case corresponds to the consolidation history of the experimental test fill at Cubzac-les-Ponts in France (8). In this case a type B prediction has been performed (See, also, 15) by using the first version of the model.

The embankment characteristics were: unit weight of 21 kN/m^3, side slopes of 1.5H:1V, height of 2.3 m and crest width of 17 m. The geotechnical characteristics of the soil are presented in Fig. 6. The values of the parameters for the model in both the overconsolidated and normally consolidated states are shown in Table 3. For this case, two sets of data were available for the coefficient of permeability. The first one, $k(I)$, was obtained indirectly from laboratory consolidation tests. The second set, $k(II)$, was obtained from laboratory permeameter tests for k_v and from field measurements for k_h. Fig. 7a which presents a comparison between the predicted and measured progress of settlement with time, shows that a better fitting is obtained when one uses the second set of data. However, when considering the pore water

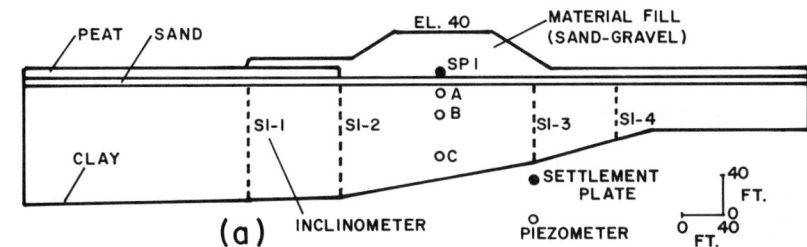

(a)

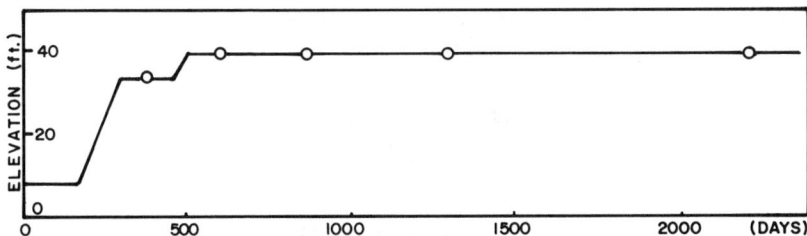

(b)

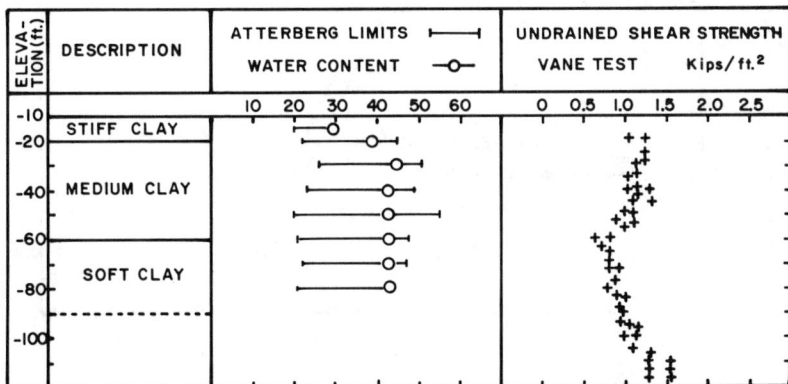

(c)

Figure 4.   M.I.T. I95 Embankment. a) Geometry; b) Construction Sequence;
             c) Soil Properties

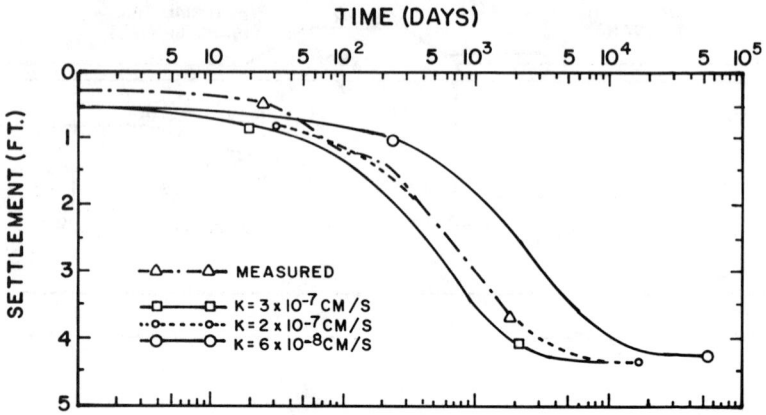

(a)

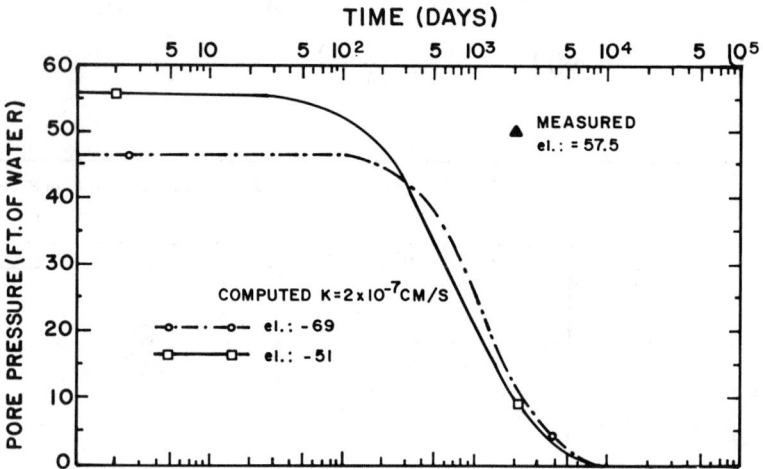

(b)

Figure 5.   M.I.T. I95 Embankment. a) Settlement; b) Pore Pressure
             Dissipation

Table 3.　Parameters used in the analysis:　Cubzac-les-Ponts

Case	Depth (m)	ν	E_v kPa	n	G_v kPa	$k_h(I)$ m/day $\times 10^{-4}$	$k_v(I)$ m/day $\times 10^{-4}$	$k_h(II)$ m/day $\times 10^{-3}$	$k_v(II)$ m/day $\times 10^{-4}$
Undrained Loading	0-1	0.550	1800	0.6	1350	-	-	-	-
	1-2	0.375	1500	1	1125	1.0	1.0	10.0	26.0
	2-4	0.375	1100	1	825	1.2	3.4	6.4	1.0
	4-7	0.375	1033	1	775	1.3	5.3	4.7	1.0
	7-8	0.375	1350	1	1012	1.0	5.7	5.0	1.0
	8-9	0.375	1350	1	1012	1.0	6.0	5.0	1.0
Consolidation	0.1	0.550	1800	0.6	1350	-	-	-	-
	1-2	0.375	250	2	125	1.0	1.0	10.0	26.0
	2-4	0.375	45	2	22	1.2	3.4	6.4	1.0
	4-7	0.375	60	2	30	1.3	5.3	4.7	1.0
	7-8	0.375	70	2	35	1.0	5.7	5.0	1.0
	8-9	0.315	70	2	35	1.0	6.0	5.0	1.0

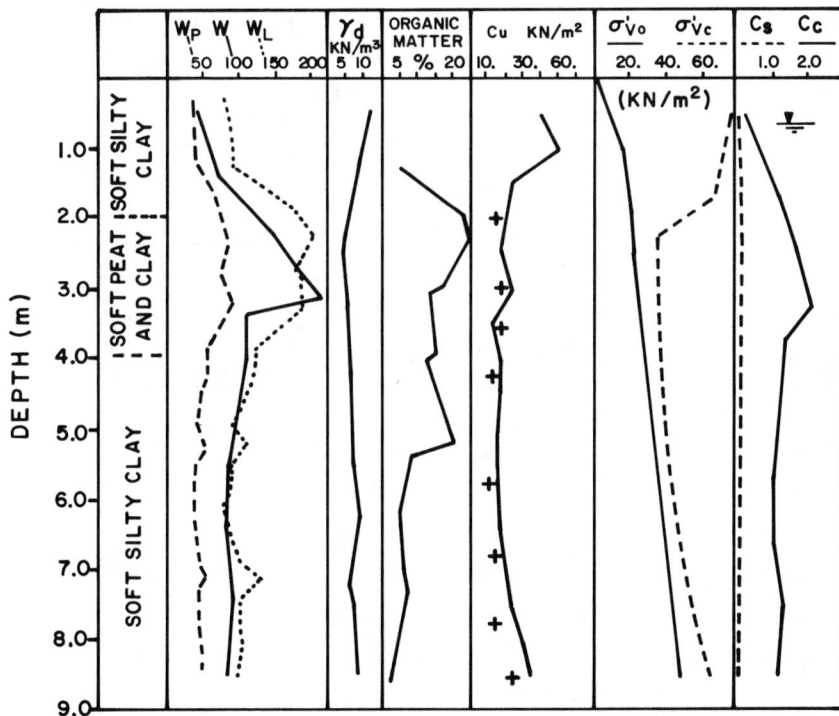

Figure 6.　Cubzac-Les-Ponts Test Fill B.　Soil Profile

pressure dissipation curve of Fig. 7b, once again a very large discrepancy exists between the observed and calculated values: the pore water pressure has been overestimated.

In fact, the slow dissipation of the pore water pressures mentioned above, in comparison with the rate predicted by the conventional theory of consolidation, has been observed on several occasions and has been attributed either to some creep phenomenon, or self-induced consolidation, or even to defective instruments (1,3,8,10).

Therefore, in order to take into account such a phenomenon, a modified version of the model has been developed (14,15) and is presented in the next section.

PHENOMENOLOGICAL CREEP

It is generally admitted that, when determining the long-term settlement of a soft clay, one must add to the settlement calculated on the basis of primary consolidation that resulting from secondary or delayed compression, or drained creep (1,2,6,9). On the other hand, very few studies have been performed concerning the phenomenon of undrained creep (4,5,6,10,12,13,16) and the settlements associated with it (3,10,11,16).

The phenomena of secondary consolidation and quasi-preconsolidation (1,2,9) may be analyzed with the help of the classical $e-\log\sigma'_v$ diagram obtained from an oedometer test, as shown in Fig. 8a. Starting from an initial state represented by point A, a typical path follows

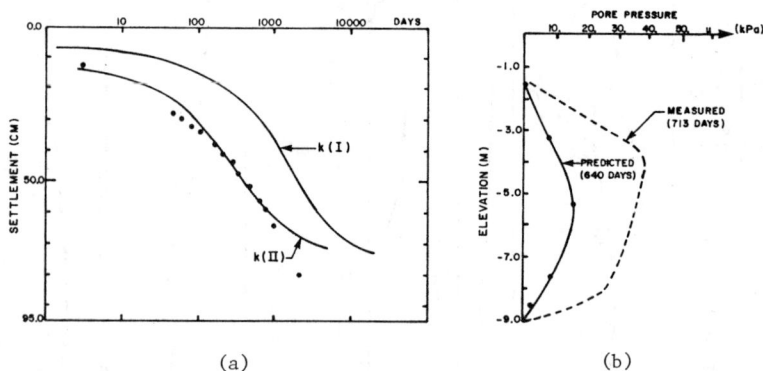

(a)   (b)

Figure 7.   Cubzac-Les-Ponts Test Fill B. a) Settlement; b) Pore Pressure Distribution

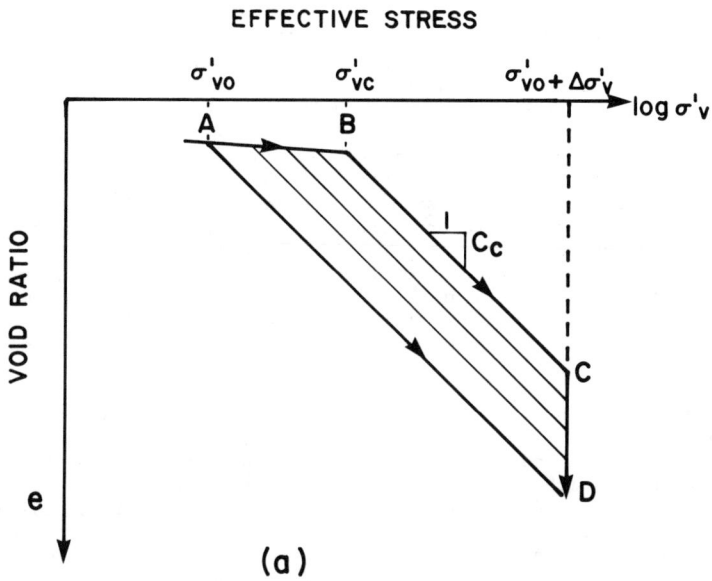

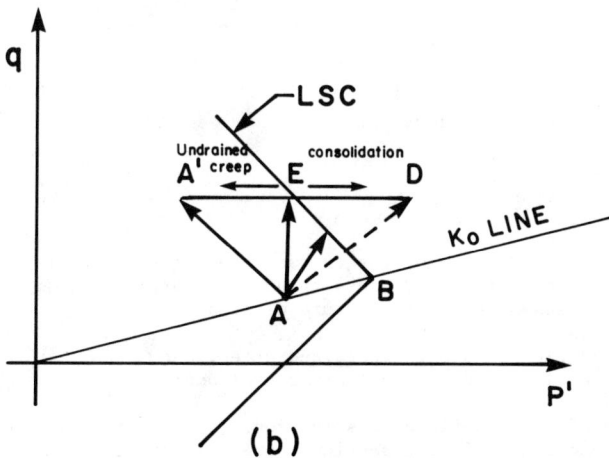

Figure 8.   Role of Creep During Consolidation. a) Oedometer Test;
             b) Undrained Creep and consolidation

ABCD for a given loading rate in the laboratory. The segment *CD* is due
to secondary consolidation. In addition, it has been shown (2,6) that
in the consolidation testing of thin samples, the duration of primary
consolidation is short and that this stage is followed by delayed compression. For this paper, the observed overconsolidation is considered
to be due to delayed compression. However, in the field, the conditions
are more complex than those existing in the laboratory. Consider, for
example, the distance separating a typical soil element from pervious
boundaries. For very short distances, the conditions being similar to
those existing in the laboratory, the pore water pressures dissipate
quite rapidly and the primary compression phase is fast. On the other
hand, for very large distances, or for very deep clay deposits, much of
the deformation that occurs at the beginning of the consolidation process may actually be undrained creep (3,10,16). It should be noted that
Chang (2) refers to undrained creep as "self-induced primary consolidation".

    Let point *E* in Fig. 8b represent the effective stress state of a
typical soil element in a very thick deposit of soft clay under an embankment, following undrained loading but for non zero lateral movements.
Undrained creep which causes an increase in pore pressures and deformations (5,12,16) will shift point *E* towards point *A'*. In fact, the a-
mount of deformation observed corresponding to the direct path *A* to *A'*
without considering undrained creep, that is, for infinitely slow rates
of loading under undrained conditions, will be the same as that following the path *AEA'*, that is for undrained loading at a given rate followed by undrained creep. Now, if at point *E*, drainage is allowed, consolidation takes place and point *E* shifts towards the corresponding total stress point *D*. Depending upon the rate of dissipation of pore
pressure due to the value of the coefficient of consolidation and the
distance to pervious boundaries, the combination of both phenomena,
i.e., primary consolidation and undrained creep, may be viewed as a superposition of two effects (14,15):

$$\Delta u_{obs} = \Delta u_{creep}\uparrow + \Delta u_{cons}\downarrow \qquad (1a)$$

and

$$\Delta \varepsilon_v = \Delta \varepsilon_{v,creep}\uparrow + \Delta \varepsilon_{v,cons}\uparrow \qquad (1b)$$

where
Δu_{obs} = observed value of the excess pore water pressure
Δu_{creep} = excess pore water pressure due to undrained creep
Δu_{cons} = excess pore water pressure dissipated
$\Delta \varepsilon_v$ = observed settlement
$\Delta \varepsilon_{v,creep}$ = settlement due to undrained creep, and
$\Delta \varepsilon_{v,cons}$ = settlement due to primary consolidation

    So, even if no significant dissipation of pore water pressure is
observed to occur in the field, settlement will continue to take place
because of undrained creep.

    And like in the oedometer test, the amount of final settlement due
to an increase in pressure can be computed by a applying a normally

consolidated behaviour directly from A to D (Fig. 8a) by using the value of C_c, the final vertical settlement in going from A to D will be
estimated by using the specified stress-strain relationship without considering creep effects along the effective stress path $A'D$.

However, as in real field situations there is no a truly undrained
loading stage, point B instead of point E represents the stress state at
the end of the loading phase. But when the effective stress path crosses the Limit State Curve, the rate of pore water pressure dissipation
due to primary consolidation will decrease considerably and the same
analysis explained above, in which the creep phenomenon will prevail,
can be applied.

In summary, the following approach has been applied for the prediction of the progress of settlement and pore water pressure with time:

i)   During the construction stage, the initial values of the pore wa-
     ter pressure (Δu_i), settlement (ρ_{vi}) and horizontal displacement
     (ρ_{hi}) are calculated by using parameters whose values depend on
     the stress level, i.e., overconsolidated and normally consolidated
     states; and
ii) During the consolidation process, pore water pressures (Δu_{nc}),
     settlements ($\rho_{v,nc}$) and horizontal displacements ($\rho_{h,nc}$) are de-
     termined at any time t by using normally consolidated parameters
     and by simulating the construction and consolidation stages as
     follows:

$$\Delta u = \begin{cases} \Delta u_{nc} & \text{if } \Delta u_{nc} < \Delta u_i \\ \Delta u_i & \text{if } \Delta u_{nc} > \Delta u_i \end{cases} \tag{2}$$

$$\rho_v = \rho_{vi} + (\rho_{v,nc} - \rho_{v,nc_o}) \tag{3}$$

$$\rho_h = \rho_{hi} + (\rho_{h,nc} - \rho_{h,nc_o}) \tag{4}$$

where ρ_{v,nc_o} and ρ_{h,nc_o} represent the initial values of the normally
                          consolidated vertical and horizontal displa-
                          cements, respectively, corresponding to the
                          stress path AA', in Fig. 8b.

It should be noted that the displacement in the brackets of Eqs.
3 and 4 correspond to the stress path $A'D$ in Fig. 8b.

CONSOLIDATION PREDICTION WITH PHENOMENOLOGICAL CREEP

A new version of the model that takes into account creep, has
been used in a type C1 prediction of the M.I.T. test fill and a type B
prediction of the Cubzac-les-Ponts embankment B.

The results for the M.I.T. embankment are presented in Fig. 9

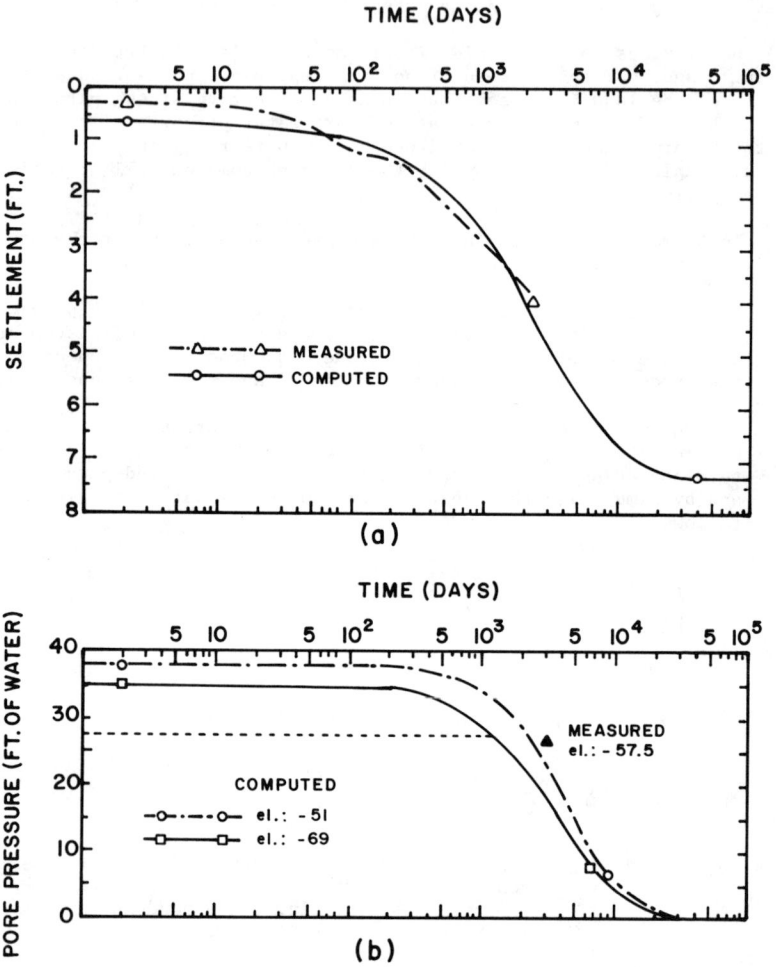

Figure 9.   M.I.T. I95 Embankment. a) Settlement; b) Pore Pressure
             Distribution; c) Lateral Displacement

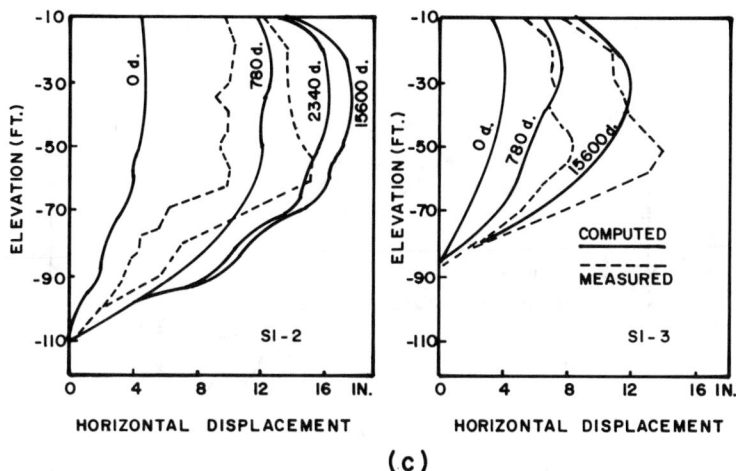

(c)

Figure 9C.

and show a much better agreement with the measured values for both the
settlements and the pore water pressures. It has to be noticed that the
final settlement has increased even though the model parameters are i-
dentical to those used in the first version of the model. This is due
to the fact that the length of the effective stress path during the consolidation stage is larger than that for the case of no creep.

In the case of the Cubzac-Les-Ponts test fill B, the results are
shown in Fig. 10. Because of the uncertainty associated with the determination of the exact values of the coefficients of permeability, the
prediction has been made by using two sets of values, k(I) and 2k(I),
where k(I) corresponds to the value obtained from consolidation tests.
The results show a good agreement for both the horizontal displacements
and the pore water pressures. Concerning the vertical settlement, a
systematic underestimation of about 10 cm is found to exist in the predicted value and has been attributed to an overestimation of the rigi-
dity of the uppermost 2m-thick layer. By taking into account undrained
creep results in an increase of about 25% in the maximum settlement as
compared to that calculated on the basis of the conventional theory of
consolidation.

CONCLUSIONS

On the basis of the content of this paper, the following conclusions are drawn:
a)   A finite element program, utilizing a simplified bilinear aniso-
     tropic elastic model and including the concept of undrained creep,
     is found to adequately describe the consolidation of soft clays.

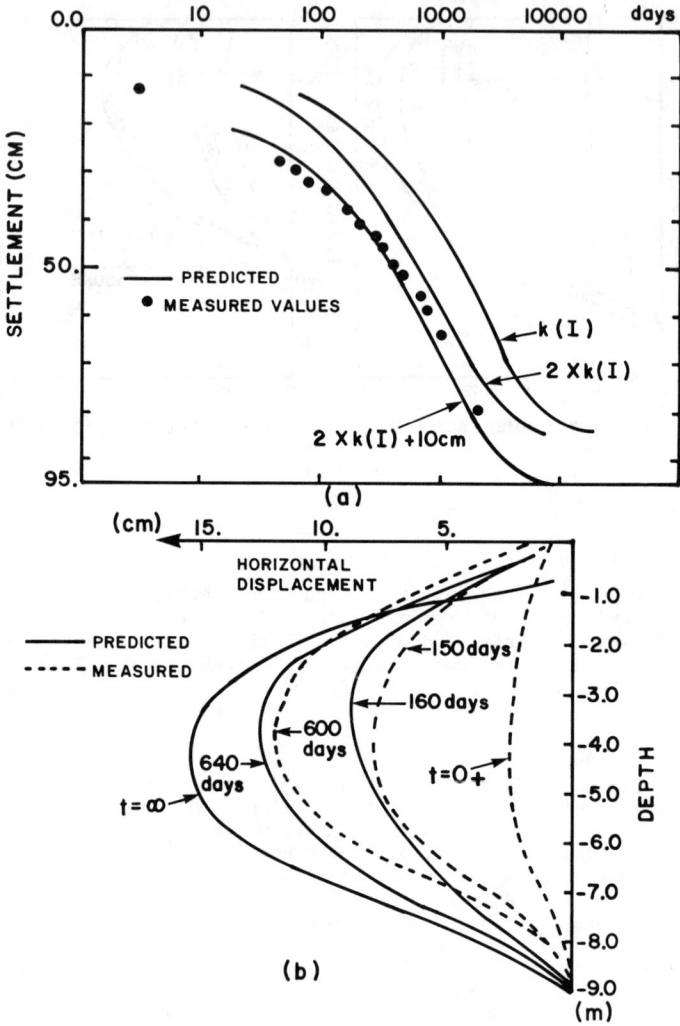

Figure 10.   Cubzac-Les-Ponts Test Fill B. a) Settlement; b) Lateral
             Displacement; c) Pore Pressure Distribution

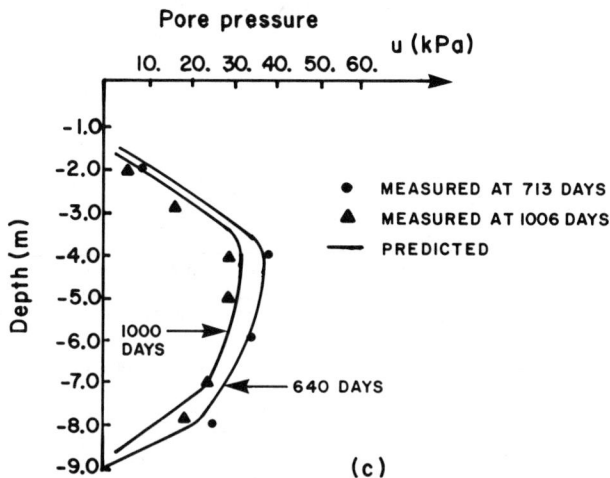

Figure 10C.

b)   The model was tested by making use of the consolidation history
     of the M.I.T. test fill.
c)   The model has been applied to the prediction of the consolidation
     performance of the Cubzac B test fill and to the prediction of the
     construction performance of the St-Alban D embankment.

ACKNOWLEDGEMENTS

      Financial support of this paper, provided by the National Research
Council of Canada and the Ministry of Education of Quebec, is gratefully
acknowledged.

APPENDIX 1 - REFERENCES

1.   Balasubramaniam, A.S. and Brenner, R.P., "Consolidation and Settle-
     ment of Soft Clay", *Soft Clay Engineering*, E.W. Brand and R.P.
     Brenner, eds., Elsevier North-Holland Inc., New York, 1981, pp.
     481-557.
2.   Bjerrum, L. "Engineering Geology of Norwegian Normally-Consolida-
     ted Marine Clays as Related to Settlements of Buildings", *Géotech-
     nique*, Vol. 17, No. 1, 1967, pp. 81-118.
3.   Chang, Y.C.E., "Long Term Consolidation beneath the Test Fills at
     Väsby, Sweden", *Swedish Geotechnical Institute*, Report No. 13,
     1981, 148 p.
4.   Finn, W.D.L. and Shead, D., "Creep and Creep Rupture of an Undis-
     turbed Sensitive Clay", *Proceedings of the Eight International
     Conference on Soil Mechanics and Foundation Engineering*, Moscow,
     Vol. 1.1, 1973, pp. 135-142.

5. Holzer, T.L., Höeg, K., and Arulandan, K., "Excess Pore Pressures During Undrained Clay Creep", *Canadian Geotechnical Journal*, Vol. 10, No. 1, Feb. 1973, pp. 12-24.

6. Ladd, C.C., Foott, R., Ishihara, K., Schlosser, F., and Poulos, H.G., "Stress-Deformation and Strength Characteristics", *Proceedings of the Ninth International Conference on Soil Mechanics and Foundation Engineering*, Tokyo, Vol. 2, 1977, pp. 421-494.

7. Leroueil, S., Tavenas, F., Trak, B., LaRochelle, P., and Roy, M., "Construction Pore Pressures in Clay Foundations under Embankments. Part I: the Saint-Alban Test Fills", *Canadian Geotechnical Journal*, Vol. 15, No. 1, Feb. 1978, pp. 54-65.

8. Magnan, J.P., Humbert, P., Belkeziz, A., and Mouratidis, A., "Finite Element Analysis of Soil Consolidation with Special Reference to the Case of Strain Hardening Elasto-Plastic Stress-Strain Models", *Proceedings of the Fourth International Conference on Numerical Methods in Geomechanics*, Edmonton, Alberta, Canada, Vol. 1, 1982, pp. 327-336.

9. Mesri, G., and Godlewski, P.M., "Time-and-Stress-Compressibility Inter-Relationship", *Journal of the Geotechnical Engineering Division*, ASCE, Vol. 103, No. GT5, May 1977, pp. 417-430.

10. Massachusetts Institute of Technology, *"Proceedings of the Foundation Deformation Prediction Symposium"*, Nov. 1974, Vols. 1 and 2, U.S. Department of Transportation, Federal Highway Administration, 1975.

11. Palmerton, J.B., "Creep Analysis of Atchafalaya Levee Foundation", *Proceedings of the Symposium on Applications of the Finite Element Method in Geotechnical Engineering*. U.S. Army Engineer Waterways Experiment Static, Vicksburg, Mississipi, U.S.A., May 1972, pp. 843-862.

12. Sekikuchi, H., "Rheological Characteristics of Clays", *Proceedings of the Ninth International Conference on Soil Mechanics and Foundation Engineering*, Tokyo, Vol. 1, 1977, pp. 289-292.

13. Singh, A., and Mitchell, J.K., "General Stress-Strain-Time Function of Soils", *Journal of the Soil Mechanics and Foundations Division*, ASCE, Vol. 94, No. SM1, Jan. 1968, pp. 21-46.

14. Siwe, A.S., *"Influence de l'Anisotropie Mécanique du Sol pendant la Consolidation"*, M.Sc.A. Thesis, Ecole Polytechnique, Montreal, 1977, 193 p.

15. Soulié, M., *"Les Eléments Finis: Outils de Prédiction"*, New Study Methods of Fills on Soft Soils, Laval University (Quebec), Ecole Nationale des Ponts et Chaussées (Paris), Dec. 1978, 50 p.

16. Walker, L.K., "Undrained Creep in a Sensitive Clay", *Géotechnique*, Vol. 19, No. 4, 1969, pp. 515-529.

APPENDIX II - NOTATION

C_c = compression index
C_v = coefficient of consolidation
E_v = vertical modulus
G_v = shear modulus
H = height of file
K_o = coefficient of earth pressure at rest
e = void ratio
e_o = initial void ratio

k_h	horizontal coefficient of permeability
k_v	vertical coefficient of permeability
n	modulus ratio
p'	$(\sigma_1' + \sigma_3')/2$
q	$(\sigma_1' - \sigma_3')/2$
u	pore pressure
Δ	increment
ε_v	vertical deformation
γ	unit weight of fill
ν	Poisson's ratio
ρ_h	lateral displacement
ρ_v	vertical displacement
σ_{vc}'	effective vertical preconsolidation pressure
σ_{vo}'	initial effective vertical pressure
σ_v'	effective vertical stress

Consolidation of Multi-layered Clay

Naotoshi Takada* and Masato Mikasa**

1. Introduction

There is no clay layer in nature that has complete homogeneity; the consolidation paremeters such as permeability k, volume compressibility m_v and coefficient of consolidation c_v generally vary along the depth. Therefore one-dimensional consolidation of non-homogeneous clay layer is an important problem from practical viewpoint. Schiffman and Gibson (6) derived an equation that generalized the Terzaghi equation for non-homogeneous cases ; Mikasa (3) also generalized his consolidation theory (2) for such cases. These equations are duly used for the case of gradual variation of the parameters along the depth. There are many cases, however, where the ground is rather considered as multi-layered, that is, composed of a number of homogeneous layers one on the top of the other. Moreover, generally speaking, any ground with gradual change of the parameters can be approximated by a multi-layered stratum (just as a differential equation can be approximated by a difference equation), which is in many cases easier than the use of the above mentioned advanced equations of non-homogeneous ground. Abbott (1) dealt with this problem and described finite difference calculation based on Terzaghi's linear consolidation theory. Here we present a series of consolidation tests of multi-layered clays and an analysis based on Mikasa's nonlinear stress-strain consolidation theory that considers large strain (2).

Three kinds of clay that have different plasticity and pre-consolidated from slurry were combined into some typical two layered systems and consolidated under 40 tf/m² with single drainage condition. Three specimens of each model type were consolidated until the degree of consolidation reached 40, 80 and 100 percent, respectively, to obtain the isochrones at each degree of consolidation.

The observed time-consolidation curves and isochrones were compared successfully with numerical solutions by Mikasa's equation.

2. Preparation of Models
2.1 Materials

Three clays were used as the unit layer : a clay taken from the reclaimed land in Osaka South Port and sieved through 0.3 mm mesh, and its mixtures with kaolinite powder and with bentonite powder at the dry weight ratio of 1 : 1 and 1 : 3, respectively. We shall call these clays Nanko clay, N-K clay and N-B clay. Their physical properties are shown in Table 1. Fig.1 shows f*-log p and f-log c_v** relations of the three clays

*Associate Professor, Civil Engineerng Department, Osaka City University, Sugimoto Sumiyoshi-ku Osaka Japan
**Professor, Civil Engineering Department, Osaka City University, Sugimoto Sumiyosh-ku Osaka Japan

obtained by oedometer test with the specimen of 10 cm diameter, 4 cm thickness and initial water content w_i shown in Table 1. The values of c_v of Nanko clay and N-K clay, in semi-log plot, increase linearly against the decrease of f, whereas N-B clay shows an almost constant c_v value for the volume ratio range of the test. The consolidation parameters Cc, p_2, f_2 ; Ccv, c_{v2}, f_2' defined in Fig.2 were obtained from the thick lines in Fig.1 and tabulated in Table 2 for the three materials. The thick f-log p

Table 1 Physical Properties of Clays

Clay	G_s	w_L	w_p	I_p	w_i*	w_{Ri}**
Nanko	2.68	87.2 %	32.3 %	55.0 %	121 %	1.61
N-K	2.69	63.6	29.6	34.0	91	1.80
N-B	2.67	108.5	38.6	69.9	137	1.40

* w_i : Initial water content before preliminaly consolidation
** w_R : Relative water content (= I_L)

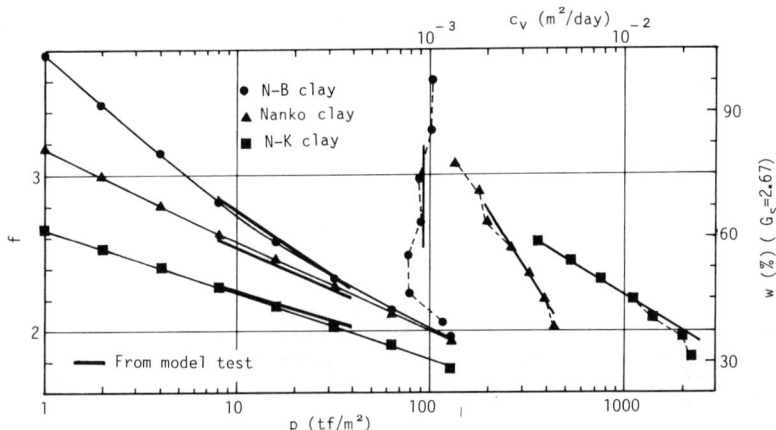

Fig.1 f-log p and f-log c_v relations

* f is the volume ratio defined as (volume of soil)/(volume of soil grains) and equals to (1+e).
** Determination of c_v is : 1) find c_v by curve rule method on the time-consolidation curve in semi-log plot, which essentially gives similar values to $\sqrt{t}$-method, 2) correct the c_v value thus obtained by multiplying the primary compression ratio r (4). This correction is necessary from the compatibility between the oedometer test and the field settlement ; in the former c_v corresponds to the compressibility in primary consolidation, whereas in the latter c_v should correspond to the volume compressibility m_v including secondary compression that is used in the field settlement calculation. Refer to Appendix 5 by Mikasa of the literature (5).

Table 2 Consolidation parameters

Clay	f-log p			f-log c_v		
	Cc	P_2	f_2	Ccv	f'_2	c_{v2}
Nanko	0.558	1000	1.409	1.977	3.380	1×10^{-3}
N-K	0.336		1.565	0.785	3.020	
N-B	0.815		1.131		9.5×10^{-4} (constant)	

unit : p in tf/m^2 and c_v in m^2/day

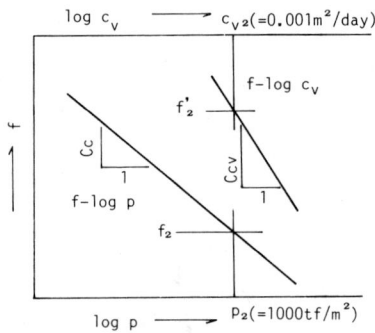

Fig.2 Definition of consolidation parameters

lines were determined from the initial and final average volume ratios of the unit layers of models, which show a slight deviation from the f-log p lines by oedometer tests. The thick f-log c_v lines were drawn on the basis of oedometer test results.

2.2 Models and Test Procedure

Three kinds of remoulded slurry clay with the water contents w_i shown in Table 1 were consolidated one-dimensionally in cylinders of 15 cm diameter (stainless steel CBR mould with a coller as a provision for the initial specimen height as much as 21 cm) under the pressure increasing stepwise to 8 tf/m^2 and lasting for 3 - 16 days to get 12 cm thick homogeneous clay blocks. The upper and lower ends of each block were cut off where the water content was found always lower than in the midst, and two 5 cm thick disks were trimmed out. Then they were combined in several ways into two layered 10 cm thick clay blocks, and were set in transparent acrylite cylinders of 15 cm diameter. In order to observe the displacement of the interface, a cross marker was set on it. Fig.3 is the setup of the test model.

Four types of combinations illustrated in Fig.4 were chosen as the two layered models, and three identical specimens were prepared for each model type. Firstly, the models were settled under the pressure of 8 tf/m², which is the same as the preliminary consolidation pressure, showing only 1 - 1.5 mm settlement. Then the models were consolidated under 40 tf/m² until the degree of consolidation in terms of settlement U_s reached 100, 80 and 40 percent, respectively. Then the models were unloaded and the distribution of water content along the depth was quickly measured to obtain the isochrones of volume ratio at each degree of consolidation. The total settlement and the displacement of the marker set on the interface were measured by means of a dial gauge and a microscope, respectively.

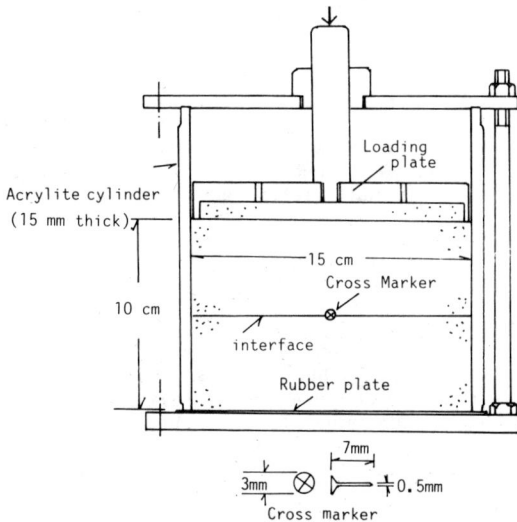

Fig.3 Illustration of test model

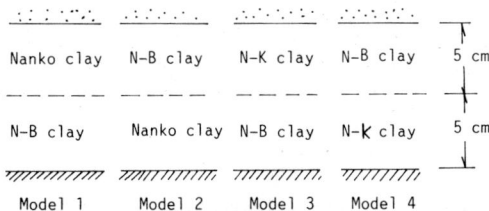

Fig.4 Models

3. Governing Equation and its Finite Difference Form

The governing equation of one-dimensional consolidation of a clay con-sidering variable permeability k, volume compressibility m_v, coefficient of consolidation c_v and consolidation pressure p together with the effect of finite strain was derived by Mikasa (2) as follows :

$$\frac{\partial \zeta}{\partial t} = \zeta^2 \left[c_v \frac{\partial \zeta^2}{\partial z_0^2} + \frac{dc_v}{d\zeta} \left(\frac{\partial \zeta}{\partial z_0} \right)^2 \right] \tag{1}$$

where t is time, ζ is the consolidation ratio $(=f_0/f$, f_0 is the original vol-ume ratio) and z_0 is the original coordinate or the coordinate in the orig-inal state in which the clay is assumed to have a certain volume ratio f_0, not smaller than the largest f value in the layer, throughout the layer. In the case of multi-layered or non-homogeneous layer, f_0 value in each ele-ment is chosen so as to have a common original effective pressure p_0 throughout the layer.

Eq.(1) has the assumption to neglect the effect of selfweight of clay, and is applied to rather hard and/or thin layers. The other assumptions of Eq.(1) are 1) homogenuity,* 2) saturated clay, 3) one-dimensional consoli-dation, 4) incompressible soil particle and water, 5) Darcy's law, 6) f-log p and f-log k relations with no time-dependency.

The finite difference form of Eq.(1), referring to Fig.5, is written as follows :

$$\Delta \zeta_{z_0} = \frac{c_{v_0} \Delta t}{(\Delta z_0)^2} \zeta_{z_0}^2 \left[\phi(\zeta)(\zeta_{z_0 + \Delta z_0} - 2\zeta_{z_0} + \zeta_{z_0 - \Delta z_0}) \right.$$

$$\left. + \frac{1}{4} \frac{d\phi(\zeta)}{d\zeta} (\zeta_{z_0 + \Delta z_0} - \zeta_{z_0 - \Delta z_0}) \right] \tag{2}$$

where $\phi(\zeta) = c_v/c_{v_0}$, c_{v_0} is the initial, and at the same time original, value of c_v, and Δt and Δz_0 are the time and space mesh, respectively. Assuming a linear f-log p relationship, the function $\phi(\zeta)$ is given as

$$\phi(\zeta) = \exp \left[\frac{f_0}{0.4343 C c_v} \left(1 - \frac{1}{\zeta} \right) \right] \tag{3}$$

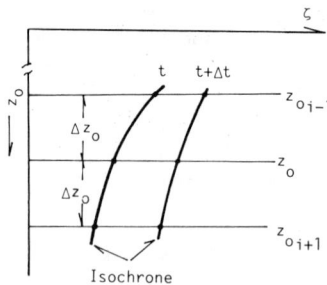

Fig.5   Schematic illustration of isochrones

* Homogenuity means here "the same sort of soil", but not "the same state".

Then from Eq.(3), $d\phi(\zeta)/d\zeta$ is

$$\frac{d\phi(\zeta)}{d\zeta} = \frac{f_0}{0.4343\,Cc_v}\,\frac{1}{\zeta^2}\,\phi(\zeta) \tag{4}$$

4 Boundary Condition at Interface

The boundary condition at the interface of multi-layered clay is determined from the continuity of effective stress and pore water flow across it. Based on a linear f-log p relationship, effective stress p and volume ratio f are related as

$$p = p_2\,10^{(f_{2A}-f)/Cc_A}$$

$$p = p_2\,10^{(f_{2B}-f)/Cc_B}$$

where f_{2A} and f_{2B} are the volume ratios of the clay A and B at pressure p_2 (here $p_2 = 1000$ tf/m^2), respectively. Since the effective stresses at both sides of the interface must take the same value,

$$\frac{f_{2A}-f_b}{Cc_A} = \frac{f_{2B}-f_c}{Cc_B}$$

where f_b and f_c are the volume ratios of the clay A and B at the interface, respectively, as shown in Fig.6. The equation is also expressed as

$$\left(f_{2A}-\frac{f_{oA}}{\zeta_b}\right)/Cc_A = \left(f_{2B}-\frac{f_{oB}}{\zeta_c}\right)/Cc_B \tag{5}$$

where f_{oA} and f_{oB} are the original volume ratios of the clay A and clay B, respectively, at a certain assumed original pressure p_o by which the original state is determined. Here p_o was taken equal to the initial pressure $p_i = 8$ tf/m^2.

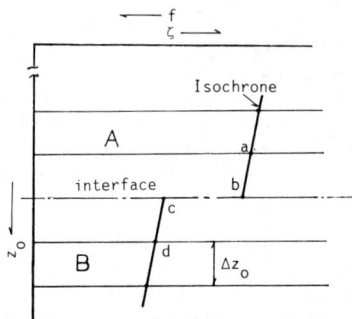

Fig.6 Isochrone and interface

The velocity of pore water flow relative to soil skeleton was expressed by Mikasa (2) in the process of derivation of Eq.(1) as follows :

$$v = c_v \frac{\partial \zeta}{\partial z_0}$$

From the continuity of pore water flow across the interface,

$$c_{vA} \left(\frac{\partial \zeta}{\partial z_0}\right)_b = c_{vB} \left(\frac{\partial \zeta}{\partial z_0}\right)_c$$

where c_{vA} and c_{vB} are the values of c_v of clay A and clay B at both sides of the interface, respectively. The equation is transformed into the following finite difference form (refer to Fig.6)

$$c_{vA} (\zeta_b - \zeta_a) = c_{vB} (\zeta_d - \zeta_c) \tag{6}$$

Eqs.(5) and (6) give the boundary conditions at both sides of the interface.

5 Numerical Calculation

In order to perform a stable calculation in the case of heat conduction type consolidation, space mesh Δz and time mesh Δt must satisfy

$$\frac{\Delta t \, c_v}{(\Delta z)^2} \leqq \frac{1}{2}$$

In the same manner, for the case where Eq.(2) stands, Δz_0 and Δt should satisfy

$$\frac{\Delta t \, c_v}{(\Delta z_0)^2} \zeta^2 \leqq \alpha$$

where ζ is the largest of the boundary value, and c_v is the largest in the given condition. α is recommended to be smaller than $1/4$ in ordinary case. For the present study Δz_0 was fixed as 0.25 cm and Δt was determined for each model to satisfy the above condition.

In order to get a good result, especially in the early stage of consolidation, the value ζ in the calculation of function $\phi(\zeta)$ and ζ^2 in Eq.(2) is recommended to take the average value $(\zeta_{z_0} + \Delta \zeta_{z_0}/2)$ in the time span of Δt, which is obtained by successive approximation as follows. First approximation $\Delta \zeta_{z_0,1}$ is obtained by using ζ_{z_0} in the function $\phi(\zeta)$ and ζ^2 in Eq.(2). Then, second approximation $\Delta \zeta_{z_0,2}$ is obtained by using $(\zeta_{z_0} + \Delta \zeta_{z_0,1}/2)$ and so on. Some preliminary calculation revealed that the second approximation in the above procedure has an adequate accuracy in the present calculation.

6 Test Results and Discussions

The observed time-settlement curves of the upper surfaces and the interfaces of Model 1 to 4 are shown in Figs.7 to 10 together with those obtained by calculation. Time-settlement curves of the models that stopped consolidation at 40 and 80 percent degree of consolidation are also presented in the figures. Model 1, regrettably, has no record of the displacement of the interface. The calculated curves coincide satisfactorily with the experimental ones. Model 2 and Model 4 are the same combinations of different clays with Model 1 and Model 3, respectively, except they are placed upside down. Therefore, the ultimate total settlement of the former two models are almost the same as the latters. The ultimate settlement of each elemental layer, also, is found to be equal for the same soil in Model 3 and Model 4 (Model 1 can not be compared with Model 2 in this respect).

From the time-settlement curves it is seen that the consolidations of

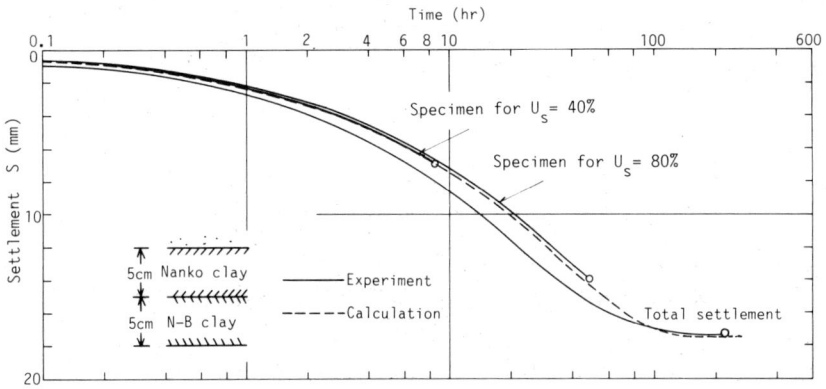

Fig.7 Time-settlement curves (Model 1)

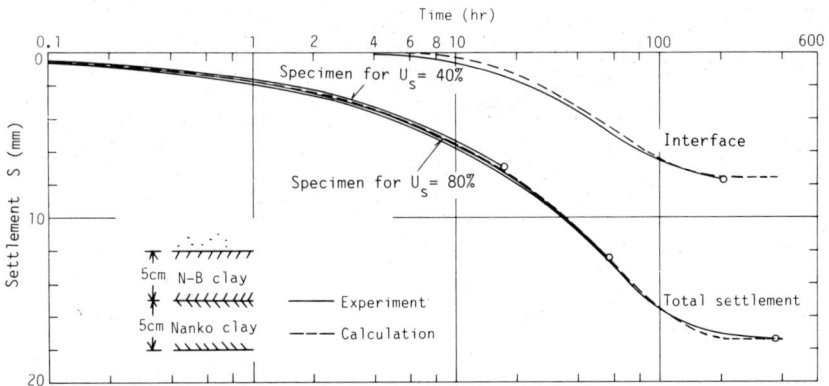

Fig.8 Time-settlement curves (Model 2)

Model 2 and Model 4 proceed slower than the other two, because N-B clay that has the lowest permeability among the three clays occupies the upper halves of the composite layers through which the total amount of pore water corresponding to the total settlement must flow out.

Figs.11 to 14 show the observed isochrones of volume ratio at 40, 80 and 100 percent degree of consolidation together with the calculated ones both plotted against the original, and at the same time the initial, coordi-

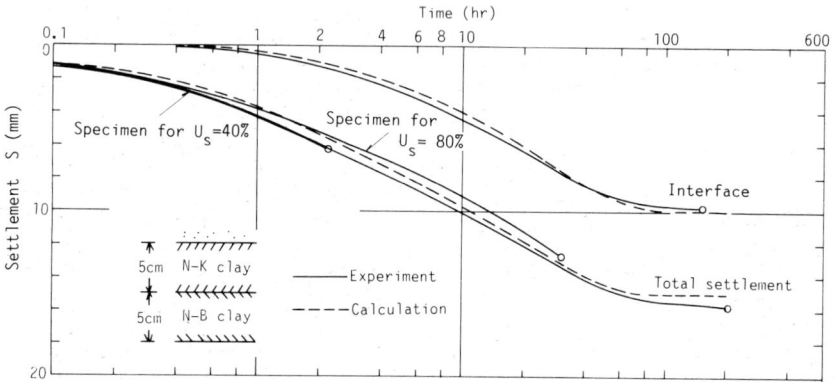

Fig.9   Time-settlement curves (Model 3)

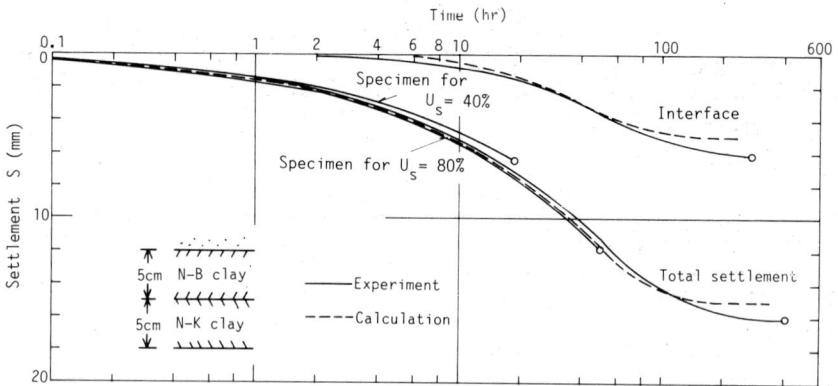

Fig.10   Time-settlement curves (Model 4)

nate z_o of the four models.[*] The isochrones show discontinuity at the inter-face due to difference of compressibility between two elemental clays. The coincidence of the ultimate volume ratio of N-B clay and the initial volume ratio of N-K clay in Figs.13 and 14 is utterly a matter of haphazard.

Theoretical solutions coincide satisfactorily with the experiments. The observed volume ratios adjacent to both the impermeable bases and inter-faces, however, show irregular small values, which suggest the existence of

* Mark ⊡ is of a piece of clay trimmed in preparing specimen.

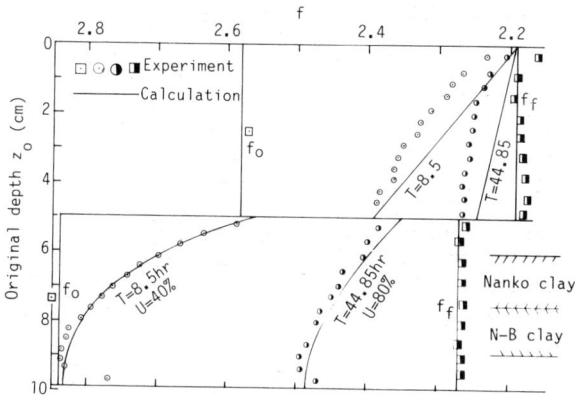

Fig.11 Isochrones of volume ratio (Model 1)

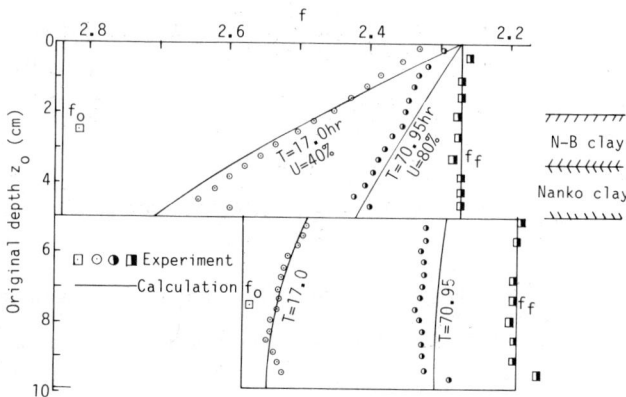

Fig.12 Isochrones of volume ratio (Model 2)

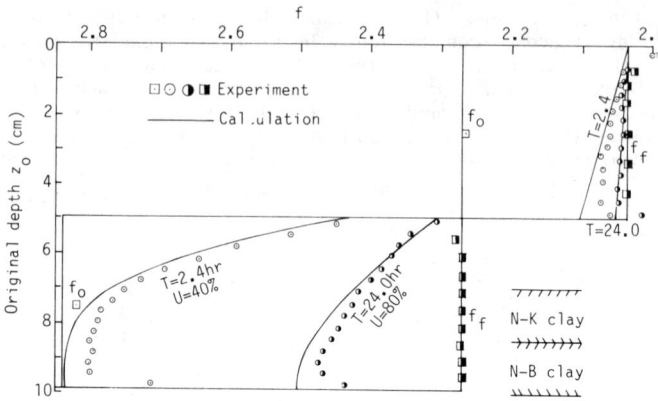

Fig.13 Isochrones of volume ratio (Model 3)

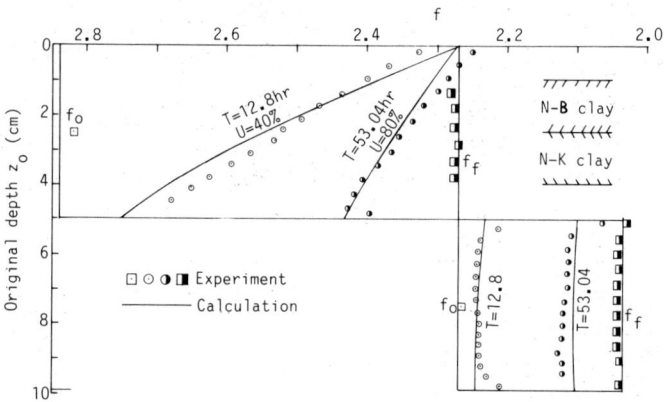

Fig.14 Isochrones of volume ratio (Model 4)

thin disturbed layers with higher compressibility and/or larger permeability due to the preparatory handling of the specimens.

The settlement S in Figs.7 to 10 was calculated by the equation

$$S = \frac{1}{f_o} \int_o^H \Delta f \, dz_o \tag{7}$$

where Δf denotes the decrease in volume ratio at each depth z_o, which is read as the difference in abscissa between the isochrones in Figs.11 to 14.

Figs.15 and 16 show some examples of theoretical isochrones of the

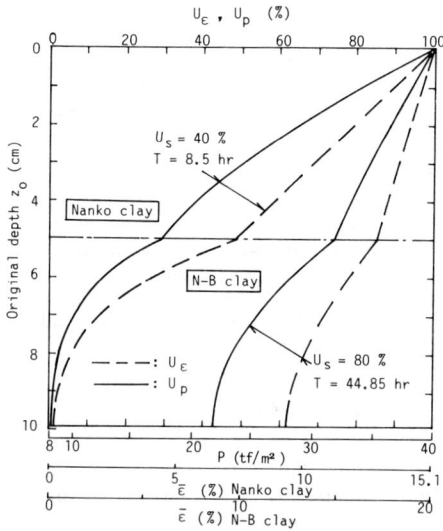

Fig.15 Isochrone of degree of consolidation in terms of strain U_ε and effective stress U_p (Model 1)

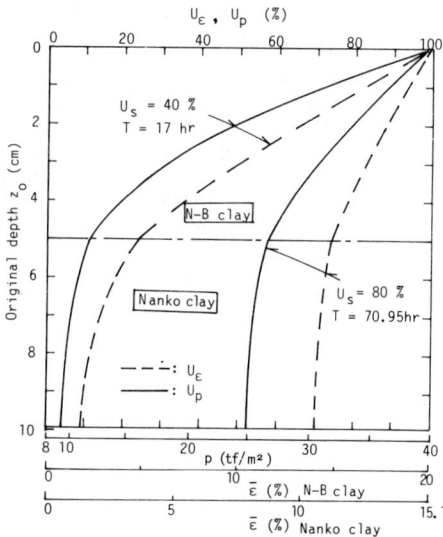

Fig.16 Isochrone of degree of consolidation in terms of strain U_ε and effective stress U_p (Model 2)

degree of consolidation both in terms of strain U_ϵ and in terms of effective stress U_p for Model 1 and Model 2, respectively. These isochrones hold the continuity at the interface, which lacks in the isochrones of volume ratio itself and strain itself. (isochrones of strain and effective stress, the latter having continuity at the interface, are not shown here.) It is seen that the degree of consolidation in terms of stress proceeds slower than that in terms of strain because of the nonlinear stress-strain relationship of the clay.

Comparing deliberately the experimental and analytical results, we may conclude that the consolidation of multi-layered clay can be analyzed by the method shown in this paper.

The following two points, however, are essential to understand the overall implication of the test results : 1) side friction between the specimen and the acrylite cylinder was small enough not to cause appreciable influence to test results ; 2) Mikasa's consolidation equation that neglects the secondary consolidation is duly applicable for the consolidation of remoulded-and-reconsolidated clay under the load increment ratio as big as four ($8 \rightarrow 40\text{tf/m}^2$), provided that c_v value was determined from the oedometer test by the method proposed by Mikasa (2), (4).

Acknowledgement

The authers wish to thank Mr. Hideo Morimoto for his help in conducting model tests and calculations.

Appendix.--- Referrences

1. Abbott, M. B., "One-dimensional Consolidation of Multi-layered Soils," Geotechnique, vol.10-151, 1960.
2. Mikasa, M., "The consolidation of Soft Clay -- A New Consolidation Theory and its Application," Kajima Shuppankai, (in Japanese) 1963.
3. Mikasa, M., "Consolidation of Non-homogeneous Ground," Proc. of Autumn Annual Convention of Japan Society of SMFE (in Japanese), 1963.
4. Mikasa, M., "Determination of Consolidation Parameters from Oedometer Test," Proc. of Annual Convention of JSCE, III-7, (in Japanese), 1964.
5. Mikasa, M. and Ohnishi, H., "Soil Improvement by Dewatering in Osaka South Port, Geotechnical Aspects of Coastal Reclamation Projects in Japan," Proc. of 9th ISSMFE, Case History Volume, pp.639-664, 1981.
6. Schiffman, R. F. and Gibson, R. E., "Consolidation of Non-homogeneous Clay Layer," ASCE, SM 5, 1964.

CONSOLIDATION BY SAND DRAIN IN ANISOTROPIC GROUND

TAKESHI TAMURA[*]

ABSTRACT

In this paper, we firstly reduce Biot's equations of consolidation into a single governing equation with the excess pore water pressure as the only unknown function and secondly derive the theory of eigenvalue problem for it, which is very similar to that of the one-dimensional Terzaghi's equation. Namely, we can find a set of eigenvalues and the corresponding eigenfunctions which compose the orthogonal bases in the function space. Therefore it is easy to write the formal solution of consolidation for any initial distribution of excess pore water pressure. Finally, we apply the present method to the problem of sand drain in the anisotropic ground and find out that the anisotropy of ground could cause a remarkable influence on the degree of consolidation under a certain condition.

INTRODUCTION

It is naturally accepted that three-dimensional consolidation is governed by Biot's theory which consists of simultaneous partial differential equations, with both displacements and pore water pressure as unknowns, and is considered to be much more complicated than the one-dimensional (Terzaghi's) theory. Indeed, we have few solutions for Biot's theory except the numerical approaches.[1]

However, we can understand that the essential mechanism of consolidation common to both one- and three-dimensional consolidation is composed of (1) the condition of equilibrium under total stresses and (2) the flow of pore water. Therefore the treatment of Biot's equations should be made similar to that of Terzaghi's equation. From this point of view we will advance the discussion on Biot's theory in the present paper.

Firstly, we try to reduce Biot's equation into a single equation of the excess pore water pressure alone, by viewing the consolidation process from its final state.

Secondly, the theory of eigenvalue problem for the above single equation, which is closely similar to the one-dimensional one, will be established, i.e., we will show the orthogonality of eigenfunctions and write the formal solution in a very compact way.

Finally, we will apply the present method to the problem of a sand drain which is installed in anisotropic ground. We will

*Associate Professor, Dept. of Civil Eng., Kyoto University , 606 Kyoto, Japan.

explain the method in detail for practically deriving the closed form solution.   The solution indicates the remarkable influence of anisotropy of ground on the effectiveness of sand drain.

EQUATION OF CONSOLIDATION [5) 6)]

Assuming Darcy's law for the flow of incompressible pore water, Biot's equations of consolidation of saturated clay are written as

$$\sum \frac{\partial \sigma'_{ij}}{\partial x_j} + \frac{\partial u}{\partial x_i} + f_i = 0 \qquad (1)$$

$$\frac{\partial \theta}{\partial t} = -\frac{1}{\gamma_w} \sum k_{ij} \frac{\partial^2}{\partial x_i \partial x_j} (u + \gamma_w \Omega) \qquad (2)$$

in which σ'_{ij} = effective stress, u = pore water pressure, f_i = body force, θ = volumetric strain, γ_w = weight of unit volume of pore water, k_{ij} = permeability, Ω = potential of height which is combined with u to form the total head $u/\gamma_w + \Omega$, x_i = cartesian coordinate, and t = time.

If the clay skeleton is assumed to behave linear-elastically, we have

$$\sigma'_{ij} = \sum D_{ijkl} \, \varepsilon_{kl} \qquad (3)$$

in which D_{ijkl} = elastic coefficient and ε_{ij} = strain tensor defined in terms of the displacement v_i :

$$\varepsilon_{ij} = \frac{1}{2} \left(\frac{\partial v_i}{\partial x_j} + \frac{\partial v_j}{\partial x_i} \right) \qquad (4)$$

Associated with Eqs.1 and 2, the following boundary conditions are set up:

(B_u) boundary condition of displacement
$$v_i = \bar{v}_i \qquad \text{on } S_v$$
(B_σ) boundary condition of traction T_i
$$T_i = \bar{T}_i \qquad \text{on } S_\sigma$$
(B_D) draining boudary condition
$$u = \bar{u} \qquad \text{on } S_D$$
(B_{UD}) undraining boundary condition
$$\sum k_{ij} \frac{\partial}{\partial x_j} (u + \gamma_w \Omega) \, n_i = 0 \qquad \text{on } S_{UD}$$

in which $\bar{v}_i$ = prescribed displacement on the displacement boundary S_v, $\bar{T}_i$ = prescribed (total) traction on the stress boundary S_σ, $\bar{u}$ = prescribed pore water pressure on the draining boundary S_D and n_i = unit outward normal vector on the boundary and S_{UD} denotes the undraining boundary.

It is noteworthy that when the body force and the boundary conditions are time-independent, Eqs.1 and 2 are equivalently replaced by the following homogeneous equations:

$$\sum \frac{\partial \sigma'_{ij}}{\partial x_j} + \frac{\partial u}{\partial x_i} = 0 \tag{1}'$$

$$\frac{\partial \theta}{\partial t} = -\frac{1}{\gamma_w} \sum k_{ij} \frac{\partial u}{\partial x_i \partial x_j} \tag{2}'$$

In the above, all the quantities are measured referring to their values at the final state of consolidation.   In other words, they are residual which are to decay along with the progress of consolidation.

In connection with the change of variables, the non-homogeneous boundary conditions are altered also to be homogeneous i.e.,

(B_{uo})	$v_i = 0$	on S_v
$(B_{\sigma o})$	$T_i = 0$	on S_σ
(B_{Do})	$u = 0$	on S_D

Since we assume the linear stress-strain relationship of the clay skeleton, it is apparent that the (residual) displacement v_i at x is represented only in terms of the distribution of (residual or excess) pore water pressure in the following form:

$$v_i(x) = \int_V V_i(x,X)u(X)dV_X \tag{5}$$

in which $V_i(x,X)$ means the displacement generated by the unit amount of concentrated excess pore water pressure at X i.e., $u(x) = \delta(x,X)$ (Dirac's delta function). To be more specific, $V_i(x,X)$ might be obtained by solving the equation of equilibrium:

$$\sum \frac{\partial \sigma'_{ij}}{\partial x_j} + \frac{\partial \delta}{\partial x_i} = 0 \tag{6}$$

with the homogeneous boundary conditions (B_{vo}) and $(B_{\sigma o})$. In Eq.5, V and X denote the region of integration and the integral variable, respectively.   Hereinafter, we use the coordinates x_i or X_i and their points x or X without any distinction.   Although it is elaborate to specify $V_i(x,X)$ for the individual case, we are allowed to write formally Eq.5 since the existence and uniqueness of $V_i(x,X)$ is guaranteed in general.

Taking the divergence operation on both sides of Eq.5 with respect to the variable x, the following relation is obtained:

$$\theta(x) = \int_V \Theta(x,X)u(X)dV_X \tag{7}$$

in which

$$\Theta(x,X) = \sum \frac{\partial V_i(x,X)}{\partial x_i} \tag{8}$$

Substituting Eq.7 into Eq.2', we finally obtain the governing equation of consolidation in terms of the excess pore water pressure alone:

$$\int_V \Theta(x,X)\frac{\partial u}{\partial t}(X,t)dV_X = -\frac{1}{\gamma_w}\sum_{i,j} k_{ij}\frac{\partial^2 u}{\partial x_i \partial x_j}(x,t) \tag{9}$$

EIGENVALUE PROBLEM OF CONSOLIDATION

Decomposing the excess pore water pressure u(x,t) into the product of functions of space variable x and time t :

$$u(x,t) = u_\alpha(x)\, e^{-\lambda_\alpha t} \tag{10}$$

and substituting it into Eq.9, we have the following homogeneous equation for $u_\alpha(x)$ with a scalar parameter λ_α:

$$\lambda_\alpha \int_V \Theta(x,X)u_\alpha(X)dV_X = \frac{1}{\gamma_w}\sum k_{ij}\frac{\partial^2 u_\alpha}{\partial x_i \partial x_j} \tag{11}$$

λ_α and u_α are, respectively, called the eigenvalue and the eigenfunction, if Eq.11 has a non-trivial solution under the homogeneous boundary conditions (B_{Do}) and (B_{UD}).

(a)Positiveness of eigenvalue

Multiplying both sides of Eq.11 by $u_\alpha(X)$ and integrating over the whole region with respect to X, we have

$$\lambda_\alpha \iint_V \Theta(x,X)u_\alpha(x)\,u_\alpha(X)dV_x dV_X = \frac{1}{\gamma_w}\int_V (\sum_{i,j} k_{ij}\frac{\partial^2 u_\alpha}{\partial x_i \partial x_j})u_\alpha dV_x \tag{12}$$

After a little manipulation with consideration of the definition of $\Theta(x,X)$ and the boundary conditions, it is easy to obtain

$$\iint_V \Theta(x,X)u_\alpha(x)u_\alpha(X)dV_x dV_X = -\int_V D_{ijkl}\varepsilon_{aij}\varepsilon_{akl}dV_x \tag{13}$$

in which ε_{aij} denotes the strain associated with the eigenfunction $u_\alpha(x)$.   Eq.(13) shows that the integral in the left hand side of Eq.12 never takes any positive value since D_{ijkl} is positive definite.   For the sake of simplicity in the sequel, we introduce the following notation and call it the inner product:

$$[u_\alpha, u_\beta] = -\int_V \Theta(x,X)u_\alpha(x)u_\beta(X)dV_x dV_X \tag{14}$$

On the other hand, the right hand side of Eq.12 is transformed into:

$$\frac{1}{\gamma_w}\int_V (\sum k_{ij}\frac{\partial^2 u_\alpha}{\partial x_i \partial x_j})u_\alpha dV = -\frac{1}{\gamma_w}\int_V \sum k_{ij}\frac{\partial u_\alpha}{\partial x_i}\frac{\partial u_\alpha}{\partial x_j}dV \tag{15}$$

Eq.15 means that it is always negative due to the positive definiteness of k_{ij}.   Therefore we can conclude that λ_α in Eq.12 is positive.

(b) Orthogonality of eigenfunctions

$u_\alpha(x)$ and $u_\beta(x)$ denote the eigenfunctions associated with two eigenfunctions λ_α and λ_β, i.e.,

$$\lambda_\alpha \int_V \Theta(x, X) u_\alpha(X) dV_X = \frac{1}{\gamma_w} \sum k_{ij} \frac{\partial^2 u_\alpha(x)}{\partial x_i \partial x_j} \qquad (16)$$

$$\lambda_\beta \int_V \Theta(x, X) u_\beta(X) dV_X = \frac{1}{\gamma_w} \sum k_{ij} \frac{\partial^2 u_\beta(x)}{\partial x_i \partial x_j} \qquad (17)$$

After multiplying both sides of two equations by $-u_\beta(X)$ and $-u_\alpha(X)$, respectively, we integrate over the domain. By subtracting them, the following relation is obtained:

$$(\lambda_\alpha - \lambda_\beta)[u_\alpha, u_\beta] = -\frac{1}{\gamma_w} \int_{S_{D+UD}} \{(\sum k_{ij} \frac{\partial u_\alpha}{\partial x_i} n_j) u_\beta - (\sum k_{ij} \frac{\partial u_\beta}{\partial x_i} n_j) u_\alpha\} dS \qquad (18)$$

Due to the homogeneity of boundary conditions, we have

$$[u_\alpha, u_\beta] = 0 \qquad (19)$$

if $\lambda_\alpha \neq \lambda_\beta$. Eq.19 means that two eigenfunctions with differenet eigenvalues are orthogional each other in the sense of the norm defined in Eq.14.

(c) Formal solution in terms of eigenfunctions

Considering the construction procedure of $u_\alpha(x)$, we can understand that each function $u_\alpha(x) e^{-\lambda_\alpha t}$ ($\alpha = 1,2,\dots$) forms a solution of Eq.9 or the governing equation of consolidation and satisfies the homogeneous boundary conditins (B_{Do}) and (B_{UD}). In order to meet the initial condition of the excess pore water pressure, it is necessary to sum up the infinite series of such solutions as follows:

$$u(x, t) = \sum_{\alpha=1}^{\infty} a_\alpha u_\alpha e^{-\lambda_\alpha t} \qquad (20)$$

in which the coefficients a_α are determined so as to satisfy the initial condition of excess pore water pressure. The initial distribution of excess pore water pressure $u_o(x)$ might be obtained by solving the equation of equilibrium under the undrained condition but it is here supposed to have been known. Putting $t = 0$ in Eq.20, we have

$$u_o(x) = \sum_{\alpha=1}^{\infty} a_\alpha u_\alpha(x) \qquad (21)$$

Calculating the inner product of both sides of Eq.21 with u_β and using the orthogonality condition of eigenfunctions, we get

$$a_\beta = \frac{[u_\beta, u_o]}{[u_\beta, u_\beta]} \qquad (22)$$

Substituting the above equation into Eq.20, we finally obtain the formal solution of consolidation in terms of the eigenfunctions of excess pore water pressure:

$$u(x,t) = \sum \frac{[u_\alpha \, , \, u_o]}{[u_\alpha \, , \, u_\alpha]} \, u_\alpha(x) \, e^{-\lambda_\alpha t} \tag{23}$$

APPLICATIONS TO THE SAND DRAIN MODEL IN ANISOTROPIC GROUND[2)7)]

(a) Assumptions

Let us consider the consolidation problem of the hollow cylinder region as shown in Fig.1, of which height, inner and outer radii are H, R_i and R_o, respectively. Along with the radial and axial directions, we set the cylindrical coordinates (r,z). The angular coordinate ϑ is suppressed because of the axisymmetry of the problem. The region is considered to be a kind of sand drain model in which the stiffness of the sand pile is ignored. The following assumptions concerning the deformation and drainage are supposed.

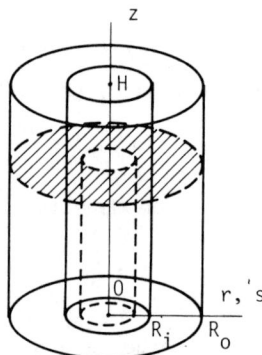

Fig.1 Hollow Cylinder Region

(1) The radial displacement v_r at $r=R_o$ is fixed, i.e., the outer cylindrical surface is the homogeneous displacement boundary S_{vo}.
(2) When the vertical pressure p_o is applied on the upper surface of the region, the same intensity of uniform excess pore water pressure p_o is initially generated in the whole region. This means equivalently that the sand pile causes the radial pressure of p_o at $r=R_i$ and keeps it unchanged in time although the stiffness of it is not directly considered. Therefore the inner cylindrical surface $r=R_i$ is assumed to be the time-independent stress boundary S_σ.
(3) The vertical strain ε_z is uniform in the whole region. As a consequence of it, the applied force on the upper surface is distributed along the radial direction except just after loading. Therefore the upper surface is a kind of mixed boundary where the parts of deformation and loading are prescribed and is noted as (B_M).
(4) The pore water flows in the radial direction alone and is drained at $r=R_i$, i.e., the inner cylindrical surface is the homogeneous draining boundary S_{Do} while the outer one is the undraining boundary B_{UD}.

(b) Stress-strain relationship[4)]

It would be more reasonable to take into consideration some kind of anisotropic of stress strain relationship of the ground, even if it is modeled to be linearly elastic. The most appropriate anisotropy for the deposit ground is the so-called transversally isotropic elasticity, of which anisotropy results from the depositing process of the ground.   As a matter of course, it is one of the problems to apply the linearly elastic model to real ground since it generally behaves non-linearly according to both strain level and loading duration.   But we can make much of the anisotropic elastic model as the first approximation.

Taking a polar coordinate (r, ϑ) arbitrary in the horizontal plane, the stress-strain relationship can be written as

$$
\begin{pmatrix} \sigma'_{rr} \\ \sigma'_{\vartheta\vartheta} \\ \sigma'_{zz} \\ \sigma'_{\vartheta z} \\ \sigma'_{zr} \\ \sigma'_{r\vartheta} \end{pmatrix} = \begin{pmatrix} c_1 & c_1 - 2c_5 & c_2 & 0 & 0 & 0 \\ c_1 - 2c_5 & c_1 & c_2 & 0 & 0 & 0 \\ c_2 & c_2 & c_3 & 0 & 0 & 0 \\ 0 & 0 & 0 & 2c_4 & 0 & 0 \\ 0 & 0 & 0 & 0 & 2c_4 & 0 \\ 0 & 0 & 0 & 0 & 0 & 2c_5 \end{pmatrix} \begin{pmatrix} \varepsilon_{rr} \\ \varepsilon_{\vartheta\vartheta} \\ \varepsilon_{zz} \\ \varepsilon_{\vartheta z} \\ \varepsilon_{zr} \\ \varepsilon_{r\vartheta} \end{pmatrix}
\tag{24}
$$

in which $c_1 - c_5$ are elastic constants.   If the ground is isotropic as a special case, we have

$$
c_1 = c_3 = \lambda + 2\mu, \quad c_2 = \lambda, \quad c_4 = c_5 = \mu
\tag{25}
$$

in which λ and μ are Lame's constants.   Although $c_1 - c_5$ are mutually independent, the following inequalities:

$$
0 < c_5 < c_1, \quad 0 < c_3, \quad c_4, \quad 0 < (c_1 - c_5)c_3 - c_2^2
\tag{26}
$$

are imposed to satisfy the positiveness of strain energy.

(c) Boundary conditions

Since each quantity in the present method is measured from its final value of consolidation process, the boundary conditions are homogeneous as well as the governing equation.   As a consequence of assumptions (1) - (4), we have the following conditions:

(B_{vo})	$v_r = 0$	at $r = R_o$
$(B_{\sigma o})$	$\sigma_{rr} = 0$	at $r = R_i$
(B_{Mo})	$\int_{R_i}^{R_o} \sigma_{zz} r\,dr = 0$	at $z = H$
(B_{Do})	$u = 0$	at $r = R_i$
(B_{UD})	$\dfrac{\partial u}{\partial r} = 0$	at $r = R_o$

in which σ_{rr} and σ_{zz} are the total stresses in the horizontal and vertical directions, respectively.

As is earlier explained, we assume further the initial distribution of excess pore water pressure is uniformly unit:

$$
u(r,0) = u_0(r) = 1
\tag{27}
$$

(d) Deduction of governing equation

Following assumptions (3) and (4), the governing equation of consolidation is simply reduced to

$$\int_{R_i}^{R_0} \Theta(r, s) \frac{\partial u}{\partial t}(s, t)ds = -\frac{k}{r_w}\left(\frac{\partial^2 u}{\partial r^2} + \frac{1}{r}\frac{\partial u}{\partial r}\right) \tag{28}$$

in which k = permeability in the horizontal direction and $\Theta(r,s)$ = volumetric strain at r caused by the excess pore water pressure $u(r) = \delta(r,s)$.

To obtain $\Theta(r,s)$, we first have to solve the equation of equilibrium in the radial direction:

$$\frac{\partial \sigma'_{rr}}{\partial r} + \frac{\sigma'_{rr} - \sigma'_{\theta\theta}}{r} + \frac{\partial \delta}{\partial r} = 0 \tag{29}$$

which corresponds to Eq.6.

The general solution of Eq.29 is written as

$$v_r = -\frac{1}{c_1}\frac{s}{r}H(r, s) + A_s r + B_s\frac{1}{r} \tag{30}$$

while the condition of uniform vertical strain leads the vertical displacement:

$$v_z = a_s z \tag{31}$$

in which $H(r,s)$ = Heaviside's step function (i.e., $\frac{\partial}{\partial r}H(r,s) = \delta(r,s)$). We have to obtain $\Theta(r,s)$ which is to be substituted into Eq.28. In this case, $\Theta(r,s)$ assumes the form:

$$\Theta(r, s) = -\frac{1}{c_1}\delta(r, s) + 2A_s + a_s \tag{32}$$

A_s, B_s and a_s which depend on the parameter s, can be determined by the boundary conditions (B_{vo}), $(B_{\theta o})$ and (B_{Mo}) as shown below.

Substituting the strain derived from the displacements v_r (Eq.30) and v_z (Eq.31) into Eq.24, we have the total stresses

$$\sigma_{rr} = 2\frac{c_s}{c_1}\frac{s}{r^2}H(r, s) + 2(c_1 - c_s)A_s - 2c_sB_s\frac{1}{r^2} + c_2 a_s \tag{33}$$

$$\sigma_{zz} = \frac{c_1 - c_2}{c_1}\delta(r, s) + 2c_2A_s + c_3 a_s \tag{34}$$

Therefore the above boundary conditions are expressed as

$$-\frac{1}{c_1}\frac{s}{R_0} + R_0 A_s + \frac{1}{R_0}B_s = 0 \tag{35}$$

$$2(c_1 - c_s)A_s - 2c_s\frac{1}{R_i^2}B_s + c_2 a_s = 0 \tag{36}$$

$$c_2(R_0^2 - R_i^2)A_s + c_3\frac{R_0^2 - R_i^2}{2}a_s + \frac{c_1 - c_2}{c_1}s = 0 \tag{37}$$

Obtaining A_S and a_S from Eqs.35 - 37 and substituting them into Eq.32, we have

$$\theta(r, s) = -\frac{1}{c_1}\delta(r, s) + \frac{c}{c_1}\frac{s}{R_0^2} \tag{38}$$

in which

$$c = \frac{2n^2\{(c_1-c_2)^2 + (n^2+1)c_5(c_1-c_3)\}}{(n^2-1)\{n^2c_3c_5 + c_3(c_1-c_5) - c_2^2\}} \tag{39}$$

with $n = R_i/R_0$.

Then the governing equation of consolidation of this model is written as follows:

$$\frac{\partial u}{\partial t} + \frac{c}{R_0^2}\int_{R_i}^{R_0} s\frac{\partial u}{\partial t}\partial s = \frac{kc_1}{\gamma_w}\left(\frac{\partial^2 u}{\partial r^2} + \frac{1}{r}\frac{\partial u}{\partial r}\right) \tag{40}$$

If $\frac{kc_1 t}{\gamma_w R_0^2}$ and r/R_0 are rewritten by t and r, respectively, Eq.41 is simplified as

$$\frac{\partial u}{\partial t} + c\int_{1/n}^{1} s\frac{\partial u}{\partial t}ds = \frac{\partial^2 u}{\partial r^2} + \frac{1}{r}\frac{\partial u}{\partial r} \tag{41}$$

which is the final equation to be solved with the initial condition ($u_0(r)=1$) and the boundary conditions:

(B_{DO}) $\qquad u|_{r=1/n} = 0$

(B_{UD}) $\qquad \dfrac{\partial u}{\partial r}\bigg|_{r=1} = 0$

It should be noted that the effect of the anisotropic stress-strain relationship is reflected merely in a parameter c and that there is no distinction in the solution method for the anistropic and isotropic cases.

(e) <u>Eigenvalue</u> and <u>eigenfunction</u>

Similarly to Eq.10, we assume the form of solution:

$$u(r, t) = u_\alpha(r)e^{-\eta_\alpha^2 t} \tag{42}$$

in which λ_α is replaced by η_α^2 since $\lambda_\alpha > 0$.
Substituting it into Eq.41, then we have the following ordinary integro-differential equation:

$$\frac{d^2 u_\alpha}{dr^2} + \frac{1}{r}\frac{du_\alpha}{dr} + \eta_\alpha^2(u_\alpha + cI_\alpha) = 0 \tag{43}$$

in which

$$I_\alpha = \int_{1/n}^{1} su_\alpha(s)ds \tag{44}$$

is unkown constant but can be taken as constant.

The general solution of Eq.43 is

$$u_\alpha(r) = AJ_0(\eta_\alpha r) + BY_0(\eta_\alpha r) - cI_\alpha \tag{45}$$

in which A and B are constants. J_m and Y_m are the m-th order
Bessel functions of the first and second kind, respectively and
both are collectively noted as Z_m in the sequel.
     Noticing $Z_0' = -Z_1$ and (B_{UD}), we have

$$u_\alpha(r) = A\left\{J_0(\eta_\alpha r) - \frac{J_1(\eta_\alpha)}{Y_1(\eta_\alpha)} Y_1(\eta_\alpha r)\right\} - cI_\alpha \tag{46}$$

Substituting Eq.46 into Eq.44 and using $\int xZ_0(x)dx = xZ_1(x)$, we
get

$$I_\alpha = -\frac{A}{n\eta_\alpha} \frac{J_1\left(\dfrac{\eta_\alpha}{n}\right) - \dfrac{J_1(\eta_\alpha)}{Y_1(\eta_\alpha)} Y_1\left(\dfrac{\eta_\alpha}{n}\right)}{1 + \dfrac{c(n^2-1)}{2n^2}} \tag{47}$$

Since the magnitude of A is immaterial, we can set A = 1 and
obtain the general form of the eigenfunction:

$$u_\alpha(r) = \left\{J_0(\eta_0 r) - \frac{J_1(\eta_\alpha)}{Y_1(\eta_\alpha)} Y_0(\eta_\alpha r)\right\} + \frac{h}{\eta_\alpha}\left\{J_1\left(\frac{\eta_\alpha}{n}\right) - \frac{J_1(\eta_\alpha)}{Y_1(\eta_\alpha)} Y_1\left(\frac{\eta_\alpha}{n}\right)\right\} \tag{48}$$

in which

$$h = \frac{2cn}{2n^2 + c(n^2-1)} \tag{49}$$

In order to determine the eigenvalue η_α, we have only to apply
the boundary condition (B_{Do}) to Eq.48 and then we obtain the
characteristic equation of η_α:

$$\frac{S(\eta_\alpha)}{T(\eta_\alpha)} + \frac{h}{\eta_\alpha} = 0 \tag{50}$$

in which

$$S(\eta_\alpha) = J_0\left(\frac{\eta_\alpha}{n}\right) - \frac{J_1(\eta_\alpha)}{Y_1(\eta_\alpha)} Y_0\left(\frac{\eta_\alpha}{n}\right) \tag{51}$$

$$T(\eta_\alpha) = J_1\left(\frac{\eta_\alpha}{n}\right) - \frac{J_1(\eta_\alpha)}{Y_1(\eta_\alpha)} Y_1\left(\frac{\eta_\alpha}{n}\right) \tag{52}$$

We define $\eta_1, \eta_2, \dots$ as the ascendant series of solutions of
Eq.50.
     Any function $u_\alpha(r)$ defined in Eq.48 with each eigenvalue is
called an eigenfunction of Eq.43.

(f) Composition of solution
     We have already explained the formal solution in terms of
eigenfunctions (Eq.23).   But we must calculate the inner
products $[u_\alpha, u_\alpha]$ and $[u_\alpha, u_0]$ for each eigenfunction if the
practical solution is required.   Let $u_\alpha(r)$, $u_\beta(r)$ be defined

through Eq.48 and η_α be an eigenvalue.   This means that both $u_\alpha(r)$ and $u_\beta(r)$ satisfy Eqs.43, 44 and (B_{UD}) and that $u_\alpha(r)$ further satisfies (B_{DO}).   Then Eq.18 is reduced in this case into

$$[u_\alpha, u_\beta] = \frac{\left(\dfrac{du_\alpha}{dr} u_\beta\right)\Big|_{r=1\ n}}{n(\eta_\alpha{}^2 - \eta_\beta{}^2)} \tag{53}$$

After taking the derivative of both the numerator and denominator with respect to η_β , we let η_β approach to η_α .   After a little manipulation, we have

$$[u_\alpha, u_\alpha] = \frac{T(\eta_\alpha)}{n}\left[\left\{\frac{1}{\pi\eta_\alpha}\frac{Y_0\left(\dfrac{\eta_\alpha}{n}\right)}{Y_1{}^2(\eta_\alpha)} - \frac{T(\eta_\alpha)}{2n}\right\} + \frac{h}{\eta_\alpha{}^2}\left\{\frac{Y_1\left(\dfrac{\eta_\alpha}{n}\right)}{\pi Y_1{}^2(\eta_\alpha)} - \frac{h+2n}{2n}T(\eta_\alpha)\right\}\right] \tag{54}$$

On the other hand, it is easy to calculate $[u_\alpha, u_0]$ by following the definition of inner product and we get

$$[u_\alpha, u_0] = -\frac{1}{n\eta_\alpha}T(\eta_\alpha) \tag{55}$$

Substituting Eqs.54 and 55 into Eq.23, we can obtain the distribution of excess pore water pressure at any place and time.

(g) Examples
        In order to show the numerical examples, we have to give the value of c and n.   c is composed of 4 elastic constants c_1, c_2, c_3, c_5 and n.   We assume for the sake of simplicity:

$$c_1 = \lambda + 2\mu,\ c_2 = \lambda,\ c_5 = \mu \tag{56}$$

which is the case just like isotropy, and

$$c_3 = \beta\, c_1 \tag{57}$$

Eq.(57) means that the vertical stiffness is β times of the horizontal one.   When $\beta = 1$, we have the isotropy case. Poisson's ratio is defined here to be the ratio of the horizontal normal strain to the vertical one when the vertical pressure is applied, i.e.,

$$\nu = \frac{\lambda}{2(\lambda+\mu)} \tag{58}$$

and we fix $\nu = 1/3$.   Then we get

$$c = \frac{2n^2\{1-(\beta-1)(n^2-1)\}}{(n^2-1)\{\beta(n^2+3)-1\}} \tag{59}$$

Therefore c is defined by two parameters n and β .
        In the sequel, we use the time factor $\dfrac{kc_1t}{r_w(2R_0)^2}$ to facilitate the comparison with Barron's results.
        Figs.2 (a) and (b) shows the initial excess pore water

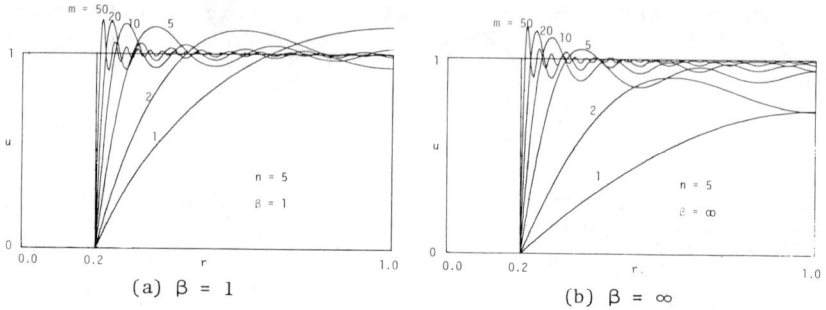

(a) $\beta = 1$ (b) $\beta = \infty$

Fig.2   Number of Summation m and Initial Distributions

pressure which is calculated through Eq.23 for the cases of n=5, $\beta=1$ and ∞, respectively.   We can observe that the initial distribution for each case converges to the prescribed value ($u_0=1$) except the portion near r=0.2 when the number of summation m is increased.

Figs.3 (a) and (b) show the isochrones for the same cases in Fig.2.   Although the curves at t=0 fluctuate with rippling, they are smoothened as soon as the consolidation starts.   It is because the component of high frequences (large η_α) decays quickly according to $e^{-\eta_\alpha^2 t}$.   As for the case of $\beta = 1$, the results agree closely with Barron's ones.   From the above, we can say that the present method has a sufficient validity even if we cut the higher components than the first several terms in Eq.23.

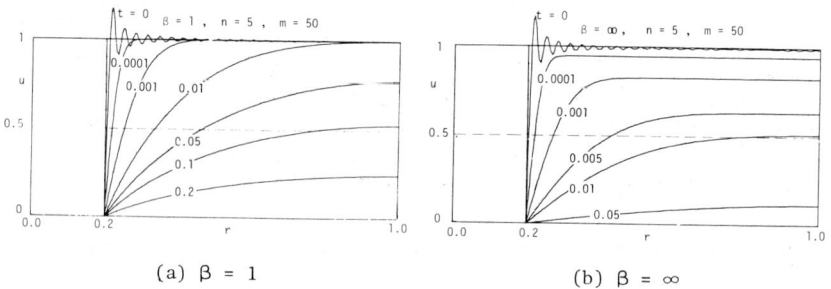

(a) $\beta = 1$ (b) $\beta = \infty$

Fig.3   Isochrones

Figs.4 (a) - (i) shows the degree of consolidation - time factor curves for several values of β in which the case of $\beta = 1$ agrees well with Barron's one.

It is considerably noteworthy that the progress of consolidation with large β be much more accerelated by the magnitude of n than the cases with small β.   Therefore as is shown in Fig.4 (i), the relative location of several curves for $\beta = \infty$ is completely reversed comparing with the usual one [Fig.4 (a)].   This paradoxical phenomenon is resulted from the fact that the part of excess pore water pressure is reduced not drainage but by mechanical effect, i.e., so-called negative

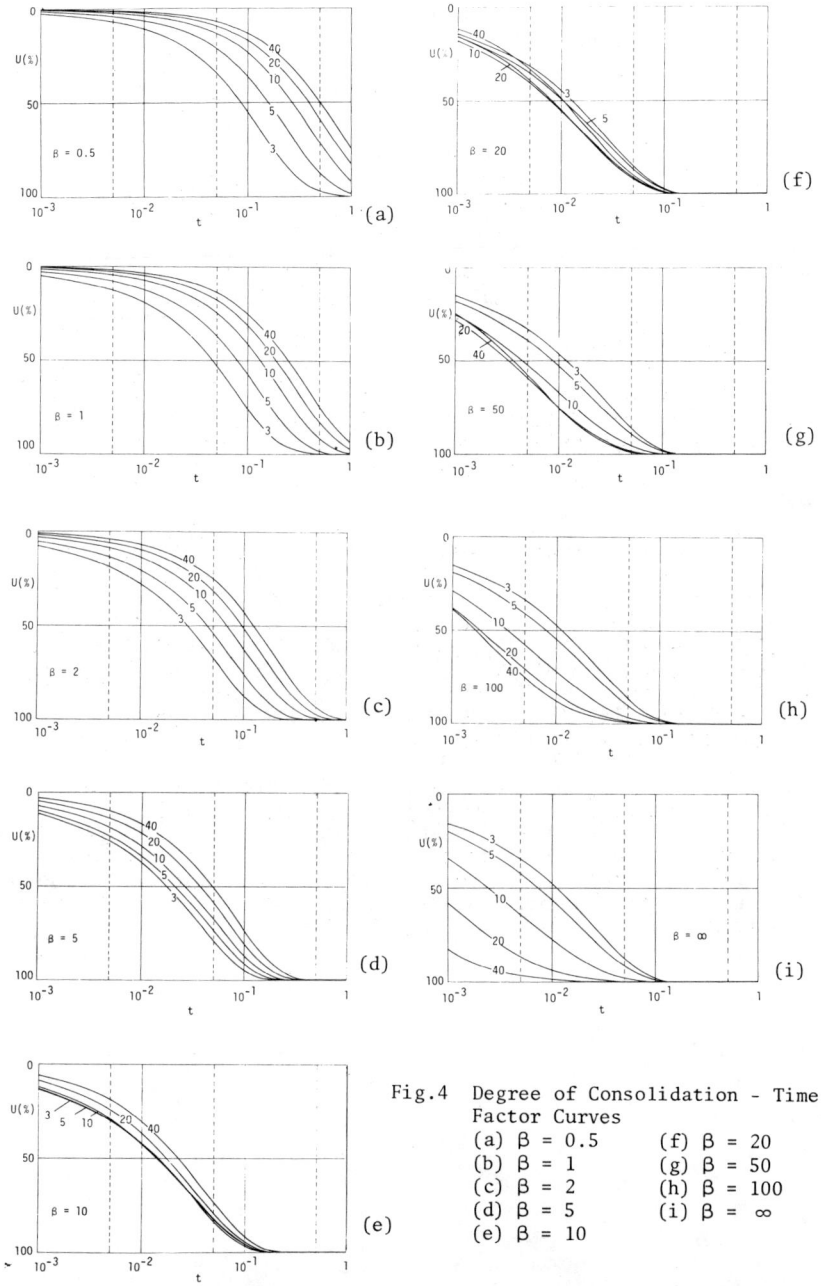

Fig.4 Degree of Consolidation - Time
       Factor Curves
       (a) β = 0.5   (f) β = 20
       (b) β = 1   (g) β = 50
       (c) β = 2   (h) β = 100
       (d) β = 5   (i) β = ∞
       (e) β = 10

Mandel-Cryer effect.[3]   Although the progress of degree of consolidation does not always mean the progress of settlement of the ground, we should take into the consideration such kind of influence due to the anisotropy in designing the sand drain.

CONCLUSIONS

We can summarize the present paper by the followoing conclusions.

1. It is possible to reduce Biot´s equations of consolidation into a single equation of excess pore water pressure.

2. There exists the theory of eigenvalue problem for the above equation quite similar to the one-dimensional consiolidation.

3. We have got the closed form solution for the problem of sand drain model in the anisotropic ground.

4.   The anisotropy of ground would have a remarkable influence on the progress of consolidation through Mandel-Cryer effect.

REFERENCES

1. Akai, K. and Tamura, T., "Numerical Analysis of Stress Path under Multi-Dimensional Consolidation," Proc. Specialty Session 12, 9th Int. Conf. Soil Mech. Found. Engrg., 1977, pp.30-53.
2. Barron, R.A., "Consolidation of Fine-Grained Soils by Drain Wells," Trans. ASCE, Vol.113, 1948, pp.811-835.
3. Cryer, C.W., "A Comparison of the Three-Dimensional Consolidation Theories of Biot and Terzaghi," Q. Journ. Mech. Appl. Math., Vol.16, Part 4, 1963, pp.401-414.
4. Lekhnitskii, S.G., "Theory of Elasticity of an Anisotropic Elastic Body," Holden-Day, 1963.
5. Tamura, T., "A Study on the Mechanism of Consolidation," Mem. Fac. Engrg., Kyoto Univ., Vol.41-4, 1979, pp.518-535.
6. Tamura, T., "Eigenvalue Problem of Consolidation," Mem. Fac. Engrg, Kyoto Univ., Vol.42-1, 1980, pp.35-52.
7. Tamura, T., "Analysis of Consolidation in Anisotropic Ground with Sand Drain," Proc. JSCE, No.338, 1983, pp.115-121 (in Japanese).

Dredge Spoil Disposal Predictions and Performance:
A Case History
Vahan Tanal,* M.ASCE

Abstract

A 146-acre (59 ha) dredge spoil disposal facility was constructed
in the Baltimore Harbor that received 3.4 million cubic yards
(2.6 million cubic meters) of dredge spoil generated from the
construction of the Fort McHenry immersed tube tunnel. The design of
the disposal site was governed by the finite size of water area that
could be enclosed by a containment structure and severe environmental
restrictions on the quality of effluent discharge that necessitated
treatment facilities comparable to those required for a large water
supply project. The design capacity of the site was established with
allowance for bulking, based on laboratory test results and reported
past experience for the different types of dredge spoil. Consolidation
of compressible in situ soils underlying the site and consolidation of
the dredge spoil during disposal were evaluated and included in
estimating the maximum available capacity. The actual capacity was
appraised during dredging, by field surveys and soundings. Disposal
procedures were modified to compensate the capacity shortfall projected
by the field monitoring results.

Introduction

A 146-acre (59 ha) dredge spoil disposal facility was constructed
in the Baltimore harbor that received 3.4 million cubic yards
(2.6 million cubic meters) of dredge spoil generated from the
construction of the Fort McHenry immersed tube tunnel. During the
design of the tunnel a search was undertaken for a site for disposal of
the excavated materials in an environmentally acceptable manner.
Numerous upland and harbor sites were identified and evaluated. The
studies concluded that upland sites did not represent feasible
alternatives because of environmental restrictions, necessary double
handling, transportation problems and costs. The studies were then
geared toward harbor sites.

Criteria for Site Selection

Engineering criteria for harbor site selection were established as
follows:

1.   The site had to have sufficient capacity.

*Professional Associate and Deputy Manager of the Geotechnical
Department, Parsons, Brinckerhoff, Quade & Douglas, Inc., One Penn
Plaza, New York, New York 10119.

2.   The site had to be within a reasonable distance from the
     tunnel and site conditions had to be such that the
     containment structure could be constructed at a reasonable
     cost and on schedule.

3.   The site had to be developed into usable upland, to support
     1,000 psf (48 kPa) in a reasonable time after it was filled
     with dredged materials.

The process of the disposal site evaluation and elimination led to
the identification of the Canton/Seagirt site as the only location
available that would likely provide sufficient capacity (depending on
the location of the containment structure), that was relatively close
to the tunnel site, and which could be developed into usable upland for
a future port facility.

Borings drilled at the Canton/Seagirt site indicated that the
subsoil consisted generally of very soft black clayey organic silt,
followed by a very soft to medium stiff gray silt and clay, underlain
by medium dense to very dense silty sand and gravel and hard red
Cretaceous silty clay. The thickness of the soft deposits increased
rapidly away from the shoreline. The borings indicated that at the
pierhead line the thickness of the harbor bottom deposits reached ten
feet (3m) and the soft silts and clays reached thickness of 80 to 100
feet (24 to 30m).

Stabilization Studies and Settlement Estimates

The requirement to stabilize the disposal site for the development
of a port related facility in the shortest possible time period after
completion of filling, was a major factor in the design of the
Canton/Seagirt site. If all the dredged materials were placed in one
area in an uncontrolled manner, the resulting fill would take many
years to stabilize at great costs. To provide a design where the land
reclamation process would be accelerated the following decisions were
taken:

1.   The softest (and hardest to stabilize) materials would be
isolated and placed in a separate area. A "muck area" was thus
designated for the very soft harbor bottom deposits. This area would
require special stabilization techniques and its development into
usable land would take approximately ten to twenty years. The better
materials would be placed in the "spoil area" and a stabilization
effort could commence soon after the completion of disposal operations.
(Photo 1.)

Photo 1.   The   27-inch   (69cm)   dredge   pipe   was   equipped   with
Y-valves to separate the spoil and muck and to fill the spoil area in a
controlled fashion.   (Photo by W.C. Grantz)

2.   The dredged materials would be placed in the disposal area in
a controlled fashion to gradually load the soft in-situ soils and
prevent deep seated failures and mud waves.   In this manner, the
stabilization process could be started during disposal by consolidating
the in-situ soft soils under the weight of the dredged materials.   The
materials would be discharged at multiple discharge points, to promote
gradual deposition in horizontal beds, starting at the structure line
and moving towards land, to prevent failure in the foundation soils.

It was estimated that under a 1,000 psf (48 kPa) surcharge load
above the elevation of +10 ft (3m), total settlements at the site would
range locally from 3 to 13 feet (1 to 4m), 2 to 8 feet (0.6 to 2.5m) in
in-situ soil plus 1 to 5 feet (0.3 to 1.5m) in dredge spoil, depending
on the thickness of the soft in-situ soils.   The lower settlements
would occur close to the shoreline where the existing compressible soil
strata   were   very   thin   or   non-existant.   The   1,000   psf   (48   kPa)
surcharge load is equivalent to about 10 to 12 feet (3 to 4m) of spoil
placed above elevation +10 ft (3m).   The settlement analyses indicated
that in all areas except close to the containment structure, 90 percent
consolidation would be achieved in one to 5.5 years, if single drainage
was assumed.   If double drainage was assumed most areas would reach 90
percent consolidation in less than two years.   The actual time required
for consolidation in the field should fall somewhere in between the two
conditions   since   only   portions   of   the   compressible   strata   were
underlain by permeable materials and the permeability of the overlying
dredged spoil would vary depending on its sand content.   Most estimates
of the time rate of consolidation are usually conservative because of
undetected lenses of permeable material[4].   Hence, it was concluded
that actual times might be expected to be closer to the double drainage
case.   Considering the eighteen months available for dredging and

disposal, some of the estimated settlements were expected to occur during this period.

Study of Dredging Methodology

A study undertaken to determine the dredging methods and equipment for this project concluded that for the 18 months allowed for dredging, the average production required would be 9000 cubic yards (7,000 cubic meters) per day, in-place measure. Reasonable production rates for equipment suitable for this project were established as follows:

Type of Dredge	Type of Material	Rate of Production	
		Cubic yards per day	(Cubic meters per day)
Hydraulic suction	Harbor bottom deposits	10,000	(8,000)
Hydraulic suction with cutterhead	Sands, soft clays and silts	15,000 to 20,000	(11,000 to 15,000)
Hydraulic suction with cutterhead	Stiff clays and very dense sands	4,000 to 10,000	(3,000 to 8,000)
Clamshell	Sands and silts	3,000 to 10,000	(2,000 to 8,000)
Clamshell	Very stiff clays	400	(300)
Clamshell	Fine dredging	1,200	(900)

The maximum dredging rates were governed by the tunnel construction schedule, the availability of equipment and very importantly, the environmental limitations on effluent discharge quality from the disposal facility.

It was determined that most of the bulk dredging could be accomplished by a 24-inch (61cm) cutter suction dredge with an effective dredging depth of about 70 feet (21m). Even though larger 27-inch (69cm) hydraulic dredges were available in the eastern United States, and these would be ideal to excavate the stiff Arundel clays in the lower zones of the trench (because of both power and operating reach) the output of these dredges would far exceed the tunnel's construction schedule requirements and they would have to be used on an intermittent basis. In addition, the high production rate of these dredges would present great difficulties in managing efficient disposal of the dredge material because of environmental limitations on the receiving capacity of the disposal site facilities. Therefore a combination of a 24-inch (61cm) cutter suction dredge for the first 60 to 70 feet (18 to 21m) and a large clamshell dredge for the remainder of the rough dredging was considered as a basis for design.

On the basis of these studies, the design rate was established as 10,000 cubic yards (8,000 cubic meters) per day for muck and 20,000 cubic yards (15,000 cubic meters) per day for spoil. The weirs and the piping at the muck and spoil sites were designed for these rates. The treatment facilities were designed for a maximum discharge rate of 50 cubic feet (1.4 cubic meters) per second, corresponding to 20,000 cubic yards (15,000 cubic meters) per day. These rates governed the maximum allowable design flow depths over the weirs of 2.5 inches (63mm) in the spoil area and 2 inches (50mm) in the muck area. The project specifications did not place a limit on the size of the equipment the contractor might elect to utilize as that would have been unduly restrictive. Depending on availability of equipment and other dredging work at the time of construction, it was conceivable that the contractor might commit a large dredge but operate it intermittently below its maximum production rate.

Operation and Effluent Treatment

It was planned that the spoil and muck arriving at the disposal site would be directed into their respective areas and the spoil would be rehandled and mounded by wide track bulldozers to utilize the available capacity to its maximum. In this manner, the dikes in the spoil area could be raised to the design elevation of +21 ft (6m) and a ponding area around the spoil area and the muck area weirs would be maintained at all times.

The treatment of the effluent from both the spoil and muck areas was achieved by adding flocculating agents and settling the suspended solids in the settling basin. The design discharge rate into the treatment basin was 50 cubic feet (1.4 cubic meters) per second based on an assumed maximum daily dredging rate of 20,000 cubic yards (15,000 cubic meters) for spoil. The water quality standards required by the permit were thus satisfied.

This is the first major dredging project to be designed for treatment of the effluent by large-scale flocculation and sedimentation. There were no precedents for the performance of such a system under conditions similar to those on the project, and it was necessary to base the design on theoretical considerations plus a considerable extrapolation of previous experience. This state-of-the-art system in dredge spoil disposal has worked remarkably well in settling the finer solids in the effluent and containing them at the disposal site. However, it has contributed to the fact that the bulking factors experienced have been higher than those previously experienced in conventionally contained disposal areas.

Disposal Site Capacity Considerations

Although the Canton/Seagirt site provided a large area between the pierhead line and the shoreline to construct a disposal site, the containment structure needed suitable subsoil for stability to meet the anticipated design loading conditions. Accordingly, an objective was established to minimize the quantities of dredged materials to be placed in the disposal site and to utilize the capacity of feasible construction to the greatest possible extent. In this manner a containment structure that did not require pre-excavation could be constructed as close to the shore as possible, in comparatively favorable subsoil conditions. The design which evolved from this was judged the best compromise between capacity and constructibility.

Dredged Material Quantities

The quantity estimate of the materials to be dredged from the tunnel trench was 3,343,000 cubic yards (2,560,000 cubic meters). The quantity breakdown of different material types that would be excavated from the tunnel was estimated on the basis of boring data obtained at the trench location. This procedure is only approximate as it requires interpolation of the soil profile between borings, both along and perpendicular to the tunnel alignment. The uncertainty in the breakdown of quantities is aggravated by the presence of several intermixed material types and the variability of their stratification between the borings. On the basis of available information, the breakdown of the material types was estimated as follows:

	Quantity	
Material Types	Cubic yards	(Cubic meters)
Harbor bottom deposits	600,000	(459,000)
Soft organic silts and clays	650,000	(497,000)
Subtotal "soft" material	1,250,000	(956,000)
Sands	1,155,000	(884,000)
Stiff to hard clays	938,000	(718,000)
Subtotal "suitable" materials	2,093,000	(1,602,000)
Total quantity requiring disposal	3,343,000	(2,558,000)

During the design phase a further adjustment was made to provide for the excavation at the east end to be performed by dredging below Elevation O.The effect of this was to increase the quantity requiring disposal by the amount of east end excavation between the elevations of 0 and -20 feet (-6m). This quantity was estimated as an additional 223,000 cubic yards (170,000 cubic meters), resulting in a total estimated quantity requiring disposal of 3,566,000 cubic yards (2,730,000 cubic meters). This quantity was the best estimate during the study and design phases and was used as the basis for the design of the disposal site.

Bulking Factors

Bulking factors represent an increase in volume that most soils experience when dredged and redeposited. The increase in volume is temporary and varies widely with the types of materials being dredged and the method and rate of dredging and disposal. The medium to which the material is being deposited (onto land or into water) and the prevailing climate also have important effects on the magnitude of the residual bulking factors and the time required to attain shrinkage.

In addition, moving the materials around during disposal and piling them up in mounds has an effect of reducing the initial bulking factors by compressing the materials. It has been reported that parameters affecting bulking factors include in-situ density, water content, void ratio, plasticity, cohesiveness, compressibility, permeability, soil particle distribution and organic material content. Other variables include the method of dredging and containment systems, water salinity, and climatic conditions in the area. In addition, the long term settlement of the dredged material as well as consolidation of the underlying soil influence the containment site capacity[2].

Although there is an abundance of published literature on dredging and disposal, there is limited direct information on bulking factors and their use in sizing disposal sites. Furthermore, the limited published data is sometimes conflicting. A literature search and consultations with experts during the project study and design phases indicated that a wide range of bulking factors could be experienced in the types of materials to be dredged from the Fort McHenry Tunnel trench. Since bulking factors associated with dredged spoil disposal are highly variable and cannot be predicted or measured accurately, it has been common practice to size disposal sites based on past experience[1, 2, 3]. Agencies and contractors involved in dredging operations rely heavily on practical experience to predict bulking factors and in sizing disposal sites.

A comparison of bulking factors reported for similar materials and considerable extrapolation of information obtained indicated that the following range of factors might be experienced by the various materials being dredged from the Fort McHenry tunnel trench:

Materials	Probable range of initial bulking factors	Probable range of residual bulking factors
Muck	1.5 - 2.0	1.0 - 1.2
Organic Clayey Silts	1.5 - 2.0	1.1 - 0.7
Sandy Materials	1.0 - 1.2	1.1 - 1.0
Stiff and Hard Clays	1.2 - 1.5	1.2 - 1.4

Application of these initial bulking factors to the excavation quantities resulted in the following range of bulked material to be accommodated at the disposal site:

Materials	In place volume		Bulked volume Lower factors		Higher factors	
	Cu. yds.	(Cu. m.)	Cu. yds.	(Cu. m.)	Cu. yds.	(Cu. m.)
Muck	600,000	(459,000)	900,000	(689,000)	1,200,000	(918,000)
Organic Clayey Silts	650,000	(497,000)	975,000	(746,000)	1,300,000	(994,000)
Sandy Materials	1,155,000	(884,000)	1,155,000	(884,000)	1,386,000	(1,060,000)
Stiff and Hard Clays	938,000	(718,000)	1,125,600	(861,000)	1,407,000	(1,076,000)
Land Excavation	223,000	(170,000)	278,750	(213,000)	278,750	(213,000)
Totals	3,566,000	(2,728,000)	4,434,350	(3,393,000)	5,571,750	(4,261,000)

The land excavation was not broken down into components and a bulking factor of 1.25 was applied. The difference in the estimated total quantity of dredged materials to be accommodated in the disposal site depending on the choice of bulking factors was in excess of one million cubic yards (765,000 cubic meters). To have conservatively sized the disposal site using the higher bulking factors would have required the construction of a containment structure further offshore in very poor subsoil conditions, where it would be impossible to meet the design loading conditions. However, if the dredged materials were piled up at the disposal site as they were being discharged, some compression would occur to reduce the initial bulking factors below their maximum values. On this basis it was assumed that the actual bulking factors would be considerably lower than the maximum reported values. The following bulking factors were therefore judged reasonably conservative for the purpose of sizing the disposal site:

Materials	In place volume Cu. yds.	(Cu. m.)	Bulking factor	Bulked volume Cu. yds.	(Cu. m.)
Muck	600,000	(459,000)	1.8	1,080,000	(826,000)
Organic Clayey Silts	650,000	(497,000)	1.8	1,170,000	(895,000)
Sandy Materials	1,155,000	(884,000)	1.0	1,155,000	(884,000)
Stiff and Hard Clay	938,000	(718,000)	1.4	1,313,200	(1,004,000)
Land Excavation	223,000	(170,000)	1.25	278,750	(213,000)
Total	3,566,000	(2,728,000)	1.4 avg.	4,996,950	(3,822,000)

The total bulked volume of the dredged materials that would be generated from the tunnel trench was thus estimated to be 4,996,950 cubic yards (3,822,000 cubic meters).

Allowance for overdredging was considered for both spoil and muck materials. As the anticipated dredging schedule indicated that muck would be removed in three stages with about three to five months in between the stages, the no-dredge periods would have an effect of reducing the bulking factor. Assuming a residual bulking factor of 1.5 for muck, the assumed bulked quantity on the basis of 1.8 bulking factor would allow for 20 percent overdredging. For the spoil materials overdredging would have to be accommodated by the inherent conservatism in the assumed bulking factors or by piling the spoil locally above the design elevations. However, because of limited capacity at the disposal site, the contract specifications were written to exercise strict control to avoid overdredging and originally included a provision that the material dredged beyond the designated trench limits would have to be deposited elsewhere at the contractor's expense. This provision was later removed to ease the limitations on the contractor and to preclude provoking excessive bid contingencies.

Disposal Site Capacity Estimates

The maximum capacity available at the Canton/Seagirt site was computed on the basis of the following assumptions:

1. The spoil could be utilized for building dikes with their toe at 50 feet (15m) behind the containment structure, using one vertical on five horizontal slopes to the elevation of +21 feet (6.4m).

2. The spoil surface inside the 21 foot (6.4m) elevation dikes could reach an average elevation of +20 feet (6m).

3. The muck area would be contained on the landside by a sheetpile structure against an existing bulkhead and would be filled to the elevation of +10 feet (3m).

4. The treatment and settling basins could be filled to the elevation of +10 feet (3m) at the end of filling operations, providing a capacity of about 86,000 cubic yards (66,000 cubic meters).

Using the above assumptions the estimated capacity of the Canton/Seagirt site, based on the original configuration and design, was 4,600,000 cubic yards (3,500,000 cubic meters). The total available capacity is computed by using the contour maps of the site and does not account for settlements of the original in-situ compressible materials and the dredged materials.

The difference between the bulked volume of 4,996,950 cubic yards (3,822,000 cubic meters) and the available maximum capacity of 4,600,000 cubic yards (3,500,000 cubic meters) is 396,950 cubic yards (322,000 cubic meters) (bulked volume). If conservative bulking factors prevailed throughout the eighteen months of filling operations and no settlement occurred during this period, it was estimated that the dredged spoil might reach Elevation +23 feet (7m) in the spoil area at termination of dredging. However, one or more of the following factors would significantly affect that conclusion:

1. Settlements would increase the receiving capacity by about 140,000 cubic yards (107,000 cubic meters) per foot of settlement in the spoil area only. Three feet (1m) of average settlement in the spoil area would create an additional 420,000 cubic yards (128,000 cubic meters) capacity. If the spoil were placed in layers without creating extensive mud waves, as specified, the achievement of one to three feet of consolidation settlements appeared to be a reasonable objective.

2. The assumed bulking factors were expected to be conservative, especially since the spoil was to be reworked at the site and would compress under its own weight. To quantify, every ten percent decrease in the average bulking factor would signify an additional 360,000 cubic yards (275,000 cubic meters) of capacity.

3. Significant reduction in the material to be taken to the disposal site could be achieved by diverting several hundred thousand

cubic yards of dredge spoil for the creation of marshlands for wetlands mitigation, a requirement established by the regulatory agencies, and backfilling of the harbor by an additional 735,000 cubic yards (562,000 cubic meters) of dredged material. Although the scheduling of the mitigation and the placement of the harbor backfill were not finalized during design, these reductions provided a potential for a substantial relaxation of the required spoil receiving capacity of the disposal site.

Taking into consideration immediate settlement of the dredged spoil and in-situ materials during disposal, it was concluded that the capacity of the Canton/Seagirt containment facility would be adequate. However, it was impossible to assess the effects of actual construction operations on the necessarily assumed factors in these computations.

Construction Methods Affecting Capacity

1. <u>Method and sequence of disposal</u>: The contract documents called for the disposal of spoil in a controlled manner to prevent failures in the in-situ compressible soils that would create uncontrollable mudwaves. The initial sand removed from the west end of the trench excavation was to be discharged under water directly behind the containment structure; the spoil would be placed by gradually moving the discharge point toward land. At no time was spoil to be deposited above the water surface until an area of 150 feet (46m) width behind the structure was filled to minimum water elevation. Subsequently, the material would be deposited in several stages on flat slopes. This procedure was planned to gradually load the in-situ compressible soils and initiate their consolidation under the spoil fill. After the designated area behind the containment structure and the settling basin sheeting was brought to above water in this manner, dike construction would commence along this perimeter to enable the raising of the ponding depth in the spoil area.

The dredging contractor proposed a filling procedure that maintained the discharge points on the containment structure until the spoil behind the structure was brought to the elevation of +10 ft (3m) in thicker lifts and less stages than indicated in the contract drawings. To expedite construction, his proposed method was accepted with the condition that the site be carefully filled and mud waves be prevented and controlled. However, the increased rate at which the spoil material was discharged and the Contractor's effort to create a dike in a single stage caused deep seated failures in the in-situ compressible soils and resulted in large mud waves. The Contractor was then unable to control the mudwaves to effectively dewater the site by creating a ponding area around the weirs and providing a gradient from the containment dikes toward the weirs. Instead, the mudwaves formed islands over the site and against the weirs and the flow toward the weirs was channeled in between the mud islands. Furthermore, it became impossible to induce settlements in the compressible soils as they have been displaced ahead of the spoil fill. Additionally, the unsuitable materials that were excavated and placed behind the cells were pushed by the incoming dredge spoil and a portion of the displaced soft materials went into suspension. The deep failures in the soft in-situ materials and the disturbance of the redeposited unsuitable materials resulted in further bulking of the materials that existed at the site before dredging started.

2. Rate of discharge: The Contractor elected to utilize one of the largest dredges available in the United States for the excavation of the entire trench, coupled with a booster pump to rapidly deliver the excavated materials to the disposal site. (Photo 2) While a large dredge provides advantages for cutting the hard clays, deep operational capability and high production rates, it has several adverse effects. First, the rate of discharge could not be accommodated by the effluent treatment facilities as they were designed for a maximum discharge rate of 50 cubic feet (1.4 cubic meters) per second (cfs). Although the contractor had redesigned and modified the hydraulics of the system to accommodate up to 80 cfs (2.2 cms) of discharge, actual discharge often exceeded that rate and reached 100 to 120 cfs (2.8 to 3.4 cms). With insufficient ponding depth, when the discharge point was relatively close to the weirs, the finer solids carried in the discharge were jetted toward the weirs without having a chance to settle in the disposal site. In addition, very high discharge velocities had pushed the deposited finer materials ahead of the discharge point in effect creating mudwaves out of the previously deposited spoil materials.

Photo 2.   The dredge spoil was pumped through a 27-inch (69cm) pipe from the tunnel site 2 miles (3.2km) away.

Reestimating Capacity During Construction

Subsequent to the award of the contract, negotiations between the owner and the Contractor resulted in shifting the alignment of the containment structure 300 feet (92m) further offshore at the western end.   The capacity of the site with the new alignment of the structure was estimated at 4,980,000 cubic yards (3,810,000 cubic meters).

In December 1981, when only about 50 percent of the trench dredging was completed, the Contractor projected a capacity shortage at the disposal site and asked for guidance. In estimating the remaining capacity the Contractor had assumed that the spoil area within the +21 foot (6.4m) dikes would only be filled to an average elevation of +19.5 feet (6m) and that the treatment and settling basins did not offer any additional capacity for operational reasons. In addition, the Contractor's design required that the dikes behind the cells be moved to a 100 foot (30m) distance from the back of the cells, instead of the originally planned 50 feet (15m), to prevent unacceptable loading of the cellular cofferdam containment structure. Furthermore, the Contractor felt that an additional one foot of ponding depth throughout the spoil area would be required at the end of the filling operation. Further the net capacity of the disposal site was decreased during construction by placing the soft silts and clays excavated from inside and from behind the 62 foot (19m) diameter cells into the spoil area. Considering all these factors, the current capacity of the site was estimated as follows:

	Volume	
	Cubic yards	(Cubic meters)
Total capacity (spoil to +20, muck to +10 excluding treatment and settling basins)	4,980,000	(3,810,000)
Loss of capacity if spoil is filled to +19.5 only	70,000	(54,000)
Loss of capacity due to pre-dredging and excavation from cells*	363,000	(278,000)
Loss of capacity by moving dikes to 100 ft behind cells	60,000	(46,000)
Loss of capacity due to one foot additional ponding depth	140,000	(107,000)
Projected current total capacity	4,347,000	(3,325,000)

The projected capacity shortage had been estimated by taking the elevation to which the site was filled as of mid-December, estimating the quantity of material that remained in the trench as of that time, and assuming that all material remaining in the trench must be accommodated in the spoil and muck areas. The trench quantity was estimated by the Contractor, broken down into "soft" and "suitable" materials and verified by the Engineer based on data supplied by the Contractor regarding limits to which dredging had been completed and based on the trench cross sections. The capacity shortage was estimated by applying two sets of bulking factors as follows:

---

*The net quantities are assumed bulked by a factor of 2 when redeposited.

	Material remaining in trench as of Dec. 10, 1981		Bulking factor high/	Range of bulked quantities in cubic yards			
	Cu. yds	(Cu. m)	low	Cu. yds.	(Cu. m)	Cu. yds.	(Cu. m)
Soft	652,000	(500,000)	2.0/1.8	1,304,000	(998,000)	1,173,600	(898,000)
Suitable	1,068,000	(817,000	1.5/1.25	1,602,000	(1,226,000)	1,335,000	(1,021,00)
Totals	1,720,000	(1,317,000)		2,906,000	(2,224,000)	2,508,600	(1,919,000)

The average bulking factor for the total volume would be a high of 1.69 and a low of 1.46.

The capacity remaining at the site was estimated using the Contractor's criteria, one foot (0.3m) additional ponding depth and dikes at 100 feet (30m) behind cells, and based on assumed elevations to which the site was filled in Mid-December 1981, as follows:

	Remaining Capacity as of 12/10/81	
Assumptions	Cubic yards	(Cubic meters)
Spoil Area full to El+8.5 ft (2.6m) Muck Area to +8 ft (2.4m)	1,481,800	(1,133,600)
Spoil Area full to El+6 ft (1.8m) Muck Area to +7 ft (2.1m)	1,857,000	(1,420,600)

Accordingly the range of projected capacity shortage was estimated as:

2,906,000 - 1,481,800 = 1,424,200 cubic yards (1,090,400 cubic meters)
2,508,600 - 1,857,000 =   651,600 cubic yards (498,400 cubic meters)

A significant assumption in estimating the projected capacity shortage was the elevation to which the site was already filled. The site at the time exhibited three zones below the weir elevation. In descending order these were: ponded water, slurry, and soft materials resting on firm bottom. On December 14, 1981 an attempt was made to measure the top elevation of each of these strata by taking soundings at ten points on a line across the spoil area. The results were as follows:

Zone	Range of Measured Elevations, ft (m)
Water surface	+9.5 (3)
Top of slurry	+4 to +9.5 (mostly +9) (1.2 to 3)
Top of soft clay	-0.5 to +7.5 (mostly +5) (0 to 2.3)
Top of firm material	below -5 to -4 (-1.5 to -1.2)

The soundings were made by three methods: (a) visually estimating the top of slurry surface, (b) noting the depth to which a sounding

basket settled under its own weight (top of soft clay) and (c) pressing
down by hand a 15 foot (5m)-long 1-1/2 inch (38mm) diameter PVC pipe to
reach firm bottom.

In general it was found that the surface of the mud slurry was
about 1/2 foot (15cm) below the water surface. Generally the basket
sank to approximately El.+5ft (1.5m). At most locations firm bottom
was not encountered with the 15-foot (5m) pipe probe putting it at an
elevation lower than -6ft (-1.8m). Unfortunately, limitations on
access to the site and mobility made it extremely difficult to survey
the entire site and obtain more accurate measurements.

Analysis of Bulking Factors

Based on the adjusted calculations of in-place quantities to be
dredged and of capacity of the containment site, plus the contractor's
estimates of remaining quantities to be dredged and average depths of
filling of the muck and spoil areas, the average bulking factor of the
material dredged up to mid-December 1981, may be derived as follows:

Quantity in Place in Trench

	Cubic yards	(Cubic meters)
Total	3,625,000	(2,773,000)
Less remaining 12/10/81	1,720,000	(1,316,000)
Completed to 12/10/81	1,905,000	(1,457,000)
% completed	52%	

Containment Site Capacity

	Cubic yards	(Cubic meters)
Original Total	4,980,000	(3,810,000)
Current adjusted total	4,347,000	(3,325,000)
Minus remaining capacity, spoil 8.5 ft (2.6m), Muck 8 ft (2.4m)	1,482,000	(1,133,600)
Capacity used to 12/10/81	2,865,000	(2,192,000)

Average bulking factor
to 12/10/81

$$\frac{2,865,000}{1,905,000} = \underline{1.50}$$

If the average level of dredged material in the spoil area is
taken at El. +6.0 ft (1.8m) and in the muck area at El. +7.0 ft (2.1m),
the remaining capacity is calculated to be 1,857,000 cubic yards
(1,420,600 cubic meters). The capacity used to 12/10/81 is then
2,490,000 cubic yards (1,905,000 cubic meters), and the derived average
bulking factor becomes 2,490,000/1,905,000 or 1.31. These derived
factors are to be compared to the originally assumed average of 1.40.

It appeared reasonably conservative to apply an average bulking
factor of 1.50 to the remaining 1,720,000 cubic yards (1,316,000 cubic

meters) to be dredged. This indicated a required containment volume for the remaining dredged spoil of 1.5 x 1,720,000 = 2,580,000 cubic yards (1,974,000 cubic meters). The capacity shortage projections would then become:

Average 12/10/81 Elevations:	Maximum		Minimum	
	Feet	(m)	Feet	(m)
Spoil	8.5	(2.6)	6.0	(1.8)
Muck	8.0	(2.4)	7.0	(2.1)
	Cu. yds.	(Cu. m)	Cu. yds.	(Cu. m)
Calculated remaining capacity	1,482,000	(1,133,600)	1,859,000	(1,422,000)
Required capacity	2,580,000	(1,974,000)	2,580,000	(1,974,000)
Projected shortage	1,098,000	(840,400)	723,000	(552,000)

These quantities could be accommodated by establishing a temporary storage area for the dredged materials that would be later used over the tunnel as harbor backfill. The capacity of the containment site would then be ample to accommodate the remaining material to be dredged.

Completion of Construction

The shortfall projections were approximate at best and could only be verified at the completion of dredging. However, given the finite size of the disposal site it was in the project's interest to accept the Contractor's projection in preparing a contingency plan to accommodate the more conservative shortfall estimates. Accordingly the following measures were taken:

1. A bucket dredge was utilized to excavate the remaining soft bottom materials and a clam shell was used to place these materials into the muck area.

2. About 100,000 cubic yards (765,000 cubic meters) of stiff clay that was later to be used as harbor backfill was temporarily stored in front of the containment area.

The disposal site was less than full when the dredging of the tunnel trench was completed in May 1982. Approximately 100,000 cubic yards (765,000 cubic meters) of suitable materials excavated from the tunnel approaches were later disposed of at the site. Presently a stabilization program is underway to develop the site into a port facility. More suitable materials will be needed to fill and grade the site as consolidation takes place. (Photo 3)

Photo 3. The 146-acre (59 ha) disposal site was almost full at the completion of dredging of the tunnel trench in May 1982.   A stabilization program is underway to develop the site into a port facility. (Photo by W.C.Grantz)

Conclusion

The capacity predictions for the Fort McHenry tunnel dredge spoil disposal site were based on assumed bulking factors and estimated disposal rates.   Although the project specifications were tight on dredging rates and disposal procedures, the contractor was allowed to discharge at substantially higher rates in order not to compromise the tunnel construction schedule.   By temporarily storing approximately 100,000 cubic yards (765,000 cubic meters) of material outside the site, and by switching to a bucket dredge for the final muck dredging, the construction was completed on schedule.   However, additional capacity was available at the site at the completion of the tunnel trench dredging, indicating that the original assumptions on bulking factors would have prevailed had the site been filled at a slower rate.

Acknowledgments

William K. Hellmann, formerly Chief of the Interstate Division for Baltimore City directed the project, Kenneth D. Merrill was Project Manager and Edward A. Terry was in charge for the FHWA. For Sverdrup-Parsons Brinckerhoff and Associates, T.R. Kuesel and E. Lemcoe were principals in charge and R.H. Hebenstreit was project manager.   V. Tirolo, H.W. Parker, G.A. Munfakh and R.J. Damigella were responsible for all geotechnical work.   For the Contractor, Kiewit/Raymond/Tidewater, Leon Heron was in charge.

References

1.   Di George F.P. and Herbich J.B. (1978) "Laboratory Determination
      of Bulking Factors for Texas Coastal Fine-Grained Materials" Texas
      A & M University Center for Dredging Studies CDS Report No. 218.

2.   Hayden T.A. (1978) "Prediction of Volumetric Requirements for
      Dredged Material Containment Areas" U.S. Army Engineers Dredged
      Material Research Program Technical Report DS-78-41.

3.   Krizek, R.J. Salem A.M. (1978) "Use of Dredging for Landfill.
      Technical Report No. 5 Behavior of Dredged Materials in Diked
      Containment Areas" Northwestern Univ. Dept. of Civil Engineering.

4.   Krizek R.J., Salem A.M. (1977) "Field Performance of a Dredgings
      Disposal Area" ASCE Geot. Practice of Disposal of Solid Waste
      Material June 1977.

CONVECTION-DIFFUSION ANALYSIS OF SEDIMENTATION
IN INITIALLY DILUTE SOLIDS-SUSPENSIONS

Raymond N. Yong[*] M.ASCE and Diaa S. Elmonayeri[**]

ABSTRACT

The settling of suspended solids in solids-suspensions is treated
from the viewpoint of the relative fluxes established as a result of the
upward diffusion of water and the downward convection of the solids. The
convection-diffusion analysis approaches the problem by first calculating
the fluid diffusion coefficients from actual laboratory solids-settling
experiments. The calculated diffusion coefficients are then used with
the convection-diffusion relationship to predict the settling rate of
the solids and also the solids concentration profile. Predictions for
various experiments and a field settling pond are compared with actual
measured values.

INTRODUCTION

The term "solids-suspension" refers to the physical phenomenon where
a suspending fluid medium contains material (generally identified as
solid particles, i.e. solids) which remain in suspension for very long
periods of time. If the suspended solids do settle at all, they will
settle at sedimentation rates not readily modelled by classical Stokesian
or modified Stokesian models. Typical situations where solids-suspension
behaviour arise include slime/sludge ponds, coastal sediments, and waste-
discharge containment lagoons.

In studying the development of sediments such as those obtained in
slime/sludge containment ponds and in many coastal areas (i.e. as coastal
sediments), their composition and physical/mechanical properties constitute major items of interest. Recognizing that the physical and mechanical properties of the sediments are both source and time dependent, it
has been found convenient to analyze the development of sediment layers
in respect to rates of settling of the solids "suspended" in the fluid
medium. By doing so, the analyses which require actual material properties input will thus implicitly factor in the resultant effects of source
material in the settling process which leads to the development of the
sediment layers [4,5,7,11].

---

[*] William Scott Professor of Civil Engineering and Applied Mechanics,
and Director, Geotechnical Research Centre, McGill University, Montreal,
Canada.
[**] Research Associate, Geotechnical Research Centre, McGill University.

When the concentration of solids in suspension is small, as is the case for many solids-suspensions derived from tailings discharge (solids concentration from 2 to 5 percent by weight), and when these solids contain surface-active material, Yong and Sethi (1978) have shown that the dispersion stability and settling behaviour of the solids are controlled more by surface-active relationships (i.e. by interactions between surface-active solids) than by gravitational mechanisms. The problem of model development which would permit one to evaluate and predict the continuous solids-settling process in initially dilute solids-suspensions, from solids sedimentation (suspension "thickening") to compact sediment, remains one of the more challenging problems facing analysts in this field. Present analytical methods using Stokesian, Kynchian and large-strain consolidation models [1,3,4,5,6,9] as separate and individual models for analysis of the various "stages" of solids sedimentation, testifies to the present state of prediction capabilities and reinforces the need to provide a better means for overall prediction of the continuous process of sediment formation.

In this study, the recently developed convection-diffusion method of analysis of solids-settling in initially dilute solids suspension [10], is used to predict solids-settling performance of mixed clay mineral suspensions. Comparisons are made between predicted settling performances and actual laboratory measured values. The method of analysis permits one to address the entire solids concentration range heretofore covered by: (a) the Stokesian model - for the more dilute solids concentration, and (b) hindered fall methods of analysis - for higher solids concentration. When the actual sediment layer is formed, large strain consolidation analyses can be applied successfully to predict consolidation performance.

CONVECTION-DIFFUSION ANALYSIS

Denoting the various known and unknown internal driving forces in the solids suspension in terms of a resultant potential difference $\Delta\psi$ acting between two arbitrary points separated by a distance $\Delta\zeta$, the relative fluid flux q_{fs}, - i.e. upward flux of fluid past the suspended solids which are themselves moving downward in the opposite direction to the fluid flux - can be expressed as:

$$\text{Relative fluid flux } q_{fs} \propto - \frac{\partial\psi}{\partial\zeta} \tag{1}$$

$$= -k\frac{\partial\psi}{\partial\zeta}$$

and
$$\frac{\partial q_{fs}}{\partial\zeta} = \frac{\partial}{\partial\zeta}(-k\frac{\partial\psi}{\partial\zeta}) \tag{2}$$

where k = proportionality constant.

The net result of the relative fluxes established because of the downward movement of the settling solids and upward movement of the fluid will be physically demonstrated in terms of the time-rate settling of the solids-liquid (fluid) interface as shown in Fig. 1. For simplicity in viewing, one might consider the total initial solids-suspension as a

stacking of an infinite number of thin "compressible" solids-suspension
layers with solids concentration varying in the fashion shown on the LHS
drawing in Fig. 1. The supernatant layer that is formed above the solids-
liquid interface at any one time is the result of the relative fluxes,
and is the accumulation of the fluid-release from the infinite number of
"compressing layers", as seen in the RHS drawing in Fig. 1.

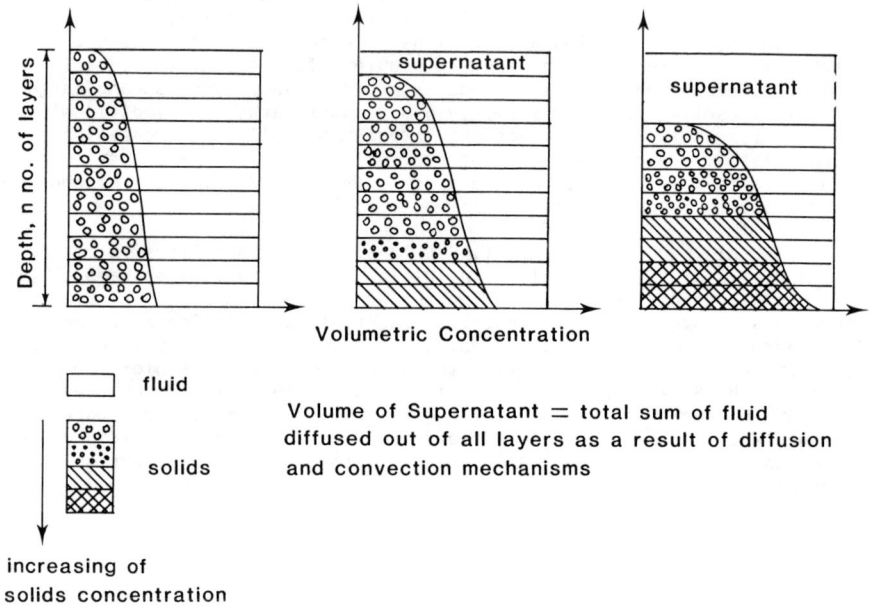

Fig. 1   Schematic representation of the solids settlement and super-
          natant development processes

        To elaborate further on the so-called "compressible" nature of the
solids-suspension layer, Fig. 2 shows a more detailed illustration of the
fluxes developed in the solids settling process in a control unit volume
in a typical solids-suspension layer. The quantity of fluid volume α_d
released in a unit time is the result of the upward fluid flux q_{fs} relative to the downward flux of solids q_s. The volume of fluid associated
directly with the solids consists essentially of two parts: (a) the part
that remains above the datum line separating layer C_{n-1} from C_n, and (b)
the part that moves in concert with the solids flux q_s in the settling
process, bringing them below the datum line into layer C_n - as shown in
Fig. 2.   The fluid associated with the solids represents that volume of
fluid held by the solids because of surface-active forces [11], and is
determined by the balance between the internal and external energy of the
local element.

Representing β as the volumetric content of the fluid associated with the convecting solids, i.e. volume of fluid directly tied in with the convecting solids in a unit solids suspension volume (Fig. 2) divided by that same suspension volume, the total flux of fluid associated with the settling solids passing below the datum separation into layer C_n will be given by $\beta q_s = \alpha_{cv}$. The total change in volumetric fluid content in the control volume after time Δt is due to (a) loss of fluid represented by the relative fluid flux q_{fs} and (b) the loss of fluid associated with the convecting solids as represented by the flux βq_s as shown in Fig. 2.

σ diffusing fluid

∴ solids flux with associated fluid

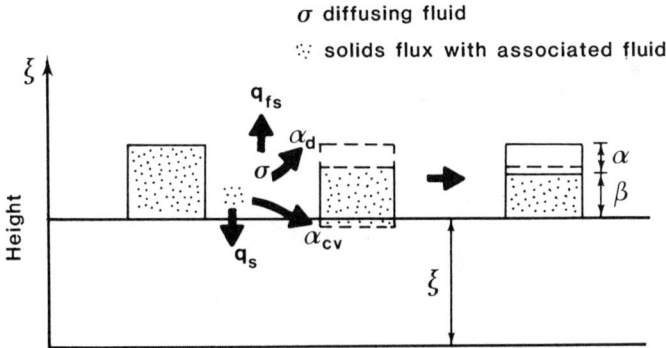

Fig. 2   Schematic representation of developed fluxes in a control volume, showing fluid flux relative to solids q_{fs} flowing in the upward side direction, and q_s solids flux leaving the control volume with associated fluid $\beta q_s = \alpha_{cv}$. Note that fluid flow due to $q_{fs} = \alpha_d$.

The total fluid volume which diffuses to the top of the settling column or pond to form the clear (solids-free) supernatant is the accumulation of all the α_d quantities from the individual layers. Because α_{cv} moves below the datum plane, the solids-liquid interface movement that will be visually evident is the result of the sum of all the α_d and α_{cv} quantities. The resultant fluid flux q_f has been previously given [10] as follows:

$$q_f = q_{fs} + \beta q_s \tag{3}$$

where

q_f = fluid flux,

q_{fs} = flux of fluid relative to the convecting solids,

q_s = flux of solids, and

β = volumetric content of fluid directly associated with the convecting solids.

Considering the movement of fluid in the ζ direction, (Fig. 2), the continuity condition can be written as:

$$-\frac{\partial \alpha}{\partial t} = \frac{\partial q_f}{\partial \zeta} = \frac{\partial (q_{fs} + \beta q_s)}{\partial \zeta}$$

$$= \frac{\partial q_{fs}}{\partial \zeta} + \frac{\partial (\beta q_s)}{\partial \zeta} \tag{4}$$

where

ζ = the upward positive spatial coordinate with origin of axis at the base,

$\alpha = \alpha_d + \alpha_{cv}$,

α_d = volumetric content of fluid lost (released) due to diffusive flow, and

α_{cv} = volumetric content of fluid lost due to convective flow.

Recalling Eq.(2):

$$\frac{\partial q_{fs}}{\partial \zeta} = -\frac{\partial}{\partial \zeta} (k \frac{\partial \psi}{\partial \zeta})$$

$$= -\frac{\partial}{\partial \zeta} (k \frac{\partial \psi}{\partial \alpha_d} \frac{\partial \alpha_d}{\partial \zeta})$$

$$= -\frac{\partial}{\partial \zeta} (D \frac{\partial \alpha_d}{\partial \zeta}) \tag{5}$$

where D = coefficient of diffusion of the diffusing fluid = $k(\frac{\partial \psi}{\partial \alpha_d})$

Combining Eqs.(4) and (5), we obtain the governing relationships previously given by Yong and Elmonayeri (1984):

$$\frac{\partial \alpha}{\partial t} = \frac{\partial}{\partial \zeta} (D \frac{\partial \alpha_d}{\partial \zeta}) - \beta \frac{\partial q_s}{\partial \zeta} - q_s \frac{\partial \beta}{\partial \zeta} \tag{6}$$

we should note that the applicability of this relationship to fully account for sediment consolidation after sediment formation and initial compression, will need to be critically examined - especially when actual interparticle contact becomes a significant issue. In its present form, experiments show that the convective-diffusive relationship can model sedimentation of the pure clays-suspensions tested - to void ratios of about 3.

DETERMINATION OF DIFFUSION COEFFICIENT

Since the convective terms on the RHS of Eq.(6) deal with the solids flux and associated effects, determination of the fluids diffusion coefficient D can be readily accomplished by considering only the diffusive term on the RHS, - so long as the Peclet number, which refers to the ratio of the convective velocity to the diffusion constant over the

domain of interest, is small. The property of maximum principle which
is well-demonstrated when the Peclet number remains small, is indeed
important if non-oscillating numerical solutions are to be sought. It
is useful to note that the standard diffusion model is also the generic
model used in the derivation of the classical consolidation relationship.

Solving Eq.(6) in the usual form of a diffusion equation, by drop-
ping off the last two convective terms on the RHS, Yong and Elmonayeri
(1984) have used a similarity solution technique to obtain the relationship between the diffusion coefficient of the fluid D with the total
fluids concentration at any one time. Note that the fluid concentration
θ is given in terms of a volume ratio - i.e. volume of fluid per unit
volume of total solids-suspension - and is also easily identifiable as
the porosity n. The diffusion coefficient D is obviously a variable
since it is directly related to the proportion of solids present (i.e.
solids concentration). The relationships for D for various suspensions
are given later in Fig. 5 - in terms of the volumetric water content.

The procedure used to obtain the necessary information for calculation of the diffusion coefficient for any one particular solids or fluid
concentration in the solids-suspension, requires experimental input
which describes the rate of fall of the suspended solids. Consequently,
laboratory settling column experiments similar to those previously conducted [10] were used. These consist of essentially producing solids-
suspensions of various initial solids concentrations, using pure clay
minerals dispersed in 15 meq/l $NaHCO_3$. The settling columns which consisted of lucite tubes measuring 208 mm diameter and 600 mm high had
sampling ports located at various heights of the column - to permit
periodic sampling of the solids-suspension for determination of solids
or fluids concentration profile. Since the volume of suspension removed
via a macro-hypodermic syringe was considered to be miniscule in relation
to the total volume of suspension material tested, no corrections were
considered necessary in measurements of the height of the solids-liquid
interface. A listing of the experiments conducted is given in Table 1.
Since it was easier to weigh the basic constituents, the fluid concentrations were reported in terms of water contents instead of the volume concentrations used in the calculation procedures. With known specific
gravities, it was an easy procedure to convert the water contents to
volumetric fluid concentrations.

The typical kind of solids-settling information obtained in experimentation is shown in the top and LHS drawings shown in Fig. 3. The
profiles drawn can be in terms of density variation or solids concentration variation with height of settling column. For convenience in calculation of the diffusion coefficients, the information presented in
Fig. 3 is given in terms of volumetric concentrations, e.g. volume of
solids divided by the total suspension volume (Fig. 3a) for the volume
element under scrutiny. Since the sampling ports provide information
with respect to fluid or solids concentration, Fig. 3b can be determined
directly. This corresponds with Fig. 3a since the sum of these two concentrations should provide one with the total solids-suspension volume.
By measuring the fluid concentrations at various times and locations, it
is possible to produce information concerning the diffused fluid concentration at any particular time period (Fig. 3c). This represents the
quantity referred to in the procedure for calculation given in Fig. 4.

TABLE 1 - Tests Conducted for Solids Settling

1	2	3	4	5	6
Experiment	Source	Mass of Solids (particles) gm	Initial Water Content w/w	Initial Height mm	Duration of Test hr
1		1012.1	1900%		
2		1222.1	1566.7%		
3	GRC	1435.1	1328.6%		
4	Experiment	1667	1150%	600	15.5
5	Set #1	2568	733.33%		
6		511.5	613.9%		
7	GRC	652	455.5%		
8	Experiment	935	316.6%	350	24.0
9	Set #2	1078.5	257%		
10		1170.5	233.3%		
		FD[*] + DSB[**]			
13		0 + 647.5	796.2%		
14		53.5 + 593.5	"		
15		107.5 + 539.5	"		
16		160.5 + 485.6	"		
17		215.8 + 431.6	"		
18	GRC	269.8 + 377.7	"		
19	Experiment	323.7 + 327.7	"	600	24.0
20	Set #3	377.7 + 269.8	"		
21		431.61+ 215.8	"		
22		485.6 + 160.5	"		
23		539.5 + 107.5	"		
24		593.5 + 53.5	"		
25		647 + 0	"		

[*] FD = Hydrite Flat D Kaolinite ("pure" kaolinite)

[**] DSB = Domtar Sealbond

To calculate or predict the settling rate of the interface established by the supernatant lying atop the solids-suspension (due to continuous accumulation of diffusing fluid), the fluids diffusion coefficient D, which is obtained as a function of α_H the diffused fluid volumetric content, is used to enter into the solution of the convection-diffusion equation - given previously as Eq.(6). At this time, the total equation (together with the convective terms) is used in the calculations. This is identified in Fig. 4.   The output information from solution of the equation permits one to view settling rate, fluid or solids concentration profiles at any specified time, and obviously sediment formation. Note that the calculation procedure used in Fig. 4 stops when the solids concentration reaches 40%.   This point is arbitrary in that we expect that particle contact forces will begin to become relevant issues and perhaps will need separate consideration. This is not to say that the governing equation will not operate beyond this region. However, much work remains to be done to further elaborate on the convective-diffusive phenomenon when particle contact stresses and excess pore water pressures become significant.

CONCLUDING DISCUSSION

The diffusion coefficients previously calculated for tests 4, 5 and 10 in Table 1, and included in calculations with other test material types used for comparison with field test results [10] have been shown to apply for all the kaolinite suspension test results obtained in this study. This is not surprising, since the tests were all concerned with the same type of clay material. The diffusion coefficients are shown in Fig. 5 - together with comparisons with other suspensions reported previously. Note that by and large, the slopes of the various D relationships are essentially similar. The relationship calculated for D using the test results reported by Been and Sills (1981) is basically identical in slope to the kaolinite test series obtained in this study. The same can be said for the D values calculated for actual field experiments conducted [10].

Using the kaolinite D values reported in Fig. 5, predictions for settling rate can be made and compared with test measurements - as for example for tests 4 and 5 in Table 1. The close agreement between measurement and prediction shown in Fig. 6 is evident for all the samples studied. A better summary is given in Fig. 7 where the slopes of the initial settling curves (i.e. settling velocity) represented in Fig. 6

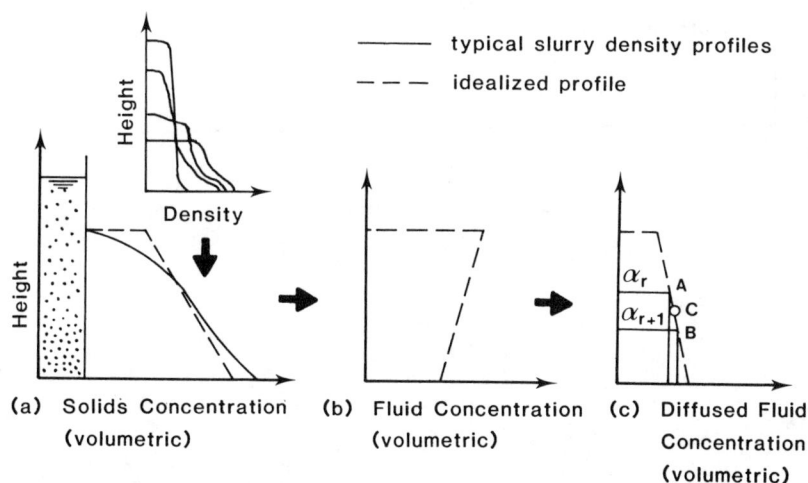

Fig. 3   Idealized representation of solids settling at any one time, together with distribution of solids, fluid and diffused fluid.

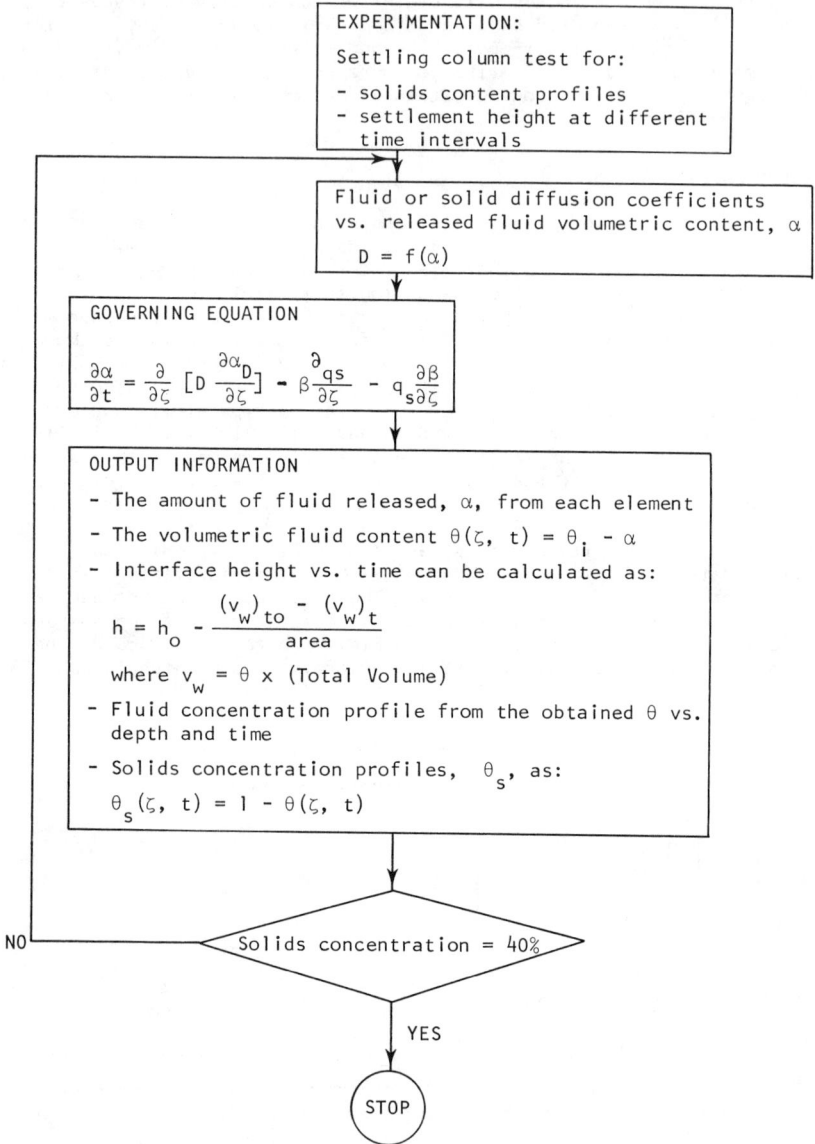

EXPERIMENTATION:

Settling column test for:

- solids content profiles
- settlement height at different time intervals

Fluid or solid diffusion coefficients vs. released fluid volumetric content, α

$$D = f(\alpha)$$

GOVERNING EQUATION

$$\frac{\partial \alpha}{\partial t} = \frac{\partial}{\partial \zeta} \left[D \frac{\partial \alpha}{\partial \zeta} \right] - \beta \frac{\partial q s}{\partial \zeta} - q_s \frac{\partial \beta}{\partial \zeta}$$

OUTPUT INFORMATION

- The amount of fluid released, α, from each element
- The volumetric fluid content $\theta(\zeta, t) = \theta_i - \alpha$
- Interface height vs. time can be calculated as:

$$h = h_o - \frac{(v_w)_{to} - (v_w)_t}{\text{area}}$$

where $v_w = \theta \times (\text{Total Volume})$

- Fluid concentration profile from the obtained θ vs. depth and time
- Solids concentration profiles, θ_s, as:

$$\theta_s(\zeta, t) = 1 - \theta(\zeta, t)$$

NO

Solids concentration = 40%

YES

STOP

Fig. 4   Schematic diagram illustrating procedure for calculation

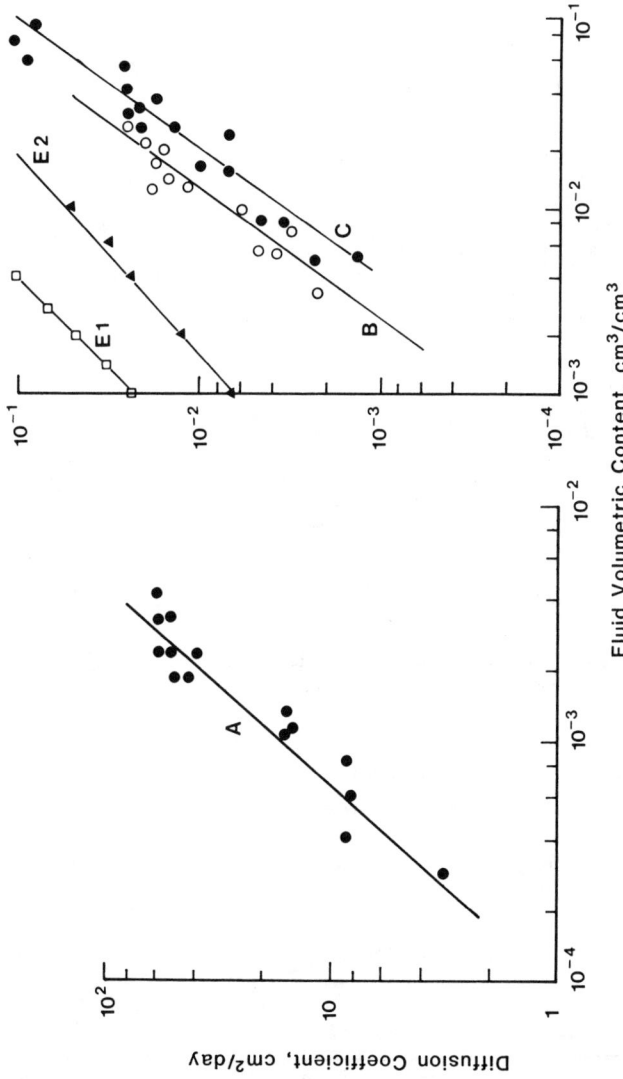

Fig. 5 Diffusion coefficients for "A" tar sand sludge, "B" kaolinite, "C" Been and Sills Tests, and E1 and E2 (in cm²/sec) for Domtar Sealbond/Kaolinite mixtures (Exp. #15 and 19 respectively)

for example, are plotted against the initial solids concentration of the solids-suspension. This shows the effect of initial solids concentration on the velocity of settling of the solids-liquids interface. Although not directly shown, the close accord represented in Fig. 6 between predicted and measured values is obviously repeated for Fig. 7.

To further test the applicability of the method of analysis, two other mixtures (reported in Table 1) were used. These consisted of a mixture of kaolinite and illite minerals. The hydrite flat D kaolinite was mixed with illite, identified as Domtar seal bond, in two proportions as shown in Fig. 8. The diffusion coefficients used were the ones calculated from experiments with the mixtures, and the resultant predictions are shown together with the measured rate of fall in Fig. 8. Whilst the agreement between prediction and measurement is not as good - in comparison with the pure kaolinite samples - it is nevertheless acceptable. The bottom boundary effect in constraining sedimentation can be seen in the lower portions of the experimental curves.

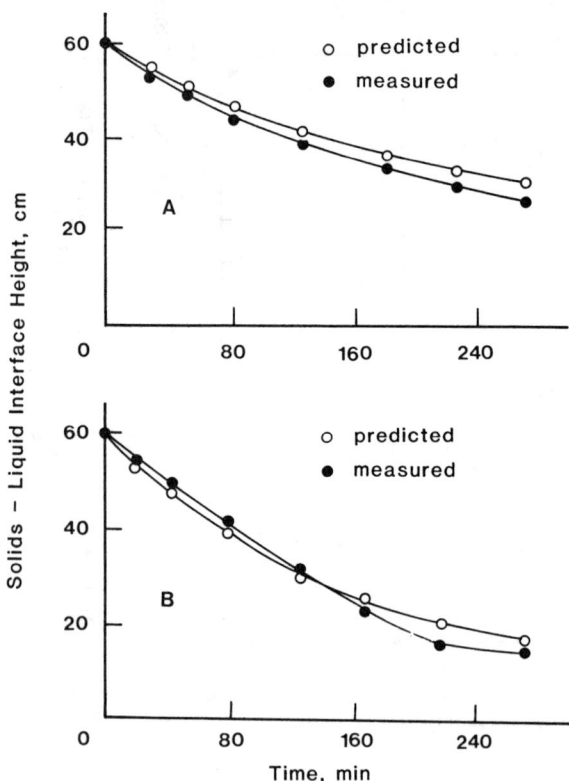

Fig. 6   Comparison between predicted and measured settling rates for Tests 4 and 5, reported in Table 1.

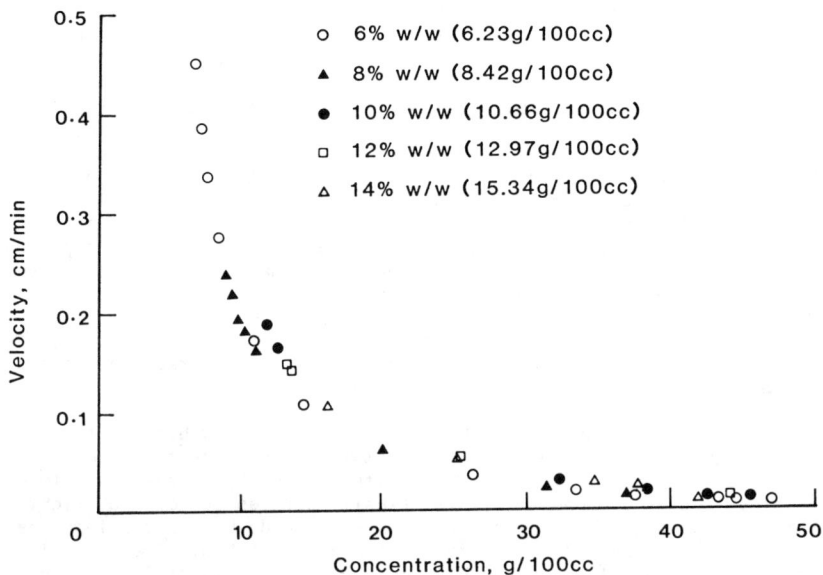

Fig. 7   Relationship between the interface settling rate and its concentration for different initial uniform solids concentration

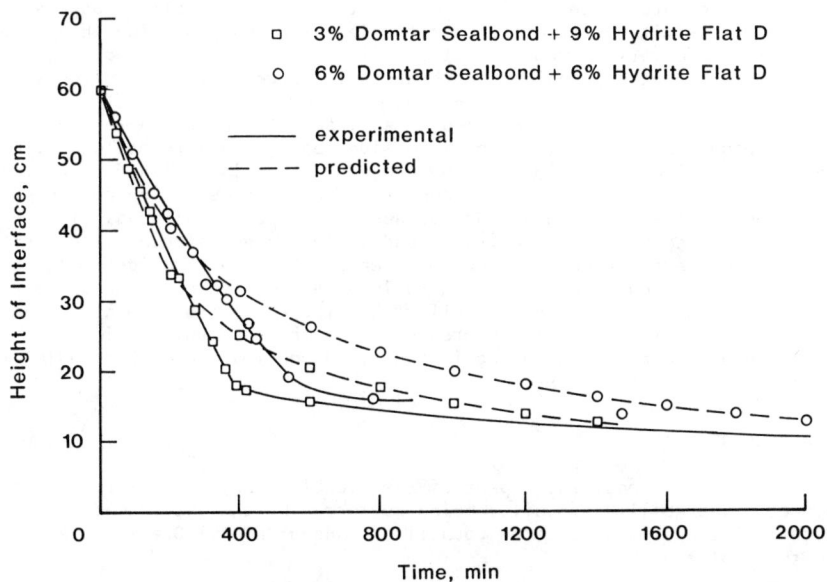

Fig. 8   Experimental and predicted height of interface for kaolinite and illite mixtures

Much work remains to be done to further develop the method of analysis - as witness the test/prediction results in Fig. 8. Because of
the addition of a more surface-active material component (illite), complete demarcation between pure supernatant and solids-suspension is not
always achieved, i.e. the solids-liquid interface is not clearly evident.
A slightly turbid supernatant is developed and some of the more active
component material remains in the supernatant. Thus visual observations
of the first part of the initial settling process tends to lag behind
predictions if visual comparisons are made with predictions - since observations can only be made by viewing the interface which is visible
because of the more dominant settling component (kaolinite). The process
is akin to the more drastic "settling-out" phenomenon seen in sand-clay-
fluid suspensions, where the granular component segregates out of the
total suspension and settles through the suspension to form the sand
sediment layer. Bottom sediments showing the general gradation of coarse
fractions at the very bottom, grading to the very fine fractions (silts
and clays) at the top testify to this phenomenon.

As time progresses, a "catching-up" process occurs because the
initial higher settling rate due to the more dominant settling component
(kaolinite and coarser fractions of illite) obviously reaches the stage
where the lagging component (illite) finally settles out of the slightly
turbid supernatant. The interface which now can be better distinguished
will be more representative of the total mixture of kaolinite and illite.
If the general modelling approach is indeed viable, the differences between prediction and measurement should become diminishingly small as
time progresses. This appears to apply for the study conducted on the
mixed system. It would indeed be interesting to see if one could predict
the settling rate of the various components in a mixed solids-suspension.
This would permit one to anticipate the layering and gradation phenomenon
in sediment formation. Further study will need to be performed to address this and other problems concerned with large convective effects.

Figure 9 shows the prediction of solids distribution in tar-sand
sludge pond (about 2 km long and 1 km wide) using the information given
by Yong and Elmonayeri (1984). The procedure used is outlined in Fig. 4,
and the values for D were determined from laboratory settling column
tests on samples obtained at various depths in the actual pond. The
mixed solids components (sand, silt and various kinds of clay minerals)
in the pond have all been lumped together as a general "solids" content.
It is useful to note that the void ratio of 4.3, which corresponds to a
solids concentration of 19% is well within the 40% cut-off point shown
in Fig. 4. Agreement between predicted and measured void ratios (as
determined from actual sampling in the field pond) shown in Fig. 9 is
indeed good.

ACKNOWLEDGEMENTS

This study constitutes part of the overall study on the Fabric and
Structural Stability of Soils, being supported by a Grant from the
Natural Sciences and Research Council of Canada (NSERC) Grant No. A-882,
to the senior author.

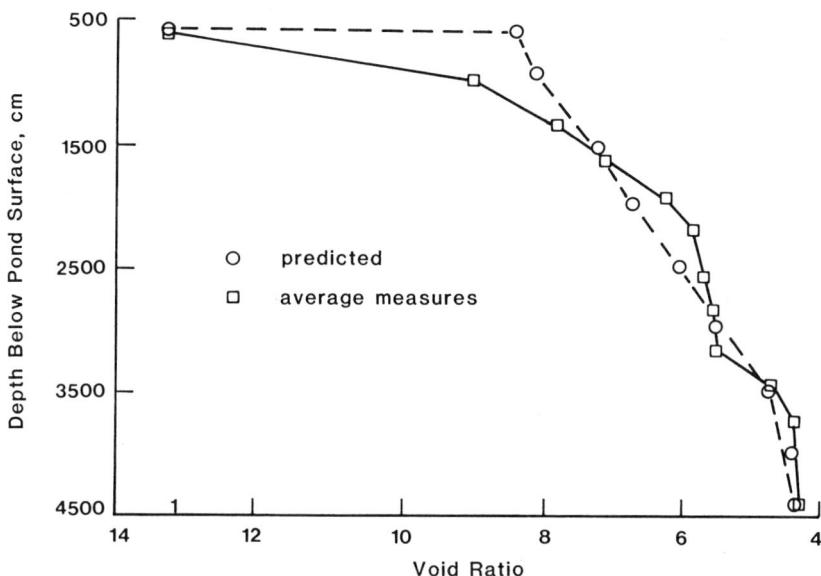

Fig. 9   Comparison between predicted and measured void ratio profile in
         tar sands tailing sludge pond

Appendix I - REFERENCES

1. Been, K. and Sills, G. C., "Self-weight consolidation of soft soils:
    an experimental and theoretical study". Geotechnique, Vol.31, No.4,
    1981, pp. 519-535.
2. Crank, J., "The Mathematics of Diffusion". 2nd Ed., Clarendon Press,
    Oxford, 1975.
3. Gibson, R. C., Schiffman, R. L. and Cargill, K. W., "The theory of
    one-dimensional consolidation of saturated clays: II. Finite non-
    linear consolidation of thick homogeneous layers". Canadian Geo-
    technical Journal, Vol.18,
4. Kynch, G. J., "A theory of sedimentation". Trans. Faraday Society,
    Vol. 48, pp. 166-176.
5. Richardson, J. F. and Zaki, W. N., "Sedimentation and fluidization".
    Transactions Institute of Chemical Engineers, Vol.2, 1954, pp.35-53.
6. Tiller, F. M., "Revision of Kynch sedimentation theory". AIChE Jour.
    Vol. 27, No. 5, 1981, pp. 823-829.
7. Yong, R. N. and Sethi, A. J., "Mineral particle interaction control
    of tar sand stability".Journal Canadian Petroleum Technology, Vol.
    17, No. 4, 1978, pp. 1-8.
8. Yong, R. N., Sethi, A. J., Ludwig, H. P. and Jorgensen, M., "Inter-
    particle motion and rheology of dispersive clays". ASCE Journal Geo-
    technical Engineering Division, Vol. 105, 1979, pp. 1193-1209.

9. Yong, R.N., Siu, S. K. and Sheeran, D. E., "On the stability and settling of suspended solids in settling ponds. Part I. Piece-wise linear consolidation analysis of sediment layer". Canadian Geotechnical Journal, Vol. 20, No. 4, 1983, pp. 817-826.

10. Yong, R. N. and Elmonayeri, D. S., "On the stability and settling of suspended solids in settling ponds, Part II. Diffusion analysis of initial settling of suspended solids". Canadian Geotechnical Journal, Vol. 21, No. 4, 1984. (In Press).

11. Yong, R. N., "Particle interaction and stability of suspended solids". ASCE Special Publication on SEDIMENTATION/CONSOLIDATION MODELS. 1984.

CONSOLIDATION OF MINING WASTES

Leslie G. Bromwell*, MASCE

ABSTRACT

Large tonnages of fine-grained mineral wastes, in dilute slurry form, are produced annually by many mining, milling, and ore processing operations. Disposal of these materials requires construction of large impoundments. Knowledge of the sedimentation and consolidation properties of the wastes is needed in order to provide adequate storage capacity (design life) for impoundments; calculate stability of the impounding dam; estimate seepage quantities; determine land reclamation and re-use options; and to evaluate alternative processes for enhancing consolidation and stabilization.

This paper describes laboratory and field tests for determining consolidation properties of these very soft sediments, presents data on several types of mining wastes, and presents the results of several case studies utilizing measured parameters and finite strain consolidation theory to predict field consolidation performance. Applications of the approach for improved waste disposal and land reclamation plans are described.

INTRODUCTION

Many mining, milling, and ore processing operations produce large quantities of fine-grained wastes, generally termed slimes, sludge, or mud. The Florida phosphate industry, for example, produces over 50 million tons (dry weight) of highly plastic waste clay slurry each year. Other mining industries that produce fine-grained plastic wastes include various metals such as copper, lead and zinc, uranium, gold, and titanium, as well as non-metals, such as trona, potash and sand and gravel operations. Coal washing operations in some areas also produce a fine-grained waste (generally non-plastic), and the chemical processing of bauxites and phosphates produce so-called ferro-silt (red mud) and phosphogypsum (gyp), respectively.

These fine-grained wastes present interesting and challenging geotechnical problems. Disposal of the wastes requires construction of impoundments, commonly termed ponds, settling areas, or stacks. Perimeter dams are generally constructed of either mining wastes (overburden or waste ore) or of coarse-grained tailings sands or silty sands.

*Principal, Bromwell & Carrier, Inc., 202 Lake Miriam Drive, Lakeland, Florida 33803

The fine-grained wastes generally report to the impoundment as a slurry with moisture contents ranging from 150% to as high as 3000%, depending largely upon (1) the amount of water added during ore processing operations and (2) viscosity requirements for slurry pumping or gravity flow. These large values of moisture content are generally not easily comprehended nor convenient to work with. Consequently, the term solids content, S, defined as (weight of solids/weight of water) x 100%, is frequently used. For saturated slurries, the moisture content (w), void ratio (e), and solids content are related as follows:

$$e \; = \; Gw \; = \; G\left(\frac{100}{S} \; - \; 1 \right) \ldots \ldots \ldots (1)$$

The consolidation behavior of fine-grained waste influences the design life of a disposal facility, the stability of the perimeter dam, quantity of seepage, land reclamation timing and land re-use options. All of these factors have become increasingly important, both from economic and environmental considerations for mining operations in the U. S. and elsewhere.

Materials that sediment to high void ratios and then consolidate slowly under self-weight stresses

- require large volumes of disposal area per unit mass of waste,
- exhibit low shear strength during filling of a disposal area,

and

- require long times for surface stabilization and reclamation.

As mentioned previously, the Florida phosphatic clays are an extreme example of poor consolidation behavior. Other, less plastic, mining wastes exhibit similar but less dramatic behavior. The following sections of this paper describe techniques developed for determining the consolidation properties of fine-grained mining wastes, and present several case studies of predicted and observed field consolidation behaviors.

SEDIMENTATION BEHAVIOR AND INITIAL VOID RATIO

After placement of a dilute slurry into an impoundment, sedimentation occurs to a condition where particles begin to transfer stress. Since sedimentation rates, even for the phosphatic clays, are on the order of several feet per day, the initial sedimentation appears to occur very quickly compared to rates of consolidation. In a matter of days, phosphatic clays typically go from a slurry void ratio of around 90 to a sedimented value on the order of 30. Other, less plastic wastes such as copper slimes, may go from a slurry void ratio of about 8 to a sedimented value on the order of 2 within a day.

The sedimentation behavior of clays can be dramatically affected by physico-chemical factors, such as mineralogy, cation type and concentration, and pH. Florida phosphatic clays are generally saturated with calcium and magnesium cations at near-neutral pH. Replacement

of these ions with sodium can produce a dispersed slurry that will
settle very slowly, if at all. An example of the effects of mineralogical composition on sedimentation behavior is shown in Figure 1.

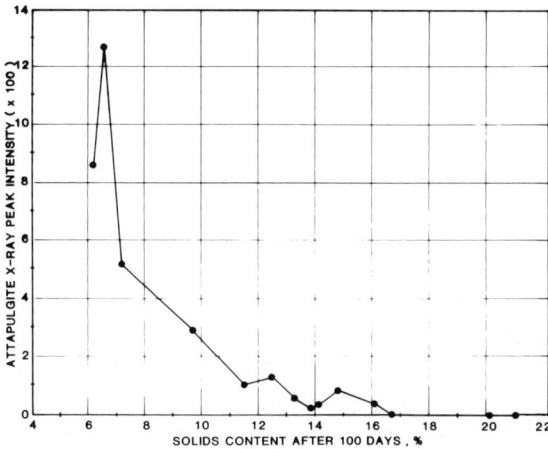

Fig. 1   Influence of Attapulgite on Sedimentation
of Phosphatic Clays

Samples of phosphatic clay slurry from 14 Florida mines were diluted
to a standard 3% suspension, using supernatant water from the plant
output, and subjected to standard 2000 mL graduated cylinder settling
tests. The samples containing higher percentages of attapulgite clay
mineral, as determined from X-ray diffraction peak intensities, sedimented to much lower solids contents (6% to 8%) than those samples
that showed no attapulgite (17% to 21%).

        In practice, the rate of sedimentation does not have a strong in-
fluence on the design capacity of a waste impoundment, since the useful life compared to sedimentation time is usually a matter of years
compared to days. Also, no correlation between poor settling behavior
and subsequent self-weight consolidation behavior has been observed.

INFLUENCE OF PLASTICITY ON CONSOLIDATION PROPERTIES

        Atterberg Limits for various fine-grained mining wastes are shown
on the plasticity chart in Figure 2. Low plasticity ML-CL materials
include most of the metals slimes, which are generally finely-ground
rock with some clay minerals; and alumina red mud, which is largely
silt-sized degraded aluminum and from silicates.

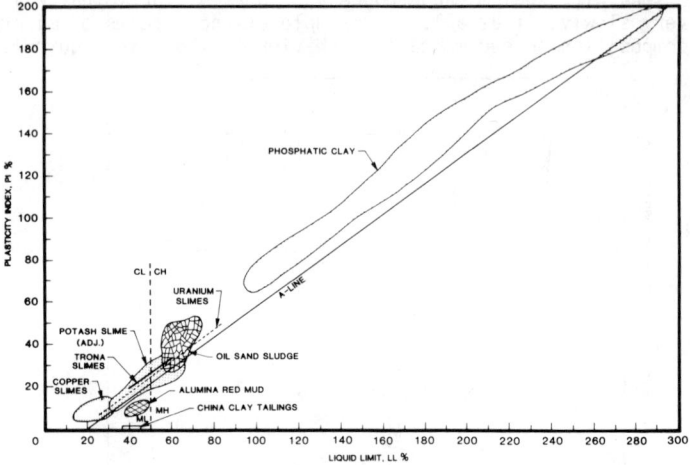

Fig. 2   Atterberg Limits of Fine-Grained Mining Wastes

The higher plasticity wastes include oil sand sludge, consisting of fine silt, kaolinite, residual bitumen, and some smectite; and phosphatic clays, composed of silt-size quartz and apatite, with principal clay minerals smectite, palygorskite (attapulgite), illite, and kaolinite. Depending upon the amount and types of clays present, the plasticity index of phosphatic clay samples can range from about 60 to 200.

The large range of plasticity for the phosphatic clays has enabled correlations to be developed between plasticity (PI) and consolidation properties. Carrier, et al (6) present power function relationships as follows:

$$e = 17.7G \left(\frac{PI}{100} \right) \bar{\sigma}^{-0.29} \quad \ldots \ldots \ldots \ldots (2)$$

and

$$k = \left[95.2G \; \frac{PI}{100} \right]^{-4.29} \frac{e}{1+e}^{4.29} \quad \ldots \ldots (3)$$

where $\bar{\sigma}$ is the vertical effective stress in pascals, and k is the permeability in meters per second.

These correlations have been found to describe fairly well the compressibility (e - $\bar{\sigma}$) and permeability (k - e) relationships for a wide range of waste slurries, and have been used extensively for preliminary design of disposal facilities and to estimate consolidation times for long-range waste disposal and reclamation planning and design.

LABORATORY AND FIELD MEASUREMENTS

Conventional soil mechanics testing equipment procedures are frequently inapplicable for these materials. Over the past ten years, several new devices have been developed that have proven useful for measuring consolidation properties of fine-grained wastes in the laboratory and the field. The most useful laboratory device is the slurry consolidometer, basically a sedimentation/consolidation cell that can apply small stress increments (on the order of 10 gm/cm^2) at high void ratios, and can accommodate large height changes of the sample during testing. Several types of slurry consolidometers that apply constant vertical loads in increments were developed during the 1970's both for testing dredged materials (10) and for phosphatic clays (9). Recently, R. L. Schiffman at the University of Colorado has developed a constant-rate-of-deformation slurry consolidometer that has proven useful for testing a variety of fine-grained wastes (15).

The results of slurry consolidation tests on several fine-grained mining wastes are shown in Figures 3 and 4.

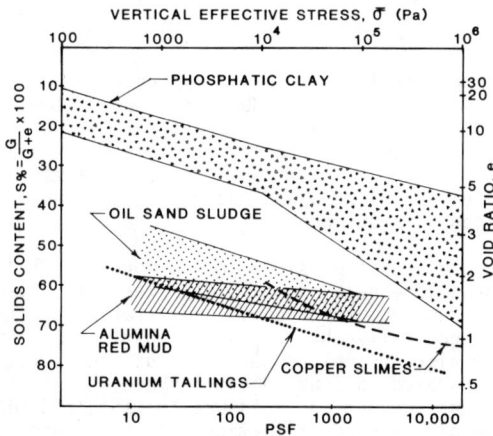

Fig. 3 Compressibility of Fine-Grained Mining Wastes

These log-log plots show a large range of compressibilities and permeabilities for these materials. The test results include much higher void ratios and much lower effective stresses than are generally encountered in natural soil deposits.

The initial points on the compressibility and permeability curves are obtained by observing sedimentation of slurry samples of differing initial heights, with no imposed load (3). Although not precise, calculation of average effective stress and void ratio from the settled volume, and of permeability from a finite strain simulation of the settling test itself, provides useful data at very low effective stresses.

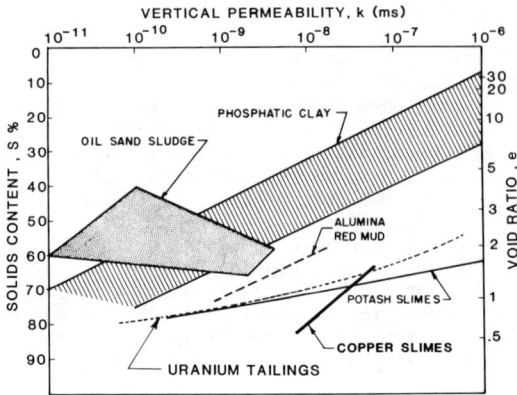

Fig. 4   Permeability of Fine-Grained Mining Wastes

FIELD SAMPLING

Obtaining representative samples of mining wastes for testing involves special considerations. Standard penetration test borings generally are of little value, as blow counts are essentially zero and the softer materials frequently cannot be retained in the spoon sampler. If a lens or pocket of coarse or dense material is penetrated, the spoon will likely be plugged and underlying soft material merely displaced as the spoon is advanced.

Undisturbed sampling is possible in the more consolidated slimes, using either thin-walled Shelby tubes or samplers with special foil or fabric liners. However, for the high void ratio materials, disturbed samples are generally obtained using bottle samplers or special small-diameter piston samplers (5). Because the samples are essentially fully saturated, disturbed samples are satisfactory for determining in situ unit weight and moisture content, and for preparing remolded slurry samples for consolidation testing. These slurries are prepared by compositing a large number of representative samples from a disposal area, using supernatant pond water for dilution. Although physico-chemical changes after deposition might make this procedure questionable for some materials, it has not proven to be a problem with recently-deposited mineral wastes.

In order to project future waste disposal requirements, representative samples of waste slurry are obtained from the plant or mill output and tested. By running characterization tests, particularly Atterberg Limits and percent minus No. 200 sieve, on a number of samples taken over a period of time, the range of consolidation behavior of the wastes being produced can be estimated, and representative composite samples prepared for slurry consolidation testing. If prospect samples of ore from future mining areas are available, they can be processed through a metallurgical laboratory or pilot plant, and the slimes thus produced collected and tested as described above for fresh plant slimes.

FIELD MEASUREMENTS

Field measurements of consolidation behavior can be made using piezometers, electrical pore pressure probes, nuclear density probes, and other devices that can accurately measure *in situ* pore pressures, moisture content, or density.

The piezometer probe (2, 14) with a small diameter extended porous tip, has proven very useful for obtaining complete pore pressure profiles in a relatively short time. Sediment density, although generally obtained from laboratory testing of recovered samples, can also be profiled with depth by use of a nuclear density probe. Such a device, with a conical tip containing a gamma source similar to that commonly used for field density testing of compacted materials, and a sensitive scintillation counter, can be pushed to a desired depth, the counter activated for a short time period (about one minute), and the probe then pushed to the next depth.

By profiling both pore pressure and density with depth, void ratio can be correlated with effective stress at a large number of points within the deposit. In this manner, a direct field determination of compressibility of the waste material can be made. Figure 5 shows the results of such measurements at a large phosphatic clay field test area in Central Florida (1). Comparisons of the field data with the results of a constant rate of deformation laboratory consolidation test, and with the compressibility relationship calculated from Atterberg Limits are also shown. The agreement is seen to be quite good, and in fact, is typical of results obtained in several types of mineral wastes.

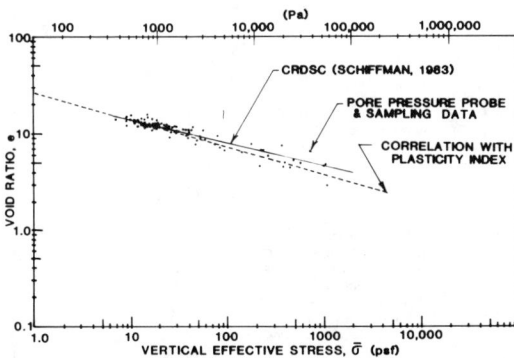

Fig. 5 Comparison of Compressibility Data -
Phosphatic Clay

The pore pressure and density profiles can also be repeated at
various times in order to measure the rate of dissipation of excess
pore pressures and the resulting consolidation. The results obtained
can then be simulated by use of finite strain computer programs (subsequently described) and the field permeability relationship thus
"back-calculated".

Direct field measurements of permeability can also be made at
various depths by use of falling head (or, preferably, rising head)
tests in wellpoints or porous tip piezometers, using very small gradients in order to avoid consolidating or fracturing the sediment.
Figure 6 shows a comparison between direct field measurements of permeability and laboratory measurements on slurry samples of phosphatic
clays. The consistency of the measurements within any data set is
quite good, and the band of results, we now realize, can be understood
in terms of variations in plasticity.

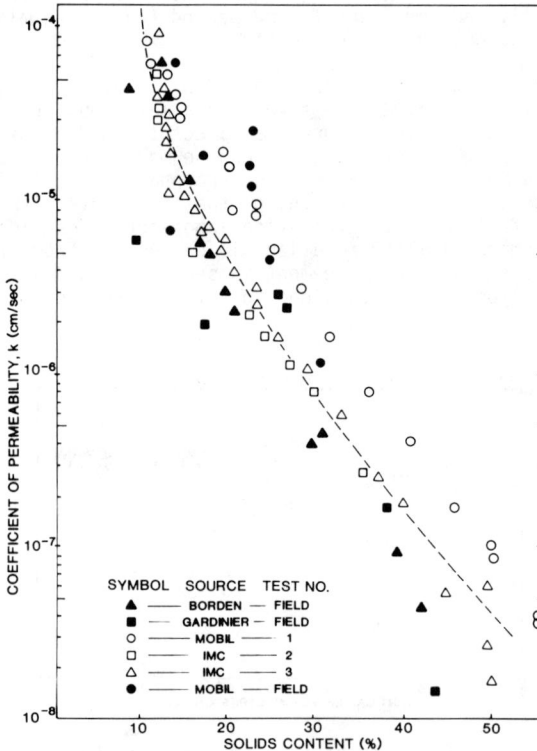

Fig. 6 Laboratory and Field Permeability Values -
Phosphatic Clays

The decrease in permeability as consolidation proceeds can be very large for these materials, as shown in Figure 7, which shows the relationship between permeability and vertical consolidation stress for a typical phosphatic clay sample. The permeability decreases by a factor of about 1000 over the effective stress range likely to be encountered in a deep disposal pond.

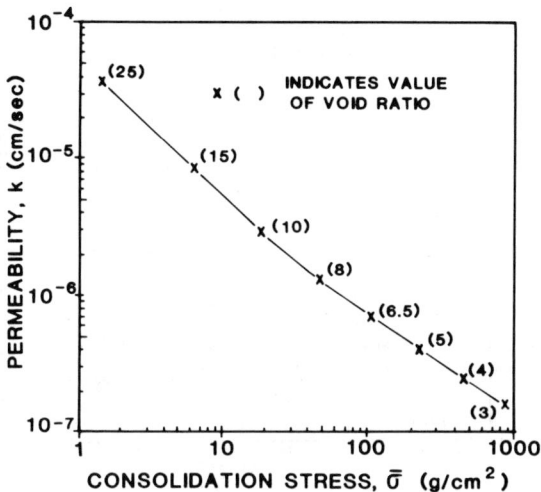

Fig. 7   Permeability Vs. Consolidation Stress - Phosphatic Clay

PREDICTIONS OF CONSOLIDATION BEHAVIOR

Conventional solutions to Terzaghi's consolidation theory assume a constant coefficient of consolidation (c_v); infinitesimal strain, and zero self-weight. For high plasticity mining wastes, these assumptions are not valid and the predicted consolidation rates are too inaccurate to be of any value.

Instead, a non-linear large (or finite) strain formulation must be used. Such a formulation was developed by Gibson, et al (8), and computer programs have been written by Somogyi (13) and Schiffman (12). Somogyi's solution involves a fully implicit finite difference method, whereas Schiffman's involves an explicit finite difference method. Both programs have been used to model a well-documented field test involving a large tank filled with a uniform slurry of Florida phosphatic clay. The initial height of clay was 6.4 m and the initial solids content was 12.6%. After completion of the test, field measurements of solids content and pore pressure distribution were used to back-calculate the best-fit power law relations among e, $\bar{\sigma}$ and k. These parameters were used by both Somogyi and Schiffman to predict settlement vs. time. Also, an average c_v value was calculated over the range of effective stresses observed in the test.

Figure 8 shows the predicted settlement by three methods: Somogyi program, Schiffman program, and conventional theory. The actual field measurements are also shown for comparison. The close agreement between the Schiffman solution and the field measurements is evident, as is the much slower, inaccurate rate of consolidation predicted by conventional theory. The Somogyi solution underestimates the rate of consolidation during early stages, but considering the fact that the writer obtained the Somogyi and Schiffman settlement predictions by scaling data from published graphs, the two computer solutions, for this case at least, give similar results.

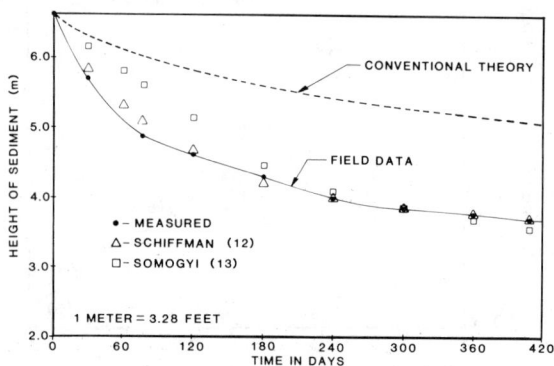

Fig. 8   Comparisons of Predicted and Measured
Settlement of Phosphatic Clay

The difference between consolidation rates predicted by finite strain theory (Somogyi program) and conventional theory is further illustrated in Figure 9, which shows the calculated time to 90% consolidation as a function of initial height. Finite strain theory predicts a much smaller influence of height on consolidation time than does conventional theory, with consolidation times proportional to about $H_o^{1.3}$ rather than H^2. For a 100-foot thick deposit having single drainage, the difference in predicted consolidation times is dramatic: 20 years by finite strain analysis vs. more than 1000 years by conventional analysis. Also, the figure shows that the predicted time difference between single and double drainage is much less using finite strain analysis than by conventional theory.

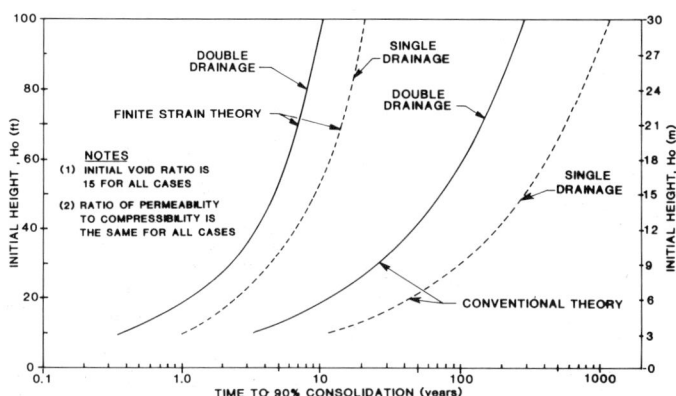

Fig. 9   Calculated Consolidation Time Vs.
Initial Height - Phosphatic Clay

MODELLING FIELD DISPOSAL SITUATIONS

The Somogyi computer programs have been available since 1978, and
have been used to model numerous disposal situations by the writer and
co-workers (3, 5, 6). The programs require the following input: (1)
compressibility and permeability parameters for the waste material;
(2) void ratio at the start of consolidation; (3) size of disposal
area; (4) slurry inflow rate (mass of solids per unit time); (5) total
inflow time; (6) surcharge, if any; and (7) boundary drainage conditions. The output from the programs include: (1) settlement vs. time;
(2) void ratio vs. depth and time; and (3) pore pressure vs. depth and
time.

A typical plot of predicted void ratio vs. depth for various
times following filling is shown in Figure 10. This case is for a
pervious base having the same boundary total head at the top and bottom; i.e., no imposed seepage gradient.

The rapid development of a high solids content layer about 1-
foot thick at the bottom drainage boundary is also shown in this example. Such "cake" formation results from the large decrease in void
ratio that accompanies the rapid increase in effective stress near the
drainage boundary. This phenomenon has been observed in many field
situations involving highly compressible slurries. The large decrease in permeability of the "cake" essentially eliminates the benefits of the free-draining boundary. This effect is automatically accounted for by non-linear finite strain solutions, and cannot be predicted by conventional theory or by any of its variations (12).

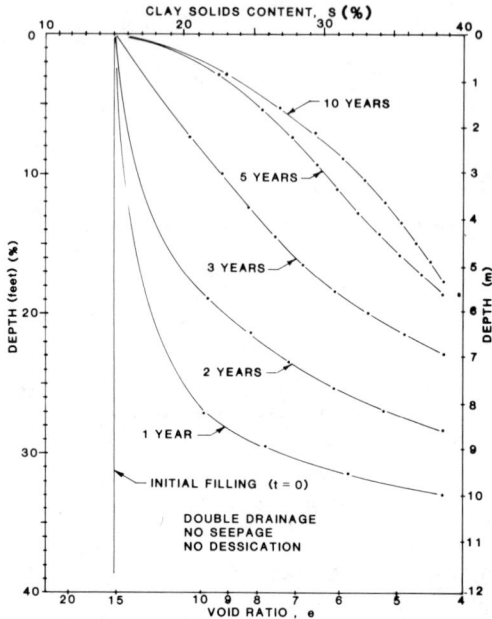

Fig. 10   Predicted Void Ratio Vs. Depth -
Phosphatic Clay

The above examples involve deposits having a uniform initial void
ratio. This is not the usual field case, where filling rates are sufficiently slow to allow partial consolidation of the waste material
during filling.

Somogyi (13) developed a series of programs that can handle virtually any sequence of filling, followed by quiescent consolidation,
then filling at a different rate, etc. These programs have been used
to determine optimum filling rates and rest periods for disposal areas,
considering all relevant mining and disposal factors. These include
cost of dam construction, stability of embankments, requirements for
water recovery, and time requirements for abandonment and reclamation.

ADVANCED WASTE DISPOSAL TECHNOLOGY

An example of a filling, resting, refilling, and reclamation sequence for a disposal area is shown in Figure 11. This example is
for a mixture of pre-thickened phosphatic clay and tailings sand. Admixing of tailings sand has been shown to enhance the consolidation
of high plasticity wastes, such as phosphatic clay, as a result of increased self-weight stresses (4). This reduces disposal volume requirements and allows more rapid land reclamation. The development of
finite strain consolidation solutions has facilitated the design of

sand/clay mixing schemes, and allows predictions to be made of the benefits obtained by such advanced waste disposal technology.

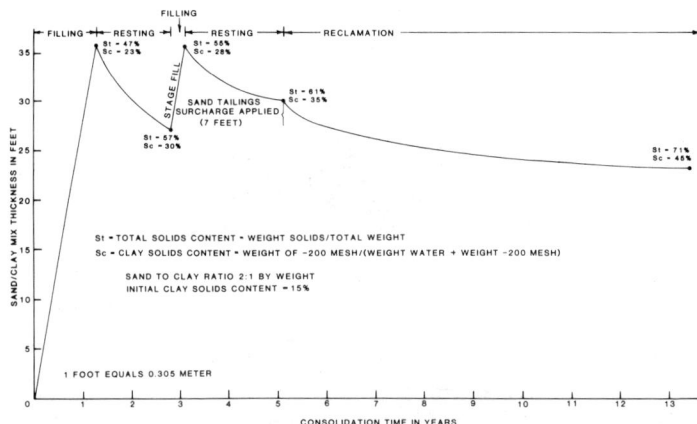

Fig. 11   Predicted Consolidation History for
Sand/Clay Mix Disposal Area

A novel method for achieving sand/clay mixing has been developed by Brewster Phosphates, a partnership of American Cyanamid Company and Kerr McGee Corporation, operating two phosphate mines in Central Florida (4). A plan view of the process, termed sand spray, is shown schematically in Figure 12. It involves placing dilute clay slurry from the beneficiation plant into mine cuts for initial settling and consolidation to an average solids content of 12% to 15%, a process that generally requires 3 to 6 months. Then spray nozzles are used to sprinkle hydraulically-delivered tailings sand over the clay. These spray nozzles can be either mounted on a floating pipeline or distributed along the edges of the overburden spoil piles.

As the sand rains down upon the pre-thickened clay, it gradually settles into the clay mass, enhancing consolidation, as shown in Figure 13, and also producing a better soil for agricultural use following land reclamation.

Other processes developed for enhancing the consolidation of phosphatic clays include use of chemical flocculants in combination with mechanical thickeners to pre-thicken the clays to a consistency such that they will suspend dewatered sand tailings in a stable sand/clay mixture that can be pumped and discharged into a disposal area without segregating.

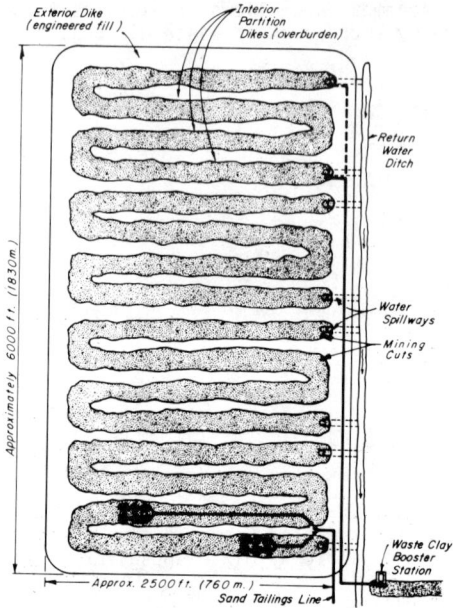

Fig. 12   Sand/Spray Process

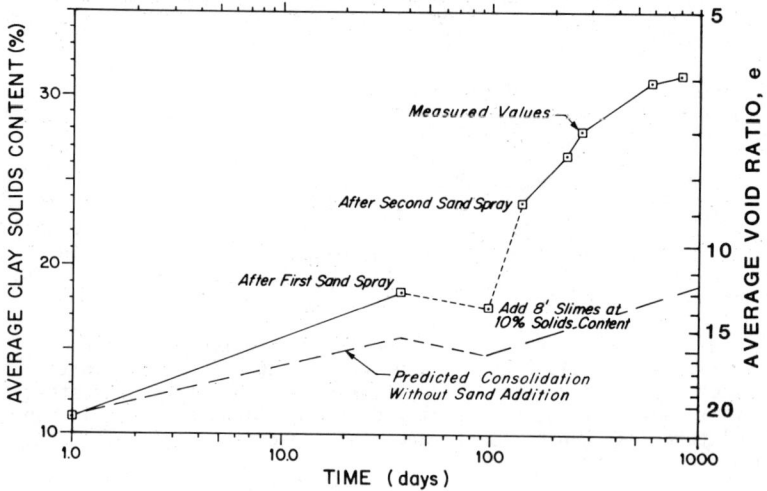

Fig. 13   Influence of Admixing Tailings Sand
on Consolidation of Phosphatic Clay

Slurry consolidation testing of flocculated clays indicates that the primary benefit of the chemical addition is to greatly reduce the time for sedimentation and initial consolidation to a solids content level of about 15%. Normal sedimentation and self-weight consolidation to 15% solids may require 3 to 6 months for unflocculated clay, and only a matter of minutes for flocculated clay. However, as shown in Figures 14 and 15, the compressibility and permeability relationships are virtually identical for both flocculated and unflocculated clay once the effective stresses start to increase. Therefore, the primary effect of flocculation, at least for the highly plastic phosphatic clays, is to decrease the initial void ratio, thereby increasing the self-weight stresses at the start of consolidation.

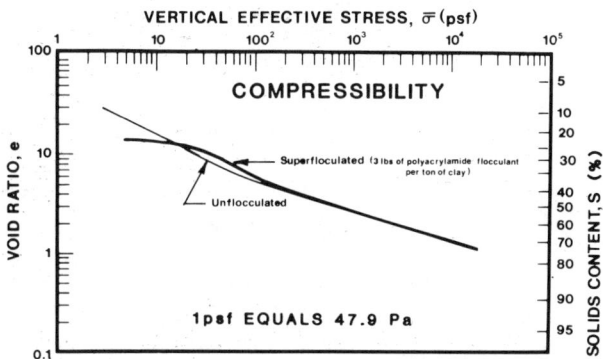

Fig. 14   Effect of Flocculant on Compressibility
          of Phosphatic Clay

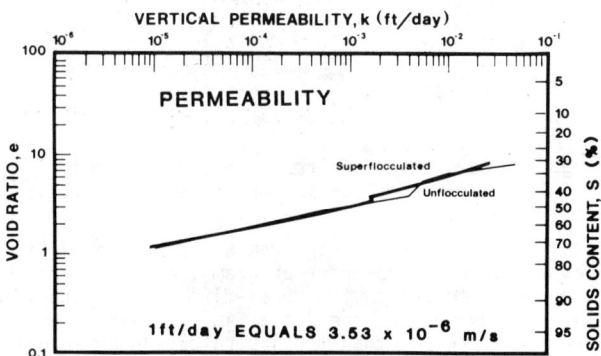

Fig. 15   Effect of Flocculant on Permeability
          of Phosphatic Clay

The same benefit can be achieved by dredging thickened clay from a large settling or holding basin, and such a process, termed dredge/mix, is also being utilized by a phosphate mine in Florida.

OTHER CONSIDERATIONS

Downward seepage gradients exist at many disposal areas, and boundary pore water pressures may be zero at both top and bottom. Such a case is shown in Figure 16, which presents pore pressure data from a 160-foot (49m) thick deposit of copper slimes. The Somogyi computer programs were modified to include seepage gradients for this analysis, and then used to model the filling history of the impoundment shown in Figure 17.

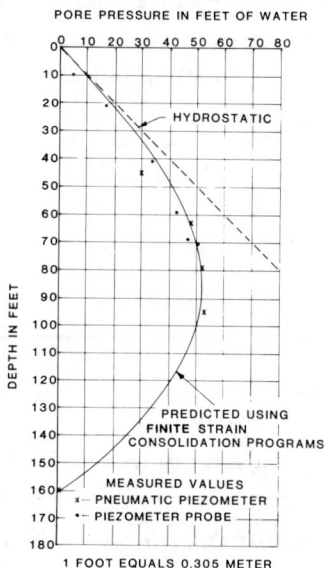

Fig. 16 Comparison of Predicted and Measured
Pore Pressures in Copper Tailings Slimes

The compressibility relationship from slurry consolidation testing was used as input to the computer programs, and the permeability parameters were varied to obtain the "best-fit" curve to the field data. A comparison showed that the "best-fit" permeability averaged approximately 2 times greater than the laboratory-determined values.

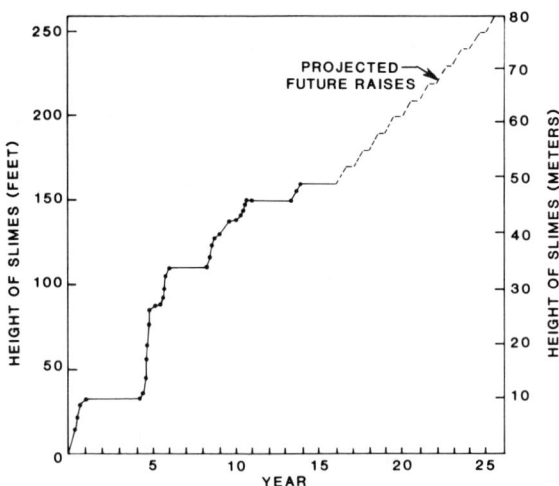

Fig. 17   Filling History - Copper Tailings Dam

The principal purpose for using the computer programs to model the existing situation was, in fact, to obtain the appropriate consolidation parameters to predict pore pressures during projected future raises. Accurate predictions of pore pressure can be used to calculate changes in shear strength of the slimes during future raises, and thereby select a raise rate that will maintain a desired factor of safety for the raised dam, which utilizes cycloned tailings sand for upstream construction.

The predicted pore pressure distribution in the slimes pond for a 100-foot raise over 10 years is shown in Figure 18. In this case, predictions were made using both finite strain (Somogyi) and infinitesimal strain methods. The infinitesimal strain analysis was done by Allen Marr of GEOCOMP, Inc., using a finite difference computer program, GEOSOL, based on Gibson's numerical approach for an increasing clay layer thickness with time (7). For this case, involving a relatively low plasticity mining waste (PI = 12), both methods gave very similar predictions, and the finite strain analysis does not appear to offer significant advantages over the simpler solution using a constant average c_v.

More research is needed to define the conditions under which finite strain solutions must be employed, and those for which conventional theory is sufficiently accurate. Also, the examples cited herein involved additional simplifying assumptions, not always met in practice. All the examples involved one-dimensional consolidation, which is satisfactory for the pond area of large disposal facilities, but not necessarily for slimes adjacent to more pervious materials, such as overburden spoil piles or the tailings sand shell of upstream embankments. For these cases, a two-dimensional analysis is needed.

Recent research (1) involving phosphatic clays in narrow mine cuts, developed a pseudo two-dimensional finite strain computer program. The results of various parametric studies using the model, combined with field tests in an actual mine cut, indicated that two-dimensional effects only became significant when the width to height ratio was on the order of 5 or less. However, this result may not be valid for other materials, particularly those less plastic than phosphatic clays and those exhibiting significant anisotropy in permeability.

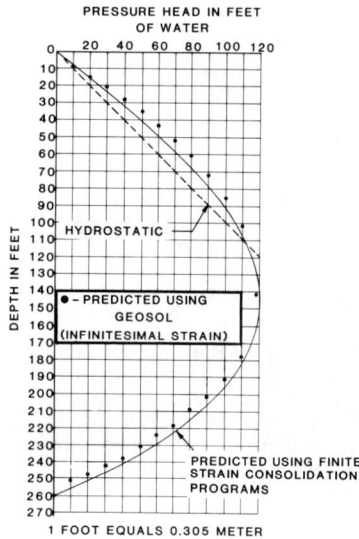

Fig. 18   Predicted Pore Pressures in Copper Slimes
Pond for 100-Ft. (30m) Raise

Also, none of the cases included herein considered effects of surface drying and desiccation. During active filling these effects can generally be ignored, but during extended rest periods or for long-term conditions, desiccation effects should be considered. Only semi-empirical methods are currently available for estimating rates of surface drying and crust development.

Finally, sedimentation has been essentially ignored in this presentation, and for most practical situations, probably is not a significant factor. However, for small disposal areas that are filled rapidly, it can be important to define better the boundary between sedimentation and consolidation. Also, for some fine colloidal materials that disperse easily, sedimentation problems can override consolidation considerations. A rational combined sedimentation-consolidation model is definitely needed, not only for the practical cases cited above, but also to advance our fundamental understanding of and ability to predict the behavior of slurried mineral wastes.

SUMMARY AND CONCLUSIONS

Mining wastes include a wide range of materials, whose consolidation properties span a very large range of compressibilities and permeabilities. The fine-grained fraction of these wastes, generally low-to-high plasticity clays, are placed as slurries into large impoundments, where they sediment and consolidate under self-weight. Understanding the consolidation behavior of these materials is important for designing adequate storage facilities, estimating seepage quantities, and planning for eventual abandonment or reclamation.

Special testing devices and procedures have been developed for measuring consolidation properties in the laboratory and for making field measurements of consolidation behavior. These techniques have been utilized extensively in research on phosphatic clays over the past decade. As a result of this work, correlations between Atterberg Limits and consolidation properties are available that are useful for making preliminary design estimates.

Improved methods for predicting the consolidation of high plasticity wastes, such as phosphatic clays, utilize computer programs based on non-linear finite strain theory. Solutions based on conventional Terzaghi theory are too inaccurate to be useful for these materials. The finite strain computer programs have been verified in numerous field tests, including large test pits and tanks, as well as full-size disposal areas of several hundred acres.

The new methodology has also proven useful for developing life-of-mine disposal plans for selecting optimum raise rates for upstream construction of tailings dams, for evaluating alternative waste disposal methods, such as sand/clay mixing, and for planning abandonment and/or reclamation of waste disposal areas.

More research is needed to define the range of materials for which finite strain solutions must be used, and those for which modified infinitestimal strain solutions are acceptable. Also, additional research is needed to develop rational procedures for incorporating sedimentation surface evaporation/dessication, and 2 or 3 dimensional drainage effects into slurry consolidation analyses.

ACKNOWLEDGEMENTS

Many individuals and organizations have contributed to various aspects of the development of the approach to waste disposal engineering described herein. The author wishes especially to acknowledge the following, who played key roles in the original formulation of the approach and the development of testing equipment and numerical procedures: Dr. R. T. Martin, Professor C. C. Ladd, Dr. W. D. Carrier, Professor Frank Somogyi, and Berg Keshian, Jr.

Essential support and encouragement was provided by David J. Raden, R. C. Timberlake, Dr. Richard McFarlin and other representatives of the Florida Phosphate Council and its member companies. A special note of appreciation is extended to Professor R. L. Schiffman, who for more than twenty years has attempted to lead the author to a better understanding of clay consolidation.

APPENDIX - REFERENCES

1. Bromwell Engineering, Inc., "Waste Clay Disposal in Mine Cuts", Publication No. 02-006-009, Florida Institute of Phosphate Research, Bartow, Fla., 1982.

2. Bromwell, L. G., "Soil Engineering Studies for Future Foundation Construction", report prepared for Climax Molybdenum N.V., Rotterdam, Holland, by T. William Lambe & Associates, Cambridge, Mass., 1967.

3. Bromwell, L. G., and Carrier, W. D., III, "Consolidation of Fine-Grained Mining Wastes", Proceedings of the Sixth Panamerican Conference on Soil Mechanics and Foundation Engineering, Lima, Peru, Vol 1, 1979, pp. 293-304.

4. Bromwell, L. G., and Raden, D. J., "Disposal of Phosphate Mining Wastes (Current Geotechnical Practice in Mine Waste Disposal)", ASCE Geotechnical Division Special Publication, 1979.

5. Carrier, W. D., III, and Keshian, B., Jr., "Measurement and Prediction of Consolidation of Dredged Material", Twelfth Annual Dredging Seminar, Texas A&M, College Station, Tex., 1979.

6. Carrier, W. D., III, Bromwell, L. G., and Somogyi, F., "Design Capacity of Slurried Mineral Waste Ponds", Journal of the Geotechnical Engineering Division, ASCE, Vol 109, No. 5, May 1983, pp. 669-716.

7. Gibson, R. E., "The Progress of Consolidation in a Clay Layer Increasing in Thickness with Time", Geotechnique, Vol 8, pp. 171-182, 1958.

8. Gibson, R. E., England, G. L., and Hussey, M. H. L., "The Theory of One-Dimensional Consolidation of Saturated Clays, 1. Finite Non-Linear Consolidation of Thin Homogeneous Layers", Geotechnique, Vol 17, 1967, pp. 261-273.

9. Roma, J. R., "Geotechnical Properties of Florida Phosphatic Clays", thesis presented to the Massachusetts Institute of Technology, at Cambridge, Mass., in 1976, in partial fulfillment of the requirements for the degree of Master of Science.

10. Salem, A. M., and Krizek, R. J., "Consolidation Characteristics of Dredging Slurries", Journal of the Waterways, Harbors, and Coastal Engineering Division, ASCE, Vol 99, 1973, pp. 439-457.

11. Schiffman, R. L., "Finite and Infinitesimal Strain Consolidation", Technical Note, Journal of the Geotechnical Engineering Division, ASCE, Vol 106, No. GT2, Feb., 1980, pp. 203-207.

12. Schiffman, R. L., Personal Communication, 1983.

13. Somogyi, F., "Large Strain Consolidation of Fine-Grained Slurries", presented at Canadian Society for Civil Engineering 1980 Annual Conference, Winnipeg, Manitoba, Canada, 1980.

14. Wissa, A. E. Z., Martin, R. T., and Garlanger, J. E., "The Piezometer Probe", In Situ Measurement of Soil Properties, ASCE Specialty Conference, North Carolina State University, Raleigh, N.C., Vol 1, 1975.

15. Znidarcic, D., and Schiffman, R. L., "Finite Strain Consolidation: Test Conditions", Technical Note, Journal of the Geotechnical Engineering Division, ASCE, Vol 107, No. GT5, May 1981, pp. 684-688.

PERSPECTIVES ON MODELLING CONSOLIDATION OF DREDGED MATERIALS

Raymond J. Krizek[1], M. ASCE and Frank Somogyi[2], A.M. ASCE

ABSTRACT--The finite strain consolidation of sedimented dredged materials presents an extremely challenging problem that encompasses new developments in mathematical modelling, new laboratory testing techniques and associated interpretations, new material property formulations, and even new perspectives in engineering judgement. The physical and mathematical aspects of this problem are highly nonlinear, and material property relationships extend well beyond the scope of standard geotechnical engineering. The potential contributions offered by each component must be skillfully synthesized to render a meaningful solution to a given problem, and the likely cause of any discrepancy must be identified to minimize the possibility that a fortuitous choice of input information might produce an unfounded "correct" solution. Addressed herein are some of the issues that must be faced when using these models to analyze or design the behavior of a containment area composed of dredged materials.

INTRODUCTION

The planning and design of disposal areas for dredged material has been and continues to be an issue of national importance because of the need to maximize economic benefits and minimize adverse environmental impacts. Simply stated, the major goals are to predict the time-dependent capacity of a containment area or to properly size the area to accommodate the anticipated volume of dredged material and to analyze settlement rates under surcharge for reclamation projects. While appearing deceptively simple on the surface, this is a particularly challenging problem because it extends beyond the borders of normal geotechnical engineering; in particular, it requires the definition of new terms, the modification of concepts formulated for other purposes, and the development of specialized tests (together with appropriate methods of interpretation).

It is instructive to examine the many components that must be synthesized to render a successful solution to this problem. First, the insitu volume of the material to be dredged must be known; this can usually be determined with reasonable accuracy from soundings or other measurements. However, this insitu volume must be adjusted by several empirical coefficients (which account for overdredging, removal efficiency, etc) to determine the actual volume of sediment that is ultimately deposited in the containment area. Quantifying the void ratio of this insitu material is a formidable task indeed; this may be

---

[1]Professor and Chairman, and [2]Associate Professor, Department of Civil Engineering, The Technological Institute, Northwestern University, Evanston, Illinois 60201

done by means of undisturbed samples (if proper equipment and expertise are available) or by use of various empirical correlations that have been established with index properties of the disturbed material.

When the material is dredged and deposited in a containment area, the bulking factor or initial void ratio of the freshly deposited sediment must be determined by means of field samples, laboratory tests, or empirical correlations. Immediately after sedimentation of the dredged material, the process of self-weight consolidation will begin; the analysis of this process requires the use of an appropriate mathematical model, together with properly defined boundary conditions and material properties stemming from laboratory tests or empirical correlations. Ideally, the predicted time-rate of settlement of the deposited materials will indicate when and what additional volumes can be placed in the site; hopefully this will be compatible with the projected time history of the dredging-disposal operation. If the area is to be reclaimed, stabilization by consolidation will probably involve surcharge loading in conjunction with methods to accelerate drainage.

OBJECTIVES

The objectives of this paper are (a) to assess briefly the advantages and disadvantages of the mathematical models that are available, (b) to present various formulations that have been developed to characterize the mechanical properties of dredged materials for input into the mathematical models, (c) to discuss some of the uncertainties that impede the application of available models, (d) to review by example the considerations involved in the implementation of models, and (e) to outline the position where the profession stands in the handling of these problems.

ANALYTICAL MODELS

While analytical models comprise the heart of most predictive techniques, we are often in a position of inadequately understanding the physical process to be simulated. Accordingly, the extreme complexity of natural phenomena usually dictates the incorporation of several simplifying assumptions to achieve mathematical tractibility (not to mention actual formulation of the governing equations in many cases). Described briefly herein are some of the models that have been developed to predict the large-deformation response of landfills composed of dredged material.

Consolidation Models

The major assumptions in classical consolidation theories (namely, infinitesimal strains and constant material properties) are generally recognized as excessively restrictive for analyzing the large volume changes exhibited by newly deposited dredged materials. The two most popular approaches to circumvent this restriction utilize either "incremental small strain" or "finite strain" formulations which are solved numerically using primarily finite difference schemes.

Incremental small strain models, such as those by Olson and Ladd (1979) and Yong, Siu, and Sheeran (1983), maintain the simplicity of the

Terzaghi-type formulation and constitute a logical "next step" to extend its applicability. The solution method usually involves dividing a deposit into a series of "numerical sublayers". Different material properties can be assigned to each sublayer, and continuity must be explicitly satisfied at the interface between each sublayer. These models can readily handle nonuniform material properties, whether caused by dissimilar materials, self-weight, or other temporal or spatial variations in effective stress. Furthermore, linearization of the consolidation process within each sublayer ensures that convergence and stability criteria can be identified for the particular numerical solution adopted. However, these models still incorporate the assumptions of infinitesimal strain and uniform material properties over small regions of space and time. In addition, the need to continuously update both the material properties and the coordinates of the numerical mesh makes the solution computationally laborious.

Finite strain formulations are invariably based on the pioneering work of Gibson, England and Hussey (1967) and result in nonlinear second-order partial differential equations. Self-weight and material nonlinearities can be directly incorporated into the equations, which are usually developed in terms of material, rather than spatial, coordinates. In terms of material coordinates, the equivalent or reduced height of a deposit remains constant in the absence of additional filling. The use of material coordinates thus has significant mathematical, as well as numerical, advantages. It transforms a boundary value problem with a moving boundary whose location is unknown into one with a fixed known boundary (no filling) or at least one in which the boundary location is always known (during filling or accretion). In addition, computational complexity is dramatically reduced by precluding the necessity for constantly redefining the numerical mesh. The actual thickness of the deposit can be readily calculated at any desired time during the computational process.

In the finite strain models, the equation describing the process of consolidation is cast in terms of either void ratio (Monte and Krizek, 1976; Schiffman, 1980; Gibson, Schiffman and Cargill, 1981) or pore pressure (Koppula, 1970; Somogyi, 1979; Koppula and Morgenstern, 1982). The void ratio based models assume that both effective stress and permeability are directly relatable to void ratio, whereas the pore pressure based models assume that void ratio depends on effective stress and permeability depends on void ratio. Monte and Krizek (1976) accomplish the numerical solution by use of a step-by-step procedure combined with a weighted residual technique which leads to a finite element discretization in the spatial variable and a finite difference discretization in the time variable. However, they report that, in the case of a particular kaolin clay, classical small strain consolidation theory can for all practical purposes adequately describe the deformation-time response after the effective consolidation stress exceeds about 8 psi. Cargill (1982) employs a void ratio based formulation and solves the resulting convection-diffusion equation explicitly. Somogyi (1979) utilizes an implicit technique to solve the self-adjoint form of a pore pressure based equation. Koppula and Morgenstern (1982) use an implicit technique to solve their pore pressure based integro-differential equation.

Explicit numerical techniques have the advantage that the solution is advanced from known (i.e. previously computed) values of the dependent variable and corresponding material properties. However, stability and convergence criteria are often unattainable for nonlinear equations and can only be deduced after appropriate linearization of the equation. Although implicit numerical techniques are computationally more efficient and unconditionally stable for linear equations, criteria for their stability and convergence may be equally difficult to prove for highly nonlinear problems, but can be conveniently demonstrated by trial. Perhaps the major limitation of implicit techniques when applied to nonlinear problems is that the finite difference equations require material properties corresponding to unknown values of the dependent variable (i.e. at the "new" time step). Those material properties must therefore be either estimated or allowed to lag one time increment behind the solution.

All of the formulations described above assume a completely saturated soil, one-dimensional movement of both fluid and soil, and primary or hydrodynamic consolidation. They have all been applied to both accreting (except Olson and Ladd, 1979) and stagnant (or quiescently settling) deposits, and solutions have been obtained for the classical boundary conditions (namely, one-way and two-way drainage). To the authors' knowledge, impeded drainage has been incorporated into only two of the formulations (Krizek and Casteleiro, 1974; Cargill, 1982). The effect of underseepage has been investigated using Somogyi's solution and assuming the somewhat unrealistic condition of a constant piezometric head across the deposit; the resulting seepage force was found to have a significant effect on the consolidation process and warrants further attention. Using Somogyi's solution, Bromwell and Carrier (1982, Private Communication) have incorporated the effects of crust formation as a time-dependent, highly permeable surcharge, where the rate of crust formation and its density are estimated or based on field observations and provided as input to the model. The results suggest that the progress and effects of crust formation deserve additional analytical and experimental effort.

Desiccation-Consolidation Model

Since the process of depositing maintenance dredgings in a containment area often takes place periodically over a number of years, a repeated sequence of desiccation and consolidation occurs with each cycle. After a certain period of time, the overlying free water either evaporates or drains through the layer into the underlying soil. If downward drainage takes place, seepage forces will add to the body forces causing consolidation. When the overlying water is removed, the surface begins to dessicate. With the associated lowering of the ground water table, the effective weight of the sediments comprising the desiccating layer increases due to its loss of buoyancy, thereby increasing the magnitude of the settlement. This process of drainage, evaporation, desiccation, and consolidation for a given layer continues until the beginning of the next cycle, at which time the site is again inundated and the same steps are repeated. The basic phenomenon involved in this process is one of water flowing through a consolidating medium in which the location of the interface between the saturated and partially saturated zones is unknown and dependent on the properties of the dredged material, weather conditions, nature of the foundation soils, and particular stage of the cycle.

Although appearing rather straight-forward, the consolidation of a partially saturated medium leads to a variety of conflicting hypotheses. On one hand, geotechnical engineers, who are interested mainly in deformations and volume change characteristics of soils subjected to external loads, have usually assumed full saturation, and the problem of defining concepts and coefficients for partially saturated media is one to which only incomplete and tentative answers exist. On the other hand, soil scientists, who have directed much effort to studying flow through partially saturated porous media, usually assume that such media possess a rigid structure and undergo no volume change during flow. The combination of these two approaches constitutes a major impediment in the development of models of this type.

The physical process of desiccation and crust formation was incorporated as an integral part of the mathematical model by Krizek and Casteleiro (1974). This process was modeled one-dimensionally by a general nonlinear partial differential equation, and simplifying assumptions were employed to deduce two nonlinear parabolic differential equations -- one for the fully saturated zone and the other for the partially saturated zone. Boundary conditions include impeded drainage at the lower boundary and evaporation plus transpiration at the upper boundary (transpiration caused by vegetation was handled by multiplying the evaporation rate by an empirical coefficient that depends on the type of vegetation). This boundary value problem was solved by use of a step-by-step finite difference technique with a correction procedure to adjust for any errors introduced by the assumptions utilized during any given small time increment. Solutions obtained from this model are able to (a) describe the water content distribution in the fill at any time after deposition, (b) predict the desiccation and consolidation behavior of the dredged material as a function of time, and (c) aid in evaluating the different techniques that are available to accelerate the dewatering of the landfill. The problems with the implementation of this model are (a) the large amount of computer time required to solve the equations, especially for a large number of cycles, and (b) the general unavailability of proper input data for the model.

MATERIAL PROPERTIES

The effective and efficient application of the various finite-strain or incremental small-strain consolidation models that have been developed to supplant the previously used empirical rules-of-thumb and to circumvent some of the unrealistic assumptions inherent to classical Terzaghi small-strain consolidation theory depends strongly on the specification of certain critical input parameters that interrelate the effective stress, void ratio, and coefficient of permeability. In general, these take the form of a compressibility relationship (which relates the void ratio to the effective stress) and a permeability of consolidation relationship (which relates either the coefficient of permeability or the coefficient of consolidation to the void ratio). In addition, most analyses of self-weight consolidation require the identification of the "initial", or "zero effective stress" void ratio. Due primarily to the large magnitude of the settlements involved, these material property relationships are highly nonlinear. In addition, their range of interest extends well beyond that normally encountered in conventional geotechnical engineering practice, thereby introducing the need for new definitions and tests together with associated interpretations.

Plasticity

Figure 1 (Carrier, Bromwell and Somogyi, 1983) shows plasticity data for a large number of dredged materials from various fresh water and salt water locations in the United States. Although the range of values is quite broad for the entire spectrum of dredged materials, experience has indicated (Krizek and Salem, 1974; Carrier, Bromwell and Somogyi, 1983) that the materials in a given locality are generally quite similar.

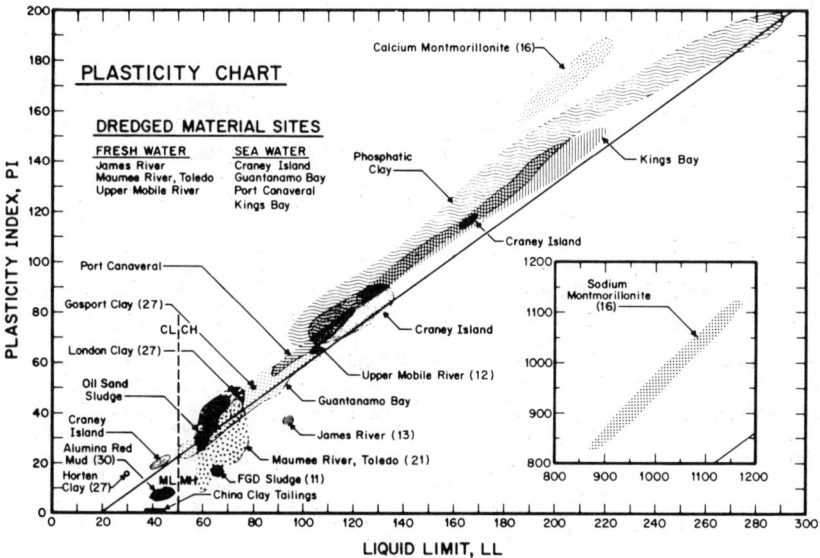

Figure 1.   Atterberg Limits of Various Dredgings and Other Mineral Wastes (after Carrier et al, 1983)

Initial Void Ratio

Conceptually, the initial void ratio represents the stress-free condition of a soil at the end of sedimentation and prior to the onset of self-weight consolidation, and its identification is required for the self-weight consolidation analysis of deposits formed from dilute slurries, as is usually the case with dredged material. Monte and Krizek (1976) have termed this condition the "fluid limit" because it represents the limiting water content at which a soil-water system behaves essentially as a viscous fluid. As such, it can be regarded as the condition of a soil at its initial formative state and provides an origin or datum for measuring the strain in the soil at any other stage during its consolidation history. Unfortunately, the quantification of this parameter is not a simple matter. Based on arguments stemming from fabric formation and permeability measurements in clay soils, Monte and Krizek (1976) suggest that the fluid limit of a soil is four to five times its liquid limit, and they showed good agreement between predicted and measured results for slurry consolidation tests where a ratio of four and a half was used.

The initial void ratio can be determined experimentally by measuring the void ratio of a sample recovered from the surface of the deposit (after sedimentation is considered to have occurred) or sedimented in a cylinder from a slurry having an initial solids content similar to that in the field. Alternatively, the initial void ratio can be estimated from empirical relationships, such as the one proposed by Carrier, Bromwell and Somogyi (1983), where they suggest that the initial void ratio be taken as seven times the void ratio at the liquid limit. Recent experimental evidence (Been and Sills, 1981; Umehara and Zen, 1982) indicates that the initial void ratio (as well as the initial compressibility) is not a unique material property, but depends on the initial solids content of the slurry. However, Carrier, Bromwell and Somogyi (1983) state that the particular value chosen for the initial void ratio usually has a minor effect on the predicted rate of consolidation and only a near-surface effect on the predicted equilibrium void ratio distribution.

Compressibility

Some of the solutions described previously (namely, those by Olson and Ladd (1979), Yong, Siu and Sheeran (1983), and Schiffman and his co-workers) can accommodate experimental compressibility data (i.e. observed void ratios at known effective stresses) directly; various linear or nonlinear interpolation schemes are then utilized to calculate the appropriate compressibility. Other solution procedures require that a mathematical relationship between void ratio and effective stress be specified. Monte and Krizek (1976) suggest the following power relationship between effective stress, $\bar{\sigma}$, and strain, ε, but actually employ a parabolic form with $N = 2$:

$$\bar{\sigma} = M \, \varepsilon^N \qquad (1)$$

Koppula and Morgenstern (1982) employ a relationship of the form

$$\frac{de}{d\bar{\sigma}} = \left[\frac{de}{d\bar{\sigma}}\right]_0 \left[\frac{\bar{\sigma}}{\bar{\sigma}_0}\right]^P \qquad (2)$$

where e is the void ratio, o is a subscript indicating an arbitrary reference state, and P is a constant that ranges between zero (linear relation) and unity (semilogarithmic relation). In the linearized form of the general finite strain equation, Cargill (1983) uses an exponential relationship of the form:

$$e = (e_{00} - e_{\infty}) \, \exp(-\lambda\bar{\sigma}) + e_{\infty} \qquad (3)$$

where e_{00}, e_{∞}, and λ are empirical parameters (e_{00} and e_{∞} are nominally the void ratios at zero and infinite effective stress). Somogyi (1979) employs a power relationship of the form:

$$e = A \, \bar{\sigma}^{\,B} \tag{4}$$

where A and B are empirical constants.

The void ratio and effective stress data required for input to any of the numerical solutions can be obtained experimentally either in the laboratory or in the field. The most useful laboratory device for measuring the compressibility of high void ratio sediments is the slurry consolidometer (Sheeran and Krizek, 1971), which is essentially a large tube (usually with rigid walls) with a loading piston. Although incremental loading has been most commonly employed to date, the long duration of these tests (typically several months) has given impetus to constant rate of deformation tests. The interpretation of this test requires accurate pore pressure measurements at the boundary and finite strain analysis, as described by Znidarcic and Schiffman (1981). Measurements of insitu density and pore pressure can also be used for estimating the void ratio-effective stress relationship, as discussed by Carrier and Keshian (1980). Finally, based on a synthesis of data from extensive laboratory and field testing programs, the empirical correlation depicted in Figure 2 (Carrier, Bromwell and Somogyi, 1983) has been

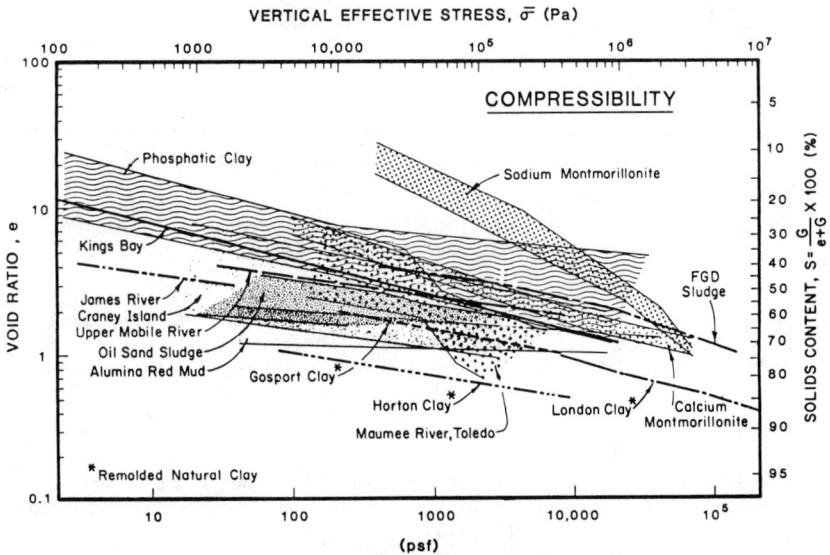

Figure 2. Compressibility of Various Dredgings and Other Mineral Wastes (after Carrier et al, 1983)

advanced for obtaining preliminary estimates of the curve-fit parameters in Equation (4). Alternatively, Carrier and Beckman (1983) have proposed the following somewhat more general version of Equation (4):

$$e = \alpha \left[\frac{\bar{\sigma}}{p_a} \right]^\beta + \xi \tag{5}$$

where p_a is atmospheric pressure and the coefficients α, β, and ξ are related to the plastic limit, plasticity index, and activity of the material.

Permeability

With the exception of Olson and Ladd's (1979) technique, which employs coefficients of consolidation, all of the previously presented analytical models require a relationship between permeability and void ratio. Of all the components that comprise the problem under consideration, the permeability relationship is probably the most important. Unfortunately, it also manifests the greatest variability, as depicted in Figure 3, and is the most difficult to quantify. Experimental data

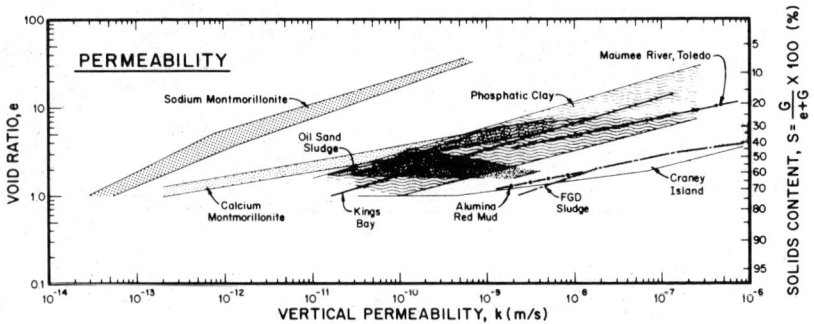

Figure 3.   Permeability of Various Dredgings and Other
Mineral Wastes (after Carrier et al, 1983)

can be input directly into the solution procedures used by Pane and Schiffman (1981) and Yong, Siu and Sheeran (1983). The procedure presented by Monte and Krizek (1976) utilizes a relationship of the form:

$$\frac{k}{1 + e} = S + Te \tag{6}$$

where k is the coefficient of permeability and S and T are empirical constants. Koppula and Morgenstern (1982) employ the relationship:

$$\frac{k}{1 + e} = \left[\frac{k}{1 + e} \right]_o \left[\frac{\bar{\sigma}}{\bar{\sigma}_o} \right]^R \tag{7}$$

where o is a subscript indicating an arbitrary reference state and R is an empirical constant. Somogyi (1979) uses a power curve of the form:

$$k = Ce^D \tag{8}$$

where C and D are empirical constants. Additional relationships which have recently been suggested, but have yet to be incorporated in the previous models, include (Samarasinghe, Huang and Drnevich, 1982; Carrier, Bromwell and Somogyi, 1983):

$$k(1 + e) = Ee^F \tag{9}$$

and (Carrier and Beckman, 1984)

$$k (1 + e) = G(e - H)^J \tag{10}$$

where E, F, G, H, and J are empirical constants.

The coefficient of permeability can be measured directly or indirectly in a slurry consolidometer. Falling head tests can be performed at the end of each loading during an incremental load test, or the permeability can be calculated from the observed consolidation rate. If falling head tests are performed, sufficiently small gradients must be used to minimize rebound or additional consolidation by the seepage forces, or the applied lead can be adjusted. Based on finite strain computer simulations of slurry consolidation tests, Carrier and Bromwell (1980) suggested the following equation for calculating permeabilty:

$$k = \frac{c_v \, a_v \, \gamma_w}{1 + e_f} \tag{11}$$

where k is the permeability of the material at the final void ratio, e_f, and

$$a_v = AB \, (\bar{\sigma}_b)^{B-1} \tag{12}$$

where A and B are the constants in Equation (4), $\bar{\sigma}_b$ is the effective stress at the base of the slurry consolidometer, and

$$c_v = Th_f^2/t_{90} \tag{13}$$

where h_f is the final height of material, t_{90} is the time for 90% of the settlement to occur, and T is an analytical parameter that depends on the void ratio (typically 0.85 to 1.20). Cargill (1984) recommended the conventional small-strain equation for computing permeability from oedometer test results at 50% consolidation and with average values of drainage distance, compressibility, and void ratio for the particular load increment. However, he also cautioned that this may be inappropriate if large changes in void ratio occur under the load increment. Znidarcic (1982) presented a method for estimating permeability from constant rate of deformation slurry consolidation test results; this technique involves the "reverse solution" of a modified, piecewise linear finite strain theory.

Permeability tests have also been conducted in the field, primarily by means of borehole or pumping tests. Field tests have invariably resulted in higher permeabilities than obtained from laboratory tests. This may be attributed to anisotropy or the less severe "caking" produced by the lower gradients typically used in field tests. The latter may, in fact, be the predominant factor, since closer agreement between predicted and actual consolidation rates have generally been obtained when using field measured permeabilities or the upper bound of scattered laboratory permeability results. Finally, Carrier and Beckman (1984) present the following empirical correlation for preliminary estimates of permeability:

$$k = \mu \frac{(e - \delta)^{\nu}}{1 + e} \tag{14}$$

where μ, ν and δ are functions of the plastic limit and plasticity index of the material.

FIELD VARIATIONS

Although material characterization presents a formidable challenge in most geotechnical engineering problems, determining the distribution of engineering properties of dredged materials deposited within diked enclosures introduces its own special set of conditions that often turns an apparently simple problem into a deceptively complicated one. In addition, there are many physical phenomena which occur in field (and sometimes laboratory) situations that are not understood in sufficient detail to enable incorporation into a mathematical model.

Heterogeneity

As a first approximation, dredged material landfills are frequently assumed to be homogeneous. This assumption is especially attractive, because the landfills are man-made under circumstances that can some-times be controlled to a limited degree. However, there is considerable post-facto evidence (Krizek, 1976) to indicate that deposits of dredged material exhibit non-negligible heterogeneity in both the horizontal and vertical planes. For example, Krizek and Salem (1974, 1977) document substantial variations in particle size distribution within the first 100 to 200 meters from the inlet pipe, beyond which the variation is negligible and the assumption of homogeneity appears reasonable. If a disposal area were operated with two or more inflow points, it follows that a "pocket" of coarser materials would form in the delta emanating from each inflow point. Since most engineering properties such as strength and compressibility are related to particle size (at least indirectly), variations in these properties can be expected and do occur. Specific examples of horizontal and vertical nonhomogeneity in dredging disposal areas have been documented by seismic surveys (Krizek, Franklin and Soriano, 1974) and electrical resistivity surveys (Giger, Franklin and Krizek, 1973). In the vertical profile there are many factors that contribute to heterogeneity. Perhaps the most obvious is whether the disposal area is filled incrementally or in a single opera-tion; in the case of an incrementally filled site, this effect is mani-fested primarily in a lowering of the water table below the surface of the spoil, desiccation and associated "alligator cracking" of the sur-face, growth of a vegetative cover, and then subsequent deposition of new material to cover the "altered" layer. Numerous variations were detected by Krizek and Salem (1974) in profiles from four different fresh-water sites in the same vicinity.

Gas Generation

Since most maintenance dredgings contain significant amounts of organic matter, the generation of gas introduces another complexity into the interpretation of test data and the prediction of settlements in a landfill. Indeed, gas generation has been observed in numerous slurry consolidation tests on dredged materials from various sources and on several occasions when sampling field sites. In some cases laboratory compression curves for low intensities of load indicate an increasing void ratio with time. At higher loads the response curves become more "regular", thus suggesting either a decrease in gas generation with time or the more probable situation whereby the effects of gas generation are suppressed by the applied load. An analysis of the gas generated in one particular sample of fresh-water maintenance dredgings with a loss on ignition of 6.4% indicated about 3 to 4 percent oxygen, 16 percent carbon dioxide, 17 percent methane, and the remainder primarily nitrogen. To date, none of the known mathematical models are capable of handling this feature, nor are appropriate interpretations available for laboratory test data. Nevertheless, the problem clearly exists and, because it has not been addressed seriously, little can be said regarding its quantitative significance. Some of the models that have been developed to predict the consolidation of sanitary landfills, such as that by Zimmerman, Chen and Franklin (1977), offer some guidance concerning how to incorporate a gas generation term in a large strain consolidation model for dredged materials.

Secondary Consolidation

"Secondary" consolidation may be considered as simply that portion of the settlement response that is not described by some appropriate theory of "primary" consolidation, and there is no implication that settlements due to secondary consolidation are of secondary or negligible magnitude. Within the context of this definition, however, there would be no secondary consolidation if a sufficiently comprehensive primary consolidation theory were available, and this should be the ultimate goal of the mathematical models that are being developed. To provide some appreciation for the magnitude and nature of the physical phenomenon that may prevail, data from two series of tests are illustrated briefly and the results are interpreted in light of the commonly accepted definitions of primary and secondary consolidation (Salem and Krizek, 1975). In the first series two slurry consolidation tests were conducted and in the second series eight conventional consolidation tests were performed.

In the slurry consolidation tests, five different loads (with a doubling each time) were applied; each of the first four loads were applied for a period between 10 and 60 days, and the fifth was maintained for 150 days. The compression-time curves for these two tests are given in Figure 4. A review of the data indicates some secondary effects for the third and fourth load increments, even though the times of application were smaller than that for the fifth load. By the time the fifth load increment was applied, the thicknesses of the two specimens were only about one-half of their original thicknesses. After a more-or-less standard response curve for the first 15 or 20 days, there is a significant increase in the rate of settlement.

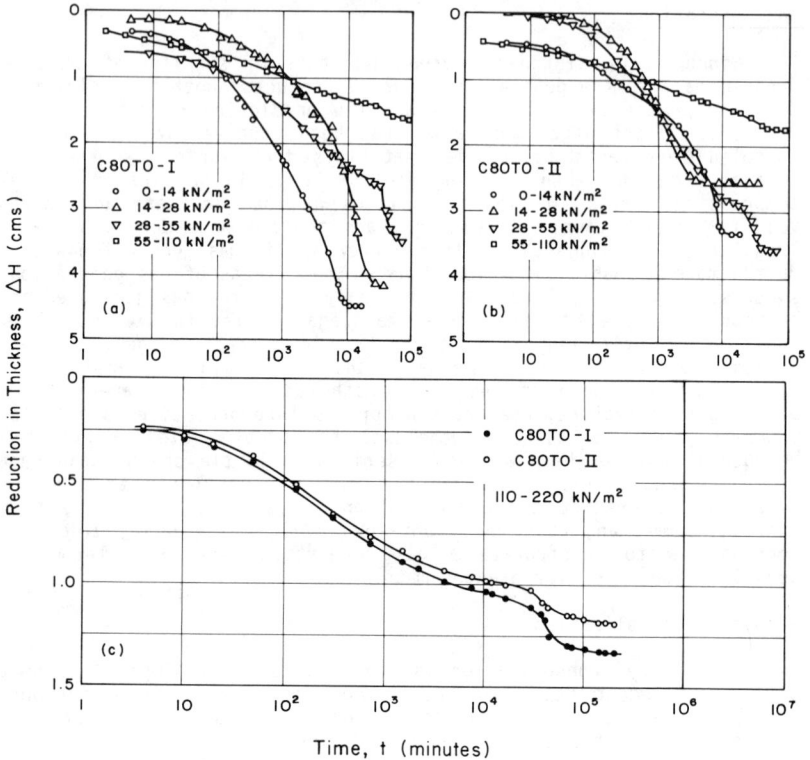

Figure 4. Slurry Consolidation of Toledo Dredged Materials
(after Krizek and Salem, 1974)

Beginning with an initial load of 31 kN/m^2, all loads on the eight specimens tested in the conventional consolidometer tests were doubled after 24 hours until a load of 248 kN/m^2 was reached, at which point this load was maintained on six specimens for 205 days and on the remaining two specimens for about 300 days. The data from these tests are given in Figure 5; the solid points in this figure represent values calculated from an empirical model developed by Krizek and Salem (1974). In all cases a "normal" primary consolidation stage was followed by a linear increase in the settlement with the logarithm of time for about 10 days, after which the rate of settlement increased substantially for about 70 days and then decreased to approach zero for longer times.

Although the unusual behavior pattern which starts at the beginning of the second S-shaped portion of the response curves is similar for both the slurry and conventional consolidation tests, certain differences between the two sets of results are apparent. First, there are different proportions of secondary to primary compressions and the times

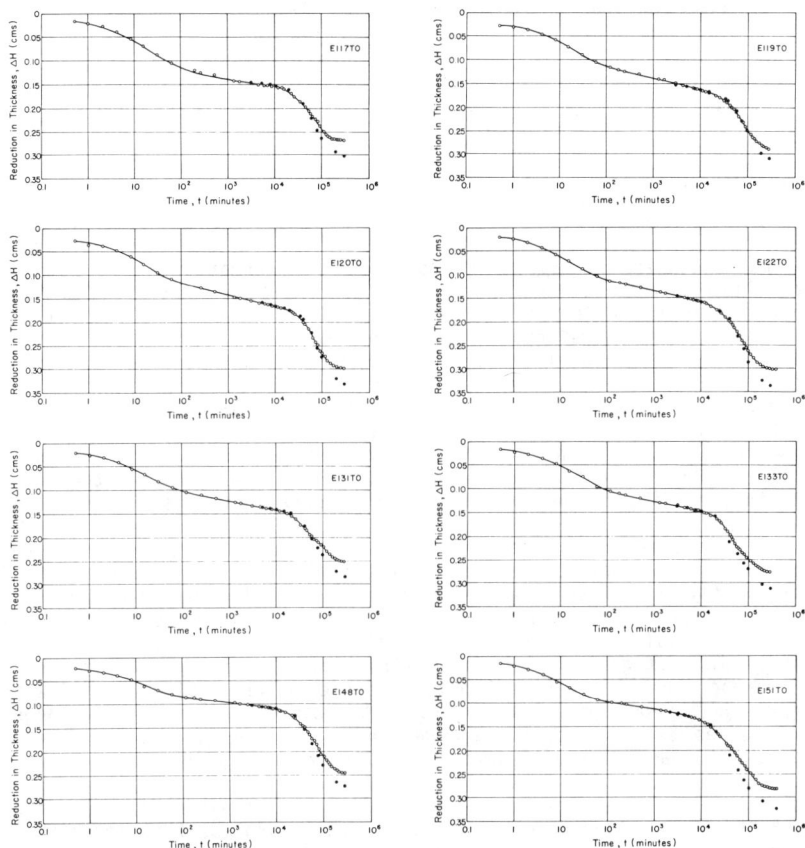

Figure 5. Secondary Consolidation (Conventional Tests) of Toledo
Dredged Materials (after Krizek and Salem, 1974)

required to reach the end of primary consolidation and ultimate settlements are different. In the slurry tests the proportion of secondary
compression to total compression is about 0.25, whereas it is on the
order of 0.60 for the conventional tests for essentially the same load
(220 kN/m² for the slurry consolidation tests and 248 kN/m² for the
conventional tests). Second, the time required to reach ultimate settlements is shorter for the slurry tests than for the conventional
tests, although the thicknesses in the former tests are much greater
than in the latter. Alternatively, the time required to complete primary consolidation in the slurry tests is about 30 times that for the
conventional tests; these times are approximately proportional to the
square of the specimen thicknesses. These observations suggest that

(a) the time required to reach ultimate settlements (including secondary compression) does not necessarily have to be longer for greater thicknesses of material and (b) the relative amount of secondary compression that occurs in the field, where the thickness of material is normally much larger than in the laboratory, may be considerably less than the value predicted from laboratory tests.

Crust Formation

In areas where the slurry is exposed, a surface crust forms as the water evaporates. While evaporative forces enhance the densification of the material at the uppermost surface, the formation of a crust is generally considered to hinder the evaporation of water from lower layers. Prompted by the conviction that crust formation impedes dewatering, engineers responsible for managing disposal areas have expended considerable effort to investigate methodologies for preventing the development of such crusts or for removing them once formed. One advantageous feature is the development of "alligator" cracking as the crust forms; this provides drainage channels for the horizontal movement of water and additional surface area for evaporation. Unfortunately, little is known about the rate of formation of such crusts, the thickness they attain, or their influence on evaporation rates.

The loss of water from a soil surface is essentially a two-stage process. In the first stage the conductivity of water in the soil is sufficiently large that water loss is controlled by the prevailing climatic conditions (radiation, air temperature, wind speed, humidity, etc), and results are comparable to those measured in pan evaporation tests. This stage has been found (Brown and Thompson, 1977) to govern the response until the water content reaches a value slightly below the liquid limit. During the second stage the converse is true and evaporative losses are essentially independent of the environment. However, due to daily and diurnal climatic variations, the governing process may alternate frequently. However, the presence of a water table near the surface may provide a sufficiently large supply of water to continuously rewet the surface and preclude the evaporative process from entering the second stage.

Within the soil itself water translocates in response to variations in the soil-water potential (gravity, matric suction, and osmotic suction) and the temperature gradient. Although the latter has little effect on the movement of water in wet or dry soils, it does contribute to the moisture flux in soils of medium water content. Richie and Adams (1974) studied the loss of water from shrinkage cracks and found that more than 80% of the water loss occured through the cracks. Indeed, in their particular study it was found that the cracks conducted water to the surface in the form of water vapor rapidly enough to supply about twice the evaporation demand per unit area. To the authors' knowledge there are no mathematical models that account for water losses through cracks. It is clear that this aspect of dredged material landfills is not well understood and offers a fruitful area for further investigation.

Vegetation

Another factor that may exert a significant influence on the behavior of a disposal site is colonization by volunteer vegetation. This influences the response of a given layer in much the same manner as the formation of a crust (except that evapotranspiration effects are even more complex than evaporation alone). In addition, when considered in the light of successive cycles of deposition-vegetation, substantial heterogeneity is introduced at the interfaces between layers. This, again, has not been incorporated into any existing models.

Other Factors

In addition to the major features (consolidation models with appropriate material characterization, boundary conditions, etc) addressed in this paper, several other factors may affect (possibly to a large degree) the capacity or sizing of a containment area. While these factors are inherently unrelated to any formal mathematical predictive model, they can be causes for substantial differences between predicted and actual response, because it is often difficult to quantify precisely the volume of solids that are placed in the site. As described by Lacasse, Lambe and Marr (1977), these factors are the (a) overdredging factor, which may be as high as 75%, (b) removal efficiency factor, which depends on such parameters as type of removal, type of sediment, pumping rate, tidal velocity, etc, (c) transport efficiency factor, which depends on the control (pipeline leaks, etc) exercised over the dredging contractor and the type of sediment, and (d) containment system efficiency, which depends on the amount of solids lost from the containment structure (e.g. over the effluent weir).

IMPLEMENTATION OF MODELS

Mathematical models of the type described herein incorporate a wide variety of conceptually new ideas and sometimes weakly documented physical relationships; in addition, there are often unknown construction irregularities and poorly understood operational problems that are not explicitly included in the formulation but nevertheless do affect the actual behavior. Many of these anomalies can be handled only by the ingenuity and expertise of an engineer who fully understands both (a) the development of the model with its inherent capability for special adaptations and (b) the pertinent factors of the field problem that are likely to influence the application of the model. In other words, just as a Winchester rifle or a Stradivarius violin require a skilled marksman or an accomplished artisan to obtain the best results, so also do these large strain models need a specially trained engineer to achieve the desired solution. This will become increasingly true as the models increase in complexity to handle a greater breadth of conditions.

The two examples described herein are intended to illustrate the types of assumptions, judgements, options, and modifications that must be addressed when using any of the analytical models. Accordingly, the objective is to emphasize the process that must be followed in the implementation of one of the typical models that are available, and not to tout the advantages of a particular model or to provide comparisons of predicted and measured behavior from case histories. The first example describes a current project where test fills are in progress to provide guidance for the future stabilization and reclamation of a newly formed dredged material landfill, and the second presents new analysis of data from a project that was undertaken over a decade ago (before large strain models became common and before the importance of the behavior at very low stresses was fully appreciated).

Seagirt Marine Terminal in Baltimore, Maryland

A field test program was undertaken to acquire data on which to base the design of a stabilization scheme (involving prefabricated wicks and surcharging) to reclaim an existing dredged material containment area, namely, the Seagirt Marine Terminal in Baltimore, Maryland. Two test fill areas, each with a different type of wick installed in a 5-foot grid pattern over the center 25-foot square zone of the fill, were constructed near one edge (due to access problems) of the 113-acre site. The centers of the two fills were about 150 feet apart, and each area was loaded with 7.5 feet of sand. The slightly crusted (on the order of 6 inches) surface of the dredgings was covered with a geofabric, followed by a 2-foot thick layer of sand, 94 feet square at the base, another layer of geofabric, and the 5.5 more feet of sand fill about 50 feet square at the base and 35 feet square at the top. The containment area had been filled in about one year and allowed to settle undisturbed for approximately two years prior to construction of the test fills.

Soil borings made after placing the 2-foot sand blanket but prior to placing the remaining 5.5-foot surcharge indicated highly non-uniform conditions, with moisture contents in the dredged material varying randomly between about 45% and 145%. The stratigraphy at the two test fill locations is shown in Figure 6. One sample taken in the black organic silt beneath the dredgings exhibited a water content of 186%, and two samples from the underlying gray clayey silt had water contents near 37%.

About one year before the fills were constructed two slurry consolidation tests were performed on samples of the dredged material recovered several hundred feet from the test fill locations. Because of the nonuniform nature of the material, combined with the desire to test a significant volume, it was decided to homogenize large portions of the samples. Material from depths of 3 to 8 feet, with an average water content of 98%, was placed in one consolidometer, and from depths between 8 and 12 feet, with an average water content of 82%, in the other. Attempts to perform permeability tests after each load increment were thwarted in most cases by the generation of gasses within the samples, and permeability was thus computed from observed settlement rates. Based on the data obtained, the following compressibility and permeability relationships were deduced:

$$e = 11.51(\bar{\sigma})^{-0.312} \tag{15}$$

$$k = 1.68 \times 10^{-4}(e)^{3.02} \tag{16}$$

where e is void ratio, $\bar{\sigma}$ is effective stress (in psf), and k is the coefficient of permeability (in feet/day).

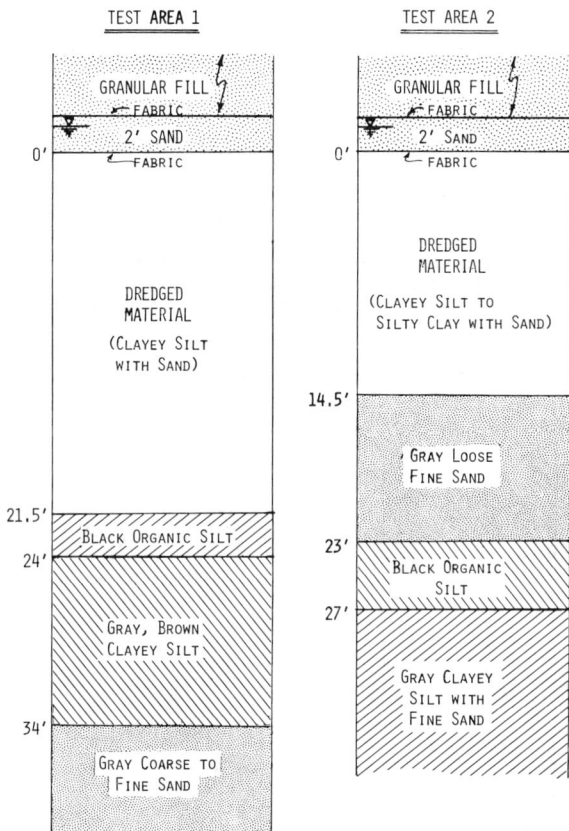

Figure 6.   Stratigraphy at Test Areas 1 and 2
(data from STV/Lyon Associates)

Test Areas 1 and 2 were idealized as initially containing 21 feet and 14 feet, respectively, of dredged material. Preliminary one-dimensional self-weight consolidation analyses were performed to estimate a likely, albeit ideal, initial void ratio distribution within the dredgings. Assuming an immediate (or "stress-free") water content of 150% (40% solids content), it was determined that 33.0 feet and 20.1

feet of unconsolidated slurry would have been required to produce equilibrium thicknesses of 21 feet and 14 feet, and that self-weight consolidation would be essentially complete within three years of instantaneous deposition.

The expected performance during and after placement of the test fills was estimated by use of a recently developed axisymmetric version of the "quasi-two-dimensional" finite strain model described by Somogyi et al (1984). Isotropic permeability was assumed, as were perfectly draining wicks. The computer analyses were performed without consideration of the crust that actually existed prior to fill placement. This simplification was deemed appropriate for two reasons. First, in light of the closely spaced drainage wicks, the crust would not significantly affect the settlement rate, regardless of its permeability. Second, the surcharging effect of the crust would be minor relative to the 7.5-foot granular fill. The crust, which was considered to be either 6 inches or 12 inches thick at a water content of 50%, was then taken into account rather simplistically by subtracting 6.9 inches or 13.8 inches from the computed settlements; these values represent the difference in height between a 6 inch or 12 inch crust and the thickness that the corresponding amount of solids would occupy at an initial water content of 150% (i.e. 12.9 inches or 25.8 inches).

The predicted and observed settlements are presented together with the actual and idealized fill placement histories in Figures 7 and 8.

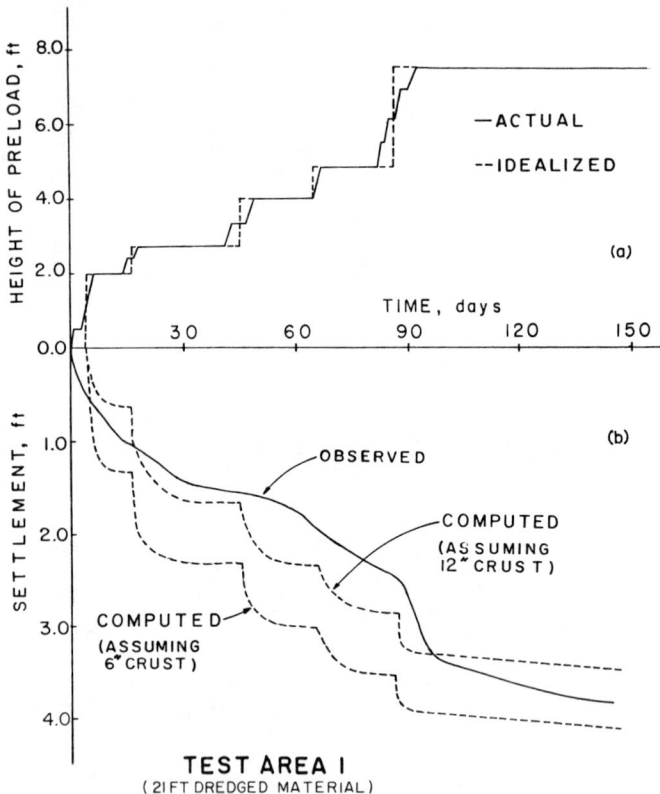

Figure 7.   Load and Settlement History at Test Area 1 (data from STV/Lyon Associates)

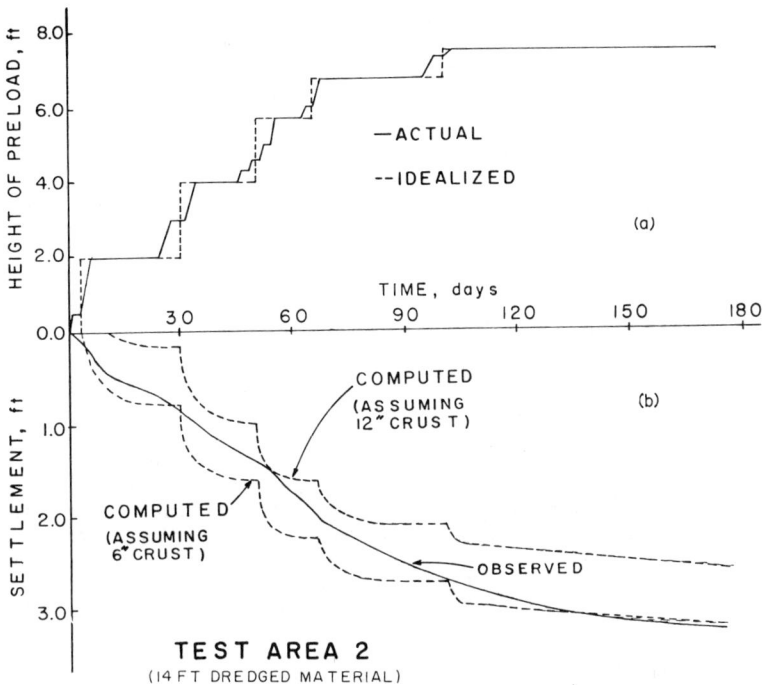

Figure 8. Load and Settlement History at Test Area 2 (data
from STV/Lyon Associates)

Very good agreement is obtained for Area 2 (Figure 8), with the observed
behavior throughout the entire history being bracketed by the predictions assuming the two different crust thicknesses, and the observed
settlement after completion of filling being well described by the
prediction using the 6-inch crust. For the case of Area 1 (Figure 7),
the agreement is not as good, with observed settlements being less
thanthose predicted until after the last surcharge lift was placed, when
the predictions are seen to bracket the observed settlement. Possible
explanations for this discrepancy include the use of inappropriate
permeability and/or compressibility relationships, significant settlement of the underlying foundation soils, secondary compression, the
effects of gas generation, reduction in effective surcharge loading
caused by submergence of bottom portions of the fill, downward seepage,
shear (rather than consolidation) movements, possible rupture of the
fabric, partial clogging of the wicks (which may have become unplugged
by the relatively high gradients produced during the rapid placement of
the large final lift), and false settlement plate readings. The last
two factors will be set aside as being improbable. The choice of inappropriate material properties is certainly possible, given the highly
variable nature of the deposit, but this would not explain the large

settlement observed under the final lift. An independent estimate
(using conventional analysis) of the settlement expected in the black
organic silt was on the order of a few inches, and this would not likely
occur only when the final lift was placed. Very little settlement was
expected from the low water content (37%) underlying clayey silt.
Secondary-type compression and/or gas generation may be responsible for
the large rapid settlement near 100 days, but this behavior was not
manifested at Area 2. A similar argument would tend to rule out the
effects of possible underseepage and reduced effectiveness of the surcharge because of bouyancy. This suggests that the most plausible
explanations for the unexpectedly large increment in the observed settlement of Area 1 at about 100 days is attributable to either shear flow
(squeezing) or fabric rupture, neither of which can be handled by any of
the models presently available.

While it would be possible at this stage to perform additional
analyses using appropriately modified material properties or more
detailed analyses incorporating a more correct simulation of crust
formation, underseepage during self-weight consolidation, and/or the
reduced effectiveness of the surcharge due to buoyancy, every effort
should be made in any situation like this to first identify the prime
cause for the existing discrepancy. There is always the danger that a
parameter study will fortuitously produce an "acceptable prediction" and
associated "peace of mind" without really identifying the true cause of
the problem. In this case, the placement of an additional small lift or
two, with special attention and appropriate instrumentation aimed at the
most likely possibilities, may reveal the cause for the discrepancy.

Finally, the basic difference between a finite strain (in its
present form) and incremental small strain approach to this problem is
worthy of mention. With an "incremental" model, the actual (experimentally determined) initial conditions (i.e. void ratio distribution prior
to surcharging) could be incorporated directly, provided the associated
material properties were known or could be estimated with confidence.
However, the finite strain technique required preliminary simulation of
the deposit formation process using averaged or homogenized material
properties to arrive at an initial condition (i.e. void ratio distribution) indicative of equilibrium under self-weight.   While it can
certainly be argued that inputting initial conditions based on actual
observations is more appropriate, especially for predicting the performance at this test location, the alternate approach may, in fact,
better simulate the overall performance of the entire site.

Disposal Sites in Toledo, Ohio

During the 10-year period from 1964 to 1973, approximately nine
million cubic yards of dredged bottom sediments were placed in four
disposal areas (Penn 7, Penn 8, Riverside, and the Island) in Toledo,
Ohio. These areas ranged in size from about 25 acres to 150 acres. In
general, the sediments were deposited periodically in each site, reaching depths on the order of 10 to 15 feet. Field sampling and testing
and associated laboratory testing was done over the four years from 1971
to 1974, inclusive, and the results have been well documented in a
series of reports and papers by Krizek and a host of coworkers. These
sediments were dredged by hopper dredge and pumped into the different

disposal areas in slurry form with a solids content between 10% and 20%, attaining a solids content between 30% and 40% after sedimentation. The ratio of sand:silt:clay was approximately 1:3:2, and the liquid limit and plastic limit of the fines were about 60 to 90 and 25 to 50, respectively. In most cases the organic content, as determined by loss on ignition, was between 4% and 8%.

A wide variety of material property tests were conducted on samples taken from the Toledo sites, but only pertinent results are presented here. In addition to the secondary consolidation data shown previously in Figures 4 and 5, a series of twelve slurry consolidation tests were performed; of these, two were started at 1 psi and the others at 2 psi. At 2 psi void ratios ranged from 2.5 to 6.0 with an average of 3.8, while at 10 psi the void ratio was between 1.0 and 2.2 with an average of 1.8. The coefficient of permeability was determined directly and indirectly (back-calculated from slurry and conventional consolidation test results) from a number of different laboratory tests, as well as two field tests, and the data are summarized in Figure 9 (Krizek, Jin, and Salem, 1974).

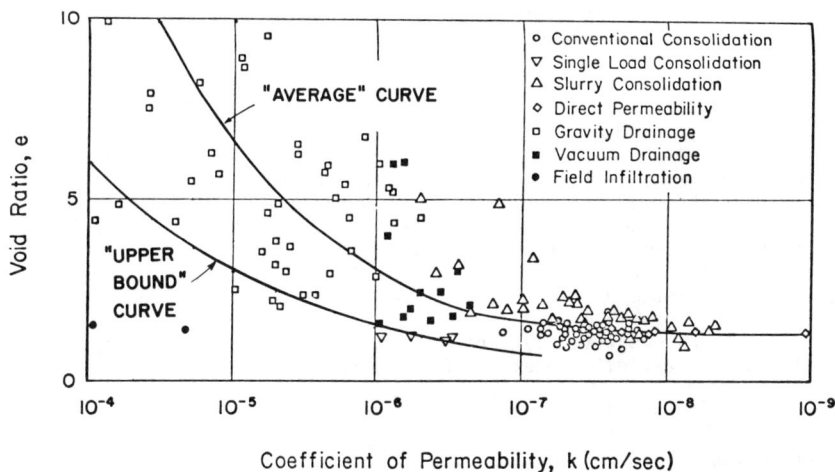

Figure 9. Summary of Permeability Data for Toledo Dredgings
(after Krizek and Salem, 1974)

Filling History-- In the case of the Penn 7 site, the filling history was documented from the beginning of the deposition process until the containment area was filled (Krizek and Salem, 1977). Between August 20, 1972 and December 20, 1972, 570,000 cubic yards of dredged material (bin volume) were pumped into this site. The pond was then left dormant until June of 1973, when the thickness varied between 4 feet and 12.5 feet (average of 8.25 feet) over the site. Surface desiccation was observed to have begun in Feburary of 1973. An additional 350,000 cubic yards (bin volume) of dredgings were pumped into the site between June 6, 1973 and August 20, 1973, and the site was monitored periodically for one year thereafter. The average solids content in the dredge hopper (bin) is not known, but the dredgings were pumped into the site at a measured average solids content of 15%. Based on contour surveys of the Penn 7 site before any dredged material was deposited (June 20, 1972) and one year after the area was filled (August 20, 1974), it was determined that 570,000 cubic yards of material were contained and, based on data from 32 undisturbed samples, the average dry density in 1974 was 52.5 pcf. Using this information, the total weight of solids contained in Penn 7 was computed to be 404,000 tons, and they occupied an average depth of 11.4 feet. The actual thickness of the deposit varied from about 7 feet near the overflow weir to about 15.5 feet in the vicinity of the inflow pipe. The natural water content of the dredgings ranged from 67% to 98%.

Input Parameters--The major parameters required to perform the settlement analysis are the size of the disposal area, solids inflow rate, initial ("stress free") solids content, and permeability and compressibility relationships. The solids inflow rate to the 31-acre site was calculated to be 0.75 x 10^6 ton/year for both the 122-day initial (1972) filling period and the subsequent (1973) 75-day period, and an initial solids content of 30% was assumed. A curve-fit of the upper bound permeability data shown in Figure 9 yielded

$$k(ft/day) = (8.9 \times 10^{-4})e^{3.5} \tag{17}$$

As previously mentioned, the slurry consolidation tests produced highly variable results. A preliminary calculation to estimate the equilibrium void ratio after self-weight consolidation indicated an average dry density of about 38 pcf using typical slurry consolidation test results. Furthermore, the equilibrium effective stress at the base of a 12-foot deposit having an average dry density equal to the observed 52.5 pcf is approximately 3 psi. Since virtually all of the slurry consolidation tests were started at a stress of 2 psi and gas generation was reported, especially at low pressures, it was decided to use the empirical curve-fit parameters suggested by Carrier, Bromwell and Somogyi (1983). For a soil with a plasticity index of 45, this relationship is

$$e = 7.0(\bar{\sigma})^{-0.29} \tag{18}$$

where $\bar{\sigma}$ must be specified in psf.

Analysis--The results of the analysis are shown in Figure 10a. The model predicts a deposit thickness of 7.0 feet just prior to the second filling, and 10.3 feet one year after the completion of filling, whereas the actual average heights at those times are 8.25 feet and 11.4 feet, respectively. The computed results are within 10 to 16 percent of the observed behavior and may represent sufficient accuracy, depending on the purpose of the analysis. If deemed necessary, improved results could be sought by first performing more detailed analyses which incorporate any actual site conditions that may be different than those assumed for the initial analysis. If sufficient accuracy is not obtained after exhausting such considerations, there is a sound basis for adjusting the assumed material properties within justifiable limits.

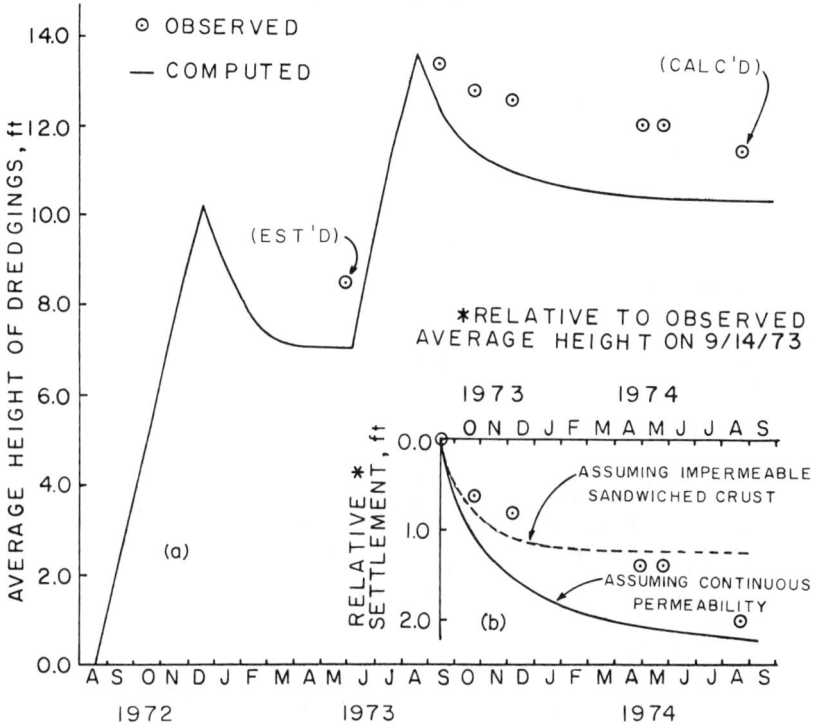

Figure 10.   Measured and Computed Settlement History at
Penn 7 (data from Krizek and Salem, 1974)

In this example, the discrepancy between computed and observed results could be attributed to the use of either a void ratio versus effective stress relationship that is "too low" (based primarily on the underestimation of the actual thickness of material at the end of the first filling) or a permeability relationship that is "too high" (based on the more rapid than actual rate of settlement computed after the second filling).   However, conditions that actually existed at the site, but which were not incorporated into the initial analysis, include the existence of a desiccated crust over most of the site prior to the second filling and a nonuniform deposit thickness (tapering from the inflow pipe to the overflow weir).

The greatest possible effect of the sandwiched crust was estimated by assuming it to be impermeable, thereby virtually halting consolidation of the underlying dredgings during and after the second filling. The resulting computed settlement relative to the observed average thickness shortly after filling is shown in Figure 10b.   Very close agreement can be seen until the summer of 1974, when surface desiccation and vegetation caused significant additional settlement.

As mentioned previously, a uniform deposit thickness was assumed in the analysis, whereas the deposit was actually tapered. Because of the nonlinearity in the compressibility relationship, the average void ratio in a tapered deposit is greater than that computed by assuming an average thickness. Although it is doubtful whether this would completely account for the indicated discrepancy (in light of the gentle taper and relatively low compressibility), at least a rough check should be made assuming equilibrium conditions. If this check shows a significant difference between calculated average void ratios, additional analyses should be performed, simulating the filling history at the thickest and thinnest sections. If this analysis does not produce sufficient accuracy, the material properties may require modification. In this case, the compressibility relation would be the prime candidate, with possible curves lying between the empirical one used initially and the best fit through the experimental data. If necessary, alternative permeabilty relationships should be chosen between the upper bound and the average of the laboratory data.

Desiccation--As described earlier, the model by Krizek and Casteleiro (1974) is capable of handling the flow of water through a vertically heterogeneous soil at any degree of saturation; accordingly, it can model desiccation due to evaporation and transpiration and consolidation for conditions of impeded bottom drainage. Prior to applying this model to the site conditions prevailing at the Penn 7 disposal area, the following empirical relationships were deduced for the compression index, C_c, and the specific water capacity, C:

$$C_c = 0.37\ e - 0.07 \tag{19}$$

$$C = -\frac{m\theta_s}{\psi_{cr}} \exp\left[-\frac{m\psi}{\psi_{cr}}\right] + \frac{n\theta_{cr}}{\psi_{cr}} \operatorname{sech}^2\left[\frac{n\psi}{\psi_{cr}}\right] \tag{20}$$

where ψ is the soil-water potential, ψ_{cr} is the limiting or air-dry soil-water potential, θ_s is the volumetric water content at saturation, θ_{cr} is the limiting or air-dry volumetric water content, and m and n are empirical material-dependent parameters). Typical comparisons of last stage (after second dredging season) settlements measured at three surface points in the landfill and those calculated by the model for various conditions are illustrated in Figure 11. For impeded bottom drainage in this example, the coefficient of permeability of the foundation soil was taken to be 10^{-5} cm/sec (based on field tests), and the coefficient of transpiration ranged from 0 to 1. The average thickness of the dredged material was taken to be 7.7 feet for the first dredging season and 4.7 feet for the second. These comparisons indicate very clearly the effect of transpiration on the settlement response and associated densification of dredged materials in a diked containment area. Within the context of the results predicted by this model, it appears that the enhancement of evapotranspiration (curves A, D, E, and F) offers greater promise than the improvement of bottom drainage (curves A, B, and C) for accelerating the rate of settlement.

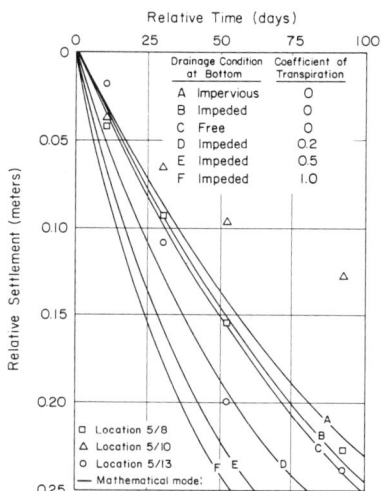

Figure 11. Last-stage Settlement Response at Penn 7 for Different Boundary Conditions (after Krizek and Casteleiro, 1974)

Storage Capacity--Although not related to the application of a mathematical model, it is worthwhile to present briefly some results of this well documented case history that (a) quantify the time-rate of densification of dredged material in a containment area and (b) relate the volume of material actually stored in a disposal area with the specified insitu volume designated to be dredged and an auxiliary bin-measure volume (which is that volume occupied by the sediment in the hopper of the dredge). During the years from 1971 to 1974 over 100 dry density measurements were made on undisturbed piston-tube samples of the dredged materials from each of the four sites, and these data were synthesized to develop a quantitative relationship for the rate of increase of dry density as a function of time. If it is assumed that the dredged material deposited in each of the sites is basically similar (this has been amply documented) and if spatial variations are handled by working with average values, the average density of a specified number of samples taken from a given site in a given year may be plotted as a function of the approximate number of years the material has undergone self-weight consolidation, as well as desiccation and/or evapotranspiration. These data, shown in Figure 12, indicate that the spoil will increase in dry density at a rate of about 4% per year. While this finding is admittedly site specific (climatic conditions, nature of dredged material, etc) and obviously can not prevail ad infinitum (the density will eventually become asymptotic to some limiting density), the trend appears to be clear for at least eight or ten years. It is significant, however, that no data are available for the first year, where considerable consolidation would be expected.

Information supplied by Boresch (Private Communication, 1974) of the U.S. Army Corps of Engineers, Detroit District, indicated that a bin-measure volume of 920,527 cubic yards was deposited into Penn 7 and that the bin-measure volume for a hopper dredge is approximately equal to the insitu volume of bottom sediment to be dredged divided by an empirical factor of 0.82. If the aforementioned 1974 disposal-area volume of 570,000 cubic yards is increased by 4% (as per Figure 12) to

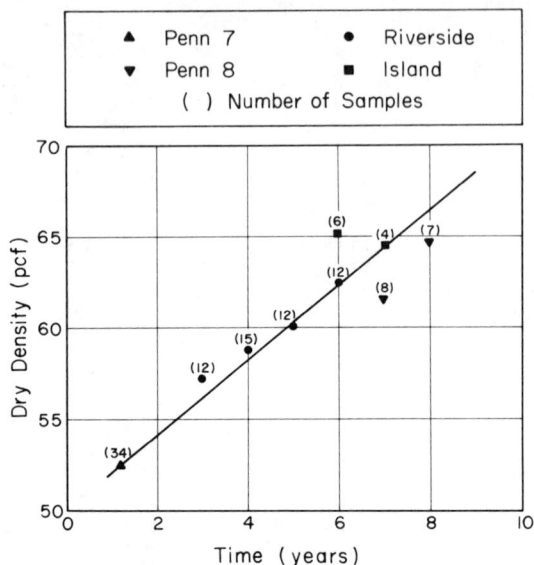

Figure 12. Measured Increase in Dry Density as a Function
of Time (after Krizek and Salem, 1974)

extrapolate a 1973 volume of newly placed sediment, the ratio between
disposal-area volume and bin-measure volume can be calculated as 0.65;
this is exactly the empirical value (Powell, Private Communication,
1973) used by the Detroit District of the U.S. Army Corps of Engineers
for this relationship. Finally, this suggests that the ratio between
disposal-area volume immediately after deposition and insitu volume is
0.8; note, however, that the insitu volume stated here has not been
adjusted by any correction factors to account for overdredging, removal
efficiency, transportation efficiency, and containment area efficiency.

QUO SUMUS ET QUO VADIMUS?

Based on the foregoing review and assessment of the present status
of our ability to analyze the response of landfills composed of slurried
dredged materials, the following questions are suggested.
1. To what degree of accuracy can we quantify the volumes and
    densities of the insitu materials to be dredged, and what is
    the effect of any inaccuracies on calculated predictions?
2. What are the advantages and disadvantages of the various mathe-
    matical models that have been developed?
3. How confidently can we specify the boundary and initial condi-
    tions and the material property relationships needed to
    exercise these models?
4. How important are the various field variations and uncertain-
    ties in predicting the behavior of a given landfill?

Volume of Material

When designing or sizing a disposal area, the insitu volumes and densities, together with the different empirical or experience factors that are used to adjust for overdredging, removal, transportation, and containment area efficiencies, are extremely important, because they provide a direct measure of the solids that are deposited in the disposal area. Without a reasonably accurate knowledge of the quantities of dredged materials that are actually placed, one cannot hope to properly predict the contained volume at any given time or to adequately size the area. Modern technology is capable of providing quite good estimates for the ideal volume of insitu material to be dredged, but various efficiency factors may introduce significant error in the volume that is actually dredged. To determine the volume of solids that will be dredged, it is necessary to know the insitu density or void ratio in addition to the gross volume. As in virtually all geotechnical engineering operations, obtaining undisturbed samples is a challenging task, and this challenge becomes even more awesome when sampling very soft material submerged under tens of feet of water. The predicted response for the landfill will be in error by an amount that is more-or-less proportional to the error in determining the volume of solids placed in the containment area. Even when analyzing the response of an already filled site, an accurate knowledge of the volume of solids is useful in conjunction with field sampling to assess whether or not equilibrium under self-weight has been attained.

Mathematical Models

Although the advantages and disadvantages of the various mathematical models and the alleged superiority of one model over another provide considerable "food" for argument, the simple fact is that most of the available models can be adapted to handle comparable cases. To some extent it may be said that each model is as good or versatile as its proponent wants it to be, with details such as computer efficiency and economy remaining privileged information. In any event, it is more than likely that the analytical capability of any chosen model exceeds our ability to specify correct boundary conditions and to input proper relationships for the material properties. Constantly improving computer technology is becoming increasingly conducive to undertaking "million dollar analyses" based on "two dollars worth of data".

Notwithstanding this criticism (which is applicable to more than large strain consolidation problems), certain advantages seem apparent for the different analytical formulations. Finite strain models are most appropriate for the analysis of conditions during and immediately after filling (i.e. while the deposit remains submerged) because the formulation utilizes material coordinates and incorporates self-weight directly; the choice of coordinates has the advantages that the position of the boundary is always known and there is no need to update the mesh. Alternatively, incremental small strain models are rather inefficient because the material properties vary rapidly and therefore dictate the need for frequent updating and mesh redefinition. Based on the models which are documented in the literature to date, incremental small strain models appear to be the most attractive for handling layered (continuous) deposits, possibly including sandwiched crusts; to handle this situation finite strain models would require the incorporation of internal boundary conditions. However, the permeability of a sandwiched crust is very difficult to quantify. The effects of surface crust formation can apparently be simulated with either type of model, but further study is required to enable an analytical (as per Krizek and Casteleiro, 1974), rather than empirical, modeling of the physical process.

Although confidence in incremental small strain models and linearized finite strain models increases as nonlinearity decreases, it is a basic fact that the actual behavior is highly nonlinear. Caution must be exercised against "over-linearizing" for sake of mathematical convenience (or necessity) and thereby obtaining a solution to a problem that doesn't really exist. Alternatively, the "real" problem can be addressed "head on" by using highly nonlinear field equations, in which case it will be necessary to remain aware of possible numerical instabilities and convergence difficulties. If well defined stability and convergence criteria cannot be obtained, replicate analyses with different mesh sizes must be performed, particularly during those early periods when the behavior is the most nonlinear. Since each of the available models contains a variety of "internal idiosyncrasies" which are best (and perhaps only) understood by its proponent, special care must be taken to avoid the possibility of giving large-strain theories a bad reputation by making particular models generally available to the profession without a proper appreciation for their limitations and uncertainties, together with the required level of material characterization. Unless the models are diligently "tuned" and "played" by the hands of their proponents, it is highly likely that the resulting "music" will not be pleasing. On the other hand, should unpleasant "music" be experienced, it is not clear whether the cause is the instrument or the player.

Material Properties

Perhaps the greatest shortcoming in applying a given mathematical model lies in specifying the boundary conditions and material properties. At the lower boundary, it is probable that impeded drainage and non-hydrostatic (relative to the upper boundary) steady-state pore fluid pressure exist in many situations, but few models account for it. Similarly, the formation of a crust almost certainly alters the normally assumed free-drainage condition at the surface, but little has been done to handle this condition in a manner that reflects the physical process that occurs. Determining the "initial" or "stress free" void ratio of a newly sedimented deposit is a difficult task; fortunately, in most practical problems its effect is manifested most strongly in the top few inches of sediment and reasonable "ball park" estimates will usually suffice. As with probably every aspect of geotechnical engineering, specifying realistic relationships for the material properties is paramount to the successful use of a mathematical model. In this case the required compressibility and permeability relationships are especially challenging because they span such a broad range and extend into regions (viz, slurries) that have not been comprehensively studied heretofore. Of these two relationships, the one for permeability is undoubtedly the most critical because it is the most difficult to determine experimentally and it influences strongly the predicted time-rate of settlement. While the proper choice of drainage conditions at the boundaries is important, the specification of an unrealistic permeability relationship can obscure boundary effects.

In a general sense, the compressibility and permeability relationships may be viewed as encompassing three phases; the behavior during each of these phases may be regarded as very highly nonlinear, highly nonlinear, and nonlinear, respectively. The first phase begins after sedimentation is complete and includes the very early stages in the formation history of a soil (immediately after particle interactions become influential). This period is particularly significant in problems involving the periodic disposal of dredging slurries in containment areas because it is the period during which large settlements will occur very quickly. Although the finite strain models probably have their greatest advantage during this phase, the very highly nonlinear material relationships are most difficult to quantify. The relationships determined during this phase will probably be highly dependent on the expertise of the personnel, the specialized equipment employed, and the methods of interpretation utilized. As such, it is not likely that universal relationships will be developed for use interchangeably with any of the available models; rather, the specific relationships formulated will be tailored to the needs and idiosyncrasies of a particular model.

The second phase may be considered as beginning after the extremely large, rapidly occurring settlements have taken place. While the origin of this phase can not be identified quantitatively, it will likely begin some weeks after deposition or when the liquidity index reaches a value on the order of two. The termination of this phase might be taken as the point when the liquidity index is about unity or when the vertical effective stress reaches about 4 or 5 psi. While the dredged materials during this phase would certainly be characterized as very soft, they are nevertheless amenable to sampling and testing by means of techniques that are modest modifications or extensions of those for which a reasonable background of usage exists and are thereby relatively free of operator bias. During this highly nonlinear phase it would appear that the finite strain and incremental strain models are both applicable, with the choice being dependent in large part on model availability and data compatibility. For example, a profile with clearly identifiable layering would suggest the use of an incremental strain model, whereas the converse would be true if the deposit was sensibly homogeneous.

If the fill were to be loaded externally (such as by surcharging or when used for storage) or if the deposit were deep, the third phase of these compressibility and permeability relationships will come into play. In this case the problem is encroaching into the realm of conventional geotechnical engineering where a broad base of experience is available. To be sure, the sediments will be very soft and the settlements will still be large, but standard sampling and testing methods are increasingly applicable, analyses and interpretations are more conventional, and extrapolations become more comfortable. The logic underlying the incremental small strain models becomes increasingly apparent during this phase because the piece-wise linear increments can be larger. It is during this stage that more conventional approaches, such as that presented by Krizek and Krugmann (1972), become more useful--at least for providing first-order estimates. However, finite strain models should not be considered inappropriate during this phase; in fact, some proponents of finite strain theory predict that it will be used routinely for all consolidation analyses within a decade. It must be appreciated that the three arbitrary phases described above may each prevail simultaneously at different points in a given landfill; in a deep deposit the lower material may be in the third phase while the freshly deposited surface material will be in the first phase.

The major factor that affects the range in which the compressibility and permeability relationships must be known is the nature of the disposal operation. If the site is planned for use in the future, the emphasis will be on the early phases when the void ratio is very high; this will be especially true if dredged material will be deposited in the containment area periodically. Such an operation provides an excellent opportunity, with proper field measurements, to evaluate the effectiveness of any particular model for predicting the response because there are several chances to apply it. Since the design of the site and its predicted storage capacity will necessarily have had to been done on the basis of empirical compressibility and permeability correlations with plasticity data from disturbed samples of insitu bottom sediments, the prediction may not be consistent with measured field response. The measured parameters should include settlements, pore pressures, and densities, together with inflow quantity, all as a function of time. If measured and calculated response do not agree, one can either "tune" the model or seek improved material property relationships from sampling programs. Periodic samples, if carefully taken, can yield data for assessing the degree of homogeneity in the deposit. Since both the model formulation and the material property determinations are relatively novel and certainly unproven to any large degree, it seems advisable to seek comparisons with as many parameters as possible; consequently, every effort should be made to measure and compare several different parameters (settlement, pore pressure, density, etc) rather than only one, even at the expense of fewer measurements of each individual parameter. Resampling and updating might be undertaken several times during the filling process.

On the other hand, if the site has already been filled or if it will be loaded (as a storage yard), the problem is quite different. Depending on the age of the fill to be reclaimed, the consolidation process maybe considered in either the second or third phases, or perhaps some combination. The situation becomes particularly critical when attempting to surcharge sediments that are in place but not yet fully consolidated under self-weight. Under these conditions the conduct of slurry consolidation or possibly conventional consolidation tests on sampled materials is recommended. Empirical correlations, while useful for characterizing material properties over a broad range, are often not sufficiently accurate for specific types of material. However, they might serve well with a particular model to bound the solution in the absence of direct experimental data.

Although empirical correlations for material properties provide a useful role in allowing preliminary estimates of the response, these relationships should ideally stem from laboratory or field test data to achieve improved predictions. In general, laboratory sedimentation and slurry consolidation tests or field sampling and measurements (densities, pore pressures, etc) offer the greatest promise for providing the most reliable material property relationships. The "stress free" void ratio and the compressibility relationship at low values of effective stress can be determined quite adequately from a series of laboratory sedimentation tests utilizing several different heights of slurry with a constant solids content that is approximately the same as that of the slurry to be deposited in the containment area. Alternatively, the initial void ratio could be obtained from a representative field sample taken within the top few inches. The compressibilty can best be determined from a slurry consolidation test; to better interpret this test, one-way upward drainage should be used and total stresses and pore pressures should be measured at the bottom to allow an assessment of side friction, piston friction, and gas generation. There is a need to better understand and quantify the effects of gas generation in these often highly organic dredgings on the compressibility, unsaturated permeability, and pore pressure measurements.

Permeability determinations are best made directly on a specimen in a slurry consolidometer by applying either a head difference across or flow through the specimen between load increments. Calculations can be made to obtain permeability values indirectly from the consolidation data, but a mathematical model must be employed to analyze the data and this leads to the classic dilemma of "circular reasoning" and introduces another potential source for error. If this latter approach is followed, it is preferred to make the calculation after most (perhaps 90%) of the consolidation has occurred, because the void ratio will tend to be more uniform throughout the sample than at a lower degree of consolidation. It is recommended that the inevitable scatter in permeability test data (if more than one test is conducted) be handled by giving added weight to the "upper bound" values, because the lowest void ratio throughout the sample exerts the greatest effect on the flow quantity, which is usually associated with the average void ratio of the sample. While compressibility data may also exhibit scatter, a "best fit" average curve is probably justifiable because the settlement response is an integrated parameter and therefore more forgiving of local nonhomogeneities. When confronted with a heterogeneous sample on which to conduct a slurry consolidation test, homogenization will usually be desirable because the analysis of the test data will be greatly simplified (indeed, it may be impossible otherwise) and the measured property will be more representative of the overall mass.

Another question of concern is whether to use actual discrete data with an interpolation scheme or a continuous function with the mathematical model. Experimental data are rarely smooth, and this is usually due to various non-material factors, whereas nature is inevitably a "smooth operator". This suggests the use of a smooth continuous "best fit" curve (perhaps tempered by engineering judgement) for material relationships to avoid the direct reflection of experimental artifacts in the analysis (for example, the compression index, rather than laboratory-measured values for void ratio and effective vertical stress, is used to estimate field behavior in conventional settlement analyses). However, since laboratory data are often described by exponential or power relationships, care must be exercised to ascertain the sensitivity of a particular fit; although virtually all data "look good" when plotted on a log-log scale, exceptionally shallow or steep "best fit" lines suggest a very sensitive relationship. Unavoidable cases of widely different, but equally plausible, material property relationships can be handled by determining alternative solutions and hopefully bracketing the real solution. A sensitivity analysis is advisable even when the property relationships are considered to be well documented.

OVERVIEW

Predicting the time-dependent large-deformation behavior of a sedimented landfill composed of slurried dredged materials is an extremely challenging problem that encompasses new developments in mathematical modeling (and even solution techniques), laboratory testing and associated interpretations, material property formulations, and indeed even engineering judgement. No standard analyses exist, and limited background experience dictates a cautious approach to empirical extrapolations. Without doubt, there is a pressing need for comprehensive and well documented case histories on which to perform post facto analyses in the quest for a logical methodology to ultimately advance a priori positions. Notwithstanding criticisms that have been leveled at the approaches taken, large strain analyses by whatever technique provide at worst a rational procedure for extrapolating observed settlement data and at best an accurate prediction of anticipated settlements with due consideration for containment area geometry, drainage provisions, crust formation, and material property variations; hence, alternative disposal strategies can be assessed with a reasonable degree of confidence.

Despite the sophistication that might be incorporated into a given model or the apparent preciseness of the formulation, it is probable that actual field observations will differ from the predicted behavior. In such cases the reasons why the prediction went awry (e.g. improper boundary conditions, inadequate material property relationships, error in the amount of material deposited, etc) must be identified; efforts must be made to obtain appropriate improvements; and the model proponent must be ready and willing to make proper revisions to update the predictions. Even a correct prediction does not necessarily validate the model or its input, because the complexity of the many factors involved allows the "right" overall behavior to be obtained by using a fortuitous combination of erroneous material properties, boundary conditions, and mathematical formulation. This possibility can be minimized by comparing theoretical and measured values for a variety of individual parameters (pore pressures, densities, water contents, depth of crust, etc). As always in engineering practice, the level of computational effort and the extent of testing and field instrumentation should reflect directly the consequences of an erroneous prediction. Throughout the trying times and seemingly endless criticisms that serve to enhance our ability to handle successfully this highly nonlinear problem frought with uncertainties, we must keep the situation in perspective; toward this end, remember that we are usually more than willing to accept a ±20% error in most of our analyses. Can we expect better for this problem?

ACKNOWLEDGEMENTS

The data for the Seagirt Marine Terminal analysis were obtained from reports by STV/Lyon Assoicates to the Maryland Port Administration. Grateful appreciation is extended to Christophe Glock and Bertrand Palmer who assisted in the analyses and preparation of this paper.

REFERENCES

Been, K., and Sills, G. C. (1981), "Self-weight Consolidation of Soft Soils: An Experimental and Theoretical Study", Geotechnique, Volume 31, Number 4, pp. 519-535.

Brown, K. W., and Thompson, L. J. (1977), Feasibility Study of General Crust Management as a Technique for Increasing Capacity of Dredged Material Containment Areas, Technical Report D-77-17, U. S. Army Engineer Waterways Experiment Station, Vicksburg, Mississippi.

Cargill, K. W. (1982), Consolidation of Soft Layers by Finite Strain Analysis, Misc. Paper GL-82-3, U.S. Army Engineer Waterways Experiment Station, Vicksburg, Mississippi.

Cargill, K. W. (1983), Procedures for Prediction of Consolidation in Soft Fine-Grained Dredged Material, Technical Report D-83-1, U. S. Army Engineer Waterways Experiment Station, Vicksburg, Mississippi.

Cargill, K. W. (1984), "Prediction of Consolidation of Very Soft Soil," Journal of the Geotechnical Engineering Division, American Society of Civil Engineers, Volume 110, Number 6, pp. 775-795.

Carrier, W. D., and Beckman, J. F. (1984), "Some Recent Observations on the Fundamental Properties of Remolded Clays", Geotechnique, vol. 34, no.2, pp. 211-228.

Carrier, W. D., and Bromwell, L. G. (1980), "Geotechnical Analysis of Confined Spoil Disposal", Proceedings of the Ninth World Conference on Dredging, Vancouver, Canada, pp. 313-324.

Carrier, W. D., and Bromwell, L.G. (1983), "Disposal and Reclamation of Mining and Dredging Wastes", Preceedings of the Seventh Panamerican Conference on Soil Mechanics and Foundation Engineering, Vancouver, Canada, pp. 727-738.

Carrier, W. D., Bromwell, L. G., and Somogyi, F. (1983), "Design Capacity of Slurried Mineral Waste Ponds", Journal of the Geotechnical Engineering Division, American Society of Civil Engineers, Volume 109, Number GT 5, pp. 699-716.

Carrier, W. D., and Keshian, B. (1980), "Measurement and Prediction of Consolidation of Dredged Material", Proceedings of the Twelfth Dredging Seminar, Center for Dredging Studies, Texas A & M University, College Station, Texas, pp. 63-105.

Gibson, R. E., Schiffman, R. L., and Cargill, K. W. (1981), "The Theory of One-Dimensional Consolidation of Saturated Clays. II. Finite Nonlinear Consolidation of Thick Homogeneous Layers", Canadian Geotechnical Journal, Volume 18, pp. 280-293.

Giger, M. W., Franklin, A. G., and Krizek, R. J. (1973), "Electrical Resistivity Survey of Two Dredging Disposal Sites," Bulletin of the Association of Engineering Geologists, Volume 10, Number 12, pp. 107-119.
Koppula, S. D. (1970), The Consolidation of Soil in Two Dimensions and with Moving Boundaries, Doctoral Dissertation, University of Alberta, Edmonton, Canada.

Koppula, S. D., and Morgenstern, N. R. (1982), "On the Consolidation of Sedimenting Clays", Canadian Geotechnical Journal, Volume 19, pp. 260-268.

Krizek, R. J. (1976), "Spatial Nonhomogeneity of Dredged Materials in Confined Disposal Areas", Proceedings of the Seventh World Congress on Dredging, San Francisco, California, pp. 779-797.

Krizek, R. J., and Casteleiro, M. (1974), Mathematical Model for One-Dimensional Desiccation and Consolidation of Dredged Materials, Technical Report No. 3, Department of Civil Engineering, Northwestern University, Evanston, Illinois.

Krizek, R. J., Franklin, A. G., and Soriano, A. (1974), "Seismic Survey of a Hydraulic Landfill Composed of Maintenance Dredgings," Bulletin of the Association of Engineering Geologists, Volume 11, Number 3, pp. 173-202.

Krizek, R. J., and Giger, M. W. (1974), "Storage Capacity of Diked Containment Areas for Polluted Dredgings", Proceedings of the Sixth World Conference on Dredging, Taipei, Taiwan, pp. 354-364.

Krizek, R. J., Jin, J. S., and Salem, A. M. (1974), "Permeability and Drainage Characteristics of Dredgings", Proceedings of the Seventh Dredging Seminar, Center for Dredging Studies, Texas A & M University, College Station, Texas, pp. 154-179.

Krizek, R. J., Karadi, G. M., and Hummel, P. L. (1973), Engineering Characteristics of Polluted Dredgings, Technical Report No. 1, Department of Civil Engineering, Northwestern University, Evanston, Illinois.

Krizek, R. J., and Krugmann, P. K. (1972), Placement Rates for Highway Embankments, Technical Report, Project IHR-602, Illinois Department of Transportation, Springfield, Illinois.

Krizek, R. J., and Salem, A. M. (1974), Behavior of Dredged Material in Diked Containment Areas, Technical Report No. 5, Department of Civil Engineering, Northwestern University, Evanston, Illinois.

Krizek, R. J., and Salem, A. M. (1977), "Field Performance of a Dredgings Disposal Area", Proceedings of the Conference on Geotechnical Practice for Disposal of Solid Waste Materials, American Society of Civil Engineers, pp. 358-383.

Lacasse, S. E., Lambe, T. W., and Marr, W. A. (1977), Sizing of Containment Areas for Dredged Material, Technical Report D-77-21, U. S. Army Engineer Waterways Experiment Station, Vicksburg, Mississippi.

Monte, J. L., and Krizek, R. J. (1976), "One-Dimensional Mathematical Model for Large-Strain Consolidation", Geotechnique, Volume 26, Number 3, pp. 495-510.

Olson, R. E., and Ladd, C. C. (1979), "One-Dimensional Consolidation Problems", Journal of the Geotechnical Engineering Division, American Society of Civil Engineers, Volume 105, Number GT1, pp. 11-30.

Pane, V., and Schiffman, R. L. (1981), "A Comparison Between Two Theories of Finite Strain Consolidation", Soils and Foundations, Japanese Society of Soil Mechanics and Foundation Engineering, Volume 21, Number 4, pp. 81-84.

Richie, J. T., and Adams, J. E. (1974), "Field Management of Evaporation from Soil Shrinkage Cracks", Proceedings of the Soil Science Society of America, Volume 38, pp. 131-134.

Salem, A. M., and Krizek, R. J. (1975), "Secondary Compression of Maintenance Dredgings", Proceedings of the Fifth Panamerican Conference on Soil Mechanics and Foundation Engineering, Buenos Aires, Argentina.

Salem, A M., and Krizek, R. J. (1976), "Stress-Deformation-Time Behavior of Dredgings", Journal of the Geotechnical Engineering Division, American Society of Civil Engineers, Volume 102, Number GT2, pp. 139-157.

Samarasinghe, A. M., Huang, Y. H., and Drnevich, V. P. (1982)," Permeability and Consolidation of Normally Consolidated Soils", Journal of the Geotechnical Engineering Division, American Society of Civil Engineers, Volume 108, Number GT6, pp. 835-850.

Schiffman, R. L. (1980), "Finite and Infinitesimal Strain Consolidation", Journal of the Geotechnical Engineering Division, American Society of Civil Engineers, Volume 106, Number GT2, pp. 203-207.

Schiffman, R. L., and Cargill, K. W. (1981), "Finite Strain Consolidation of Sedimenting Clay Deposits", Proceedings of the Tenth International Conference on Soil Mechanics and Foundation Engineering, Stockholm, Sweden, Volume 1, pp. 239-242.

Sheeran, D. E., and Krizek, R. J. (1971), "Preparation of Homogeneous Soil Samples by Slurry Consolidation", Journal of Materials, American Society for Testing and Materials, Volume 6, Number 2, pp. 356-373.

Somogyi, F. (1979), Analysis and Prediction of Phosphatic Clay Consolidation: Implementation Package, Technical Report, Florida Phosphatic Clay Research Project, Lakeland, Florida.

Somogyi, F., Carrier, W. D., III, Lawver, J. E., and Beckman, J. F. (1984) "Waste Phosphatic Clay Disposal in Mine Cuts," Proceedings of the Symposium on Sedimentation/Consolidation Models, American Society of Civil Engineers, San Francisco, California.

Umehara, Y., and Zen, K. (1982), "Consolidation Characteristics of Dredged Marine Bottom Sediments with High Water Content", Soils and Foundations, Japanese Society of Soil Mechanics and Foundation Engineering, Volume 22, Number 2, pp. 40-54.

Yong, R. N., Siu, S. K. H., and Sheeran, D. E. (1983), "On the Stability and Settling of Suspended Solids in Settling Ponds. Part I. Piece-wise Linear Consolidation Analysis of Sediment Layer", Canadian Geotechnical Journal, Volume 20, pp. 817-826.

Zimmerman, R. E., Chen, W. W. H., and Franklin, A. G. (1977), "Mathematical Model for Solid Waste Settlement", Proceedings of the Conference on Geotechnical Practice for Disposal of Solid Waste Materials, American Society of Civil Engineers, pp. 210-226.

Znidarcic, D. (1982), Laboratory Determination of Consolidation Properties of Cohesive Soil, Doctoral Dissertation, University of Colorado, Boulder, Colorado.

Znidarcic, D., and Schiffman, R. L. (1981), "Finite Strain Consolidation: Test Conditions", Journal of the Geotechnical Engineering Division, American Society of Civil Engineers, Volume 107, Number GT5, pp. 684-688.

Reporter's Summary

Prediction and Validation of Consolidation for Hydraulically-Deposited Fine-Grained Soils

John E. Garlanger, M., ASCE*

Introduction

The ability to predict the magnitude and rate of consolidation of hydraulically-deposited fine-grained soils is important to those engineers who are responsible for the design and reclamation of mine tailings impoundments, industrial waste disposal areas, and dredged spoil containment areas. The size of the disposal areas, the long-term settlement of the reclaimed surface of the disposal area, and the shear strength of the sediment within the areas are among the design factors which depend upon consolidation of the impounded material.

Five papers dealing with the sedimentation and consolidation of fine-grained soils have been submitted to this symposium. Four of the papers deal primarily with the consolidation of slurries under their own weight, while the remaining paper deals with the consolidation of dredge spoil and normally consolidated river sediment under surcharge loading. Two of the papers describe and analyze the sedimentation and consolidation of slurries in laboratory column tests, one of the papers discusses the use of finite strain consolidation theory in the planning and design of mine tailings impoundments, and one paper discusses the development of a graphic tool which can be used to analyze and describe numerous properties and behavior of slurries composed of mixtures of coarse- and fine-grained materials. The last paper describes the design, instrumentation and performance of test fills constructed over a dredge spoil area.

Summary of Paper by Lin and Lohnes

Lin and Lohnes in their paper entitled "Sedimentation and Self Weight Consolidation of Dredge Spoil" analyze the settling/consolidation behavior of a highly plastic silt obtained from a proposed dredging site in Guthrie County, Iowa. The tests were performed in 5.5-inch diameter by 72-inch high plexiglas columns. Lin and Lohnes' analysis of the test results suggest that self weight consolidation begins when the slurry reaches a critical concentration. At the critical concentration a distinct interface forms in the settling column. When the slurry was introduced into the settling column at an initial concentration greater than the critical concentration, the interface formed immediately.

*Principal and Chief Engineer, Ardaman & Associates, Inc., P.O. Box 13003, Orlando, Florida 32809.

Analysis of the data indicated that the rate of consolidation of the sediment could be characterized by a coefficient of consolidation and that the consolidation coefficient could be determined graphically from the time-settlement curve of the sediment interface. The finite strain consolidation theory, as modified by Been & Sills (1), was used to develop the proposed curve-fitting procedure.

The procedures for determining the coefficient of consolidation from the test results is described in the paper. The coefficient of consolidation for the test soil had a minimum value at an initial concentration close to the critical concentration.

The paper by Lin and Lohnes provides experimental evidence that what previous researchers referred to as zone settling is actually the result of self weight consolidation. However, additional work is needed to demonstrate whether or not the coefficient of consolidation determined by the zone settling tests and the Been & Sills analytical model can be used to predict consolidation performance of actual disposal areas. The results of the analytical work described by Carrier et al. (3), which suggest that the time-settlement curve for a given height of solids is independent of the loading rate, could prove useful in interpreting the results of full-scale field tests utilizing the Been & Sills model.

Summary of Paper by Elder and Sills

Elder and Sills in their paper entitled "Time Dependent Stress-Strain Behavior of Hydraulically-Deposited Soft Soils" analyze the creep (or delayed compression) of a high plasticity silt during sedimentation/consolidation in 4-inch diameter plexiglas cylinders. The sample heights varied from approximately 8 inches to approximately 138 inches. Density and pore water pressure were measured with depth in the columns and total stresses were measured at the base of the columns.

Elder and Sills also documented a critical concentration, similar to that reported by Lin and Lohnes, at which the slurry behaved as a consolidating soil. Of much greater importance, however, was the finding that the settlement behavior of a gradually deposited sediment was very different from that of an instantaneously deposited slurry. This finding suggests that the stress and depositional history of a sediment may be of major importance in evaluating the sedimentation/ consolidation behavior of hydraulically-deposited fine-grained soils.

Analysis of the data from several experiments indicated that the void ratio of the consolidating sediment remained fairly constant as the effective stress increased during the early stages of consolidation and then decreased substantially at nearly constant effective stress during later stages of consolidation. No unique relationship between void ratio and effective stress followed by all elements of the consolidating sediment could be derived. The behavior of the sediment in the column experiments was similar to that described by Bjerrum (2) for normally consolidated, highly plastic marine clays. Under field conditions, volume change will occur as a result of changes in effective stress as well as by creep and analytical models used to predict the sedimentation/consolidation behavior of hydraulically–deposited fine–grained soils may have to include the superposition of both effects.

Summary of Paper by Caldwell et al.

The paper by Caldwell et al. entitled "Application of Finite Strain Consolidation Theory for Engineering Design and Environmental Planning of Mine Tailings Impoundments" describes three case histories where finite strain consolidation theory and laboratory–determined soil parameters were used to predict the consolidation behavior of hydraulically–deposited mine tailings. The results of the laboratory testing and the computer simulation of the tailings deposition are presented for four low plasticity silts: a massive sulfide tailings and gold tailings from three different ore bodies.

In each case, the computer simulation showed that the tailings will be essentially fully consolidated by the time deposition has ceased and, consequently, the impoundment embankment could be designed for the consolidated sediment thickness. The effect of desiccation on the surface of the tailings deposit was not investigated. Use of conventional consolidation theory with the unconsolidated sediment height resulted in significantly longer consolidation times than predicted using the finite strain theory.

Summary of Paper by Scott and Cymerman

The paper by Scott and Cymerman entitled "Prediction of Viable Tailings Disposal Methods" proposes the use of a sand–fines–water diagram as a tool for analyzing and describing the properties and behavior of slurries composed of both coarse- and fine–grained soils. The boundaries between sedimentation and consolidation states, segregating and non–segregating mixtures, pumpable and non–pumpable mixtures, and fluid and solid states can be shown on the diagram for the complete range of mixtures from predominantly coarse to predominantly fine. The triangular–shaped diagram shows the total solids content on one side of the triangle, the percent fines in the total dry solids (or sand–clay ratio) along the base of the diagram, and the water content and/or void ratio on the remaining side of the triangle. The paper also describes experimental programs used to determine these boundaries for a particular set of waste materials and then uses the diagram to illustrate various disposal options for an oil sand mine where 90 million tons of oil sand are mined and processed annually.

This paper is well presented and the use of the ternary diagram should prove quite useful in planning, designing and controlling tailings disposal operations. The tests developed to locate the segregating–nonsegregating boundary should prove to be particularly useful for designing and evaluating sand–clay mix disposal areas.

Summary of Paper by Morin et al.

The paper by Morin et al. entitled "Stabilization of Dredged Slurry for the Seagirt Marine Terminal, Baltimore, Maryland" describes the design, construction, instrumentation and performance monitoring of a test fill program carried out within a dredged spoil area. The dredged spoil consists predominantly of silt mixed with variable amounts of clay and fine sand and contains a significant amount of organic matter. The natural moisture contents of the dredged spoil ranged from 50 to over 200 percent. The purpose of the test program was to evaluate the performance of two types of prefabricated wick drains.

The instrumented test fill program includes a control section without artificial drains and two identical drained sections on either side of the control section. Drains were installed on 5-foot centers. The two types of drains evaluated were Mebra drain, which consists of a plastic ribbed core surrounded by an envelope of bonded filter fabric, and Desol, which is a needle-punched geotextile filter with hollow interior channels.

Instrumentation consisted of slope inclinometers to monitor lateral displacement of the foundation soils, piezometers and settlement platforms. Time-settlement curves for the wick sections, which flattened out after 120 to 125 days, indicated ultimate measured settlements between 80 and 90 percent of the calculated values. Both types of wick drains performed satisfactorily in accelerating the rate of consolidation. The value of the horizontal coefficient of consolidation backfigured from the measured time-settlement curves for both wick sections were essentially the same as the vertical value determined from laboratory testing. Comparison of the time-settlement curves for the drained sections with that of the control section indicate that the drained sections settled approximately 3.7 times faster than the control section. The paper did not discuss the data obtained from the piezometers.

Appendix-References

1. Been, K. and Sills, G., "Selfweight Consolidation of Soft Soils: An Experimental and Theoretical Study," Geotechnique, Volume XXXI, 1981, pp. 519-535.

2. Bjerrum, L., "Engineering Geology of Norweigian Normally-Consolidated Marine Clays as Related to Settlements of Buildings," Geotechnique, Volume XVII, Number 2, June 1967, pp. 83-118.

3. Carrier, D. W., Bromwell, L.G., Somogui, F., "Design Capacity of Slurried Mineral Waste Ponds," Journal of Geotechnical Engineering, Volume 109, No. 5., pp. 699-716.

GENERAL REPORT FOR SESSION II: VALIDATION AND FIELD STUDIES
by

J.H.A. Crooks*

The papers submitted to this session deal with a wide variety of material types - organic soils, permafrost, hydraulic clay fill and
in-situ soft clays. Further, the treatment of the subject matter is
quite different with varying emphasis on field instrumentation techniques, empirical constitutive relationships for compressible soils
and a qualitative approach to interpreting the field behaviour of soft
clays.

The papers by Crory, Edil and Mochtar, and Carrier and Bromwell describe useful empirical approaches to prediction of settlements in
permafrost, organic soils and hydraulic clay fill respectively.
Crory's approach for permafrost relates unit strains to initial frozen
dry unit weight and discusses various aspects of laboratory consolidation testing for these materials. While this presentation is instructive, comparison of predicted and observed field behaviour is not
included.

Carrier and Bromwell provide an interesting overview of the uncertainties faced by Casagrande in the 1940's in his assessment the behaviour
of hydraulic clay fill used in the construction of the Logan Airport.
Their presentation concentrates on the magnitude and time rate of
settlement of the fill which basically consisted of clay balls in a
"slurrified" clay matrix. Casagrande's approach assumed that the behaviour of the clay mass would be controlled by the clay balls whereas
the authors' analysis assumes that the clay matrix controls overall
behaviour. Their analysis is based on empirically derived relationships for compressibility and permeability of slurrified clays and
finite strain consolidation theory. The input parameters for these
relationships are related to activity'and Atterberg limits. Assuming
liquidity indices of 2 and 3, their predicted settlements bracket the
observed settlement behaviour and the predicted time rate of settlement is close to that observed in the field. This agreement for what
is considered to represent a "difficult" material is encouraging.
Further work is warranted to more specifically define the conditions
under which the clay matrix controls the fill behaviour as opposed to
clay balls.

Problems associated with predicting the behaviour of hydraulic fills
are not restricted to dredged clays. Similar difficulties exist for
hydraulic sand fills. For example, hydrocarbon exploration in the

---

* Golder Associates, Calgary, Canada

Beaufort Sea is frequently carried out from artificial islands constructed using dredged fine to medium sands. Knowledge of the in-situ
state of these materials is necessary for rational selection of borrow
materials and placement techniques together with prediction of behaviour under large caisson loads. Unfortunately there is little information available for this purpose and expensive failures can occur.
One such case is reported by Mitchell (1984) who described major failures during construction of a massive subsea berm at the Nerlerk site
in the Canadian Beaufort Sea. These failures appear to have been the
result of liquefaction caused by high residual porewater pressures
and/or inadequate fill density. Given the pace of exploration in the
Alaskan North Slope area where borrow materials consist of even finer
sands probably with a higher silt content, further research is warranted into the sedimentation and consolidation aspects of hydraulic
sand fills.

Edil and Mochtar discuss settlement prediction in peat and organic
soil deposits and note that both primary and secondary compressibility
characteristics are stress level dependent. They also note that both
the magnitude of secondary compression strains and rate factor for
secondary compression are also strain rate dependent. Although stress
level dependency can be readily accounted for in laboratory testing,

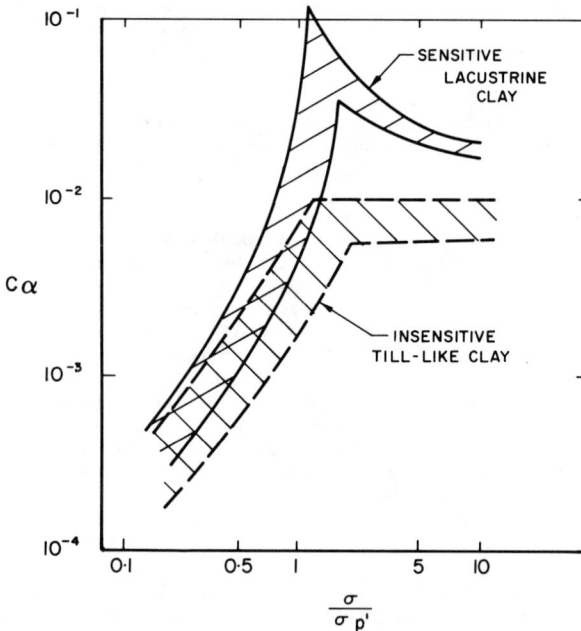

Figure 1: Stress Level Dependence of $C\alpha$ for Inorganic Clays

this is not the case for strain rate effects. The authors' approach
involves the use of the Gibson and Lo model but includes empirical
correction factors based on laboratory and field data. Good agreement
between observed and predicted settlements is obtained for two field
cases where correction factors which account for strain rate and
stress level effects are included.

Although the problem of secondary compression prediction is clearly
more significant for compressible organic soils, the authors' comments
are also at least partially applicable to less compressible inorganic
soils. For example, the coefficient of secondary compression for soft
clays is dependent on stress level with the significance of this
effect increasing with increasing sensitivity, (Figure 1).

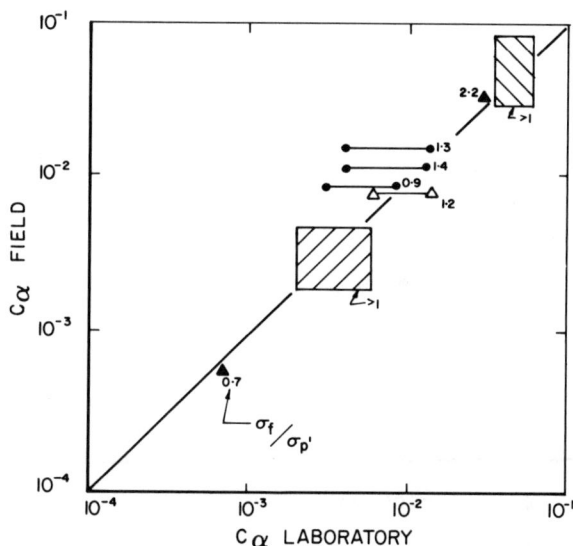

LEGEND:

▲ TIMBER CRIB ON LACUSTRINE CLAY (BREZENSKI, 1968)
● EMPRESS HOTEL - VICTORIA CLAY (CRAWFORD & SUTHERLAND, 1971)
△ FILL ON LEDA CLAY (SAMSON & GARNEAU, 1973)
▨ 7 STOREY BUILDING ON LACUSTRINE CLAY (GOLDER ASSOC.)
▧ INDUSTRIAL PLANT ON ST. CLAIR TILL (GOLDER ASSOC.)

Figure 2:   Comparison of Field and Laboratory Secondary Compression
            Rates for Inorganic Clays

However, contrary to the authors' experience with compressible soils, there appears to be better correlation between laboratory and field secondary compression settlements for inorganic soils provided stress level effects are recognized, (Figure 2). Further evidence of the significance of stress level on secondary compression settlements is shown on Figure 3 for major structures founded on the Fraser River deposits in Vancouver. In this area, secondary compression is concentrated in an extensive micaceous marine silt deposit which underlies the upper sand unit. Secondary settlements of large structures in this area can be of the same order of magnitude as primary settlements.

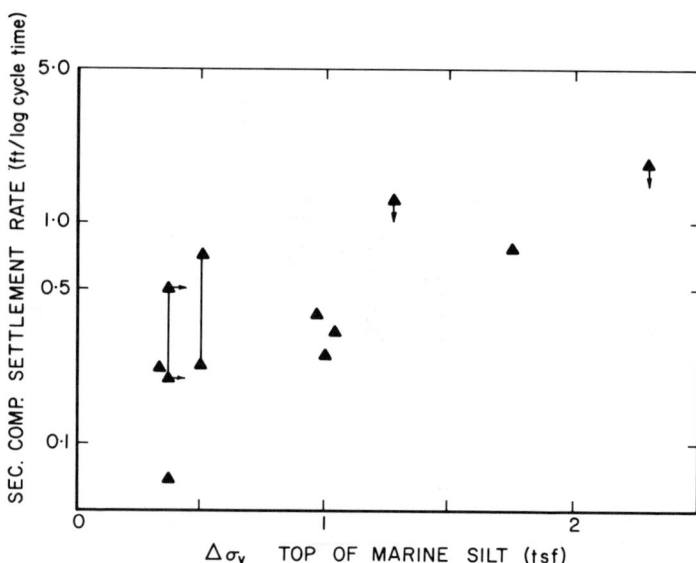

Figure 3:   Field Secondary Compression Settlements of Major
            Structures;   Frazer River Delta Area

The paper presented by Laier et al focusses mainly on techniques used to monitor the behaviour of a 60 ft. thick lightly overconsolidated clay deposit subjected to a wide areal loading of relatively high intensity. Descriptions of the various types of instruments (vertical and horizontal inclinometer casings, Sondex casing, pneumatic and standpipe piezometers), installation procedures and performance are provided. The failure rate for piezometers and the Sondex casing installations was high as a result of the large deformations which occurred in the clay, (up to 4 ft. vertical settlement and 0.5 ft. lateral displacement).

While this paper does not discuss in detail the behaviour of the loaded clay, sufficient information is presented to permit some observations to be made. The porewater pressure response in the central area of the preload is interesting in that a piezometer (SP-10) recorded increasing porewater pressure for a period of about 3-4 weeks following completion of filling, (authors' Figure 12). A similar but less marked response is evident in SP-12. Although this behaviour could be attributed to lag (i.e. slow piezometer response) since SP-10 was a standpipe-type installation, similar behaviour was not observed in the remaining standpipe piezometers at this location. Therefore, it appears that this phenomenon is related to soil behaviour. Similar occurrences are described in the paper by Crooks et al also submitted to this session. They attributed the phenomenon of continued porewater pressure increase following completion of loading to redistribution of total stresses in a critically stressed soil mass. The Effective Stress Path/Yield Envelope approach (ESP/YE) described by Crooks et al was used to determine the effective stress path at SP-10 and the results are shown on Figure 4. Following this interpretation, the porewater pressure response during initial loading (10 ft. fill) was small indicating a stiff soil response and/or significant drainage. Continued loading resulted in a higher porewater pressure response until the effective stress state reached the assumed yield envelope (based on $\sigma_p' = 2$ tsf) at the maximum fill height. Despite a modest reduction in loading at the end of construction, the excess porewater pressure increased dramatically with an abrupt change in direction of the ESP indicating strain softening. This behaviour

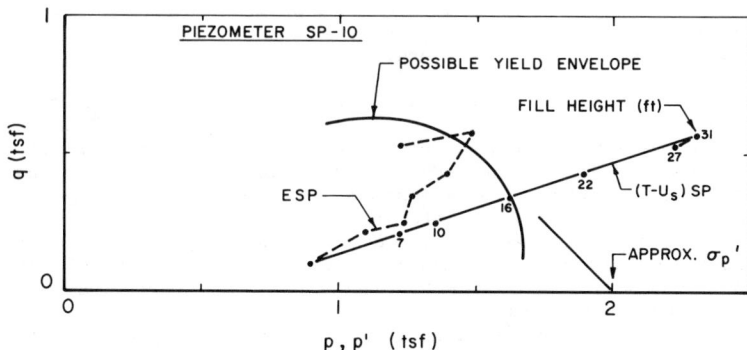

Figure 4:   Effective Stress Path Below Center of R.L.S.A. Fill (Laier et al)

suggests that at the completion of loading, the clay mass at this
piezometer location was critically stressed (i.e. the imposed shear
stress equalled the shearing resistance of the soil). It is noted
that the inferred shear strength (Su) based on this interpretation is
close to 60 kPa. The inferred Su/σ_p' ratio of 0.3 is common for compression failure which is pertinent to the area at shallow depth below
the centre of the fill. As noted by Crooks et al, similar conditions
prevailed at other sites where similar behaviour was observed. The
lateral deformation data (authors' Figure 11) is also interesting in
that it indicates concentration of deformation in a narrow zone close
to the surface of the clay deposit. Lesser deformations were recorded
in the underlying materials. This deformation pattern is consistent
with the above interpretation of the piezometric response in the clay
mass.

The phenomenon of continued porewater pressure increase following completion of loading is of considerable interest again with respect to
the performance of offshore exploration structures in the Arctic.
Many exploration sites are underlain by soft unfrozen clays and imposed stresses are high. Although the foundation clays are critically
stressed, the overall geometry of these structures (i.e. very wide
area loading and relatively thin clay deposits) is such that there is
adequate confinement of the clay and opportunity for lateral stresses
to increase. Since stress redistribution is a time dependent process,
associated porewater pressure response is also time dependent. Based
on observations at two Arctic caissons, the magnitude of porewater
pressure increases under these conditions can be significant. For
example, at one caisson island site operated by Esso Resources Canada
Limited, the excess porewater pressure in a thin (2m) soft clay deposit at seabed level was 110 kPa immediately following construction. During
During a three month period following completion of construction, the
excess porewater pressure tripled before dissipation occurred. Clearly
large porewater pressure increases of this nature below widely loaded
areas give rise to concern and further research into this phenomenon
is warranted particularly in view of potential soft clay foundation
conditions on the Alaskan North Slope.

The paper submitted by Mikasa and Takada describes the behaviour of an
upper alluvial soft clay deposit under one stage of a multi-staged
fill loading operation during the construction of Kobe Island. The
data presented relates to the 8th loading stage and describes the
settlement and porewater pressure response in the soft clay deposit
under that load increment. Settlements were predicted based on a
finite strain approach in which self weight of the clay and variation
in c_v magnitude could be taken into account. Good agreement between
predicted and observed behaviour for the upper soft clay layer was
achieved. However, the total settlements recorded were greater than
anticipated because of compression of diluvial sand and clay deposits
underlying the upper alluvial clay. The authors have concluded that
this settlement could not have been predicted but insufficient information is presented to demonstrate this conclusion.

The porewater pressure response in the alluvial clay is interesting in
that the initial excess porewater pressure due to the final loading

stage appears to have dissipated very quickly for a short period following construction. The dissipation rate then changed dramatically and little or no dissipation occured for several months. About 6 months after completion of construction, the dissipation rate increased but remained relatively slow. Crooks et al observed similar changes in dissipation behaviour for a number of cases. Unfortunately, insufficient information is given in the Mikasa and Takada paper to permit ESP/YE analysis and to compare the Kobe Island conditions with those pertaining to other sites.

Mikasa and Takada also describe various site investigation and instrumentation techniques developed during the course of the project. Because of the low clay strength, a CPT device was developed which incorporated a larger than usual cone tip and a friction reduction The non-standard cone tip makes it difficult to adopt conventional correlations. A similar friction reduction system has been successfully used in offshore Arctic investigations, (Jefferies and Funegard, 1983). A new type of settlement anchor is described which requires an above ground tensioning system. This type of system may produce acceptable settlement data but the requirement for a permanent above ground installation has obvious drawbacks. The porewater pressure gauge described in the paper incorporates many of the elements included in conventional pneumatic piezometers but does not appear to incorporate the facility to flush the instrument.

References:

Jefferies, M.G. and Funegard, E. (1983) "Cone Penetration Testing in the Beaufort Sea". Proceedings ASCE Speciality Conference on Geotechnical Practice in Offshore Engineering.

Mitchell, D. (1984) "Liquefaction Slides in Hydraulically Placed Sands", Proc. Int. Symposium on Landslides, Toronto, 1984.

LOGAN AIRPORT RE-VISITED

W. David Carrier, III* FASCE
Leslie G. Bromwell,* MASCE

ABSTRACT

The first detailed evaluation of self-weight consolidation of a large hydraulic clay fill was performed by Arthur Casagrande during the construction of Logan Airport in Boston nearly 40 years ago. The purpose of this paper is to re-analyze the settlement of the fill by means of finite strain consolidation theory, coupled with correlation relationships for compressibility and permeability. These new techniques allow a designer to quickly evaluate possible deposition alternatives; in particular, to consider the effect of the filling rate, which was not previously possible.

INTRODUCTION

Many geotechnical engineers are familiar with the design and construction of the Logan Airport dredged fill. Arthur Casagrande was the soil mechanics consultant on the project and in 1949 he published a classic paper in the Journal of the Boston Society of Civil Engineers (7). Casagrande's work on hydraulically placed clay fill had a significant influence on the development of soil mechanics and thus it is worthwhile to review this historic project in light of current knowledge and practice.

Logan Airport was constructed by the State of Massachusets in Boston Harbor nearly 40 years ago. The fill consisted primarily of hydraulically placed clay, approximately 7 metres deep and approximately 30 million cubic metres in volume. It was recognized at the time that a clay fill would present very difficult engineering problems, but the state authorities were very desirous of having a convenient, centrally located airport. Idlewild Airport (now John F. Kennedy) in New York was already under construction and the competition for transatlantic flights was intense.

*   Principal, Bromwell & Carrier, Inc.
    202 Lake Miriam Drive
    Lakeland, FL 33807

In Casagrande's words, *there was no reliable, quantitative information on the consolidation characteristics of hydraulic clay fills*** and thus development of the final design continued throughout the project, based almost exclusively on observations made during construction. In order to monitor the performance of the clay fill, a new, greatly improved piezometer was developed, the details of which were presented in an extensive appendix to the paper. This piezometer is still known as the "Casagrande-type." Surprisingly, not a single laboratory soil measurement is mentioned in the paper, not even a water content.

Prior to construction of the fill, there were four major technical issues that had to be addressed. These are summarized as follows:

1. In the project area, the harbor bottom was just below mean low water level and consisted of 1.5 to 3.0 m of very soft, organic silt. Should this silt be removed prior to placing the clay fill? The dredging contractors said that it would be displaced by the fill, but Casagrande was not certain and recommended that it be removed from beneath the planned runways. Casagrande was overruled and the silt was left in place. Borings later showed that the silt, in fact, had not been displaced. Casagrande felt that the silt contributed significantly to differential settlement, but ultimately concluded that removal of the silt was not justified for this particular project, because the layout of the runways was changed later during construction.

2. What would be the slope of the clay fill at the end of construction? Based on experience, the contractors estimated stable slopes of 20 horizontal to 1 vertical. The actual slopes were about 50 to 1. *This resulted in a considerably larger volume of fill placed than was included in the estimates.*

The differences between the initial design cross section and the final design are presented in Figure 1 (based on Figure 4 in Casagrande's paper).

** Quotes from Casagrande's paper are in italics.

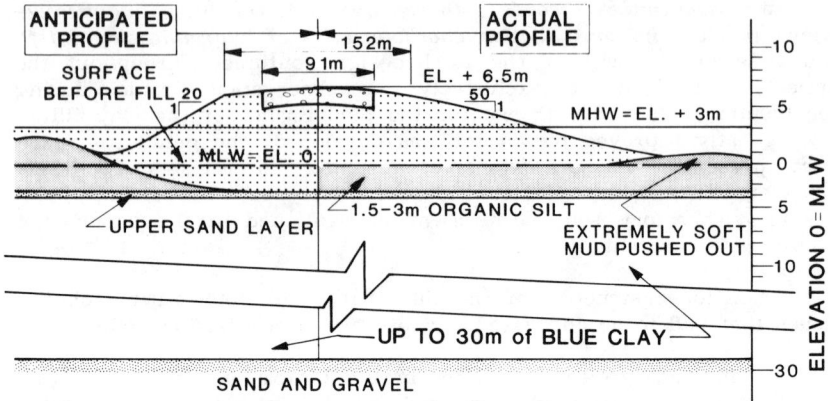

Fig. 1 Comparison of Anticipated and Actual Cross Sections
Through Logan Airport Runway Embankments (based on
Figure 4 in Casagrande (7)).

3. It was known that the fill would consolidate under its own
weight, and thus it was necessary to place it higher than the
final design elevation. How much and how fast would the fill
settle? Prior to construction, Casagrande estimated 60 to 120 cm
(*largely a guess*) and that most of this would occur within one
year. The actual measurements were 46 cm after one year and 52
cm after two years.

4. How quickly would the surface of the fill stabilize and how
strong would it be for design and construction of the runways?
It had been anticipated that the clay fill would strengthen in a
reasonable period of time and that the runways would be designed
in accordance with accepted procedures then available based on
measured CBR values of the fill. Just the opposite occurred.
Except for a dried crust a few centimeters thick, the clay
remained very soft even after a year, with a CBR of less than
1%, thereby rendering a conventional design impossible. Instead,
it was necessary to develop a design procedure which took into
account the increase in shear strength that would occur as the
fill consolidated beneath the imposed load of the pavement and
base. This was successfully accomplished, and a combined
thickness of 1.5 m was selected for the runways, and 1.7 m for
the touchdown areas and taxiways. Based on field tests,
Casagrande was able to predict that for a 1.5-m thick runway,
the clay fill would develop a CBR value in excess of 60% after 5
years. This implied a single-wheel bearing capacity of 500 kN,
more than enough for the aircraft of that era.

In the present paper, just one of the four major issues described above will be reviewed in detail: self-weight consolidation of the clay fill.

CONSOLIDATION PROPERTIES OF THE HYDRAULIC CLAY FILL

Prior to construction, undisturbed boring samples had been recovered from the dredging areas, and the borrow material had been found to consist of overconsolidated Boston Blue Clay. Although Casagrande did not report the Atterberg limits of the clay, such data have been well-documented in other literature and are presented in a plasticity chart in Figure 2. Also shown for comparison are the Atterberg limits from other dredged material sites in the United States. These latter data were not available to Casagrade, because until recently, most fine-grained dredged material was disposed of in unconfined areas, such as out to sea. Regulations now require that much of this material be stored on land within confining dikes. Hence, data on the properties of these materials are now being collected. With the perspective of current knowledge, it can be seen that the clay fill of Logan Airport, in fact, has a lower plasticity index (and, thus, better consolidation properties) than many of today's dredged material disposal sites.

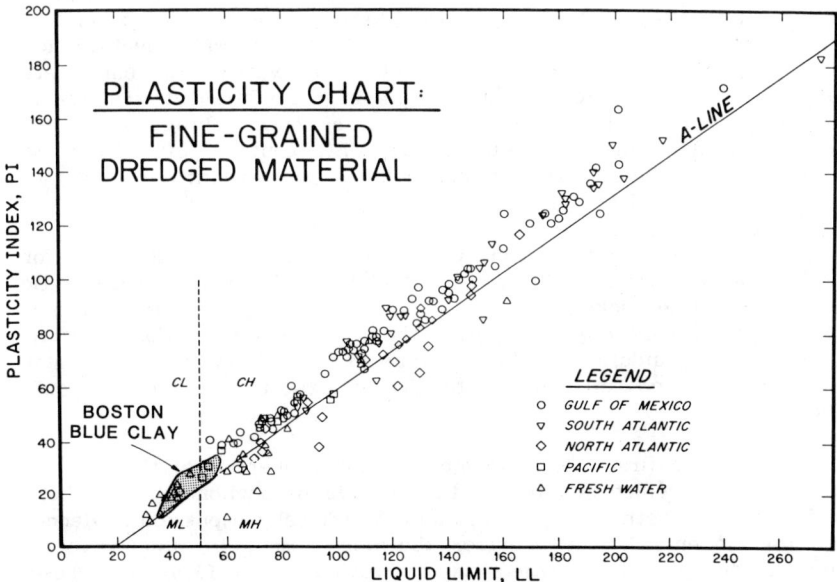

Fig. 2   Plasticity Chart of Dredged Materials (includes data from Ref. 1).

Because the borrow material was overconsolidated, and the pumping distance was short, the clay was not completely fluidized when it was deposited on the airport site. Instead, in Casagrande's words, the fill consisted of *balls of clay varying from pebble to head size, which are laid down in a matrix of semi-fluid clay.*

Casagrande *concluded that compression of the clay fill would be largely due to plastic deformation of the clay pebbles and balls rather than to consolidation of the clay itself, and that this adjustment would develop at a considerably faster rate than the primary consolidation of a natural clay deposit because of the relatively pervious matrix.*

Casagrande evidently felt that initially, the self-weight of the hydraulic fill would be carried by the semi-fluid matrix, which would result in excess pore pressure. As this pore pressure dissipated, the stress would be transferred to the clay balls, which would squeeze together at the points of contact. This process would occur very rapidly because the matrix had a comparatively high permeability. Thus, in Casagrande's view, the clay balls controlled the amount of consolidation and the matrix controlled the rate.

However, since the excess pore pressure was high at the end of filling, it is clear that the clay balls were not in intimate contact at that moment, even among the balls near the bottom which had already been in place for a month. If the interstices between the balls were large, it is possible to imagine that all of the settlement was due to re-arrangement of the slippery balls into a denser configuration. In this case, the balls might never come into contact, but simply be suspended in the matrix. Consequently, the amount and rate of consolidation would be controlled exclusively by the matrix.

For example, the maximum void ratio (face-centered packing) for uniform spheres in contact is 0.92. Whereas the minimum void ratio (hexagonal close-packing) is 0.35. Thus, in going from the minimum to the maximum relative density, without plastic deformation, uniform spheres would undergo a volume change of 30%. This is much greater than the observed settlement of the Logan Airport fill, which was only about 7%.

Thus, the first step in evaluating the properties of the clay fill would be to predict in advance the relative proportions of clay balls and semi-fluid matrix. This was, and is, virtually impossible. Hence, for the moment, let us consider the worst case, in which the fill consists entirely of slurrified clay. Previous work (5,6) has shown that the compressibility and permeability can be expressed as follows:

Compressibility
$$e = \alpha \left(\frac{\bar{\sigma}}{P_{atm}} \right)^{\beta} + \varepsilon \quad \text{(Equ. 1)}$$

Permeability
$$k = \mu \frac{(e - \delta)^{\nu}}{1 + e} \quad \text{(Equ. 2)}$$

where e = void ratio

$\bar{\sigma}$ = effective stress

P_{atm} = atmospheric pressure

k = coefficient of permeability (m/s)

and $\alpha, \beta, \varepsilon, \mu, \delta, \nu$ = empirical factors

If the initial water content of the slurry is high, corresponding to a liquidity index of about 3 or more, then the empirical factors are related to the Atterberg limits and activity of the clay. The correlations are given by:

$\alpha = 0.0208 \, (PI) \, [1.192 + (ACT)^{-1}]$

$\beta = -0.143$

$\varepsilon = 0.027 \, (PL) - 0.0133(PI) \, [1.192 + (ACT)^{-1}]$

$\mu = 0.0174 \, (PI)^{-4.29} \, m/s$

$\nu = 4.29$

$\delta = 0.027 \, [(PL) - 0.242(PI)]$

where   PL = plastic limit (%)
            PI = plasticity index (%)
            ACT = activity = PI ÷ fraction finer than 2 microns

Approximate average values for Boston Blue Clay are:

Liquid Limit = 40
Plastic Limit = 20
Plasticity Index = 20
Activity = 0.4

Substituting into Equations 1 and 2 yields:

$$e = 1.54 \left(\frac{\bar{\sigma}}{P_{atm}} \right)^{-0.143} -0.442$$

and

$$k = 4.56 \times 10^{-8} \frac{(e - 0.409)^{4.29}}{1 + e} \quad m/s$$

These relationships are plotted in Figures 3 and 4, respectively. Thus, it is possible to estimate the important consolidation properties of the clay slurry before it is deposited, based solely on index tests.

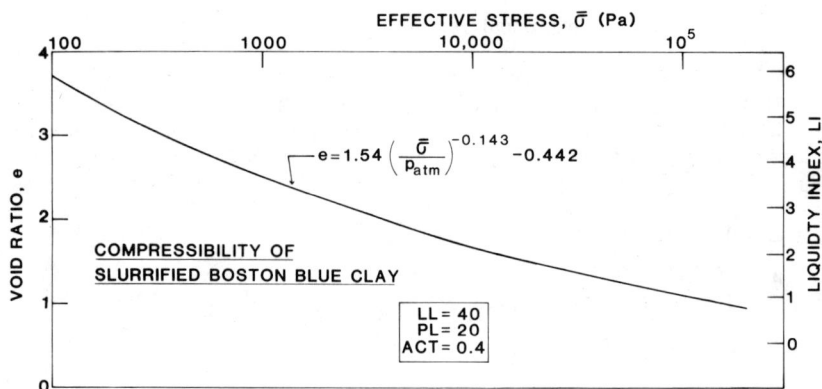

Fig. 3   Compressibility: Based on Atterberg Limits and Activity

ANALYSIS OF SELF-WEIGHT CONSOLIDATION

Numerical modeling of the settlement of the clay slurry has been performed by means of a non-linear, finite strain consolidation computer program written by Prof. Frank Somogyi of Northwestern University. This program utilizes a finite difference technique to solve the Gibson et al.(8) differential equations. Details regarding the program and its use are described elsewhere (2,3,4,5,9).

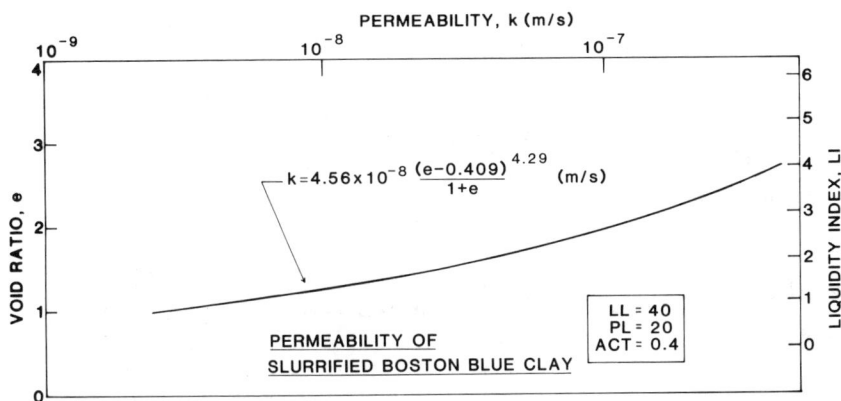

Fig. 4   Permeability: Based on Atterberg Limits

It was first assumed that the slurry was deposited at an initial water content of 80%, corresponding to a liquidity index of 3 for Boston Blue Clay. The deposition occurred over a one-month period and then the fill was allowed to rest for nearly 2 years. The results of the numerical analysis are shown in the upper half of Figure 5, along with Casagrande's predictions and measurements. It can be seen that for this "worst case" analysis, in which the fill is assumed to be entirely slurrified, the calculated amount of consolidation is approximately 72 cm. This is somewhat greater than the measured value of 52 cm, although within the 60 to 120-cm range originally predicted by Casagrande. Hence, the worst case represents a reasonable upper bound on the expected amount of settlement. More important, the calculated rate of settlement is in close agreement with the measured rate: approximately 90% complete within one year, and essentially 100% within two years. This confirms two points: first, the matrix does indeed control the rate of consolidation; and second, the permeability relationship shown in Figure 4, which is based on the Atterberg limits of Boston Blue Clay, must be reasonably accurate.

Note that Casagrande did not predict or measure any consolidation during the filling period. The numerical analysis shows that approximately 38 cm of consolidation occurred during filling; that is, if the fill were placed instantaneously, instead of over the course of a month, then 38 cm of additional settlement would occur during the resting period, or a total of 110 cm. On the other hand, if the filling period were extended from one month to, say, six months, most of the consolidation would occur during the filling period (87 cm), and only a small amount of settlement would occur during the resting period.

This is shown in the bottom half of Figure 5. Note that the total settlement (filling plus resting) is essentially the same regardless of the rate of filling.

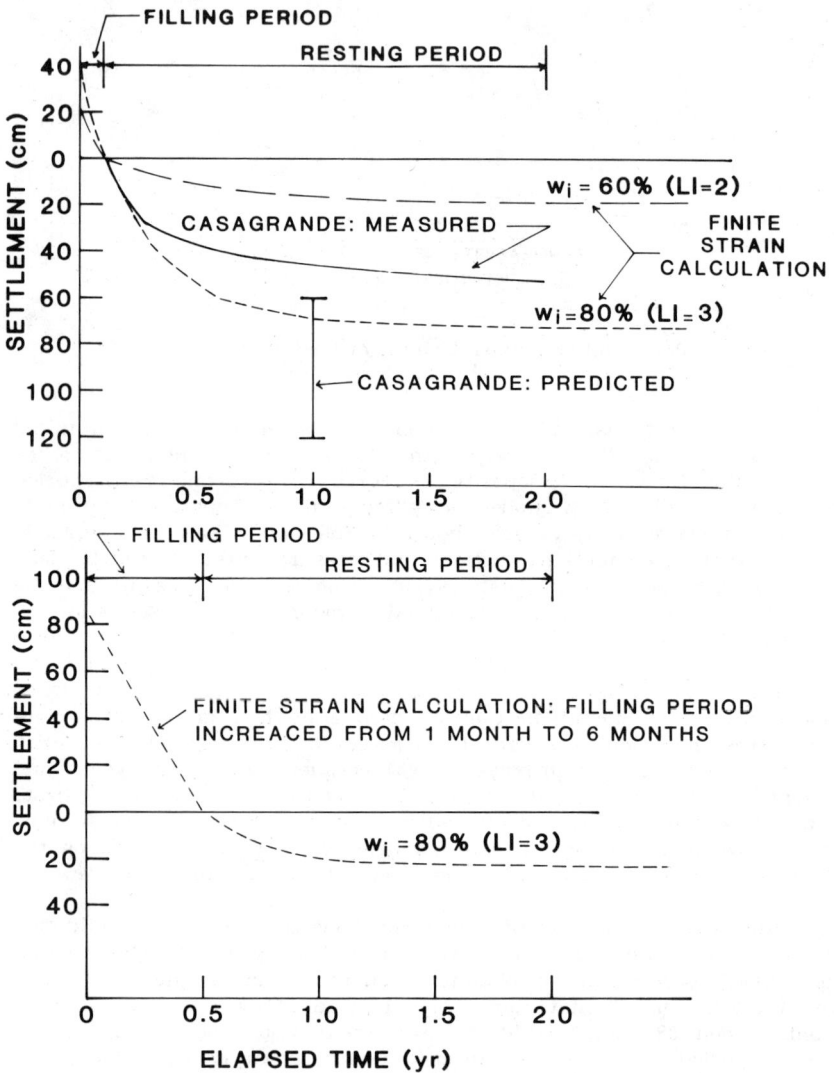

Fig. 5 Logan Airport Hydraulic Fill: Comparison of Predicted and Measured Settlement.

It was next assumed that the slurry was deposited at an initial water content of 60%, corresponding to a liquidity index of 2. This value is likely greater than the weighted average value of the fill obtained by combining the high water content of the semi-fluid matrix and the low water content of the dense clay balls. Hence, considering the fill to be completely remolded at this water content should result in a greater calculated settlement than was actually measured, if the settlement were primarily due to deformation of the clay balls. In point of fact, the compressibility of Boston Blue Clay, remolded at a liquidity index of 2, would be less than the compressibility presented in Figure 3, which is based on a liquidity index greater than 3. Thus, using the same compressibility curve would tend to further over-estimate the amount of settlement. The results of this analysis are also presented in Figure 5. It can be seen that in spite of the conservative assumptions, the calculated settlement is less than the measured settlement, although the rate is still very comparable. Thus, in order to account for the greater measured settlement, it is necessary to consider consolidation of the semi-fluid matrix, as well as the clay balls.

The following observations can be made:

1. If the fill is simply assumed to behave as a clay slurry, the calculated amount of settlement "brackets" the measured value. Furthermore, the rate of settlement is very similar.

2. Although settlement of the clay fill is due to consolidation of both the clay balls and the matrix, the numerical analyses suggest that the matrix has the most effect on the amount and rate.

The finite strain consolidation computer model can also be used to analyze other possible filling scenarios. For example, if the height of the clay fill were doubled, and the time of filling were increased to two months (i.e., the filling rate held constant), then for a liquidity index of 3, the calculated consolidation during filling is 79 cm and the consolidation during resting is 204 cm. Thus, compared to the earlier analysis, the settlement during filling would be more than doubled and during resting, nearly tripled. Casagrande apparently felt that the settlement would be approximately proportional to the fill height, and probably would have doubled his estimate.

On the other hand, the finite strain analysis shows that the settlement would be approximately 90% complete within two years, or only twice as long as before. According to the classic Terzaghi consolidation theory, which was all that was available to Casagrande, the time of consolidation is proportional to the square of the stratum thickness. Thus, if the fill were doubled, Casagrande would probably have quadrupled his estimate of the time required for settlement. (For the same soil properties and conditions, finite strain theory always predicts a faster settlement rate than Terzaghi theory).

CONCLUSIONS AND RECOMMENDATIONS

Interestingly, Casagrande recognized the influence of the matrix on the shear strength: *The mass as a whole was rather unstable due to the matrix of semi-fluid clay, and even a light caterpillar tractor would often sink and mire down.* However, he apparently did not consider that the matrix could also control the amount of consolidation. And, yet, he predicted the amount and rate of consolidation of the clay fill within satisfactory limits, and completed the project successfully. In this case, incorrect assumptions appear to have led to an acceptably accurate prediction, not an uncommon occurrence when new engineering techniques are being developed.

On the other hand, if the fill had been twice as high, Casagrande would presumably have doubled his settlement estimate and would have quadrupled his estimate of the time for most of the settlement to occur. However, non-linear, finite strain analysis indicates that the amount of consolidation would have nearly tripled, but the time would have only doubled.

Casagrande stated that *the desirable procedure would have been to build first a full-size test section and keep it under observation for whatever period might be needed*, but because of the press of time, this was not possible. Finite strain consolidation theory, coupled with correlation relationships, allows a designer to quickly evaluate possible alternatives, and begin to optimize the disposal plan. In particular, the influence of the filling rate can now be evaluated, whereas previously it was not considered. With a slower filling rate, it would have been easier to obtain the final design elevation, since the subsequent settlement would have been much less.

It is still not possible to quantify in advance the relative proportions of clay balls and matrix that might be produced in a given dredging project. In this particular case study, ignoring the clay balls and considering the liquidity index at placement to fall between 2 and 3 gave a reasonable range for design estimates of consolidation behavior.

REFERENCES

(1) Bartos, M.J. (1977) "Classification and Engineering Properties of
     Dredged Materials," Dredged Material Research Program Technical
     Report, D-77-18, Army Corps of Engineers, Waterways Experiment
     Station, Vicksburg, Mississippi.

(2) Bromwell, L.G., and Carrier, W.D., III, (1979) "Consolidation of
     Fine-Grained Mining Wastes," Proceedings, Sixth Panamerican
     Conference on Soil Mechanics and Foundation Engineering, Lima,
     Vol. 1, pp. 293-304.

(3) Bromwell Engineering, Inc. (1979), "Analysis and Prediction of
     Phosphatic Clay Consolidation: Implementation Package,"
     Lakeland, Florida.

(4) Carrier, W.D., III and Bromwell, L.G., (1980), "Geotechnical
     Analysis of Confined Spoil Disposal," Proceedings, Ninth World
     Dredging Conference, Vancouver, pp. 313-324.

(5) Carrier, W.D., III and Bromwell, L.G. (1983), "Disposal and
     Reclamation of Mining and Dredging Wastes," Proceedings,
     Seventh Panamerican Conference on Soil Mechanics and Foundation
     Engineering, Vancouver, Vol. II, pp. 727-738.

(6) Carrier, W.D., III, and Beckman, J.F. (1984), "Correlation
     Among Index Tests and the Properties of Remolded Clays,"
     Geotechnique, Vol. 34, No. 2, June, pp. 211-228.

(7) Casagrande, A. (1949), "Soil Mechanics in the Design and
     Construction of the Logan Airport," Journal of the Boston Society
     of Civil Engineers, re-printed in Contributions to Soil Mechanics,
     1941-1953, BSCE (1953), pp. 176-205.

(8) Gibson, R.E., England, G.L., and Hussey, M.H.L. (1967), "The
     Theory of One-Dimensional Consolidation of Saturated Clays, 1.
     Finite Non-Linear Consolidation of Thin Homogeneous Layers,"
     Geotechnique, Vol. 17, pp. 261-273.

(9) Somogyi, F. (1980), "Large Strain Consolidation of Fine-Grained
     Slurries," presented at Canadian Society for Civil Engineers 1980
     Annual Conference, Winnipeg.

YIELD BEHAVIOUR AND CONSOLIDATION. I: PORE PRESSURE RESPONSE
by

J.H.A. Crooks, M.ASCE[1], D.E. Becker, [1]
M.G. Jefferies, M.ASCE [2] and K. McKenzie[2]

INTRODUCTION

It is not unusual that observed field consolidation behaviour of soft clays is not as predicted or expected. A review of approximately 50 case records related to field consolidation of soft clays indicates that anomalous behaviour, which is not accurately predicted by conventional consolidation models, occurred in a significant number of cases. These anomalies include:

° Magnitude and distribution of initial excess porewater pressure, including continued porewater pressure generation following completion of loading

° Differences between field consolidation rates and those predicted based on laboratory measurements

° Changes in rates of dissipation during and following construction

One of the reasons for inaccurate prediction of consolidation behaviour is that the constitutive models used often implicitly assume effective stress paths which do not reflect those which occur in the field. Similarly, input parameters for these models are often obtained from laboratory tests which incorporate unrealistic effective stress paths. Finally, many models do not account for yielding of lightly overconsolidated clays which can control field behaviour.

This paper discusses the frequency of occurrence of anomalous consolidation behaviour. The behaviour of soft clay is discussed briefly in terms of Effective Stress Path/Yield Envelope method of analysis proposed by Folkes and Crooks (1984) [16]. About 20 case records have been analyzed to determine the field effective stress paths and associated yielding of the foundation clays. The case records analyzed include a wide variety of soil types, loading and stratigraphic geometries, and boundary drainage conditions. Sites at which vertical drains/wicks were and were not used to accelerate consolidation are also included*.

---

(1) Golder Associates, Calgary, Canada
(2) Gulf Canada Resources Inc., Calgary, Canada

*In this paper, the term "non-drained" is used to identify sites which did not employ vertical drains and "drained" denotes cases in which vertical drains were installed.

These analyses demonstrate the significance of effective stress path/
yield behaviour in controlling various aspects of soft clay consolidation behaviour. A companion paper (Becker et al, 1984) [3] discusses
changes in strength associated with consolidation of soft clays.

FREQUENCY OF OCCURRENCE OF ANOMALOUS BEHAVIOUR

Porewater Pressure Response to Load

A total of 36 case records were reviewed to determine the frequency of
occurrence of anomalous or unusual porewater pressure response to load
(Table 1). In 11 of the 31 cases for which data is available, the
excess porewater pressure in the foundation materials continued to increase following completion of loading. In some cases (Asrum, Belfast,
Bangkok, Sandwich, Saint-Alban and Gloucester) this phenomenon was not
significant in terms of either the magnitude of additional porewater
pressures generated or the period during which the increases took place.
Not surprisingly, three cases in which vertical drains were used
(Belfast, Sandwich and Bangkok) fall within this category. Where
anomalous behaviour is not significant (i.e. in terms of magnitude and
time), it could be attributed to factors other than soil behaviour such
as lag in piezometer readings and time dependent redistribution of
excess porewater pressures.

Extraordinary increases in excess porewater pressures following completion of fill placement were observed at three sites, namely, Natal,
Mission Bay and New Liskeard. In the Natal and Mission Bay cases, the
magnitude of porewater pressure increase following completion of filling
(40 - 60 kPa; 835-1250 psf) was 50 - 100% of that which occurred during
construction. Further, these increases occurred during an extended
period of up to three months following completion of construction. In
the New Liskeard case, the excess porewater pressure which developed
following construction (5 - 30 kPa; 100-625 psf) varied throughout the
clay deposit and occurred in a two week period.

Of the remaining cases, the piezometers near the base of the clay layer
underlying the San Fransisco fill dissipated but other piezometers in
the clay showed increasing porewater pressures for periods up to two
months. In the Trinidad tanks case record, the porewater pressure generated during the first loading stage together with subsequent dissipation behaviour was as expected. When the second stage load was applied,
the porewater pressure response was modest but increased with time following completion of loading. Again, the magnitude of additional porewater pressure generated was about 30 kPa (625 psf). Other aspects of
porewater pressure response to load are indicated by the summaries contained on Table 1. For example, characterization of porewater pressure
response to load has often concentrated on relating excess porewater
pressure to applied load using parameters such as $\overline{B}$, $(\Delta u/\Delta \sigma_1)$. As shown
on Table 1, there is a wide range of porewater pressure response as de-
fined by the shapes of the Δu–Δp relationships during load application.

Porewater Pressure Dissipation Behaviour

A total of 34 case records were reviewed to determine the frequency of
occurrence of unusual dissipation behaviour. The results of this review

SEDIMENTATION/CONSOLIDATIONS MODELS

TABLE 1: SUMMARY OF LITERATURE REVIEW

CASE/REFERENCE	PLASTICITY INDEX (%)	$\Delta u - \Delta p$ BEHAVIOUR	INCREASE IN Δu AFTER LOADING	CHANGE IN DISSIPA- TION RATE	VERTICAL DRAINS USED
Asrum Fill (19)	15–25	Bilinear	Yes	–	No
MIT–195 (Stage 1) (12)	20–30	Bilinear	No	No	No
Natal Embankment (21)	20–35	Bilinear	Yes	No	No
Matagami Embankment (13)	30–45	–	(Failure)	–	No
Muda Embankment (20)	30–45	Bilinear	No	Yes	No
Launceston Levee (34)	100	–	No	No	No
New Hampshire 195 (23)	10–20	Bilinear	(Failure)	–	No
Rt.17 Fills and Embankments (26)	20–25	Linear Linear	No No	– Yes	No Yes
Bund Loading (31)	40–50	Linear	No	Yes	No
San Francisco Fill (38)	25	Linear	Yes	–	No
Tees Tank (35)	50	Linear	No	Yes	No
Huntsville Embankment (30)	20–25	Curved	No	–	No
Shellmouth Dam Fill (37)	37	Curved	No	–	No
Mission Bay Dykes (16)	40	Multi-linear	Yes	No	No
Trenton Embankment (6)	3–10	–	No	No	No
Wallaceburg Silos (2)	17–32	Bilinear	No	Yes	No
Queenborough Embank- ment (32) (9)	30–80	Bilinear Bilinear	No No	No Yes	No Yes
Bangkok Embankment (4)	55–65	Linear Linear	No Yes	Yes Yes	No Yes
Belfast Embankment (14)	10–50	Bilinear	Yes	Yes	Yes
Perth Fill (28)	65–150	Linear	No	No	Yes
New Hampshire 195 (24)	10–20	Bilinear	No	Yes	Yes
Iraq Fills (Golder Associates)	22–35	Bilinear	No	–	Yes
Porto Tolle Tank (17)(18)	25–37	Bilinear	No	Yes	Yes

Cont'd

TABLE 1: SUMMARY OF LITERATURE REVIEW (Cont'd)

CASE/REFERENCE	PLASTICITY INDEX (%)	$\triangle u - \triangle p$ BEHAVIOUR	INCREASE IN $\triangle u$ AFTER LOADING	CHANGE IN DISSIPA- TION RATE	VERTICAL DRAINS USED
Trieste Fills (17)	37-57	Linear	No	No	Yes
Sandwich Embankment (9)	30-60	Linear	Yes	Yes	Yes
Ulverston Preload (8)	-	Curved	No	Yes	Yes
Taiwan Preload (40)	-	Curved	No	Yes	Yes
Napa River Embankments (7)	37	-	-	Yes	Yes
Launceston Embankment (10)	110-160	-	-	Yes -	Yes No
Yarra Test Embankment (41)	20-56	-	-	No	No
Fiddler's Ferry Embankment (1)	22	-	-	Yes	No
King's Lynn Fill (43)	30-40	-	-	Yes	Yes
Orebro Fill (15)	25-50	-	-	-	Yes
New Liskeard Embankment (16)	10-40	Multi-linear	Yes	No	No
Trinidad Tanks (42)	15-35	Bilinear	Yes	-	No
CNR-Don River Embankment (16)	12-30	Multi-linear	No	Yes	No
Gloucester Test Fill(16)	15-35	Bilinear	Yes	Yes	No
Thames Tidal Defense Dykes (29)	12-44	Bilinear	(Failure)	-	No
Saint-Alban Test Fills (27)	10-37	Multi-linear	Yes	-	No
Dutch Hollow Dam (22)		-	-	Yes	No
Izmir Fill (33)	40-75	-	-	Yes	No
CFS Gloucester (11)	15-35	-	No	Yes	No
Kars Embankment (16)	18-26	-	No	Yes	No
Thunder Bay Preload (SWPCP) (16)	40	-	No	Yes	No
Trinidad Preload (42)	15-35	-	No	Yes	No

are also summarized on Table 1. This summary indicates that a significant change in dissipation rate following completion of loading is a
frequent occurrence being observed in 25 of the 34 case records for
which data is available. In some cases, mainly involving highly
sensitive soils, the change in dissipation rate was marked. In the Muda
embankment case, James (1970) [20] concluded that this behaviour was
caused by disruption of lateral drainage layers. Crawford and Burn
(1976) [11] attribute this behaviour at the two Gloucester sites and the
Kars bridge overpass embankment to generation of excess porewater
pressures as a result of soil structure breakdown. In other cases, the
change in dissipation rate was less marked although still distinct and
somewhat surprisingly, some of these cases also involved relatively
sensitive soils. Reasonably distinct changes in cv also occurred at
sites where the clay deposit contained numerous sandy layers (e.g. Trinidad) and at "vertically drained" sites. Therefore, it appears that
although internal drainage elements increase the overall dissipation
rates, the fundamental soil behaviour which leads to reduced cv values
with increasing stress levels, is still significant and in some cases is
the controlling factor with respect to field performance.

The prediction of rate of dissipation of excess porewater pressure is
often erroneous because of inaccurate modelling of drainage boundary
conditions. In particular, thin layers of permeable silts and sands
which are present within the clay deposits are frequently undetected
during site investigations and not taken into account in analyses. Even
if their presence is appreciated, it is difficult to quantify their
effect on the rate of consolidation of the overall clay mass.

THE EFFECTIVE STRESS PATH/YIELD ENVELOPE (ESP/YE) APPROACH

The behaviour of lightly overconsolidated clays is strongly influenced
by yielding at small strains. At the stress levels normally associated
with structures founded on soft clays, yield behaviour is more relevant
to the understanding and prediction of field performance than are large
strain failure concepts. Based on analyses of several embankments on
soft clays, Folkes and Crooks (1984) [16] developed an approach for
examining the yield behaviour of these materials. This approach
involves determining the effective stress paths (ESPs) at specific
(piezometer) locations within the foundation soils based on computed
total stresses and measured porewater pressures. The observed soil
behaviour (i.e. excess porewater pressure and deformation response) can
be related to the locations of these effective stress paths relative to
the yield envelope (YE). To date the effective stress path/yield
envelope (ESP/YE) approach has only been applied to piezometer locations
where principal stress rotation has not occurred (e.g. below the centre
of a loaded area). Further, since the approach has not been developed
in terms of a formal constitutive model, it has not been used herein to
make quantitative predictions of soil behaviour. Nevertheless, it does
provide a rational framework for interpreting observed behaviour and is
used in this context. The following is a brief description of the more
salient features of the ESP/YE method.

The yield envelope (YE) is the locus of stress states in q-p stress
space which cause yield of the soil, where yield is indicated by a
change from small strain to large strain response to loading. The yield

envelope is approximately symmetrical about the Ko line and its apex is defined by the preconsolidation pressure (Figure 1). Although normally determined using drained triaxial tests, the upper and lower portions of the yield envelope are approximately coincident with undrained shear strengths in compression and extension respectively. For effective stress states lying below the critical state line and within the yield envelope, the soil behaves in an overconsolidated manner with small strain response to loading and relatively rapid dissipation of excess porewater pressures. If the effective stress state lies within the yield envelope, but above the critical state line, the soil is in a metastable condition. Additional loading or increase in porewater pressure which causes yield results in strain softening, with the effective stress path "tracking down" along the critical state line. Large increases in total horizontal stresses, lateral strains and porewater pressures are associated with strain softening stress paths of this nature. Even if the soil is not strain softening, large increases in horizontal stress can occur when the applied shear stress equals the shear strength. Thus, the effective stress state in the critically stressed soil is constant and during additional loading, the total horizontal stress (and porewater pressure) increases as the total vertical stress increases to maintain this effective stress state. Finally, for effective stress paths located outside the yield envelope, below the critical state line, normally consolidated behaviour is evident with large strain and porewater pressure response to loading together with slow porewater pressure dissipation rates.

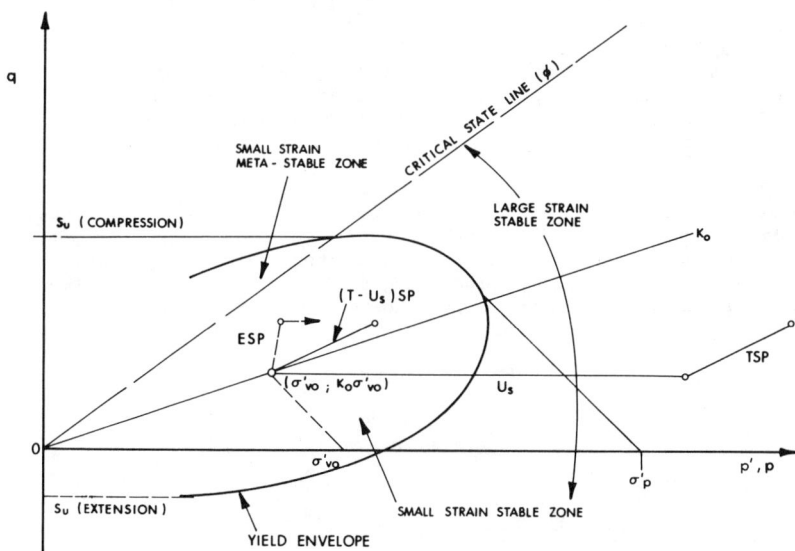

Figure 1. Major features of yield envelope (YE).

The effective stress path at a point beneath a loaded area is defined by the initial in situ effective stress state, the total stress changes resulting from loading and the total porewater pressure at that location. The initial vertical effective stress is readily computed from measured densities and initial piezometric pressures. However, the accuracy with which the initial horizontal effective stress state can be defined is limited by difficulties associated with the measurement of Ko. In most of the case records analyzed, Ko data was not provided. Since these soils are generally moderately plastic, lightly overconsolidated clays, a Ko value of 0.7 (Brooker and Ireland, 1965) [5] was assumed for ESP/YE analyses. It is noted that minor inaccuracies in the assumed Ko magnitude would alter only the location of the computed effective stress path in stress space, but would not change its shape.

Total major and minor principal stress changes were predicted using conventional elastic methods (e.g. Poulos and Davis, 1974) [36] and an assumed Poisson's ratio of 0.5. This approach is reasonable provided that significant strain softening does not occur and the "elastic" properties of the foundation materials do not change significantly because of drainage.

Strain softening results in stress redistribution and, as previously noted, increased total horizontal stresses. A modified elastic approach was used to account for strain softening behaviour as shown on Figure 2. This approach assumes that when the peak shear strength of a strain softening soil is reached, continued loading results in an effective stress path which moves down along the critical state line. It is also assumed that the major (vertical) principal total stress remains unchanged (i.e. arching is not significant). The measured porewater pressure is fitted between these two lines and defines the minor (horizontal) principal total stress. A similar approach is adopted for non-strain softening soils, except that the effective stress state in the critically stressed soil remains constant. The shear stress remains constant and the total stress state is defined by the measured porewater pressure.

The value of Poisson's ratio can change if significant porewater pressure dissipaton occurs. However the degree of dissipation which occurred during construction at most piezometer locations examined is unlikely to have resulted in large changes in Poisson's ratio. Therefore, the assumed value of 0.5 is considered to be reasonable.

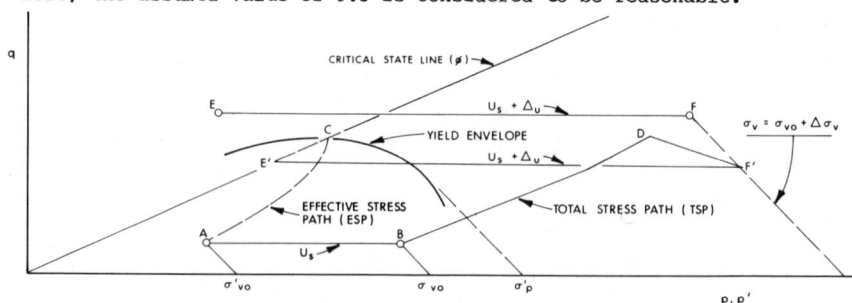

Figure 2.   Modified elastic analysis.

The total piezometric pressure at a point within a loaded soil mass consists of the excess pore pressure generated by the applied load and the "static" porewater pressure. In calculating effective stresses, only the total stress state and the total piezometric pressure are required. If the "static" porewater pressure (U_s) remains constant, the total stress path (TSP) can be conveniently described by the "total minus static porewater pressure" stress path, $(T-U_s)SP$, (Figure 1).

POREWATER PRESSURE RESPONSE TO LOAD

Accurate prediction of consolidation behaviour depends on the magnitude and distribution of initial excess porewater pressures. In this section, the results of ESP/YE analyses are used to identify some of the factors which affect excess porewater pressure development.

Drainage during Construction

Porewater pressure response to load is reflected in the direction/shape of the ESP. At stress states within the YE, the initial ESP direction can be quite variable depending on factors such as degree of overconsolidation, in situ stress conditions, rate of construction, load/stratigraphy geometry and stiffness of the soil skeleton. One of the more significant factors affecting the ESP direction is the degree to which drainage takes place during construction.

Significant drainage during construction is typified by the Natal Embankment case record. This site is underlain by about 14 m of soft clay which underlies a 4 m thick surface sand deposit. The ESP for a piezometer installed about 4 m below the clay surface is shown on Figure 3 and indicates that during initial loading (points 1 - 3), only modest excess porewater pressures were induced suggesting that significant drainage occurred. However, when the ESP reached the YE, the soil yielded and large porewater pressures were generated by further loading causing the ESP to follow the YE. This behaviour was observed in a minority (6) of the 20 case records analyzed. A more common ESP which is representative of an essentially undrained response is shown on Figure 4 for the Trinidad Tanks case record. In this case, significant excess porewater pressures were generated from the onset of loading and as a result the ESP rises almost vertically. In 60% of the cases analyzed the initial ESP direction varied between 30 degrees to the left and right of the vertical in q-p stress space indicating minimal drainage during construction.

The radically different porewater pressure responses exhibited by those two case records lead to similar shapes of $\Delta u - \Delta p$ plots; in both cases they are bilinear. However, the interpretation of the "break point" in the $\Delta u - \Delta p$ relationships is quite different. In the Natal case where significant drainage occurred, the "break point" occurred when the vertical effective stress approximately equalled the preconsolidation pressure (Figure 3). However, in the Trinidad Tanks case (Figure 4), the "break point" in the $\Delta u - \Delta p$ plot occurred when the ESP crossed the line which represents an effective stress ratio of 0.5 (Watson et al, 1984) [42].

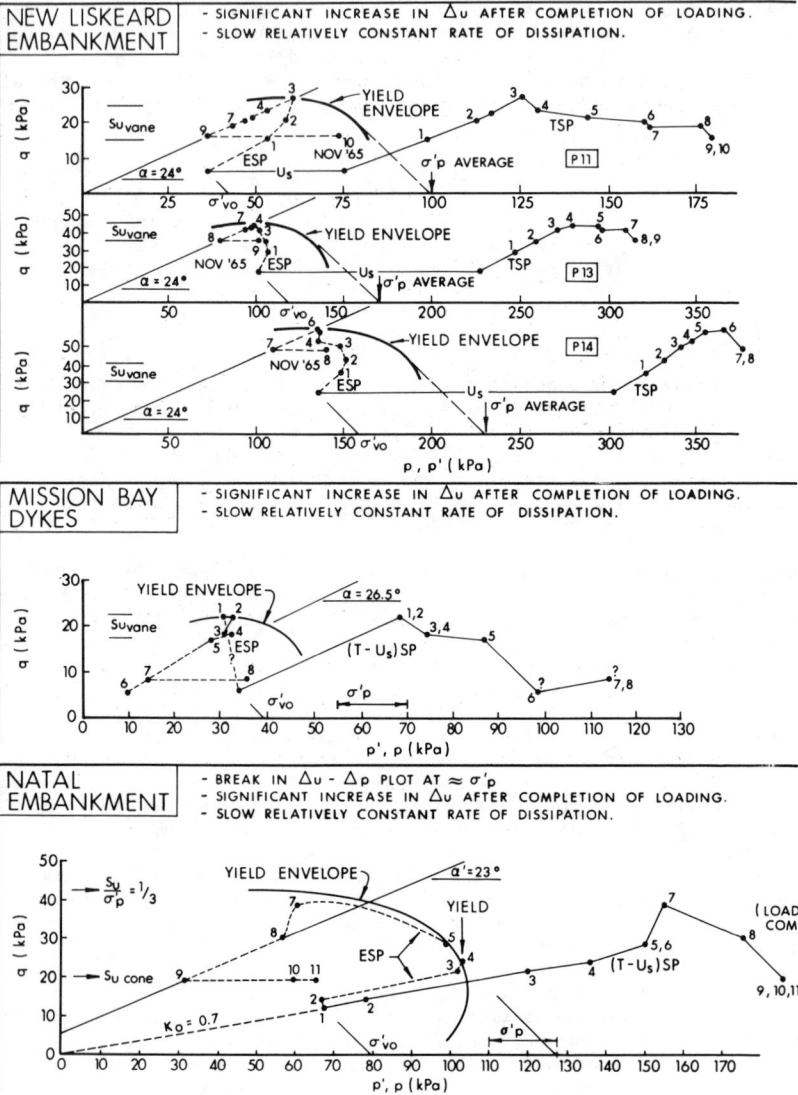

Figure 3.   ESP/YE analysis of the New Liskeard, Mission Bay and
            Natal case records - strain softening behaviour.

Strain Softening and Critically Stressed Soils

In a significant number of the cases analyzed, the imposed shear stresses were close to or equalled the shear strength of the soil. Strain softening was observed in seven cases and the Mission Bay Dykes case record also shown on Figure 3 is typical. Significant strain softening is also clearly evident in the EPSs for Mission Bay, New Liskeard, Natal and Asrum II cases (Figures 3 and 5 respectively). Less marked strain softening behaviour is evident in the Launceston (with vertical drains), Asrum I, Sandwich and Fiddler's Ferry cases (Figures 4 to 7 respectively).

In those cases where significant strain softening occurred, excess porewater pressures continued to rise following completion of the load increment which caused yield. While the ESP/YE analysis provide a rational explanation with respect to the final magnitude of the excess porewater pressure, it does not explain the time dependent nature of the porewater pressure increase following completion of loading. It is thought that the reason for this phenomenon is that the associated total horizontal stress increase is time dependent. Thus, as yielding occurs, lateral straining takes place. Lateral straining and associated development of horizontal stress is also a time dependent phenomenon. Clearly, this behaviour will be most marked in relatively thin layers of foundation clay below widely loaded areas. On the contrary, where narrow loads are imposed on deeper deposits of clay, the potential for confinement, and hence horizontal stress increase, is reduced. In these cases, the phenomenon of excess porewater pressure development following completion of loading is less likely and for the same conditions (i.e. strength and loading intensity) outright failure and/or increased deformations can occur.

The degree to which strain softening occurs is a function of the soil type, provided that stress conditions are similar. For example, the upper portion of the YE for highly sensitive, cemented or bonded soils lies well above the critical state line. When yield of these materials occurs, the ESP drops dramatically down onto the critical state line. In less sensitive soils the upper portion of the YE lies closer to the critical state line and strain softening in these materials is less dramatic or does not occur. Similarly, the type of strain softening behaviour which takes place also depends to some extent on the initial ESP direction. For example, in the New Liskeard case (Figure 3), yield occurred at about the intersection of the critical state line and the YE. Strain softening resulted in an ESP which immediately followed the critical state line without experiencing a dramatic decrease in shearing resistance (i.e. near vertical ESP).

When non-strain softening soils are critically stressed, the ESP lies on the critical state line and on further loading this effective stress state remains unchanged. Porewater pressures and associated total stresses continue to increase as load is applied. This behaviour was observed in a total of seven out of the 20 cases analyzed with the Asrum I case being typical as shown on Figure 5.

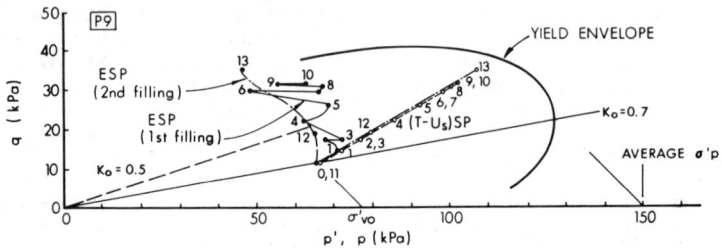

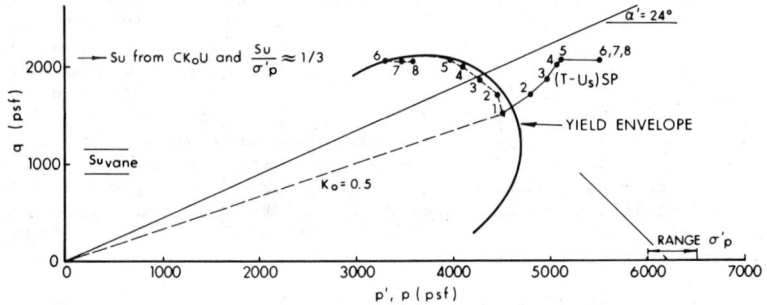

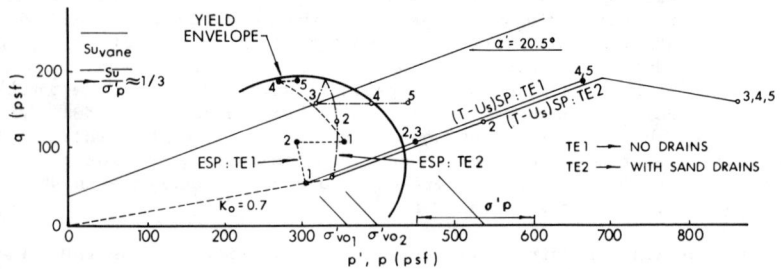

Figure 4.   ESP/YE analysis of the Trinidad Tanks, MIT and
             Launceston case records.

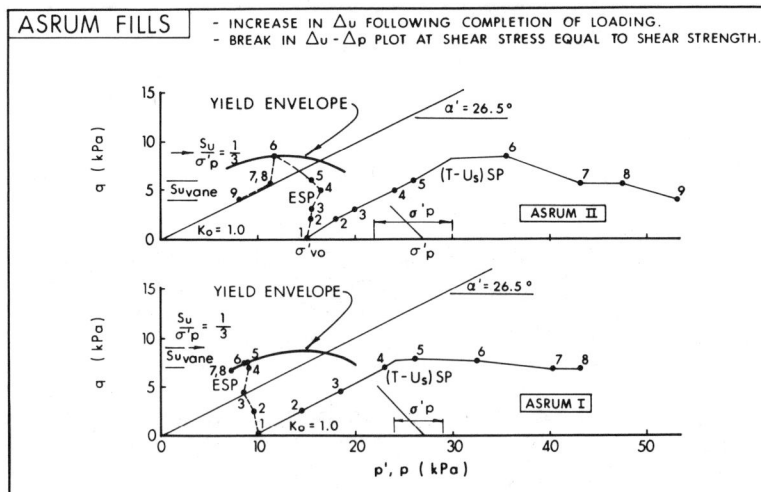

Figure 5. ESP/YE analysis of the Asrum Fills case record

Normally Consolidated Clay Behaviour

In only one case, the MIT test fill (I95 Embankment) was normally consolidated behaviour observed. The ESP for a piezometer (P9) installed in
the lower clay deposit is shown on Figure 4. Since oedometer test
results indicate that this material is normally consolidated, the
initial effective stress state lies on the YE. Loading results in
immediate yielding and an ESP which follows the YE. In the MIT test
fill case, loading was continued until the soil at the piezometer
location was critically stressed (i.e. the imposed shear stress equalled
the shearing resistance). This point represents the "break point" in
the bilinear Δu –Δp plot. Only modest further loading was imposed and
the material did not appear to strain soften.

Dilatant Behaviour

Analyses which indicate ESPs tracking upward along the critical state
line indicate dilatant behaviour. This response was observed in the
following cases:

- ° in the layered clay/silt under Stage III loading at the Mission
  Bay site, (Figure 7).

- ° under Stage III loading at the Fiddler's Ferry site, (Figure 7).

- ° during construction of the Iraq fill, (Figure 7).

- ° (Possibly) during the final load increment at the CNR - Don River
  site, (Figure 8).

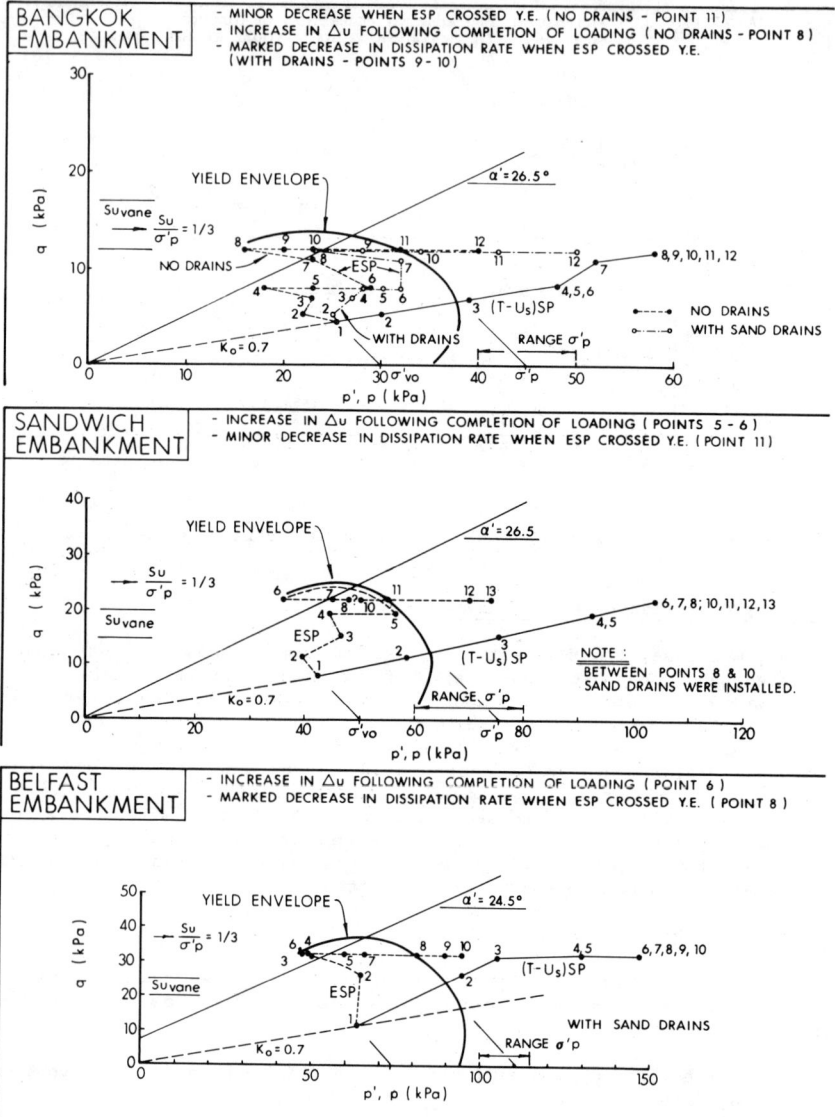

Figure 6.   ESP/YE analysis of the Bangkok, Sandwich and Belfast embankments case records.

In the latter three cases, significant drainage occurred during the relevant loading stage. Following yield at about the apex of the YE, the ESP followed the YE to the critical state line. On further loading, the ESP tracked upward along the critical state line indicating dilation. In the layered clay and silt underlying the Mission Bay site, a similar behaviour was observed except that the initial response to load was essentially undrained.

CONSOLIDATION BEHAVIOUR

As stated previously, significant differences in laboratory and field cv values are reported in many cases. This difference is usually attributed to inadequate modelling of boundary drainage conditions and in particular, the effect of undetected thin continuous permeable layers in the clay deposit. This rationale is undoubtedly true. However, in cases involving very wide areal loads (i.e. close to one-dimensional loading situations) lateral drainage through thin "permeable" layers is less significant. For example, the presence of permeable layers below the wide Trinidad and SWPCP preload fills (Figure 8) did not greatly increase the rate of dissipation over that observed at "non-layered" sites.

A further significant factor in the accurate prediction of consolidation rates is the dependence of cv on stress level. The results of oedometer tests on a wide variety of soft clays indicate that the magnitude of cv decreases significantly as the stress level increases up to and beyond the preconsolidation pressure. In highly sensitive soils, cv is a minimum in the σ_p' stress range as a result of breakdown of the soil structure. As the stress level increases beyond σ_p' in these materials, the cv value increases again to a constant value, although this value is still about one order of magnitude lower than at stress levels less than σ_p'. A similar pattern of behaviour occurs in less sensitive soils, although the magnitude of variation in cv is less. The significance of cv dependence on stress level is discussed in the following sections.

Dissipation During Construction

The variation in cv with increasing stress level in the overconsolidated stress range is well illustrated by the CNR-Don River case (Figure 8). This fill was constructed in two stages. Stage I construction (points 1-9) was relatively slow with 6 m of fill being placed over a period of three months. Excess porewater pressure generated during this first filling stage was allowed to dissipate during the subsequent eight months. Stage II construction (points 10-13) raised the fill to a final height of 9.9 m over a period of about 6 weeks. The stage I ESPs indicate a paritally drained porewater pressure response to loading and are close to $(T-U_s)SP$. The rate of dissipation changed with stress level, as reflected by back-calculated cv values of $2 \times 10^{-2} cm^2/s$ at the 5m height and $5 \times 10^{-3} cm^2/s$ at a fill height of 6 m. There was negligible dissipation of excess porewater pressures in piezometers installed in the layered clay and silt during the 4-5 months prior to stage II filling, although modest excess pressures still existed in this deposit.

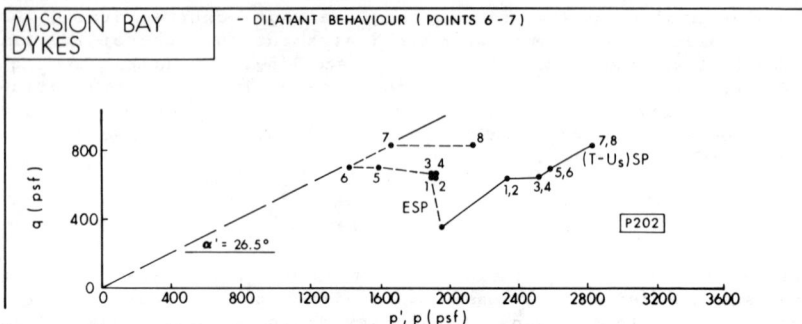

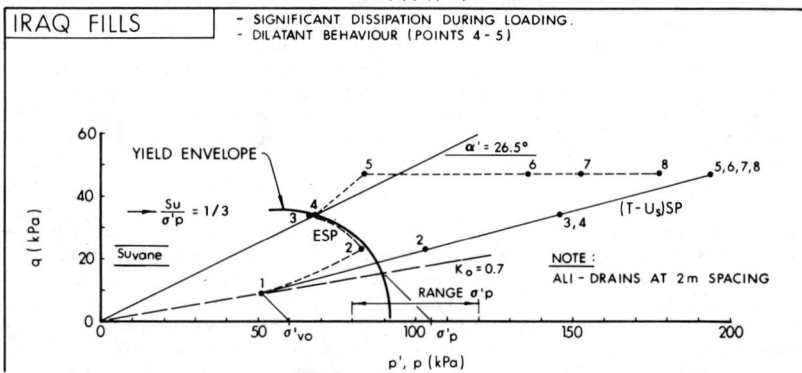

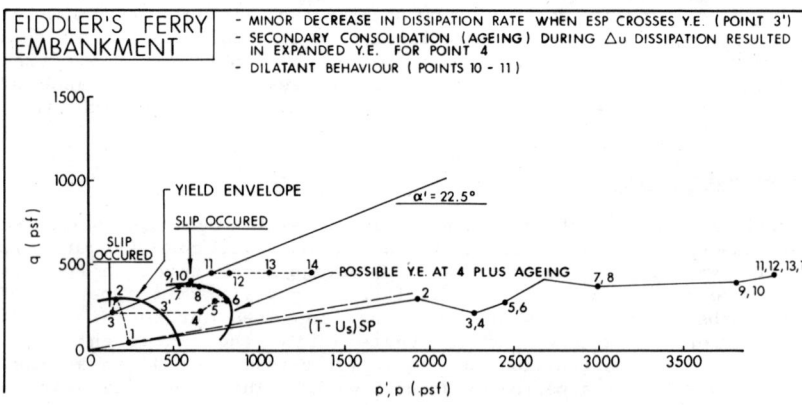

Figure 7. ESP/YE analysis of the Mission Bay, Iraq and Fiddler's Ferry
          case records.

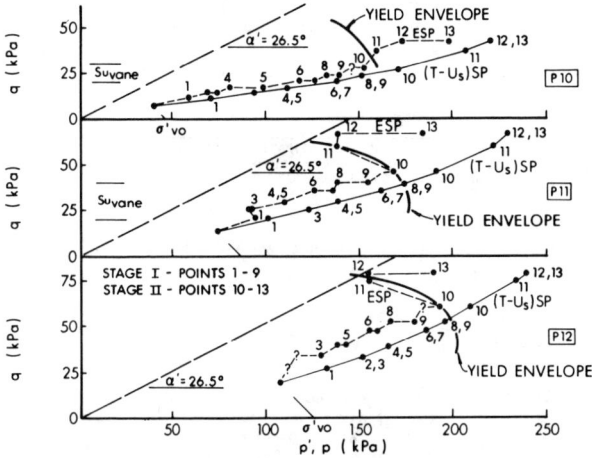

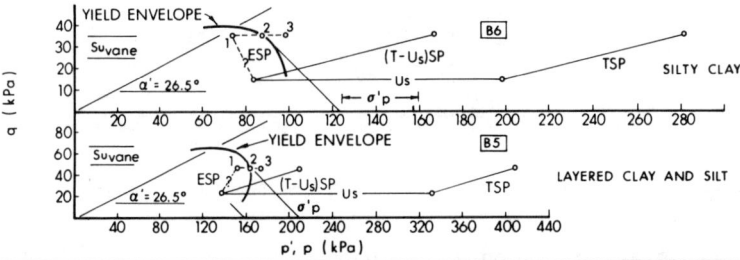

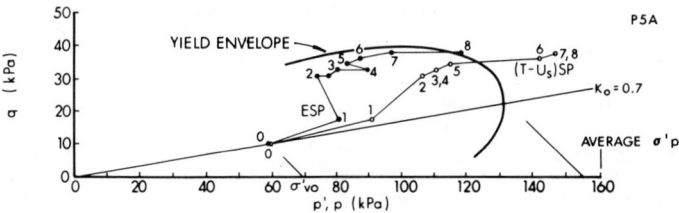

Figure 8. ESP/YE analysis of the CNR-Don River, Thunder Bay (SWPCP) and Trinidad preload case records.

The first stage II load increment resulted in only modest excess porewater pressures in all piezometers. The next load increment caused large excess porewater pressures, particularly in the lower portion of the layered clay and silt and in the sandy clayey silt deposit. The resulting effective stress paths between points 10 and 11 appear to define yield envelopes, since the vertical effective stresses defining the apex of the YEs fall within the range of $\sigma p'$ values from laboratory oedometer tests.

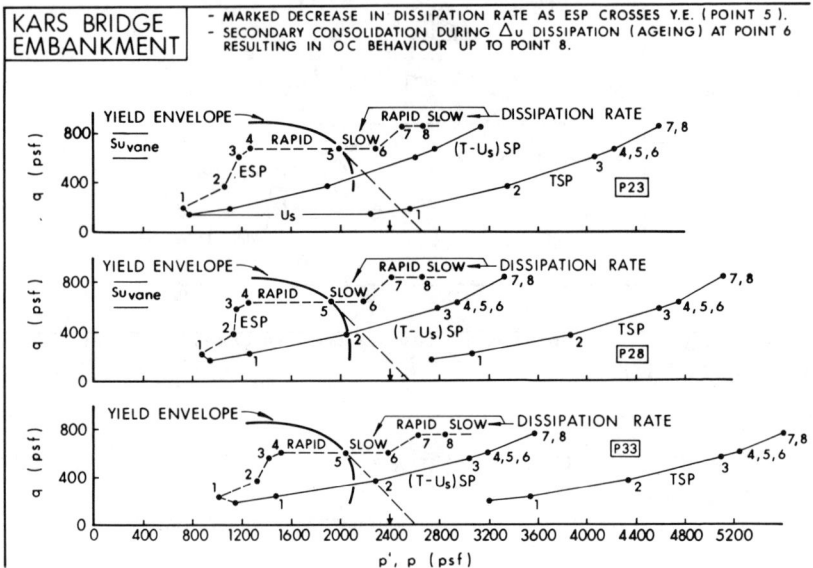

Figure 9.   ESP/YE analysis of the Kars embankment case record.

Dissipation Behaviour Following Completion of Loading

With porewater pressure dissipation following completion of loading, the effective stress in the soil increases. The increase in effective stress is associated with a change in dissipation rate and this behaviour is evident from the results of ESP/YE analyses of a number of case records. However, the manner in which cv varies depends on the stress level and soil type. Basically, four types of dissipation behaviour were observed as discussed below:

Marked change in dissipation rate: As indicated on Table 1, this behaviour occurred in a significant number of cases including Bangkok, (with drains), Belfast, Trinidad Preload, Thunder Bay (SWPCP) and Kars (Figures 6, 8 and 9, respectively). In these cases, the effective stress state following completion of loading was within the YE. Initial dissipation was rapid (cv typically $10^{-2} cm^2/s$) while the clay was overconsolidated (i.e. the effective stress state remained inside the YE). However, when yielding occurred as the ESP reached the YE, the dissipation

rate decreased abruptly to about the value associated with normally consolidated clay (10^{-3}cm^2/s). It is noted that this behaviour was clearly evident even in those cases where the effective stress state following completion of load was close to the upper portion of the YE.
Although explanations of this behaviour similar to that given above have
been published (e.g. Crawford and Burn, 1976) [11] this behaviour is
still not widely discussed in the literature.

Very slow or insignificant porewater pressure dissipation for long
periods following construction: This behaviour occurred at six sites.
ESP/YE analyses were carried out for three of these cases (MIT,
Launceston (undrained area) and Ska Edeby (undrained area) (Figures 4
and 10). In each case, the effective stress state following construction coincided with the upper portion of the YE but the imposed stresses
were not sufficient to cause strain softening. Therefore, continued
soil structure breakdown resulted in continued generation of porewater
pressures following completion of load. While some dissipation of
excess porewater pressure was taking place, the net effect was a very
slow or negligible rate of net porewater pressure dissipation. ESP/YE
analyses were not carried out for the remainder of the case records
(Yarra, Napa and Porto Tolle Tank). However, based on comments included
in these publications, it is likely that shear strain induced porewater
pressures were the cause of very slow dissipation rates.

Slow, relatively constant rate of dissipation following construction:
This behaviour occurred at the three sites at which significant strain
softening took place (Mission Bay, New Liskeard and Natal, (Figure 3).
The effect of significant strain softening is to cause major disturbance
to the soil structure and associated rates of dissipation are similar to
those observed for normally consolidated clays. No change in cv would
be expected as the effective stress level increased since the current
effective stress state defines the current YE. As a result of strain
softening, the original YE no longer controls the soil behaviour.

Minor change in dissipation rate following completion of construction:
A total of four cases exhibited this behaviour (Launceston (vertically
drained area), Bangkok, Sandwich and Fiddler's Ferry. The results of
ESP/YE analyses for these cases are shown on Figures 4, 6 and 7
respectively and indicate consistent ESP behaviour. In the Fiddler's
Ferry, Launceston and Sandwich cases, minor strain softening occurred
while in the Bangkok case, the effective stress state following
construction coincided with the top of the YE. Following completion of
loading, dissipation occurred and the cv value changed when the ESP
reached the original YE. It is noted that the magnitude of change in cv
was modest but distinct in all cases. This interpretation appears to be
contradictory in that yield is inferred to have occurred twice; once
when the ESP coincided with the top of the YE and once during
dissipation. A possible explanation of this apparent contradiction is
that since significant strain softening did not occur, the initial soil
structure was retained to a sufficient extent that the influence of the
original YE was not completely obliterated. However, some disturbance
of the soil structure occurred, causing yield during subsequent
dissipation to be less marked than would be expected.

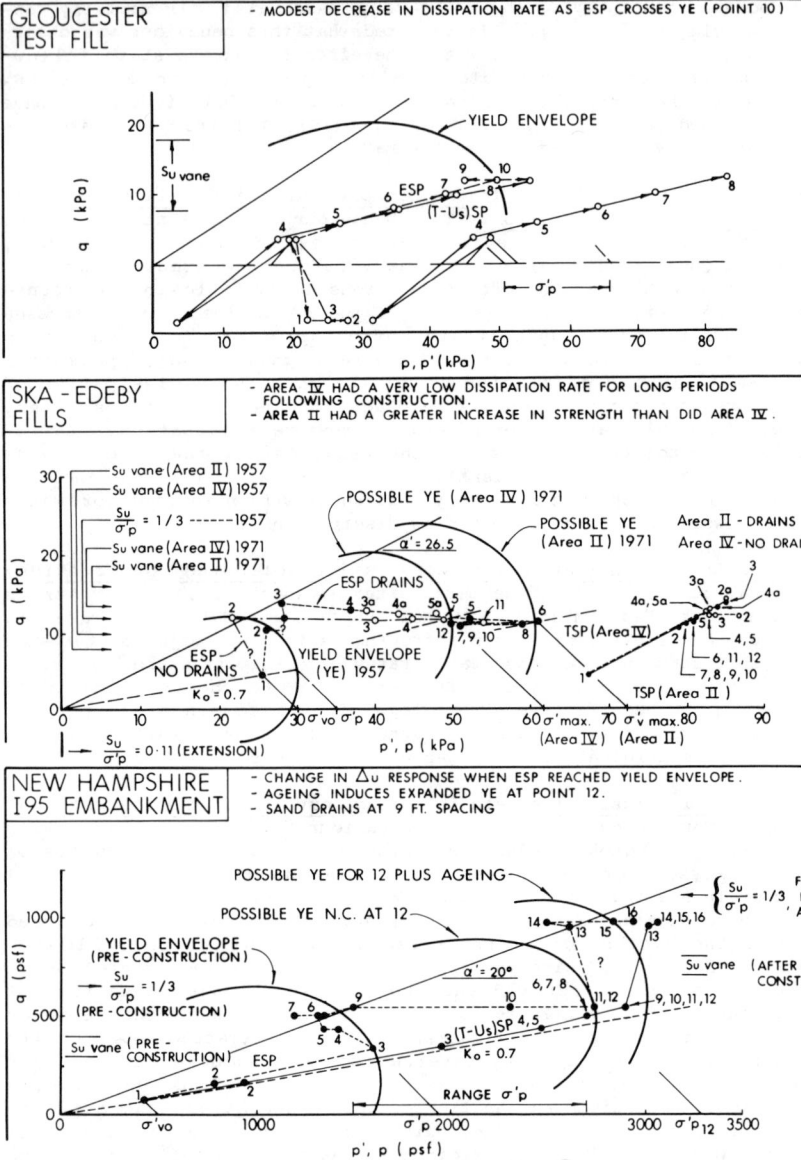

Figure 10: ESP/YE analysis of the Gloucester, Ska-Edeby and New Hampshire I95 case records.

The Gloucester test fill case (Figure 10) also showed a modest change in dissipation rate following construction. In this case the shear stress level at which initial yield occurred was well below the top of the YE. Nevertheless, this material is highly sensitive and some disturbance to the soil structure occurred at yield. This disturbance influenced the magnitude of change in cv during subsequent dissipation.

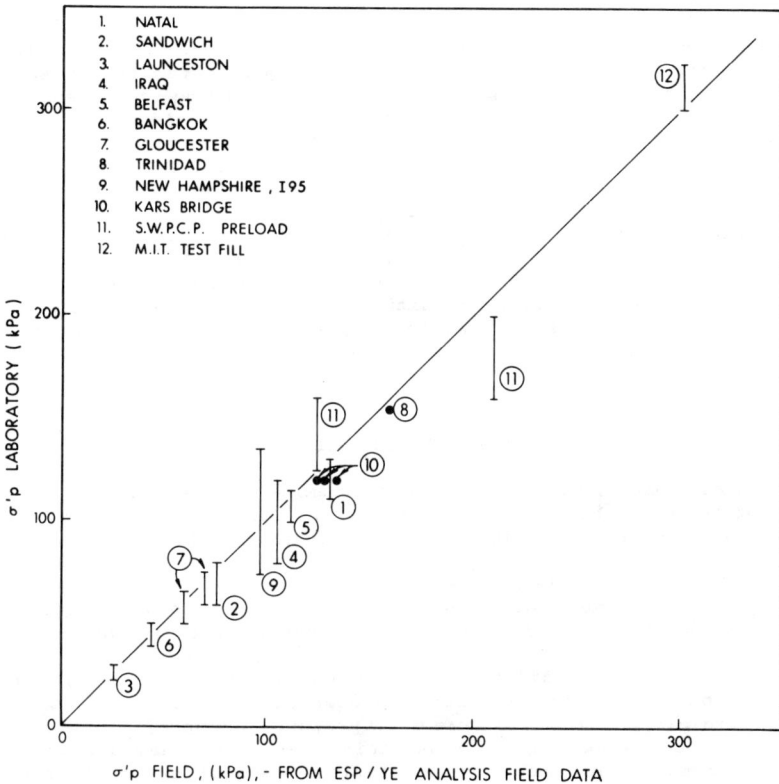

Figure 11. Comparison of "field" and laboratory preconsolidation pressure.

Settlement Magnitude

In predicting the magnitude of settlement of a soft clay, it is necessary to account for the "average" ESP in the soft clay. This aspect of settlement prediction has been discussed in detail by Folkes and Crooks, (1984) [16]. In summary, the model used to predict settlement should be based on the "average" ESP in the soft clay. For those cases where the effective stress state immediately following construction is well within the YE, the parameter which largely controls the predicted settlement magnitude is the preconsolidation pressure. A comparison of the $\sigma p'$ values predicted from ESP/YE analyses and reported laboratory $\sigma p'$ values of these cases is shown on Figure 11. It is evident that the comparison is good despite the fact that $\sigma p'$ measured in oedometer tests can be significantly affected by test procedures (Leonards and Girault, 1961) [25]. The likelihood that "normal" testing procedures provide a reasonable estimate of $\sigma p'$ is further confirmed by comparison of predicted and observed settlements which generally shows good agreement (Tavenas and Leroueil, 1980) [39].

Good agreement between predicted and observed settlements is less likely in cases where significant strain softening occurs for two reasons. Firstly, the original preconsolidation pressure is of no consequence in settlement analyses since strain softening will obliterate the memory of this $\sigma p'$ value; virgin consolidation will take place from the strain softened effective stress state. Secondly, for normal fill/foundation clay geometries, strain softening will be associated with large lateral deformations which are not readily taken into account in conventional settlement predictions. Strain-softening is also difficult to account for using available constitutive models in finite element analyses. However, in the case of wide fills or tanks on relatively thin layers of strain softened soil, boundary conditions tend to decrease the effect of lateral deformations except at the edges of the loaded area where excessive deformations/local failures can occur.

SUMMARY AND CONCLUSIONS

A tabulated summary, of the case records examined, indicating the occurrence of the above behaviour is presented. Based on these data and ESP/YE analyses, the following key observations are noted:

- ° The initial porewater pressure response to load, often charac-. terized by plots of porewater pressure versus load, included a range of linear, bilinear, multi-linear and curved relationships

- ° In 11 cases for which data was available, the porewater pressure continued to increase after the completion of loading. Extraordinary increases in porewater pressure occurred in three of the cases examined. Thus, this anomalous behaviour is not uncommon. In general, this behaviour was attributed to creep/ breakdown of the soil structure with the more extreme cases associated with strain softening. Critically stressed soils undergo time dependent redistribution of total streses and associated time dependent porewater pressure increases.

- ° In some cases, the analyses indicate that dilation occurs in the field.

° In the majority of cases, the field porewater pressure dissipation rate was significantly faster than that predicted. This
was most commonly attributed to inaccurate modelling of drainage
boundary conditions and in particular, the presence of undetected layers of more permeable silts and sands in the consolidating
clay stratum. The difference between predicted and observed
rate decreases as loading conditions become more one-dimensional.

° In the majority of cases examined, the dissipation rate changed
dramatically during the consolidation process subsequent to
completion of loading; field consolidation rates are strongly
dependent on stress level. For example, a marked decrease in
the dissipation rate takes place as the field effective stress
state crosses the original in situ yield envelope. However,
very slow or insignificant excess porewater pressure dissipation
for long periods following construction occurs when the
effective stress state following construction lies on the upper
portion of the yield envelope. Continued soil structure
breakdown causes continued generation of excess porewater
pressure. Consequently, while some dissipation of excess
porewater pressure is taking place, the net effect is a very
slow or negligible rate of net porewater pressure dissipation.
Finally, if the effective stress state at the end of
construction lies outside of the original in situ yield
envelope, a slow relatively constant rate of dissipation follows
construction, similar to that observed for normally consolidated
clays.

ACKNOWLEDGEMENTS

The authors wish to express their appreciation to Gulf Canada Resources
Inc. for permission to publish this paper. The drawings were prepared
by G. Sundholm and the manuscript was typed by K. Koch.

NOTATION

The following symbols are used in this paper:

cv	= coefficient of consolidation
p', p	= $(\sigma_1' + \sigma_3')/2$, $(\sigma_1 + \sigma_3)/2$
q	= $(\sigma_1 - \sigma_3)/2$
su	= undrained shear strength
$\bar{B}$	= porewater pressure parameter $(\Delta u/\Delta \sigma_1)$
CKoU	= Ko consolidated - undrained traixial compression test with porewater pressure measurements
ESP	= effective stress path
Ko	= effective stress ratio (σ_3'/σ_1')
TSP	= total stress path
(T-Us)SP	= total minus static porewater pressure stress path
Us	= static porewater pressure
YE	= yield envelope
α'	= angle of effective internal friction in p'- q plot

$\sigma'p$ = preconsolidation pressure
σ_{vo}' = original effective vertical stress
σ_{vo} = original total vertical stress
σ'_1, σ'_3 = major and minor principal stresses
Δu = excess porewater pressure
Δp = applied surface loading (pressure)
$\Delta\sigma_1$ = increment in major principal stress

REFERENCES

1. AL-DHAHIR, Z.A., KENNARD, M.F. and MORGENSTERN, N.R. (1970). Observations on pore pressures beneath the ash lagoon embankments at
   Fiddler's Ferry power station. Proc. of Conference on In Situ
   Investigations in Soils and Rocks. British Geotechnical Society,
   pp. 265-276.

2. BECKER, D.E. (1981). Settlement analysis of intermittently loaded
   structures founded on clay subsoils. Ph.D. thesis. Faculty of
   Engineering Science, University of Western Ontario, London,
   Canada.

3. BECKER, D.E., CROOKS, J.H.A., JEFFERIES, M.G., and McKENZIE, K.
   (1984). Yield behaviour and consolidation. II: Strength Gain.
   ASCE Symposium on Prediction and Case Histories of Consolidation
   Performance, San Francisco.

4. BRENNER, R.P. and PBEBAHARAN, N. (1983). Analysis of sandwick performance in soft Bangkok clay. Proc. 8th European Conference on
   Soil Mechanics and Foundation Engineering, Helsinki, Vol. 2,
   pp. 579-586.

5. BROOKER, E.W. and IRELAND, H.O. (1965). Earth pressures at rest
   related to stress history. Canadian Geotechnical Journal, Vol.
   2, p.1.

6. BURWASH, W.J. and MATICH, M.A.J. (1981). Stage loading of a highway
   embankment on tidal flats. Canadian Geotechnical Journal,
   Vol. 18, No. 4, pp 535-542.

7. CASAGRANDE, L. and POULOS, S. (1969). On the effectiveness of sand
   drains. Canadian Geotechnical Journal, Vol. 6, No. 3,
   pp. 287-326.

8. CHALMERS, A. and HARRIS, A.B. (1981). Six-storey building on soils
   improved by sand drains. Proc. 10th International Conference on
   Soil Mechanics and Foundation Engineering, Stockholm, Vol. 3,
   pp. 611-616.

9. COLE, K.W. and GARRETT, C. (1981). Two road embankments on soft
   alluvium. Proc. 10th International Conference on Soil Mechanics
   and Foundation Engineering, Stockholm, Vol. 1, pp. 87-94.

10. COOPER, I.D. and MEYER, P.A. Road embankments on soft foundations - Launceston, Tasmania. Australian Road Research Board, Adelaide (Date of publication unknown)

11. CRAWFORD, C.B. and BURN, K.N. (1976). Long-term settlements on sensitive clay. NGI Publication, Oslo, Laurits Bjerrum Memorial Volume - Contributions to Soil Mechanics, pp. 117-124.

12. D'APPOLONIA, D.J., LAMBE, T.W. and POULOS, H.G. (1971). Evaluation of pore pressures beneath an embankment. Proc. ASCE Journal of Soil Mechanics and Foundations Division. Vol. 97, SM. 6, pp. 881-898.

13. DASCAL, O., TOURNIER, J.P., TAVENAS, F. and LAROCHELLE, P. (1972). Failure of a Test Embankment on Sensitive Clay. Proc. of ASCE Specialty Conference on Performance of Earth and Earth-Supported Structures. Purdue University pp. 129-158.

14. DAVIES, J.A. and HUMPHESON, C. (1981). A comparison between the performance of two types of vertical drains beneath a trial embankment in Belfast. Geotechnique Vol. 31, No. 1, pp. 19-32.

15. ERIKSSON, L. and EKSTROM, A. (1983). The efficiency of three different types of vertical drain-results from a full-scale test. Proc. 8th European Conference on Soil Mechanics and Foundation Engineering, Helsinki, Vol. 2, pp 605-610.

16. FOLKES, D. and CROOKS, J.H.A. (1984). Effective stress paths and yielding in soft clays below embankments. (Submitted to Canadian Geotechnical Journal).

17. HANSBO, S., JAMIOLKOWSKI, M. and KOK, L. (1981). Consolidation by vertical drains. Geotechnique, Vol. 31., No. 1, pp. 45-66.

18. HEGG, U., JAMIOLKOWSKI, M.B., LANCELLOTTA, R. and PARVIS, E. (1983). Behaviour of oil tanks on soft cohesive ground improved by vertical drains. Proc. 8th European Conference on Soil Mechanics and Foundation Engineering, Helsinki, Vol. 2, pp. 627-632.

19. HOEG, K., ANDERSLAND, O.B. and ROLFSEN, E.N. (1969). Undrained behaviour of quick clay under load tests at Asrum. Geotechnique, 19, No. 1, pp. 101-115.

20. JAMES, P.M. (1970). Behaviour of a soft recent sediment under embankment loadings. The Quarterly Journal of Engineering Geology. Vol. 3, pp. 41-53.

21. JONES, G.A. and RUST, E. (1981). Design and monitoring of an embankment on alluvium. Proc. 10th International Conference on Soil Mechanics and Foundation Engineering, Stockholm, pp. 151 - 156.

22. KASTMAN, K.H. (1972). Analysis and Records - Dutch Hollow Dam. Proc. of ASCE Specialty Conference on Performance of Earth and Earth-Supported Structures. Purdue University, pp.285-298.

23. LADD, C.C. (1972). Test embankment on sensitive clay. Proc. of
    ASCE Specialty Conference on Performance of Earth and Earth-
    Supported Structures. Purdue University, pp. 101-128.

24. LADD, C.C., RIXNER, J.J. and GIFFORD, D.G. (1972). Performance of
    embankments with sand drains on sensitive clay. Proc. of ASCE
    Specialty Conference on Performance of Earth and Earth-Supported
    Structures, Purdue University, pp. 211 - 242.

25. LEONARDS, G.A. and GIRAULT, P. (1961). A study of the one-dimens-
    ional consolidation test. Proc.,5th ICSMFE (Paris), Vol.1,p.213.

26. LEATHERS, F.D. and LADD, C.C. (1978). Behaviour of an embankment
    on New York varved clay. Canadian Geotechnical Journal, Vol. 15,
    No. 2, pp. 250-268.

27. LEROUEIL, S., TAVENAS, F., TRAK, B., LaROCHELLE, P. and ROY, M.
    (1978). Construction pore pressures in clay foundations under
    embankments. Part I: The Saint-Alban test fills. Canadian
    Geotechnical Journal, Vol. 15, No. 1, pp. 54-65.

28. MARSH, J.G. (1963). A test embankment with sand drains on a weak
    and compressible foundation at Perth, Western Australia.
    Proc., 4th Australian-New Zealand Conference on Soil Mechanics
    and Foundation Engineering, pp. 149-154.

29. MARSLAND, A. and POWELL, J.J.M. (1977). The behaviour of a trial
    bank constructed to failure on soft alluvium of the River Thames.
    International Symposium on Soft Clay, Bangkok, Asian Institution
    of Technology, pp. 505-525.

30. MILLIGAN, V. and SODERMAN, L.G. (1962). Experience with Canadian
    varved clays. Proc. ASCE Journal of Soil Mechanics and Founda-
    tions Division, Vol. 88, No. SM4, pp. 31-67.

31. MOH, Z., BRAND, E.W. and NELSON, J.D. (1972). Pore pressures under
    a bund on soft fissured clay. Proc. of ASCE Specialty Conference
    on Performance of Earth and Earth-Supported Structures. Purdue
    University, pp. 243-272.

32. NICHOLSON, D.P. and JARDINE, R.J. (1981). Performance of vertical
    drains at Queenborough bypass. Geotechnique, Vol. 31, No. 1,
    pp. 67-90.

33. ORDEMIR, I.M. and BIRAND, A.A. (1981). Settlement of a test fill.
    Proc. 10th International Conference on Soil Mechanics and
    Foundation Engineering, Stockholm, Vol. 2, pp. 541-544.

34. PARRY, R.H.G. (1968). Field and laboratory behaviour of lightly
    overconsolidated clay. Geotechnique, Vol. 18, No. 2, pp.151-171.

35. PENMAN, A.D.M. and WATSON, G.H. (1965). The improvement of a tank
    foundation by the weight of its own test load. Proc. of the 6th
    International Conference on Soil Mechanics and Foundation
    Engineering, Vol. 2, pp. 169-173.

36. POULOS, H.G. and DAVIS, E.H. (1974). Elastic Solutions for Soil and
    Rock Mechanics. John Wiley & Sons, Inc., New York.

37. RIVARD, P.J. and KOHUSKA, A. (1967). Shellmouth Dam test fill.
    Canadian Geotechnical Journal, Vol. 2, No. 3, pp.198-211.

38. TAYLOR, H.T. and BUCHIGNANI, A.L. (1972). Field test of debris fill
    over soft soil. Proc. of ASCE Specialty Conference on
    Performance of Earth and Earth-Supported Structures. Purdue
    University, pp.395-414.

39. TAVENAS, F., and LEROUEIL, S. (1980). The behaviour of embankments
    on clay foundations, Canadian Geotechnical Journal, Vol. 17,
    No. 2, pp. 236-260.

40. TSAI, K.W., LEE, C.C. and CHAD, C.S. (1981). Site improvement by
    preloading with sand drains. Proc. 10th International Conference
    on Soil Mechanics and Foundation Engineering, Stockholm, Vol. 3,
    pp.781-783.

41. WALKER, L.K. and MORGAN, J.R. (1977). Field performance of a firm
    silty clay. Proc. of 9th International Conference on Soil
    Mechanics and Foundation Engineering, Tokyo, Vol. 1, pp.341-346.

42. WATSON, G.H., CROOKS, J.H.A., WILLIAMS, R.S., and YAM, C.C. (1984).
    Performance of preloaded and stage-loaded structures on soft
    soils in Trinidad, W.I. (To be published in June 1984, Geotech-
    nique).

43. WILKES, P.F. (1972). An induced failure at a trial embankment at
    King's Lynn, Norfolk, England. Proc. of ASCE Specialty
    Conference on Performance of Earth and Earth-Supported Struct-
    ures, Purdue University, pp. 29-64.

YIELD BEHAVIOUR AND CONSOLIDATION. II: STRENGTH GAIN
by

D.E. Becker[1], J.H.A. Crooks, M.ASCE[1]
M.G. Jefferies, M.ASCE [2] and K. McKenzie[2]

INTRODUCTION

Accurate prediction of strength gain due to consolidation is a prerequisite for successful staged-loading and preloading operations. However, it is not unusual that observed field consolidation behaviour of soft clays is not as predicted or expected. In Part I of this paper a number of aspects of the field consolidation behaviour of soft clays was discussed (Crooks et al, 1984) [8]. It was shown that in situ yielding of soft foundation clays significantly affected consolidation behaviour and that yield could be reasonably well defined using the Effective Stress Path/Yield Envelope (ESP/YE) approach proposed by Folkes and Crooks, (1984)[11]. This approach is used in this paper to examine strength gain associated with consolidation of soft clays.

A literature review was carried out to determine the frequency of occurrence of reported lack of strength gain with consolidation. Where sufficient data were available, ESP/YE analyses were carried out to examine the influence of yielding on strength gain. The case records analyzed include a variety of soil types, loading and stratigraphic geometries, and boundary drainage conditions. As in the Part I paper, analyses were carried out for sites at which vertical drains were and were not employed.*

LITERATURE REVIEW

Information related to strength gain with consolidation was available in 19 of the case records examined. A summary of these case records is given on Table 1. In ten of the cases reviewed, the strength of the foundation clay deposits is reported to have increased as a result of consolidation. In six cases, no strength gain was observed with the majority of these cases being at sites where vertical drains were not employed. Lack of strength gain in these cases can be attributed to insufficient consolidation as opposed to anomalous soil behaviour. In the remaining cases, the results of post construction strength testing were not conclusive. For example, at the Muda embankment site, only

---

(1)   Golder Associates, Calgary, Canada
(2)   Gulf Canada Resources Inc., Calgary, Canada

*Throughout this paper the term "non-drained" is used to identify those sites at which vertical drains were not employed, and "drained" denotes cases in which vertical drains were installed.

marginal strength increases were recorded by field vane tests. At the Kars embankment and Gloucester test fill sites, no strength increase was recorded by field vane tests after 16 and 8 years respectively; however, in both cases triaxial compression test data on relatively undisturbed samples, obtained after construction, indicated increased strengths. No strength gain was measured in similar foundation soils at the Boundary Road site by field vane tests five years following completion of loading.

TABLE 1: SUMMARY OF LITERATURE REVIEW

CASE/REFERENCE	PLASTICITY INDEX ($\%$)	STRENGTH CHANGE	VERTICAL DRAINS USED	REMARKS
Asrum Fill [13]	15–25	No	No	Su measured after 8 months essentially same as pre-construction. Effective vertical stress had not exceed $\sigma'p$.
Matagami Embankment [9]	30–45	No	No	No Su increase measured below failure surface.
Muda Embankment [15]	30–45	?	No	Tests not conclusive. Slight increase (7 kPa) recorded by field vane tests.
Launceston Levee [22]	100	Yes	No	Su increased from 10 to 17 kPa at 3 m depth. Greatest Su increase occurred at shallow depths.
Izmir Fill [21]	40–75	No	No	Su unchanged 70 days after fill constructed.
Visag Fill [24]	25–50	No	No	No Su increase measure below "non-drained" fill.
		Yes	Yes	Su increase occurred ($Su/\sigma'p = 0.3$). Su increase greater in sand drain (360 mm dia.) area than in sand wick area – attributed to effect of lateral deformmation.
New Hampshire I95 [17]	10–20	Yes	No	No strength gain observed in centre of clay layers after 35$\%$ consolidation complete. Su increase recorded at top and bottom of clay.
		Yes	Yes	Significant Su vane increase in drained area at 95$\%$ consolidation. Expanded yield envelope and associated Su increase evident.
New Liskeard Embankment [20]	10–40	Yes	No	Su increase with time following fill completion.

TABLE 1: SUMMARY OF LITERATURE REVIEW (Cont'd.)

CASE/REFERENCE	PLASTICITY INDEX (%)	STRENGTH CHANGE	VERTICAL DRAINS USED	REMARKS
Ska Edeby Fills [14]	40	Yes	No	Su increase measured by vield vane, and by lab tests 14 years after fill construction. Expanded yield envelope and associated strength gain evident.
		Yes	Yes	Significant Su increase recorded below 3 fills 14 years after construction.
Kars Embankment [19]	18-26	No-Vane Yes-Triax	No	Su vane did not increase after 16 years. Expanded yield envlope and associated strength gain evident from ESP/YE analysis and from triaxial tests.
Gloucester Fill [19]	16-35	No-Vane Yes-Triax	No	Su vane did not increase after 7-8 years Expanded yield envelope and associated strength gain evident from ESP/YE analysis and detailed laboratory testing.
Boundary Road Fill [19]	22-40	No	No	Su vane did not increase after 5 years.
Hampton Roads Embankment [23]	-	Yes	Yes	Dramatic increase in Su in upper clay in less time than was anticipated.
King's Lynn Fill [26]	30-40	Yes	Yes	Su increased by up to 25-50%. Rate of Su increase decreased with time.
Middle East Fill (Golder Associates)	22-35	Yes	Yes	Up to 50% increase in Su vane after one month with rate of Su gain decreasing with time thereafter.
Porto Tolle Tanks [12]	25-37	Yes	Yes	-
Norfolk Airstrip [4]	42-85	No	Yes	Su may have decreased as a result of disturbance to sensitive soil caused by sand drain installation. Based only on SPT results.
Taiwan Preload [25]	-	Yes	Yes	Su increased by a factor of 2 to 5. Water content decreased and N-values increased.
Fiddler's Ferry Embankment [1]	22	Yes	No	-

Most soft clay deposits are anisotropic with respect to undrained strength as a result of anisotropic in situ stress states imposed on the soil since deposition. Thus the measured strength depends on the manner in which the sample is sheared in the laboratory. As indicated on Figure 1, the degree of strength anisotropy decreases with increasing plasticity. Also shown on Figure 1 is the reported typical range for field vane tests (Larsson, 1980) [18]. Other factors such as strain rate effects and disturbance are also known to affect vane strength values. Based on these data, it is evident that field vane tests may not reflect either triaxial compression or extension, or direct simple shear strengths. Therefore, the usefulness of the field vane in providing an absolute measurement of strength is dubious and this may be a factor in the Kars, Gloucester and Boundary Road case records.

In the majority of ESP/YE analyses of case records where the ESP reached the top of the YE, the shear strength indicated by ESP behaviour agreed reasonably well with the interpretation shown on Figure 1 (i.e. Su compression is approximately equal to $1/3\,\sigma\mathrm{p'}$). It is noted that analyses were only carried out for piezometer locations below the centre of the loaded areas where compression loading occurs. However, the degree of correspondence between the strength indicated by ESP behaviour and field vane strengths varied greatly. Based on the trends indicated on Figure 1, it would be expected that vane strengths for soils with plasticity index of 30 - 50% would be reasonably consistent with ESP/YE behaviour. On the other hand, for soils with a plasticity index less than 30%, the vane strengths would be expected to be less that that indicated by ESP analyses. The vane and triaxial compression shear strengths are indicated on the figures illustrating the results of the ESP/YE analyses of case records (Figures 2 to 8 in this paper and Figures 3 to 10 in the Part I paper by Crooks et al, 1984) [8]. The relationships between these strengths and the top of the YE are evident from these figures and are summarized on Table 2. With two exceptions out of a total of 14 relevant case records, the ESP/YE analyses confirm the trend indicated on Figure 1.

TABLE 2:

CORRESPONDENCE BETWEEN VANE SHEAR STRENGTH AND THE UPPER PORTION OF THE YIELD ENVELOPE

CASE/REFERNCE	PLASTICITY INDEX (%)	VANE STRENGTH wrt TOP OF YE	TREND AS EXPECTED (FIG. 1)
Asrum FIII [13]	15-25	less than/approx. equal	yes
New Hampshire I95 [17]	10-20	less than	yes
Ska Edeby FIlls [14]	40	less than	no
New Liskeard Embankment [20]	10-40	approx. equal/less than	yes
Kars Embankment [19]	18-26	less than	yes
Gloucester FIII [19]	16-35	approx. equal/less than	yes
Iraq FiIls [GA]	22-35	less than	no
MIT Test FIII [7]	20-30	less than	yes
Launceston Embankment [6]	110-150	greater than/approx. equal	yes
Thunder Bay Preload [11]	40	approx. equal	yes
Bangkok Embankment [3]	55-65	approx. equal/greater than	yes
Sandwich Embankment [5]	30-60	less than/approx. equal	yes
Belfast Embankment [10]	10-50	less than	yes
Mission Bay Dykes [11]	40	approx. equal	yes

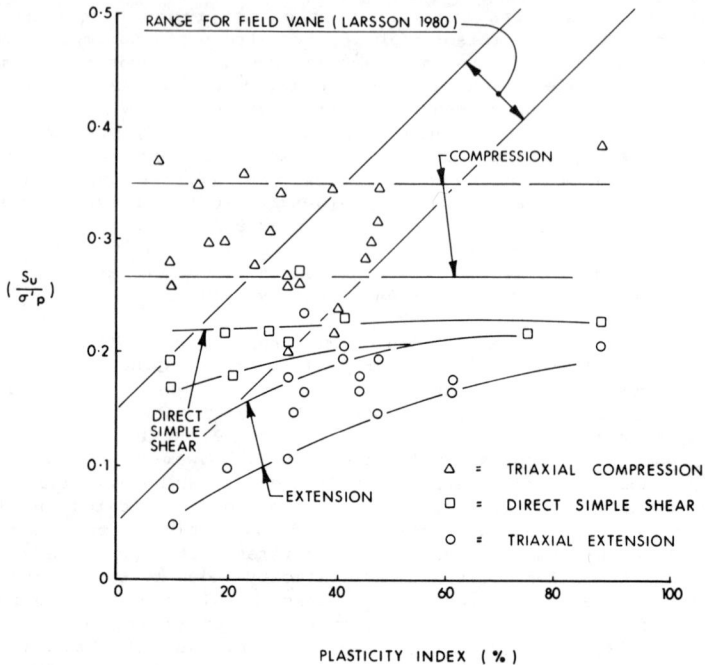

Figure 1. Undrained strength anisotropy in clays.

APPARENT LACK OF STRENGTH GAIN

Unfortunately, in most of the cases where lack of strength gain is reported, insufficient information was presented for detailed ESP/YE analyses. However, based on comments and other information provided in these publications, the reported lack of strength gain can generally be related to insufficient consolidation having taken place when post construction strength measurements were made. Therefore, these case records do not necessarily indicate lack of strength gain only that a sufficient degree of consolidation is required for measurable strength gain to occur. These individual case records are considered separately below.

Asrum: The induced shear stress levels were high below both of the Asrum fills and strain softening occurred in both cases as indicated by the effective stress path analyses on Figure 2. Therefore, porewater pressure dissipation rates following construction would have been low. The strength tests were carried out only eight months following completion of construction, and consequently it is likely that effective stress increases were not sufficient to cause a measurable strength increase.

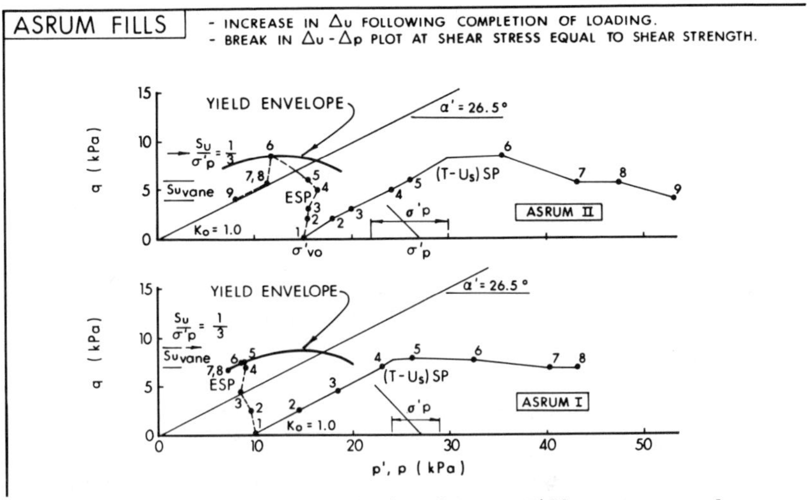

Figure 2.   ESP/YE analysis of Asrum Fills case record.

Matagami: This was a test fill loaded rapidly to failure. Insufficient consolidation would have occurred during construction to permit strength gain.

Izmir: Strength tests were conducted 70 days following completion of construction. It is expected that little consolidation and strength gain would have occurred during this period in the relatively impermeable foundation clays.

Visag: No strength gain ocurred in the non-drained area at this site. However, significant strength gain was recorded in the vertically drained area at the same time. Therefore it appears that insufficient consolidation had occurred in the area with no vertical drains.

New Hampshire I95: In the centre of the clay layer below the non-drained fill, consolidation was only 35% complete which would not be sufficient to cause measureable strength gain. It is noted that strength increase was evident in the upper and lower portions of the clay layer which were closer to drainage boundaries and therefore had experienced a greater degree of consolidation. Also, in the vertically drained area where consolidation was 95% complete at the time strength tests were carried out, significant strength gain was recorded.

Norfolk: In this case record, strength was inferred from the results of SPT tests which are not sufficiently accurate to properly assess changes in the undrained strengths of soft clay.

In two cases (Kars and Gloucester), no post construction strength increase was indicated by field vane tests although increased strength was measured in triaxial compression tests carried out on samples of "stressed" soil (i.e. obtained from below the loaded area). The increases in strength measured in these laboratory tests are discussed further in a subsequent section of this paper. In the Boundary Road case involving similar materials, no strength gain was recorded in field vane tests. The post construction strength measurements were carried out 5 - 16 years following completion of loading when it would be expected that strength gain from consolidation should have developed.

CASES IN WHICH STRENGTH GAIN WAS MEASURED

ESP/YE analyses have been carried out for six case records (Kars, New Liskeard, Fiddler's Ferry, Iraq, New Hampshire and Ska Edeby). Data are also available from Bozozuk (1984) [2] for the Gloucester test fill. These case records are considered individually below.

New Liskeard: Lo (1973) [20] reported strength gain in the soft clays underlying the New Liskeard embankment. Actual measurements of vane strength are presented on Figure 3b at different times after construction. For the 1.5 year period it is seen that, above El. 187 m strength increase does occur but there appears to be a slight decrease between El. 187 to 185 m in the region where the strength is minimum. This was attributed by Lo to breakdown of the soil structure (i.e. strain softening). Below El. 185 m the results showed no overall consistent change. After 2.5 years, the initial strength loss in the El. 187 to 185 m depth range appears to have been recovered.

The above observations are consistent with the results of ESP/YE analyses. As shown on Figure 3b, piezometer P11 was located in the original depth range of minimum strength and subsequent decrease in strength after 1.5 years. This region had strain softened considerably as indicated by point 9 for piezometer P11 (Figure 3a) upon completion of loading. Consequently, the influence of the original YE had been obliterated and the available shearing resistance in the clay was controlled by the effective stress state at point 9. Because excess porewater pressure dissipation occurs slowly in strain softened soils, the effective stress state can lie within the original YE for a considerable period. This interpretation is consistent with the observed loss in strength as indicated by the measurements 1.5 years after construction. Point 10 for piezometer P11 (Figure 3a) corresponds to a time of approximately 2.5 years following construction (July '63 to Nov. 65). This effective stress state is close to the original YE and the original strength should be restored. Again, this is consistent with the strength measurements indicated on Figure 3b. Finally, 10 years after construction, it follows that the effective stress state would lie outside (to the right) of the original YE and that a new YE would be formed with an associated increase in strength. From Figure 3b it is seen that at this time the measured strength is consistently greater than the original strength and generally greater than that measured after 2.5 years.

Effective stress paths for piezometers P13 and P14 installed at greater depth in the foundation clays are also shown on Figure 3a. It is noted that at these greater depths, the foundation clay also strain softened. However, the rate of excess porewater pressure dissipation (represented by points 8-9 for P13 and points 7-8 for P14) is, as expected, lower than that observed at shallow depths. This observation is consistent with the drainage boundary conditions at the site (i.e. the drainage path for P13 and P14 is longer than that associated with P11).

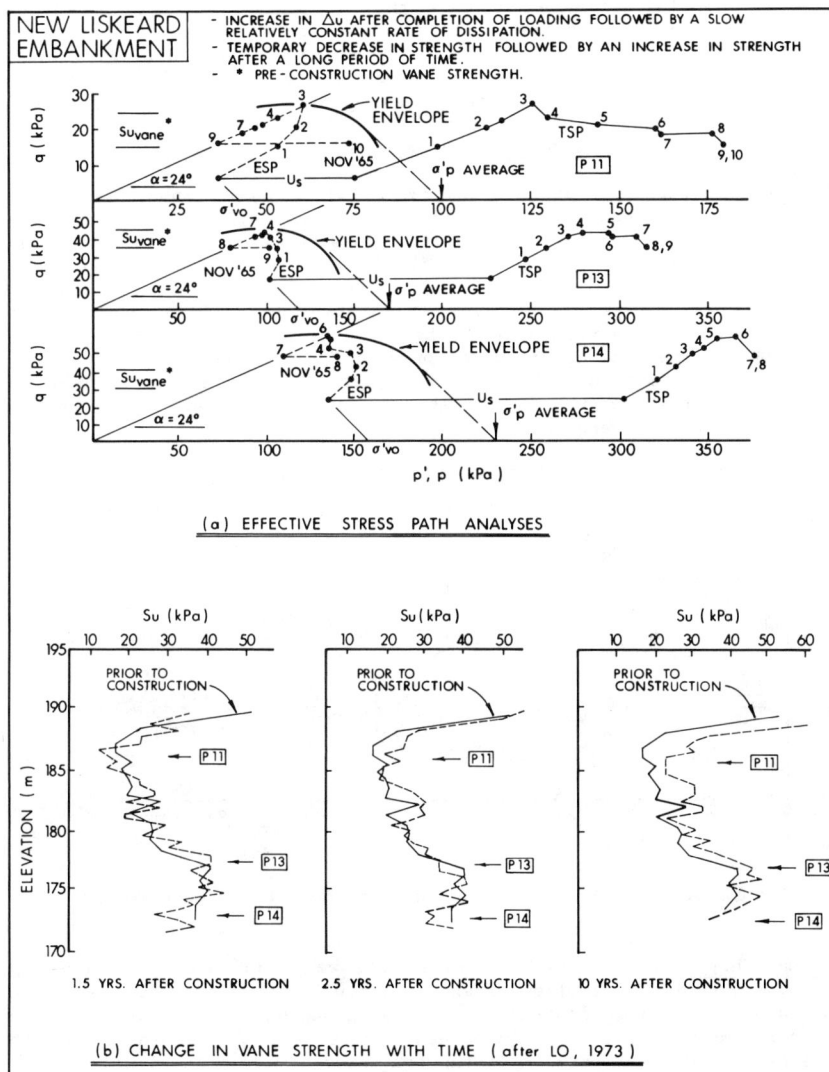

(a) EFFECTIVE STRESS PATH ANALYSES

(b) CHANGE IN VANE STRENGTH WITH TIME (after LO, 1973)

Figure 3. ESP/YE analysis and strength gain for the New Liskeard embankment case record.

The degree to which strain softening occurred is also less at the deeper piezometer locations. This is attributable to the initial strength being greater at depth and the induced shear stress level decreases with depth. It is interesting to note, however, that the "residual" strength following strain softening (i.e. strength at large strain, La Rochelle et al, 1974) [16] increases with depth. This raises the interesting question as to whether the failure which occurred at this site was associated with the "weak layer" at shallow depth (i.e. a predominantly translational failure) or whether a deep seated rotational failure occurred. Since the deformations in the weak layer between El. 189 and 183 m were significantly greater than at depth (Lo, 1973) [20], it is likely that a translational failure occurred.

Ska Edeby (Figure 4): In both the vertically drained and non-drained areas (II and IV respectively), excess porewater pressures had fully dissipated 14 years after completion of construction (i.e. 1957 to 1971) and new YEs associated with the final effective stress states had developed. This is confirmed by the increased vane strengths recorded in both areas in 1971. However, the magnitudes of vane strength increase were not as high as would be indicated by the new YEs shown on Figure 4 for the reasons discussed previously. It is also noted that based on ESP/YE analysis the strength gain beneath Area II should be greater than that under Area IV. This is also confirmed by the trends in measured vane strengths.

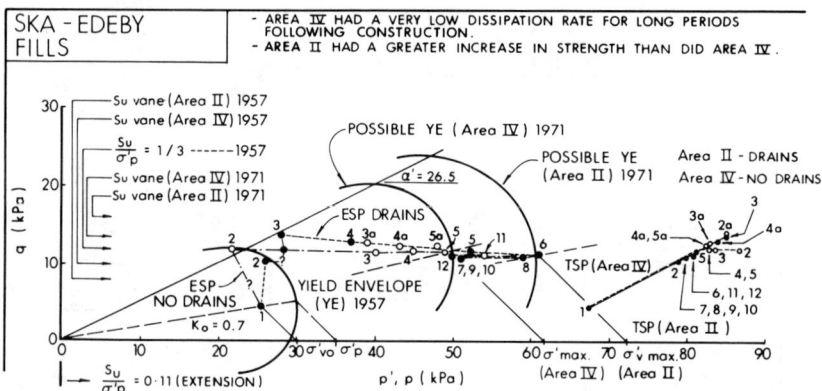

Figure 4. ESP/YE analysis of the Ska-Edeby Fills case record.

New Hampshire I95 (Figure 5): This site was vertically drained and relatively rapid dissipation of excess porewater pressure occurred both during and following construction. Vane strength increase was recorded and confirms the development of a new YE. However, the vane strength lies below the upper portion of the estimated new YE. It is noted that the magnitude of the porewater pressure response to Stage II filling was lower than would be expected for a YE passing through the final effective stress state associated with Stage I loading (point 12). Further, during subsequent porewater pressure dissipation, there was a modest change in dissipation rate at point 15. Also, the shear stress level associated with the Stage II load was higher than the shear strength indicated by the estimated yield envelope passing through the final effective stress state (point 12). These factors suggest that the YE at completion of the Stage I loading does not pass through the final effective stress state but lies beyond it. Expansion of the YE in this manner could result from secondary consolidation or ageing as discussed below.

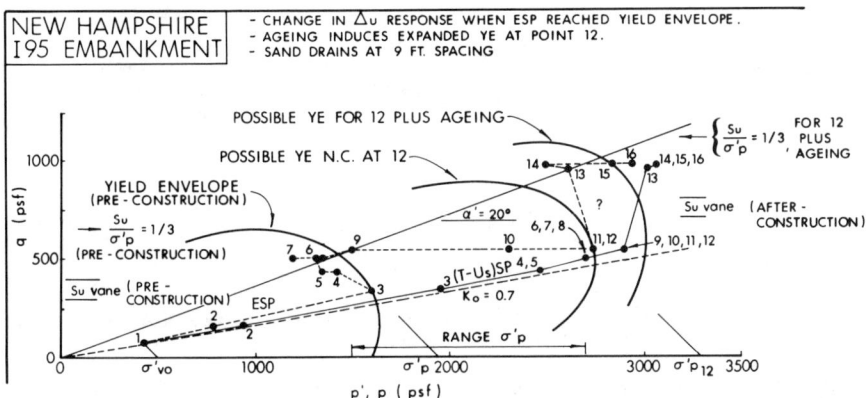

Figure 5.   ESP/YE analysis of the New Hampshire I95 embankment case record.

Kars: (Figure 6): This case is basically similar to the New Hampshire case described above. However, unlike New Hampshire, no increase in vane strength was observed although triaxial compression tests indicated strength gain consistent with the conclusions based on the ESP/YE analyses. In this case, the porewater pressure response under the second loading stage is of interest. As shown on Figure 6, the rate of porewater pressure dissipation following yielding under the first stage loading was relatively slow and when the second stage load was applied, consolidation was not complete. The soil was in a "normally consolidated" condition under the first stage loading for about 16 months before the second loading stage was applied. The excess porewater pressure induced by the second stage loading was modest and associated vertical deformation was small. Further, following completion of the second load increment, the porewater pressure dissipation rate was rapid for a short period before changing abruptly to the slow rate which prevailed following yield under the first loading stage. These observations indicate that when the second stage loading was applied, the soil response was initially overconsolidated, despite the fact that consolidation under the first loading stage was not complete (Crooks et al, 1984) [8]. Therefore it is tentatively concluded that during dissipation of excess porewater pressure to point 6, (Figure 6), secondary consolidation also occurred and resulted in a YE which lay beyond the current effective stress state. This new expanded YE appears to pass through point 8, (Figure 6).

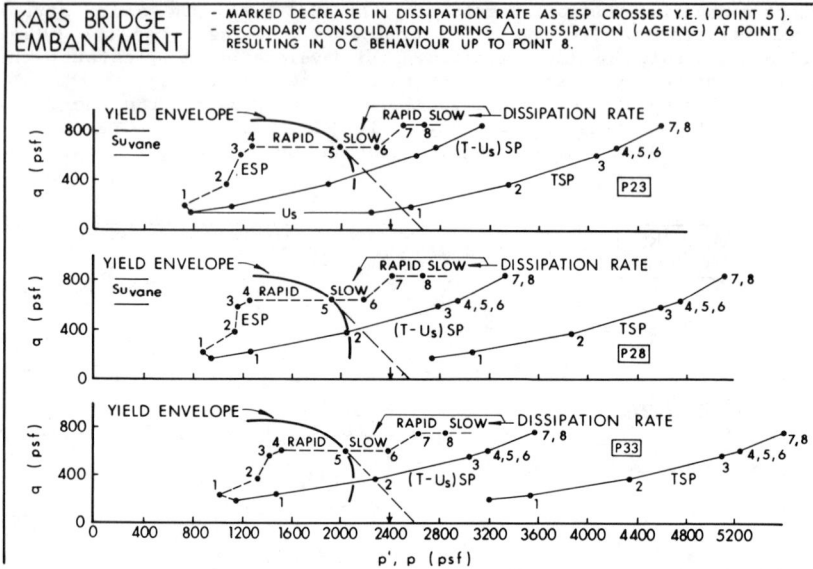

Figure 6.   ESP/YE analysis of the Kars bridge embankment case record.

Similar behaviour was observed in the Fiddler's Ferry embankment case
(Figure 8).   In this case, the effective stress state was at point 4
(i.e. outside the initial yield envelope) when the second stage loading
was applied.   The soil had been in a "normally consolidated" condition
(i.e. points 3' to 4) under the first loading stage for a period of
about 4 - 5 months.   The initial porewater pressure response to second
stage loading was modest.   However, when the loading was further
increased, the ESP followed the expanded YE to point 7.   This expanded
YE lies well beyond the effective stress state at point 4 and again
suggests that significant secondary consolidation occurred during excess
porewater pressure dissipation.

Gloucester (Figure 7):   The final effective stress state within the
overstressed zone following loading lies outside the initial YE.   Laboratory oedometer and triaxial tests were carried out on samples from this
zone before the start of loading and about 13 years following completion
of loading.   The authors' interpretation of these test results (Figure
7) indicates the development of a new YE which lies well beyond the
final effective stress state.   Further, there is a distinct difference
in the shapes of the initial and current YEs probably because of a loss
of natural cementing/bonding.   Also field vane strengths had not increased 7 - 8 years following completion of filling which is inconsistent with the results of the laboratory tests.   Finally, significant
ageing effects are evident from the results of the laboratory tests
since the YE 13 years after construction lies well beyond that which
would be expected based on the final effective stress state.

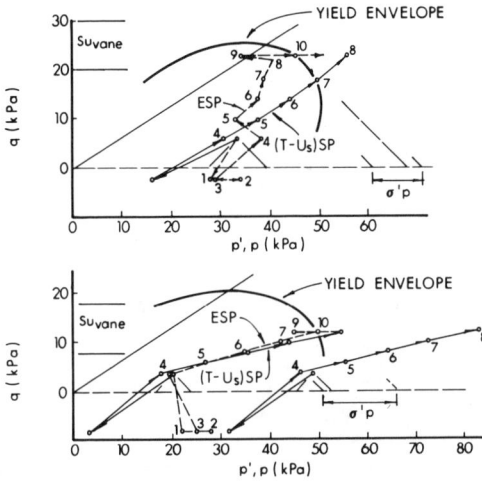

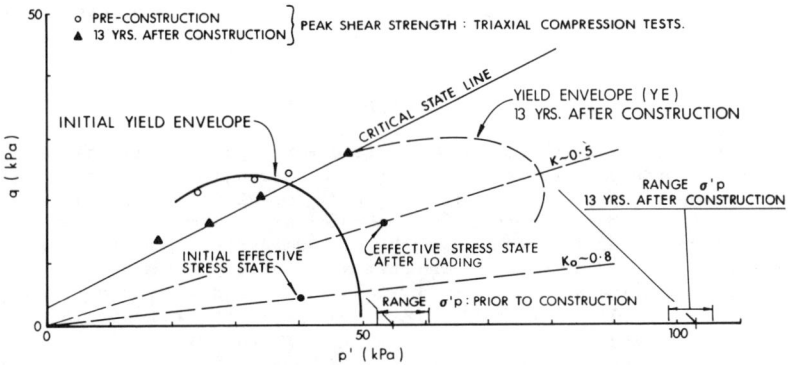

Figure 7. ESP/YE analysis and strength gain for the Gloucester Test Fill case record.

Fiddler's Ferry (Figure 8): The first of three loading stages in this fill caused overstressing and minor strain softening which was associated with a failure. During dissipation of Stage I porewater pressures, there was a modest change in dissipation rate as the ESP crossed the

initial YE (point 3'). On stage II loading the porewater pressure response was initially small but rapidly increased when the ESP encountered
the new YE with the ESP following the YE to the critical state line.
Modest dissipation took place at this effective stress state and when
the Stage III load was applied, the ESP quickly returned to the critical
state line. At this time, a second failure occurred. Continued loading
resulted in the ESP "climbing" the critical state line followed by
dissipation (i.e. dilatant rather than strain softening behaviour
occurred).

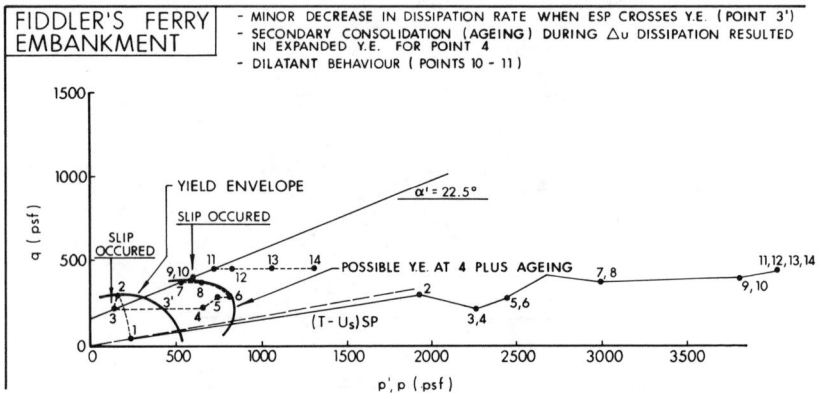

Figure 8. ESP/YE analysis of the Fiddler's Ferry embankment case record

SUMMARY AND CONCLUSIONS

The results of Effective Stress Path/Yield Envelope (ESP/YE) analyses of
the case records examined above confirm that strength gain occurs provided there is sufficient consolidation to result in a new effective
stress state which lies outside of the original in situ yield envelope.
In all of the cases in which strength gain was evident, the final effective stress states lay outside the initial in situ yield envelope (i.e.
the final effective vertical stress was greater than the preconsolidation pressure). If the effective stress state at the time of sampling
or in situ testing lies inside the original yield envelope (i.e. non-
strain softened cases), an increase in strength will not be observed
since the strength at this time is still governed by the original
preconsolidation pressure. Further, if significant strain softening
occurs, the new yield envelope will be based on the current effective
stress state. Because dissipation occurs slowly for significantly
strain softened stress paths, the effective stress state can lie within
the original YE for a considerable length of time. Assuming that ageing
effects are insignificant in the early stages of dissipation, lower
post-construction strengths can exist for lengthy periods, (i.e. at
least until the effective stress path again reaches the location of the
initial YE).

Conventional understanding would suggest that during excess porewater pressure dissipation, the foundation clay can be considered to be in a "normally consolidated" condition with respect to its current effective stress state. In fact, the ESP/YE analyses reported in this paper consistently indicate that even though consolidation is not complete, lightly overconsolidated behaviour occurs under additional loading. Insufficient data is available at this time to quantify the effect of secondary consolidation on strength gain during dissipation of excess porewater pressures. However, it appears that it may provide some measure of conservatism in design.

A further important consideration which is difficult to quantify is dilation. ESP/YE analyses clearly indicated that significant dilation does occur in the field in some soil types. Dilation only occurs at relatively large strains; however, depending on the deformation tolerances associated with individual projects, it is possible to rely on dilatant strength gain in staged-loading and preloading operations.

The observations noted in this paper are relevant in the design of staged-construction and preloading operations which require that accurate predictions of consolidation rate and strength increase be made. Thus, effective stress paths at piezometer locations should be plotted as construction proceeds to ascertain where the current effective stress state lies with respect to the original in situ yield envelope. Similarly, the relatively low rate of dissipation at effective stress states outside the original YE and in strain softened materials should be taken into account. The manner in which strength increase is measured is also important; field vane tests may underestimate strength gain in some clays.

ACKNOWLEDGEMENTS

The authors wish to express their appreciation to Gulf Canada Resources Inc. for permission to publish this paper. Dr. M. Bozozuk provided post-construction strength data for the Gloucester Test Fill case record. The drawings were prepared by G. Sundholm and the manuscript was typed by K. Koch.

NOTATION

The following symbols are used in this paper:

cv	= coefficient of consolidation
p', p	= $(\sigma_1' + \sigma_3'/2, (\sigma_1 + \sigma_3)/2$
q	= $(\sigma_1 - \sigma_3)/2$
Su	= undrained shear strength
$\bar{B}$	= porewater pressure parameter $(\Delta u/\Delta\sigma_1)$
CKoU	= Ko consolidated - undrained traixial compression test with porewater pressure measurements
ESP	= effective stress path
Ko	= effective stress ratio (σ_3'/σ_1')
TSP	= total stress path
(T-Us)SP	= total minus static porewater pressure stress path

Us	= static porewater pressure
YE	= yield envelope
α'	= angle of effective internal friction in p'- q' plot
σ'_p	= preconsolidation pressure
σ_{vo}'	= original effective vertical stress
σ_{vo}	= original total vertical stress
σ'_1' σ'_3	= major and minor principal stresses
Δu	= excess porewater pressure
Δp	= applied surface loading (pressure)
$\Delta\sigma_1$	= increment in major principal stress

REFERENCES

1. AL-DHAHIR, Z.A., KENNARD, M.F. and MORGENSTERN, N.R. (1970). Observations on pore pressures beneath the ash lagoon embankments at Fiddler's Ferry power station. Proceedings of Conference on In Situ Investigations in Soils and Rocks. British Geotechnical Society, pp. 265-276.

2. BOZOZUK, M. (1984). Personal communication.

3. BRENNER, R.P. and PBEBAHARAN, N. (1983). Analysis of sandwick performance in soft Bangkok clay. Proceedings 8th European Conference on Soil Mechanics and Foundation Engineering, Helsinki, Vol. 2, pp. 579-586.

4. CASAGRANDE, L. and POULOS, S. (1969). On the effectiveness of sand drains. Canadian Geotechnical Journal, Vol. 6, No. 3, pp. 287-326.

5. COLE, K.W. and GARRETT, C. (1981). Two road embankments on soft alluvium. Proceedings 10th International Conference on Soil Mechanics and Foundation Engineering, Stockholm, Vol. 1, pp. 87-94.

6. COOPER, I.D. and MEYER, P.A. Road embankments on soft foundations - Launceston, Tasmania. Australian Road Research Board, Adelaide (Date of publication unknown)

7. D'APPOLONIA, D.J., LAMBE, T.W. and POULOS, H.G. (1971). Evaluation of pore pressures beneath an embankment. Proceedings ASCE Journal of Soil Mechanics and Foundations Division. Vol. 97, SM. 6, pp. 881-898.

8. CROOKS, J.H.A., BECKER, D.E., JEFFERIES, M.G. and McKENZIE, K. (1984) Yield Behaviour and Consolidation. I: Pore pressure response. ASCE Symposium on Prediction and Case Histories of Consolidation Performance. San Francisco.

9. DASCAL, O., TOURNIER, J.P., TAVENAS, F. and LaROCHELLE, P. (1972). Failure of a test embankment on sensitive clay. Proceedings of ASCE Specialty Conference on Performance of Earth and Earth-Supported Structures. Purdue University pp. 129-158.

10. DAVIES, J.A. and HUMPHESON, C. (1981). A comparison between the performance of two types of vertical drains beneath a trial embankment in Belfast. Geotechnique Vol. 31, No. 1, pp. 19-32.

11. FOLKES, D. and CROOKS, J.H.A. (1984). Effective stress paths and yielding in soft clays below embankments. (Submitted to Canadian Geotechnical Journal).

12. HANSBO, S., JAMIOLKOWSKI, M. and KOK, L. (1981). Consolidation by vertical drains. Geotechnique, Vol. 31., No. 1, pp. 45-66.

13. HOEG, K., ANDERSLAND, O.B. and ROLFSEN, E.N. (1969). Undrained behaviour of quick clay under load tests at Asrum. Geotechnique, 19, No. 1, pp. 101-115.

14. HOLTZ, R.D. and BROMS, B. (1972). Long-term loading tests at Ska-Edeby, Sweden. Proc. of ASCE Specialty Conference on Performance of Earth and Earth-Supported Structures, Purdue University, pp. 435-464.

15. JAMES, P.M. (1970). Behaviour of a soft recent sediment under embankment loadings. The Quarterly Journal of Engineering Geology. Vol.3, pp. 41-53.

16. La ROCHELLE, P., TRAK, B., TAVENAS, F. and ROY, M. (1974). Failure of a test embankment on a sensitive Champlain clay deposit. Canadian Geotechnical Journal, Vol. 11, No. 1, pp.142-164.

17. LADD, C.C., RIXNER, J.J. and GIFFORD, D.G. (1972). Performance of embankments with sand drains on sensitive clay. Proceedings of ASCE Specialty Conference on Performance of Earth and Earth-Supported Structures, Purdue University, pp. 211 - 242.

18. LARSSON, R. (1980). Undrained shear strength in stability calculation of embankments and foundations on soft clays. Canadian Geotechnical Journal, Vol. 17, pp. 591-602.

19. LAW, K.T., BOZOZUK, M. and EDEN, W.J. (1977). Measured strengths under fills on sensitive clay. Proceedings 9th International Conference on Soil Mechanics and Foundation Engineering, Tokyo, Vol. 1, pp. 187-192.

20. LO, K.Y. (1973). Behaviour of embankments on sensitive clays loaded close to failure. Report to Ontario Ministry of Transportation and Communications, OJT & CRP L-2.

21. ORDEMIR, I.M. and BIRAND, A.A. (1981). Settlement of a test fill. Proceedings 10th International Conference on Soil Mechanics and Foundation Engineering, Stockholm, Vol. 2, pp. 541-544.

22. PARRY, R.H.G. (1968). Field and laboratory behaviour of lightly overconsolidated clay. Geotechnique, Vol. 18, No. 2, pp.151-171.

23. RAFAELI, D. (1972). Designs of the south island for the second Hampton Roads crossing. Proceedings of ASCE Specialty Conference on Performance of Earth and Earth-Supported Structures. Purdue University, pp. 361-378.

24. SUBBARAJU, M.H., NATARAJAN, T.K., and SHANDARI, K.K. (1973). Field performance of drain wells designed expressly for strength gain in soft marine clays. Proceedings 8th International Conference on Soil Mechanics and Foundation Engineering, Moscow, Vol. 2, Part 2, pp. 217-220.

25. TSAI, K.W., LEE, C.C. and CHAD, C.S. (1981). Site improvement by preloading with sand drains. Proceedings 10th International Conference on Soil Mechanics and Foundation Engineering, Stockholm, Vol. 3, pp. 781-783.

26. WILKES, P.F. (1972). An induced failure at a trial embankment at King's Lynn, Norfolk, England. Proceedings of ASCE Specialty Conference on Performance of Earth and Earth-Supported Structures, Purdue University, pp. 29-64.

CONSOLIDATION OF PERMAFROST UPON THAWING

By Frederick E. Crory,[1] M. ASCE

Abstract: An assessment of the consolidation associated with the thawing of permafrost is essential
in the design of facilities in cold regions. Simple,
yet accurate, methods of predicting the initial consolidation upon thawing of permafrost and the subsequent consolidation with time of the thawed soils
under increasing pressures are presented in terms of
changes in unit weight. Procedures for conducting
one-dimensional consolidation tests on originally
frozen specimens are described, and the analysis of
the laboratory tests with respect to total and
differential settlement analysis are discussed.

Permafrost, or permanently frozen ground, underlies about one-
fifth of the land mass of the Earth. Permafrost is defined as soils
or rock that remains frozen for more than a year (11). While the
presence and extent of permafrost is a direct reflection of near-
surface thermal conditions, influenced by drainage and surface cover
changes, the deeper frozen conditions are the result of thermal
changes that have occurred over hundreds or thousands of years (5).
In Alaska, Canada and the Soviet Union permafrost can extend to
depths of 2000 ft (610 m) or more. In northern areas permafrost is
considered continuous, being present everywhere but under large lakes
and rivers, or near warm springs or other areas with unfavorable
geothermal conditions (5). In subarctic areas permafrost may be discontinuous, or sporadic, and when present is often no more than 100
or 200 ft (30-60 m) thick. Farther south, and even into the temper-
ate zone, permafrost may be found on the top or north side of the
higher mountains (e.g. Mt. Washington, N.H.).
    Permanently frozen soils often contain volumes of ice that ex-
ceed those normally occupied by the pore water associated with the
same soil type when thawed and consolidated by the effective overburden pressure. These excessive volumes of ice may have been formed
during the frost action associated with the formation of the permafrost (the redistribution and migration of water through the frozen
or partially frozen soil) or by the periodic infiltration of water
into thermal contraction cracks associated with the formation of ice
wedges (6). The ice may exist as distinct layers many feet thick, as
distinct or limited lenses ranging from inches to fractions of an
inch thick, or as crystals barely discernible to the naked eye. While
the strains associated with the thawing of thick layers or lenses of

---

[1]Research Civil Engineer, U.S. Army Cold Regions Research and
Engineering Laboratory, Hanover, N.H.

Figure 1.   Photographs of frozen soils (8).

ice can equal 100%, the thaw-consolidation of frozen ground and thin
ice lenses will be much less; however, it will be much greater than
the potential strains of normally consolidated thawed soils.   The ice
lenses need not be horizontal, for they can be inclined at any angle,
including vertical.   The number and distribution of the ice lenses
often vary with depth and can vary with soil type, as shown in Figure
1.   Some permanently frozen soils may contain clearly visible layers
of ice, but the soil between the ice lenses may be very dense; apparently the pore water migrates to the ice, causing the intervening
soil layer to consolidate.   Under such conditions the pore water in
the soil layers may be only partially frozen.   Unfrozen pore water may
also be prevalent in other situations, particularly if ground temperatures are only slightly below the normal freezing point, especially in
soils having a high clay content (1).

     Problems associated with construction on permafrost can be avoid-
ed, or at least minimized, if the permanently frozen conditions are
maintained.   However, this passive approach is often not feasible or
economical, and consideration is given to the active approach wherein
thawing is permitted or actually induced.   At least three categories
of construction techniques involve the settlement associated with the
thawing of permafrost.   The first category is where artifical heating
and thawing is purposely induced, so that the major settlements occur
before construction (17, 19).   The ground is thawed to depths in excess of that required to avoid zones in which intolerable or undesira-
ble consolidation would occur.   The additional settlement associated
with thawing the low-ice-content soils at greater depths, due to the
progressive expansion of the initial thawing effort and the long-term
heat flow from the structure, become the focal point of such foundation designs.   The second category includes linear facilities, such
as roads, airfields and pipelines, where no attempt is made to thaw

the ground before construction (9, 12, 14, 15, 18). Under such conditions the permafrost often thaws and settles with time because of
the new surface heat balance or the heat from the pipeline. Engineering analyses for this category are normally concerned with both total
and differential settlements. For pipelines the settlement analysis
forms the basis for decisions on whether to bury or elevate the line.
The third category involves facilities that have been built directly
on permafrost, creating thawing and subsequent settlement (2, 3, 4,
14). This paper describes an engineering approach to all of these
categories by addressing the consolidation associated with the thawing
of initially frozen soils.

Basic Relationships

    In general a core sample of permafrost, as depicted in Figure 2a,
is considered to have two constituents; ice and soil. In some samples
there may also be air, unfrozen water and even organics. Any attempt
to analyze the potential strains associated with the thawing of a
specific sample must provide for all of these constituents. In a previous paper (3) the author derived several equations for calculating
the strains associated with the thawing of permafrost.
    The dry density of a frozen sample, γ_{df}, can be expressed in
terms of specific gravity, G, the unit weight of water, γ_w, and the
water content, w_f. When the sample is fully saturated with ice, the
equation is

$$\gamma_{df} = \frac{\gamma_w}{\frac{1}{G} + 1.09\ w_f} . \tag{1}$$

    While equations are available for calculating the dry density of
samples having both ice and variable amounts of unfrozen water (3, 17),
such refinements are seldom necessary. Conversely the relationship between water content and dry density after the samples have been completely thawed and subjected to increasing consolidating pressures is
of great importance. The expression for these conditions is

$$\gamma_{dt} = \frac{\gamma_w}{\frac{1}{G} + w_t} \tag{2}$$

where w_t is the water content of the thawed soil and γ_{dt} is the dry
density of the sample upon thawing or the high densities associated
with the subsequent applied pressures. The water content, w, in equations 1 and 2 is the ratio of the weight of water to the weight of dry
soil, and is expressed as a decimal.
    The unit strains, $\Delta H/H$, associated with the thawing of a frozen
soil sample are directly related to the thawed density:

$$\Delta H/H = (\gamma_{dt} - \gamma_{df})/\gamma_{dt}. \tag{3}$$

Equation 3 has been shown (3) to be identical to the Terzaghi (16)
equation:

$$\Delta H/H = (e_0 - e_1)/(1 + e_0) = (\gamma_{dt} - \gamma_{df})/\gamma_{dt} \tag{4}$$

where e_0 is the initial void ratio of the frozen sample, and e_1 is the final void ratio of the thawed sample; the void ratios are defined as the ratio of the volume of voids to the volume of solids. Thus the unit strains in equations 3 and 4 are a direct function of the void ratio or dry density when the frozen sample has been thawed or consolidated under different pressures.

Equation 1 can be extremely useful in defining the conditions with depth of the entire length of an undisturbed core obtained from a drill hole. Water content can also be determined from augering cuttings. Comparisons of the frozen dry density, as determined by direct volumetric displacement methods on selected samples, will confirm the degree of saturation and the usefulness of the water content data for defining initial densities. Normally, frozen soils below the active layer (the zone that freezes and thaws annually) will be saturated, although there can be exceptions.

Consolidation Tests

Tests to define the consolidation characteristics of a frozen sample upon thawing and under the subsequent application of loads are not unlike conventional tests of thawed samples (7). Perhaps the biggest difference is that undisturbed frozen specimens are used and the test must be prepared and started in a below-freezing environment. Trimming frozen specimens and mounting samples in the consolidation rings in a coldroom can be difficult for technicians unfamiliar with frozen samples. Special care must be exercised to avoid or minimize any significant voids between the specimen and the consolidation ring, and between the frozen sample and the porous stones. Such voids can produce objectionable errors in the intial height and subsequent strains upon thawing. One should carefully consider the amount and distribution of significant ice lenses in the section of the core being tested, avoiding the possibility of excessive tilting of the upper porous stone during the initial thawing period. The relatively large strains that can be encountered, regardless of the amount of visible ice, will also require special attention with respect to the duration of each applied load, because such samples have normally lost all or most of their normal consolidation or preconsolidation in the frozen state. It may be necessary to maintain each applied load increment for 1,000 to 10,000 minutes to clearly define the completion of the primary consolidation.

Although some investigators (10, 13) have used special oedometers to control the rate and direction of the thawing of the specimens, such equipment is not required for defining the strains associated with consolidation. Accordingly consolidation tests usually ignore the more complicated strains that occur during thawing; they measure only the change in height of the specimen at the end of the thawing period and during each successive load application.

Since the test specimen is constrained laterally by the consolidation ring, only the one-dimensional consolidation, or axial change in height (ΔH), is recorded. As depicted in Figure 2 the volume of the sample can increase, show no change, or decrease on thawing (3). The testing procedures for thaw-consolidation tests should therefore be

c. Expands on thawing

d. No volume change on thawing

e. Consolidates on thawing

Water

Water | Soil ΔH

Water | Soil H

Void

Water | Soil H 1 cm²

ΔH

Ice | Soil 1 cm²

H (1 cm)

b. Frozen core in terms of soil mechanics units

a. Original frozen core

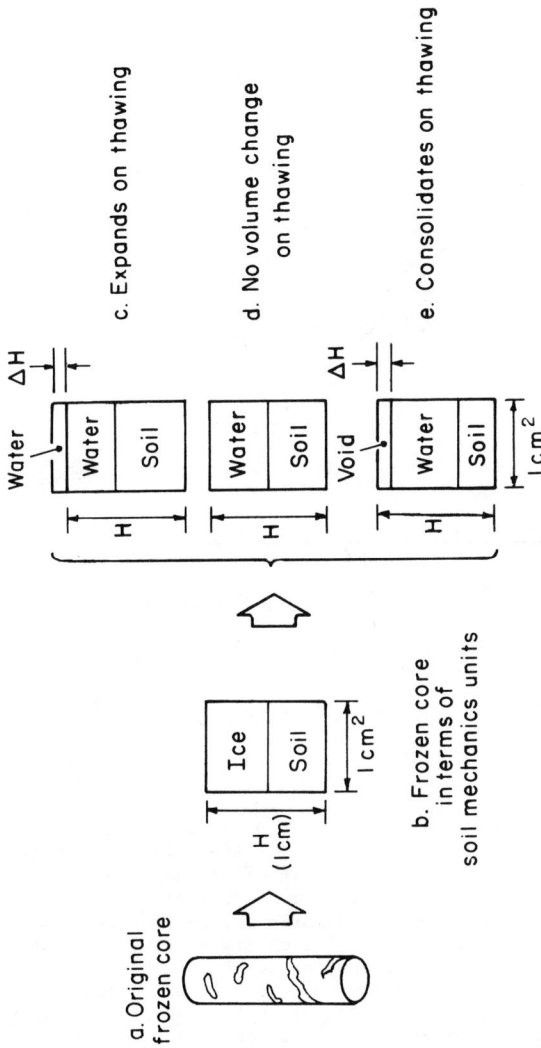

Figure 2.   Basic volume relationships of frozen cores upon thawing (3).

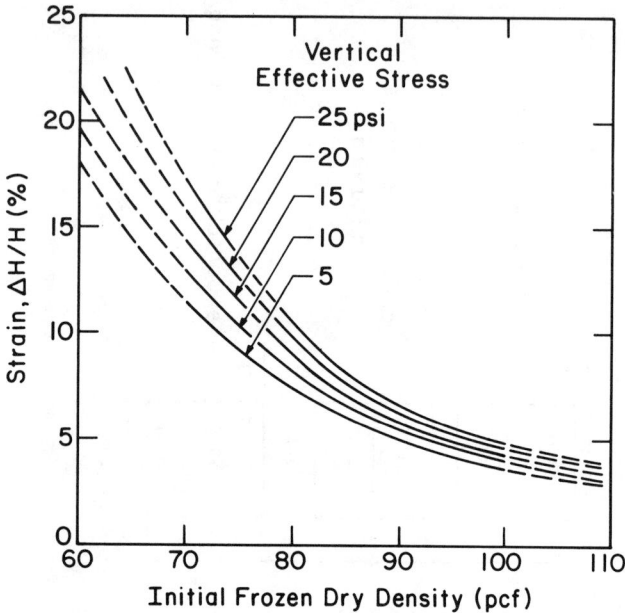

Figure 3.   Strain versus frozen dry unit weight for
samples from Kotzebue, Alaska (2).

equally prepared for heave or consolidation, as in similar tests of unfrozen soils (7).

Analysis of Consolidation Test Results

The analysis of the data from the consolidation tests and the
presentation of the results should reflect the inputs and outputs
necessary for a settlement analysis.   The field data that will constitute the major parameter in the settlement analysis will be the frozen
dry unit weights or densities.   Similarly the basic input data in the
consolidation test consists of the frozen dry unit weight.   During the
consolidation test the only measured parameter is the deflection associated with thawing the specimen under little or no load and under
each subsequently applied load.   Time is not a factor, provided the
duration of each load increment is sufficiently long to define the complete strains.   Although continuous plotting of the deflection, ΔH,
versus the log of time is essential for analyzing the strain under each
load increment, only the strains at the end of these periods are used.
The results from a single test represent only one soil type and one
initial frozen dry density; a series of tests is required for defining
the strains associated with a range of frozen dry densities for each
soil type.   Following a statistical analysis the output can be displayed on plots of strain versus initial frozen dry density (Figure 3)
by a family of curves representing the strains associated with the

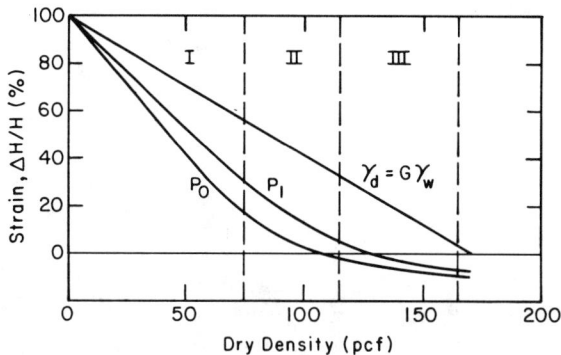

Figure 4.   Strain versus full range of frozen dry
unit densities for silts at different pressures.

initial thawing and all subsequent pressures (2). Similar plots would
be prepared for each soil type encountered in the settlement study.
     Plots of strain versus frozen dry density for different pressures
have three regions as shown in Figure 4. The first region encompasses
the low dry densities, from 0 to 50 or 75 pcf (800-1200 kg/m^3); here
the initial strains on thawing are large, but the subsequent consolidating strains under increasing pressures are relatively small and increase proportionally with increasing dry density. Frozen specimens
in this low-density region turn to mud (silts) or otherwise disinte-
grate (sands) upon thawing. Given sufficient time after thawing, however, these low-density samples settle out, the excess water drains off,
and dry densities reach 80 to 85 pcf (1281-1361 kg/m^3) or more. If one
assumes that the same soil type has a similar dry density on thawing
(Figure 5), the strains can be represented linearly by equation 4. Thus
the strain during the thawing of a specimen having an initial frozen
density of 12.5 pcf (200 kg/m^3), for example, would be twice that of
a sample that had an initial density of 25 pcf (400 kg/m^3). Conversely
the strains under each load increment would be twice as great in the
25-pcf sample as in to the 12.5-pcf sample, since the consolidation
occurs over a new sample height that is twice that of the 12.5-pcf sample. In this first region, then, both the initial thaw strain and the
strains associated with increasing pressure should be linear and should
converge at 100% strain at a frozen dry density of zero.
     The second region is characterized by frozen dry densities in the
medium range. Having less ice per unit volume than in region I, samples
in this region have smaller initial thawing strains but larger strains
under each applied pressure. Most consolidation tests are normally in
this region, since only a few tests are needed to define the potential
strains in the first region. The strains in this region are assumed to
be nonlinear and to decrease with increasing initial dry density, as
illustrated in Figure 3.
     The third region is perhaps the most difficult to understand. It
is a region of high initial dry densities and low strains on thawing
and subsequent applied pressures. This is also the region in which
thawing samples may exhibit heave, expansion or no significant strains
(3). If the sample expands or heaves on thawing, the thawed dry density

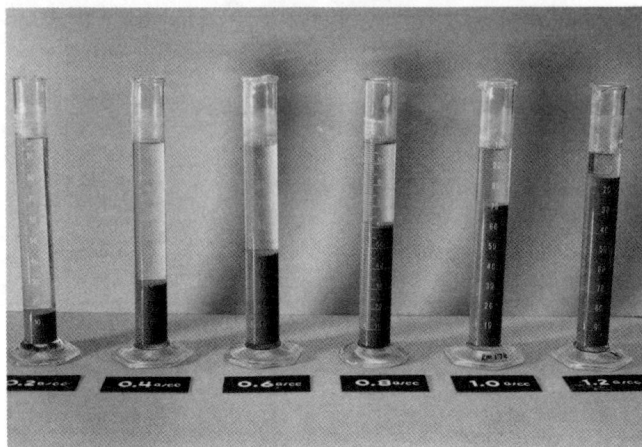

Figure 5. Simulation of thaw-consolidation of low density
silt samples, density increasing from left at 12.5 pcf
(200 kg/m^3).

will be lower than the frozen density, and additional pore water will be
taken up by the specimen during the consolidation test (Figure 2c).
After expanding during the initial thawing period, the sample may consolidate under increasing applied loads, with strains eventually passing through zero and reflecting a consolidation or settlement (Figure
2e). In tests of high-density samples, especially silts and clays,
heaving and expansion pressures during thawing should be anticipated,
and much higher load increments should be used. These increments would
be selected on the basis of the intergranular pressures expected at the
depth from which the samples were obtained and the pressures from any
superimposed surface loadings. High-density granular soils normally
will not expand on thawing.

    Another approach to analyzing the potential consolidating strains
of soil samples, whether initially frozen or thawed, is to relate the
strains to the associated changes in dry density, as defined by equation 4 and the physical parameters measured in the laboratory consolidation test. This is done by using the basic definition of dry unit
weight or density:

$$\gamma_d = \frac{W_s}{V} \tag{5}$$

where γ_d represents the initial frozen (or thawed) density or any density during the consolidation tests, W_s represents the total dry weight of
the soil or solids, and V equals the total volume at any time during the
test. The volume, V, is a function of the cross sectional area, A, of
the sample or consolidation ring, and the height of the sample:

$$V = A(H - \Delta H) \tag{6}$$

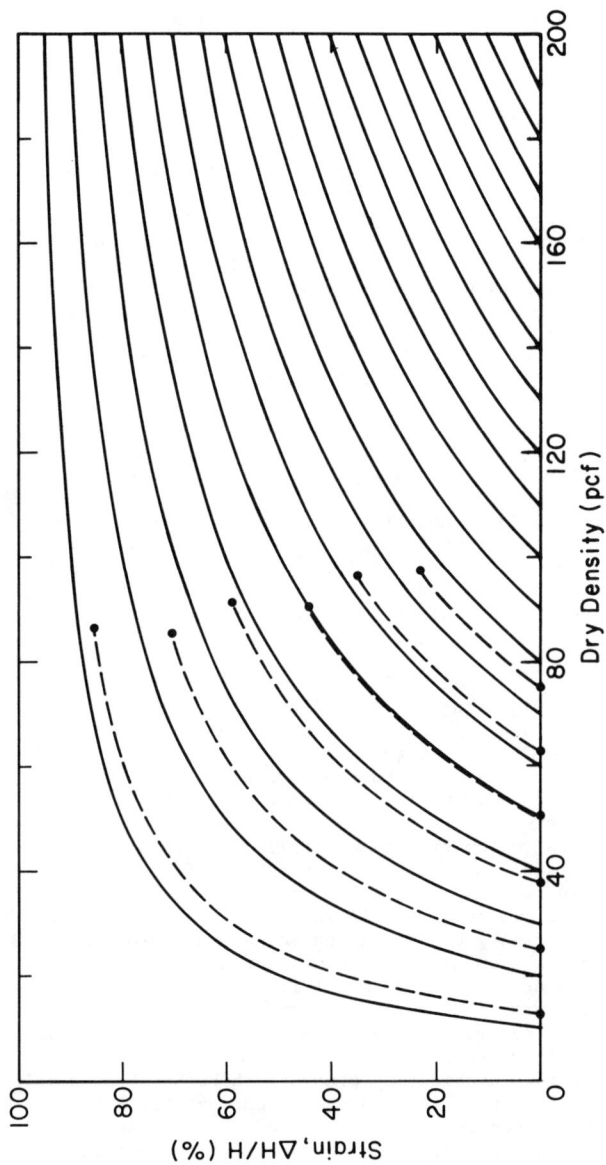

Figure 6. Strain versus associated changes in dry density for the silt samples shown in Figure 5, plotted as dashed lines.

where H is the original height and ΔH is the deformation during the test (Figure 2). Substituting equation 6 into equation 5, we have:

$$\gamma_d = \frac{W_s}{A(H - \Delta H)} \cdot \quad\quad (7)$$

By setting H in equation 7 at one and ΔH at zero, one can solve for the theoretically constant value of W_s/A for each initial dry density. One can then vary ΔH in small increments from 0 to 1 and calculate the associated dry density for the corresponding strain value of ΔH/H. When this process is repeated for different initial densities, a family of curves can be developed, as shown in Figure 6. The curves can also be extended to include the strains associated with heave or expansion of the samples, but are omitted here for simplicity.

To illustrate the use of Figure 6, the consolidating strains of the silt samples shown in Figure 5 have been plotted to reflect the associated changes in the density of each sample. Although these samples were undoubtedly influenced by different wall friction and vertical effective stress values from their own weight, they illustrate the overall strains associated with the initial thaw-consolidation of a wide range of initial dry densities at little or no applied vertical stresses. The curves in Figure 6 also illustrate the linear and nonlinear aspects discussed above with respect to Figures 3 and 4. When data from a series of actual consolidation tests are plotted on Figure 6, another family of curves, which are nonlinear and nearly vertical, can be constructed to reflect the effect of increasing vertically applied stresses (Figure 3).

Summary and Conclusions

The extremely wide range in ice volumes and initial dry densities of permanently frozen ground presents an interesting challenge to geotechnical engineers faced with a settlement analysis. The approach taken in this analysis has been to concentrate on the use of the initial frozen dry unit weight as the input parameter, because it is unaffected by the degree of saturation. The one-dimensional laboratory thaw-consolidation tests in temperature-controlled environments are assumed to represent the same consolidating strains that frozen soils of the same type will experience in the field upon thawing. Although the analysis of the laboratory data will involve an interpretation of the time required for primary consolidation and may use plots of void ratio changes with time, the emphasis is on unit strains. The thaw-consolidation data are presented as a family of curves representing the strains associated with initial thawing and increasing pressures versus the original frozen density for each soil type or the change in dry density.

The approach to consolidation presented here has been simple and straighforward with respect to laboratory testing, but there are other factors that must be considered in a settlement analysis involving frozen soils. In the laboratory thaw-consolidation test it was assumed that the frozen sample would be thawed, with little or no regard for the rate of thawing and the dissipation of the excess pore pressure. In field situations, however, a thermal analysis is necessary for indicating the extent of thawing with time. Such thermal changes are seldom uniform, being particularly sensitive to variations in ice distribution and to sources of heat and convection associated with moving water. Differential thawing and differential settlement, compounded by both the complex thawing and the variable unit thaw-strains associated with different ice contents, force the designer to consider both total and differential settlement with regard to the proposed structure. Differential settlement also involves consideration of arching effects, non-vertical displacement and remolding. The excess water released during thawing also poses potential problems with instability and tends to delay the consolidation process because of excess pore pressure at the thawing interface. Thus the thaw-consolidation process in the field is more complex than the laboratory analysis of consolidation of individual samples.

Acknowledgment

Although much of our thaw-consolidation testing of frozen soils has been done on reimbursable projects for other agencies, this paper reflects the sustained program of research on foundations in areas of permafrost which is conducted by the Cold Regions Research and Engineering Laboratory under the sponsorship of the U.S. Army Corps of Engineers.

Appendix: References

1. Anderson, D.M., Tice, A.R. and McKim, H.L., "The Unfrozen Water and the Apparent Specific Heat Capacity of Frozen Soils." In Permafrost: The North American Contribution to the Second International Conference, Yakutsk. NAS-NAE-NRC, Washington, D.C., 1973, p. 289-295.
2. Crory, F.E., "Foundation Investigation of Alaska Native Health Service Hospital, Kotzebue, Alaska." U.S. Army Cold Regions Research and Engineering Laboratory, Hanover, N.H., Internal Report, 1967.
3. Crory, F.E., "Settlement Associated with the Thawing of Permafrost." In Permafrost: The North American Contribution to the Second International Conference, Yakutsk. NAS-NAE-NRC, Washington, D.C., 1973, p. 599-607.
4. Crory, F.E., "The Kotzebue Hospital - A Case Study." Proceedings on Applied Techniques for Cold Environments, ASCE, Anchorage, AK, 1978, p. 342-359.
5. Gold, L.W. and Lachenbruch, A.H., "Thermal Conditions in Permafrost - A Review." In Permafrost: The North American Contribution to the Second International Conference, Yakutsk. NAS-NAE-NRC, Washington, D.C., 1973, p. 3-25.
6. Lachenbruch, A.H., "Mechanics of Thermal Contraction Cracks and Ice-Wedge Polygons in Permafrost." Geol. Soc. Am. Spec. Paper 70, 1962, 69 p.
7. Lambe, T.W., "Soil Testing for Engineers." Wiley, New York, 1951.
8. Linell, K.A. and Kaplar, C.W., "Description and Classification of Frozen Soils." Technical Report 150, U.S. Army Cold Regions Research and Engineering Laboratory, Hanover, N.H., 1966.
9. Luscher, U. and Afifi, S.S., "Thaw Consolidation of Alaskan Silts and Granular Soils." In Permafrost: The North American Contribution to the Second International Conference, Yakutsk. NAS-NAE-NRC, Washington, D.C., 1973, p. 325-334.

10. Morgenstern, N.R. and Nixon, J.F., "One-Dimensional Consolidation of Thawing Soils." Can. Geotech. J., 8(4), p. 558-565, 1971.
11. Muller, S.W., "Permafrost or Permanently Frozen Ground and Related Engineering Problems." J.W. Edwards, Inc., Ann Arbor, Michigan, 1947.
12. Nelson, R.A., Luscher, U., Rooney, J.W. and Stramler, A.A., "Thaw Strain Data and Thaw Settlement Predictions for Alaskan Soils." In Proceedings of Permafrost: Fourth International Conference, National Academy Press, Washington, D.C., 1983, p. 912-917.
13. Nixon, J.F. and Morgenstern, N.R., "Thaw Consolidation Tests on Undisturbed Fine-Grained Permafrost." Can. Geotech. J., 11 (1), 1974, p. 202-214.
14. Smith, N. and Berg, R., "Encountering Massive Ground Ice During Road Construction in Central Alaska." In Permafrost: The North American Contribution to the Second International Conference, Yakutsk. NAS-NAE-NRC, Washington, D.C., 1973, p. 730, 736.
15. Speer, T.L., Watson, G.H. and Rowley, R.K., "Effects of Ground-Ice Variability and Resulting Thaw Settlements on Buried Warm-Oil Pipelines." In Permafrost: The North American Contribution to the Second International Conference, Yakutsk. NAS-NAE-NRC, Washington, D.C., 1973, p. 746-752.
16. Terzaghi, K., "Theoretical Soil Mechanics." John Wiley and Sons, Inc., N.Y., 1965.
17. Tsytovich, N.A., Zaretsky, Y.K., Grigoryeva, V.G. and Ter-Martirosyan, Z.G., "Consolidation of Thawing Soils." In Proceedings, 6th International Conference on Soil Mechanics and Foundation Engineering, Montreal, 1965, I, p. 390-394.
18. Watson, G.H., Rowley, R.K. and Slusarchuk, W.A., "Performance of a Warm-Oil Pipeline Buried in Permafrost." In Permafrost: The North American Contribution to the Second International Conference, Yakutsk. NAS-NAE-NRC, Washington, D.C., 1973, p. 759-766.
19. Zaretskii, Y.K., "Calculations of the Settlement of Thawing Soil," Soil Mechanics and Foundation Engineering, May-June, (3), 1968.

PREDICTION OF PEAT SETTLEMENT

Tuncer B. Edil,* M.ASCE and Noor E. Mochtar**

Abstract

Compression of peat and organic clay deposits is characterized using a simple rheological model in which the structural viscosity is assumed to be linear. The model utilizes three empirical parameters pertaining to the primary compression, the secondary compression, and the rate of secondary compression, respectively. It has been found that the model can represent the laboratory and field compression curves adequately for soils like peat which exhibit large strain and significant secondary compression behavior. Based on an analysis of numerous laboratory and field compression-time data, it is shown that the parameters for the primary and secondary compressibility depend on the stress level. Therefore, they can be determined in the laboratory by simulating the field stress changes on representative specimens. However, the secondary compressibility and the rate factor for secondary compression are found to be non-linear. In view of the vast difference between the field and laboratory strain rates, the values obtained in the laboratory tend to overestimate the field rate of compression and underestimate the magnitude of secondary compression. Because of the non-linearity these parameters must be corrected for accurate field predictions.

Introduction

Preloading has been employed with some success as a means of improving the engineering properties of peat deposits which are otherwise considered among the worst of foundation materials in their natural state. Two important design parameters in preloading schemes are the safe and economic magnitude of surcharge to be applied, and the duration of surcharging. In order to employ preloading effectively, it is of prime importance to have a reasonable quantitative understanding of the stress-strain-time response of peat soils.

One of the analytical difficulties with laboratory and field data is the unusual shape of the settlement-time curves and the consequent inability to apply the conventional methods of analysis used with inorganic soils. In peat soils, the primary consolidation is ill-defined and takes place relatively quickly and the secondary compression is significant both in rate and accumulated magnitude. The separation of primary consolidation and secondary compression and their treatment

*Professor, Department of Civil and Environmental Engineering, University of Wisconsin-Madison, Madison, WI 53706.
**Graduate Student, Department of Civil and Environmental Engineering, University of Wisconsin-Madison, Madison, WI 53706.

by the Terzaghi consolidation theory (8) and the Buisman secondary
compression method (1), as is often done with inorganic soils, appears
to be inappropriate for peats. In an attempt to represent the
settlement-time curve as a whole, a rheological model proposed by Gibson
and Lo (4) has been used to represent the compression behavior of peat.
Using a curve-fitting procedure, this model was applied to 43 laboratory
and 10 field compression curves involving peats and organic soils from
16 sites. The results were satisfactory. Because the basic model
assumes the structural viscosity of soils to be linear, its use requires
certain corrections when differences in stress levels and strain rates
are encountered. These corrections, developed from the analysis of
numerous case histories are described in this paper together with the
proper use and application of the model in field compression of peat.

Method of Analysis

    A rheological model as proposed by Gibson and Lo (4) and shown in
Fig. 1 has been found to give satisfactory results in representing the
one-dimensional compression of peat under a given increment of stress
(3,5). For large values of time, the time-dependent strain, $\varepsilon(t)$, may
be written as

$$\varepsilon(t) = \Delta\sigma \left[a + b \left(1 - e^{-(\lambda/b)t} \right) \right] \tag{1}$$

where $\Delta\sigma$ = stress increment, t = time, a = primary compressibility, b =
secondary compressibility, and λ/b = rate factor for secondary
compression. A convenient method of analysis of a given set of vertical
strain-time data in order to determine the empirical parameters (a,b
and λ) was described by Edil and Dhowian (3). The method uses a plot of
logarithm of strain rate versus time ($\log_{10}(\Delta\varepsilon/\Delta t)$ versus t). This
should result in a straight line for the time range corresponding to the
secondary compression if the soil conforms with the basic assumptions

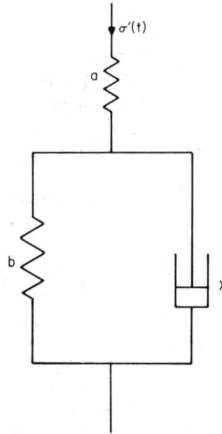

Figure 1. Rheological Model for Secondary Compression

made in the model. The slope and intercept of this best-fit line yield
the values of a, b and λ as follows:

Slope of the line = -0.434 (λ/b)   (2)

Intercept of the line = $\log_{10}(\Delta o \ \lambda)$ (3)

$$a = \frac{\epsilon(t_k)}{\Delta o} - b + be^{-(\lambda/b)t_k}$$ (4)

where t_k is the last time a reading of compression is taken. In
computing the strain rate ($\Delta\epsilon/\Delta t$) from a given laboratory or field data,
the time interval, Δt, can be kept constant at any convenient value. If
this is not practical, it can be computed incrementally for the
consecutive readings available. Fitting a straight line by linear
regression for the time range of secondary compression tends to smooth
out the variation of the strain rate caused by use of unequal time
intervals.

The model described above is based on a linear theory and does not
account for the non-linearities which may result from finite strains,
stress level, strain rate, etc. It appears that it is preferable to
retain the simplicity of a linear theory from which the parameters could
be readily determined. The variation of these parameters may then be
investigated, the non-linearities studied, and the proper corrections
can be introduced. This is the approach adopted herein in investigating
the compression of peat soils.

Application of the Model to Peats

In order to verify the applicability of the model for peats, plots
of the logarithm of strain rate versus time were constructed for 43
laboratory and 10 field sets of compression-time data of peats and
organic soils. Linearity was obtained for a substantial time interval
in each of these plots, which involved varying stress increments. This
supports the applicability of the theory and the empirical parameters
were determined by use of the method described above.

The fitted curves determined by means of the model are shown in Fig.
2 for two sets of field and laboratory data. A reasonably good
agreement with experimental data is obtained in both cases over much of
the time range indicating that the overall shape of the long-term
compression curve can be represented adequately by this rheological
model. In curve-fitting, emphasis is placed on long-term compression,
therefore, the fit is better for secondary compression and often there
is some discrepancy between the theoretical fit and the measured
compression for early times.

Non-Linearity of Behavior and Corrections

In order to evaluate preloading as a viable option in construction,
it is desirable to determine, in advance, the level and duration of
surcharging required to eliminate most of the settlement expected under
the planned construction load. This requires settlement-time
relationships for the planned construction load and the various levels

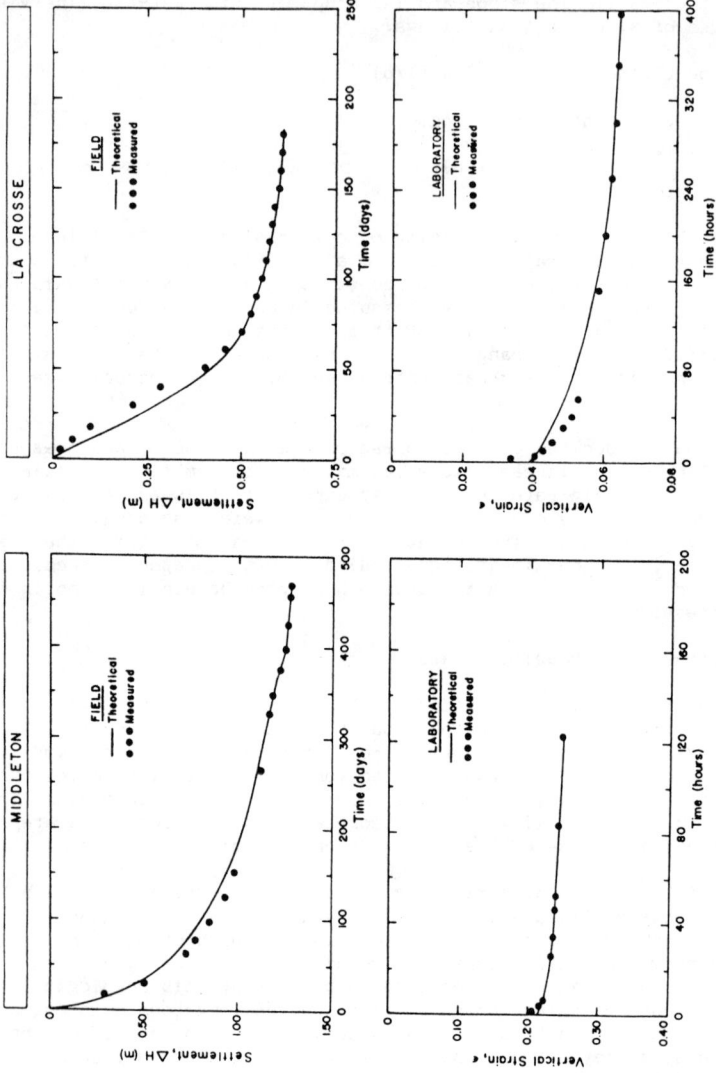

Figure 2. Measured and Curve-Fitted Compression of Peat in the Field and Laboratory.

of surcharge. These relationships often have to be predicted on the basis of laboratory or field model tests. Previous research indicated that the empirical parameters a, b, and λ/b determined from such tests display a certain amount of dependency on the magnitude of stress increment, final stress level (initial stress plus stress increment), and the average strain rate (2,3,6). For relatively small variations in stress, the parameters do not vary significantly and the effect of the magnitude of stress increment can be directly accounted for in Eq. 1 by adjusting $\Delta\sigma$.

In order to investigate the dependencies of the model parameters on the stress level and strain rate, the back-calculated values of the model parameters from all of the laboratory and the field settlement-time curves obtained for 16 peat and organic soil deposits are compared. Part of the data was obtained from the local engineering firms. The available properties of these soils are summarized in Table 1. Peats and organic soils were distinguished primarily on the basis of visual classification. Nine of these deposits were from preloading sites from which field data were collected. The characteristics of these sites are summarized in Table 2. There is a general dependency of primary compressibility, a, on stress level (final consolidation stress) as shown in Fig. 3. For comparable levels of stress, the laboratory test results provide a reasonable estimate of the field primary compressibility within the expected inherent soil property variation range. Fig. 3 includes data from both peats and organic soils indicating that the primary compression is quite comparable in magnitude for both types of soil. The average curve fitted through the data points can be used as a correction curve in adjusting the value of primary compressibility, a, measured at a given stress level to another desired stress level.

Fig. 4 gives secondary compressibility factor, b, as a function of stress level. Because of the significant difference in the magnitude of secondary compression for peats and organic soils, only the peat data are included in Fig. 4. General dependency of secondary compressibility on stress level is displayed. Correlation between the laboratory and the field values of b is not as good as it is for a. In general, the field values of b are higher than the laboratory values for the same peat at a given stress level. Translating the laboratory secondary compressibility to field conditions requires the use of the average curve drawn in Fig. 4 for stress level differences and Fig. 5 for the discrepancy between the laboratory and the field values. Fig. 5 was constructed from Fig. 4 by computing the ratio of b values as given by the field and laboratory curves as a function of stress.

Lo et al. (6) reported that the rate factor for secondary compression, λ/b, is strongly non-linear with time for an inorganic clay. This factor indicated no simple correlation between the laboratory and the field data for the peats and organic soils analyzed herein. This is expected in view of the possible non-linearity of λ which represents the structural viscosity of soil during the secondary compression. The average strain rate in the laboratory specimens and in the field vary considerably. Fig. 6 gives the rate factor for secondary compression as a function of the average strain rate for the peats and organic soils. The average strain rate was computed by dividing the

Table 1.   Average Properties of Soil Deposits

Site of Deposits	Type of Soil	Fiber Content (%)	Ash Content (%)	pH	Water Content (%)	Specific Gravity (%)	Unit Weight (kN/m³)	Liquid Limit (%)	Plastic Limit (%)
Fond du Lac	Peat	20	39.8	6.2	240	1.94	10.2	--	--
Portage	Peat	31	19.5	7.3	600	1.72	9.6	--	--
Waupaca	Peat	50	15.0	6.2	460	1.68	9.6	--	--
Middleton	Peat	64	12.0	7.0	510	1.41	9.1	--	--
Noblesville	Peat	48-52	6.9-8.4	6.4	173-757	1.56	8.4	--	--
Disney World	Peat	--	5.0-50.0	--	140-1200	--	11.0	--	--
Madison Mosque	Peat	--	25.4	--	181-206	--	9.6	--	--
La Crosse	Peat	--	--	--	115-301	--	11.2	246	133
Rhinelander	Peat	--	--	--	723	--	10.3	820	--
Solberg	Peat	--	--	--	234	--	11.0	204	77
Middleton C.S.	Org. Silty Clay	--	--	--	64	--	15.9	--	--
Middleton W.W.	Org. Silty Clay	--	--	--	44-95	2.30	15.0	--	--
Grignano	Org. Silt	--	--	--	52	--	16.5	55	27
Olin Avenue	Org. Silty Clay	--	--	--	40	--	17.5	--	--
MG & E	Org. Clay-Silt	--	--	--	39	--	18.0	--	--
Crocker Res.	Org. Silt	--	--	--	40	--	17.5	22	14

Note: These properties are listed as reported in the actual source reports. They were determined largely in accordance with the available ASTM standards. Fiber content is the rubbed fiber content based on the volume of wet soil. Ash content and water content are based on oven-dry weight.

Table 2. Characteristics of Preloading Sites

Site	Type of Deposit	Thickness of Deposit (m)	Depth of Compressible Deposit (m)	Depth of Groundwater (m)	Mid-plane Effective Overburden Stress (kPa)	Fill Stress (kPa)	Surcharge Stress (kPa)
Middleton	Peat Org. Soil	4.0 3.7	0.4.0 4.0-7.7	0	7.0	70.0	11.3
La Crosse	Peat	3.0	3.2-6.3	0.2	37.0	78.5	46.2
Rhinelander	Peat	1.7	0.8-2.4	0.6	13.1	23.5	22.0
Solberg	Peat	2.6	5.6-8.2	2.7	72.5	24.1	
MG & E	Org. Soil	5.8	0.3-6.1	0.2	25.5	37.4	29.9
Grignano	Org. Soil	4.1	3.4-7.5	3.2	80.5	78.5	
Olin Avenue	Org. Soil	5.8	3.0-8.8	1.5	55.1	27.8	25.2
Crocker Residence	Org. Soil	2.6	3.0-5.6	0	30.1	6.9	50.4
Disney World Peat		3.4	0-3.4	0	2.0	51.5	20.5

Note: The final preloading stress is the sum of the fill stress which remains permanently (to achieve a certain grade elevation) and the surcharge stress which is removed after preloading.

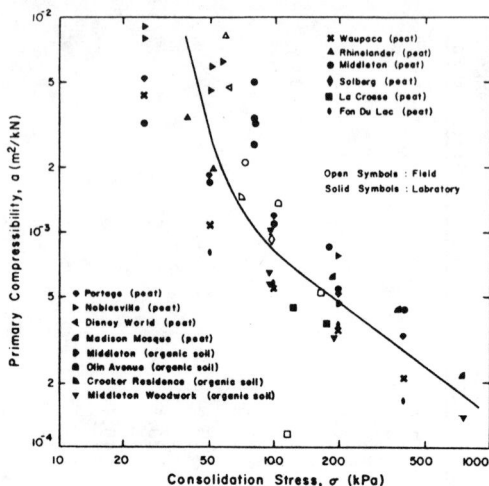

Figure 3. Primary Compressibility versus Stress Level.

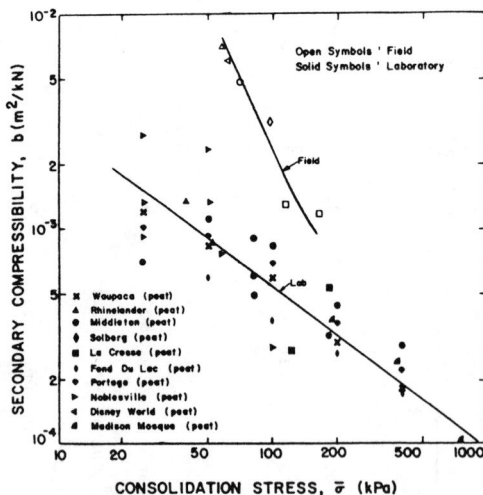

Figure 4. Secondary Compressibility versus Stress Level.

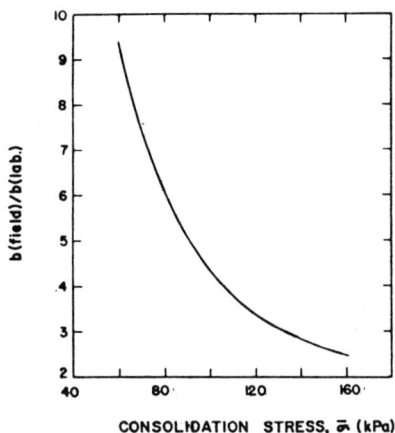

Figure 5. Correction Curve for Laboratory Values of
Secondary Compressibility.

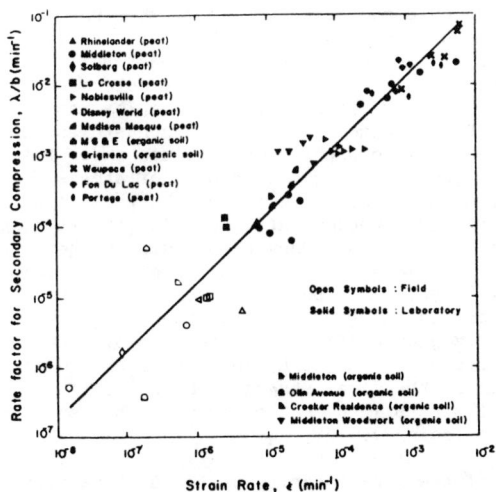

Figure 6. Dependency of Rate Factor for Secondary
Compression on Average Strain Rate.

total strain by the time required to complete it. Considering the data
points in Fig. 6, this relationship is seen to be basically linear on a
logarithmic plot. λ/b values vary considerably between the laboratory
and the field. Therefore, it is necessary that λ/b must be extrapolated
to the estimated range of the average field strain rate. The field
strain rate is often 2 to 3 orders of magnitude slower than the
laboratory value during secondary compression.

Prediction of Field Compression from Laboratory Tests

Ideally, one would like to be able to predict the field behavior on
the basis of laboratory tests performed on samples of the soil deposits
in question. There are basically two difficulties with this approach in
predicting the settlement-time behavior of peat soils. The first
difficulty is in obtaining undisturbed samples which are representative
of the deposit and in simulating all aspects of the field boundary
conditions in the laboratory. This problem can be minimized by
extensive exploration, testing and possibly by performing model tests in
the field. The second difficulty, which is the subject of this paper,
is the problem of translating the soil parameters determined in the
laboratory to the field.

The effects of stress level can be mostly eliminated by performing
the laboratory tests at stress levels and with the stress increments
anticipated in the field. This is a generally recognized procedure
since c_v and C_α in the conventional analysis are picked from those
laboratory stress increments corresponding as closely as possible to the
field stress changes. It is proposed herein to go one step further and
to perform laboratory tests by applying the anticipated field stresses
rather than the conventional consolidation test loading schedule of
doubling the applied stress incrementally. The range of initial
effective overburden stress and the anticipated range of stress
increment in the field (due to the final construction load and for a
number of design surcharge loads) should be determined and a sufficient
number of representative laboratory peat samples should be subjected to
stresses within these ranges. The overburden stress is applied first
and after the vertical compression stabilizes under this stress (1 to 2
x 10^{-3} mm/day or less), a single increment of stress corresponding to
the final construction load or design surcharge is applied. The stress
increment is left for a period sufficient for the vertical compression
to proceed well into the secondary compression range. This often is in
excess of the conventional 24 hours used for inorganic soils and it may
approach 7 to 10 days. Most peats are surficial deposits with the
ground water close to the ground surface and with the low unit weight of
peat, this results generally in low effective overburden stress (see
Table 2). Therefore, the construction and surcharge loads represent a
major proportion of the final stress. The conventional laboratory
loading ratio of stress increment to previous stress of one is seldom
encountered (see Table 2).

The effect of the non-linearity of the soil parameters due to the
vast differences in the strain rate of a thin laboratory specimen and a
much thicker deposit in the field, cannot be eliminated. This effect is
basically ignored in the conventional analysis of secondary compression
and as shown previously may introduce significant errors in the field

predictions. The parameters determined in the laboratory must be
corrected for this effect before using them in analyzing field
situations. It is proposed to use Figs. 5 and 6 for correcting,
respectively, b and λ/b obtained from the laboratory tests (performed at
stress levels comparable to those expected in the field) for use in the
field.

Two cases of field compression of peat are analyzed with and without
making the corrections described above. Fig. 7 shows the actual field
settlement-time data along with the predicted curves based on the
corrected and the actually measured laboratory parameters (a, b, and
λ/b). The laboratory tests were performed at stress levels similar to
those encountered in the field. The drastic effect of correcting for
non-linearities is seen. Correction of b affects the magnitude of the
final settlements. The two curves uncorrected for this effect give
nearly half of the final settlement observed in the field. On the other
hand, λ/b affects the rate at which this final settlement is achieved.
The two curves which are corrected for the strain rate effect on λ/b
exhibit a gentle slope comparable to that of the field data at large
times. The curves uncorrected for this effect become nearly horizontal
at large times. Fig. 7 shows the sensitivity of the model to variations
in the model parameters.

Use of Field Data in Determining Soil Parameters

Because of the usual difficulties associated with correlating the
laboratory behavior to the field, it is desirable to use field tests in
determining the soil parameters governing the settlement-time behavior.
This can be achieved by placing a test fill and monitoriong settlements
with time. Alternatively, the early settlement data obtained from the
actual surcharge application can be used to make adjustments in the
preloading scheme and in the design predictions. In either case, the
field settlement-time data are used in back calculating the operating
soil parameters governing the settlement behavior of the deposit. One
of the questions in such computations is the duration over which the
field data to be used in the curve-fitting procedure for the
determination of the compression parameters. As indicated previously,
successful application of Eq. 1 requires that the compression proceeds
well into the time range corresponding to the secondary compression.

In order to establish a practical rule for guidance in determining
the necessary field test duration, the field settlement data from four
separate sites were analyzed. Portions of the field data corresponding
to increasing test durations were used in determining the compression
parameters. The settlement-time curves obtained using these sets of
parameters are shown along with the actual field data in Fig. 8 for one
of the sites. In each case there is a certain duration beyond which the
field data yield parameters which approximate the field curves
reasonably well. If the test duration used is less than this critical
duration, t_c, the computed primary compression factor becomes very small
or negative and the secondary compression factor becomes very large.
The critical test duration was compared with t_{90} that would be obtained
from the conventional Taylor construction involving a plot of settlement
versus square root of time (7). The results are summarized in Table 3.
Considering the four cases, the critical test duration is found to be

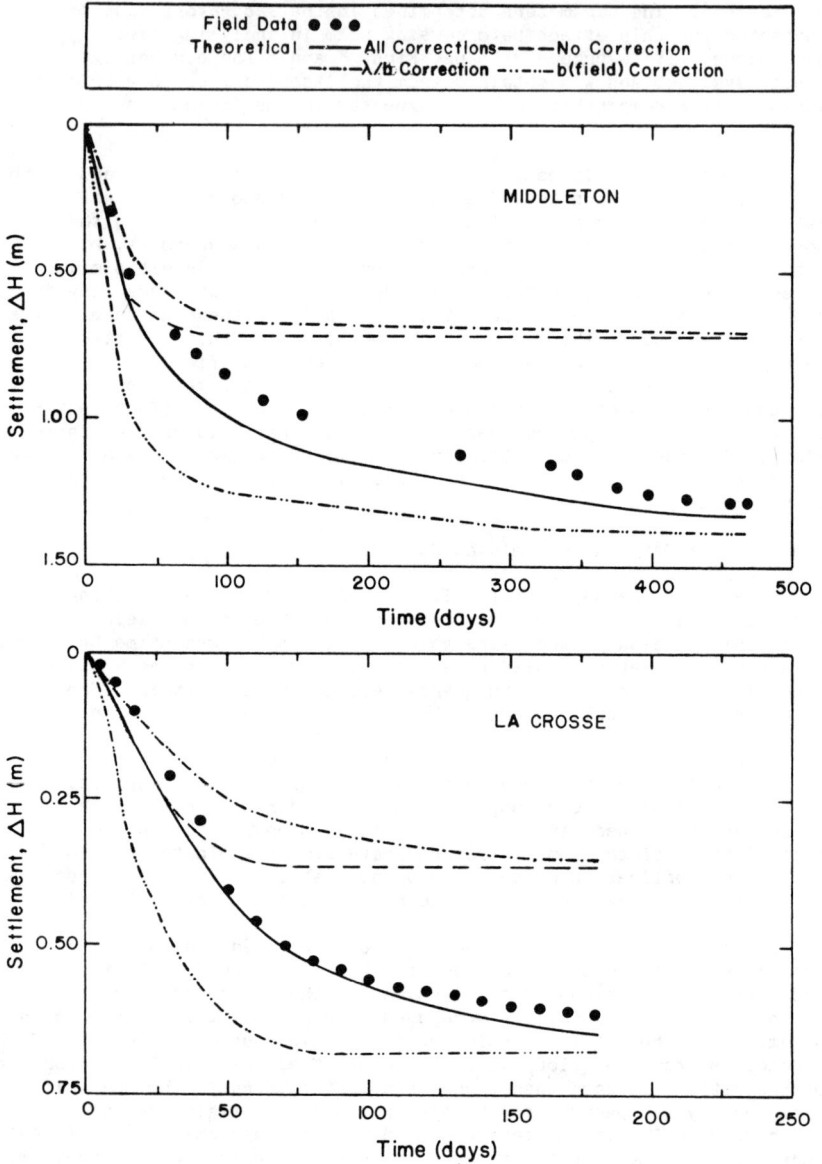

Figure 7. Influence of Corrections on Theoretical Predictions.

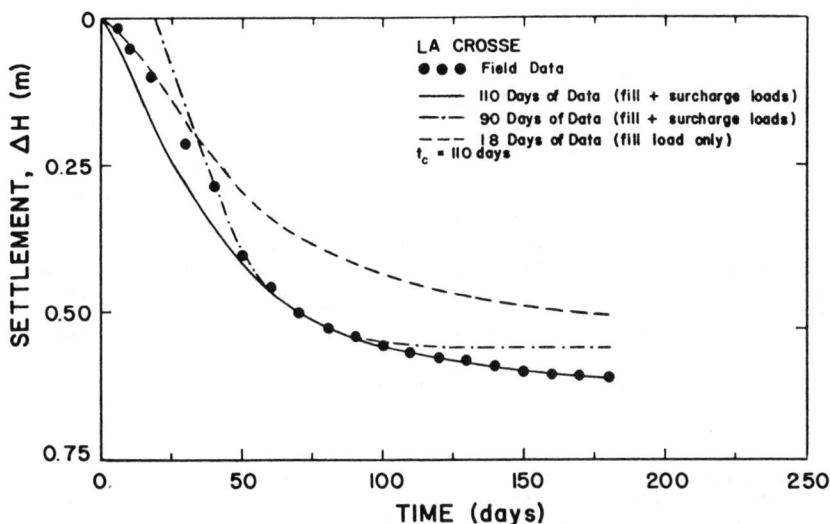

Figure 8.   Influence of Test Duration Used in Determining the Model
Parameters.

Table 3.   Critical Field Test Durations

Site	t_{90} (days)	t_c (days)	t_c/t_{90}
Rhinelander	37	55	1.49
Disney World	47	65	1.38
La Crosse	90	110	1.22
Middleton	264	264	1.00

$$t_c = (1.0 \text{ to } 1.5) \, t_{90} \tag{5}$$

Once t_c is exceeded, the field settlement data can be used in
determining the compression parameters with acceptable accuracy.   In
order to update the preloading analysis, i.e., recalculating the
anticipated life-time settlement under the final construction load and
the surcharge magnitude and duration required to eliminate this
settlement, the necessary corrections to the parameters for stress level
effects can be made if significant differences are involved.   The
differences in stress increment (surcharge magnitude) are directly
accounted for by the Δo term in Eq. 1.   An example of the successful
application of this technique was reported by Edil (2).

Conclusions

Based on the analysis of numerous laboratory and field compression data involving 16 deposits of peats and organic soils, the described procedure for settlement-time analysis is found to be useful and reasonably accurate. The procedure utilizes a rheological model in describing the vertical strain versus time in one-dimensional compression and involves three empirical parameters reflecting, respectively, the magnitudes of primary and secondary compression, and the rate of secondary compression. The two parameters involving secondary compression (magnitude and rate) are found to be non-linear. This requires the correction of parameters determined in the laboratory (under stress levels comparable to those encountered in the field) with respect to the strain rate expected in the field. Correction curves based on the cases analyzed are provided. Furthermore, recommendations for the prediction of field settlement of peat deposits based on laboratory and field model test data are advanced.

Acknowledgements

The assistance of Mr. Earl H. Reichel of Soils and Engineering Services, Inc. of Madison, Mr. Clifton Lawson, formerly of Warzyn Engineering, Inc. of Madison, and Professor C. William Lovell of Purdue University in providing the compression data for some of the deposits is gratefully acknowledged.

References

1. Buisman, A.S.K., Results of Long Duration Settlement Tests, Proceedings of the First International Conference on Soil Mechanics and Foundation Engineering, (1) (1936) p. 103-106
2. Edil, T. B,. Improvement of Peat: A Case History, Proceedings of the Eighth European Conference on Soil Mechanics and Foundation Engineering, Helsinki, 2 (1983), p. 739-744.
3. Edil, T. B. and Dhowian, A. W., Analysis of Long-Term Compression of Peats, Geotechnical Engineering, Southeast Asian Society of Soil Engineering, 10 (2). (1979) p. 159-178.
4. Gibson, R. E. and Lo, K. Y., A Theory of Soils Exhibiting Secondary Compression, Acta Polytechnica Scandinavica, Ci10296 (1961) p. 1-15.
5. Gruen, H. A. and Lovell, C. W., Compression of Peat Under Embankment Loading, TRB Symposium on Classification and Properties of Peats and Highly Organic Soils, Washington, D.C. (1984).
6. Lo, K.Y., Bozozuk, M. and Law, K. T., Settlement Analysis of the Gloucester Test Fill, Canadian Geotechnical Journal, 13 (4), (1976) p. 339-354.
7. Taylor, R. W. and Merchant, W., A Theory of Clay Consolidation Accounting for Secondary Compression, Journal of Math and Physics, 19 (1940), 167-185.
8. Terzaghi, K., Principles, of Soil Mechanics, Engineering News-Record, 25 (19-23) 25-27 (1925).

Time and stress dependent compression in soft sediments

D. McG. Elder* and G. C. Sills*

ABSTRACT

Results are presented for laboratory experiments using a typical estuarine mud settling from a slurry in columns of 100mm diameter. Measurements have been made of density and pore water pressure, from which
total and effective stress distributions are calculated. Volumetric
strains for very soft soils are shown to be considerable and are
affected both by effective stress changes and by the passage of time,
the latter giving rise to significant degrees of apparent overconsolidation for these soft soils.

1. INTRODUCTION

A wide range of cohesive material is deposited from suspensions and
slurries, some natural and some man-made. Initial suspension/slurry
densities vary from the dilute condition found in rivers, estuaries and
seabeds to comparatively dense industrial waste sludges. Despite this
range of initial conditions and types of material, the soft sediments
formed by deposition have many features in common, including consolidation due to the self weight of the sediment and the occurrence of large
strains during the consolidation process. A study of the behaviour of a
typical estuarine sediment covering a range of densities could therefore
have many applications.

One such application is the deposition of sediment in dredged channels
and harbours. Once the energy available to support sediment in the
river has fallen below a critical level, the sediment will drop through
the water. On reaching the bed, it then undergoes a process of consolidation, with compression occurring until the pore water pressures in
the bed have dissipated. An understanding of this process and an ability to predict surface position and shear strength of the settling mud
is important for an efficient maintenance dredging programme. Another
application is the disposal of industrial or mining waste by hydraulic
pumping into settling ponds or tailing dams. The rate of consolidation
is significant, affecting the capacity of the settling ponds and the
development of shear strength, governing slope stability, of the tailing
dams.

In these applications, then, the important characteristics of behaviour
are the magnitude and rate of compression and the development of shear
strength. This paper describes the results of laboratory based research
on the compression behaviour of an estuarine sediment deposited through
water. A parallel study into the development of shear strength will be
described elsewhere.

* Department of Engineering Science, University of Oxford, U.K.

2. LABORATORY SIMULATION

In many field situations, the environmental conditions on the sediment are complex and very difficult to measure. However, the deposition and consolidation process will generally occur substantially in the vertical direction, with downward sediment movement and upward flow of water. It is therefore reasonable to model this consolidation process under one-dimensional conditions in the laboratory, where the controlled boundary conditions and ease of measurement allow the fundamental parameters to be identified.

In the present experiments, a slurry of sediment at any required initial density is pumped into settling columns of internal diameter 100mm and height up to 3m, either very quickly (in less than a minute), or at a controlled preset rate. As it settles through the column, various measurements can be made. One of the most important of these is density, using a non-destructive, X-ray method that provides a spatial resolution better than two millimeters and an accuracy of measurement of the order of 0.01 Mg/m^3. The apparatus consists of an X-ray source on one side of the column which produces a narrow, horizontally collimated beam and a detector on the far side to record the attenuated intensity. The column is traversed vertically by remote control and the continuous record is converted to a density profile with height using samples of known density for calibration. The method is described in detail in Been (2). From the density profile a distribution of total vertical stress can be obtained by integration. Another important measurement is pore water pressure, obtained using pressure transducers mounted on the side of the column, behind porous sand/araldite filters let into the column wall. The pore water pressure values, taken with the total stress, allow the distribution of effective stress through the column to be calculated. A transducer mounted with its face flush with the column base measures total stress.

3.   PREVIOUS CONCLUSIONS

Some basic features of behaviour of these soft consolidating sediments have emerged from previous work and these are summarised here before the new work is presented. Detailed accounts may be found in Been (1), Been and Sills (3), Sills and Been (7) and Sills and Thomas (8). Fig. 1 shows typical results of an experiment. The sediment, an estuarine silty clay from Combwich, Somerset, was pumped quickly into the column at an initial density of around 1.09 Mg/m^3, as demonstrated by the 20 minute curve. Later, as the sediment flocs fall through the water, the level of the sediment/water interface drops and the density at and near the base of the column increases. The $4\frac{3}{4}$ hour profile illustrates this behaviour: the top part of the profile shows a value near the initial 1.09 Mg/m^3 while a shallow comparatively dense layer of density up to 1.6 Mg/m^3 has built up near the bottom. Between these two regions, a third region is growing upwards from the base layer. In this case, the intermediate density is around 1.13 Mg/m^3. With further time, the upper suspension disappears, and, as the top of the intermediate region becomes the interface between sediment and water, the subsequent rate of surface settlement is reduced. With still further time, the density of the intermediate layer continues to increase.

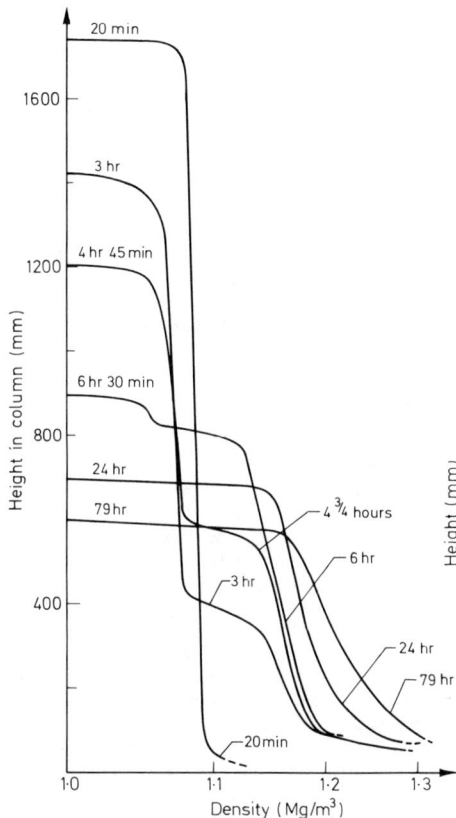

Fig.1. Typical density profiles

Fig.2. Total stress & pore pressure distributions. Experiment KB7, initial density ρ_i = 1.09 Mg/m³.

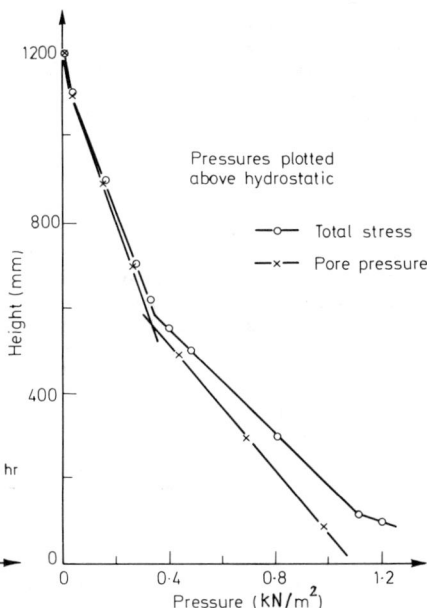

The measurement of pore pressure in the columns allows the effective stresses to be estimated and this is shown for the $4^{3}/4$ hour profile in Fig. 2. The density profile in Fig. 1 has been integrated to give the total vertical stress distribution through the column and this information is shown as the difference between total vertical stress and the hydrostatic pressure plotted against height in the column. Also plotted on this graph is the excess pore pressure against height (i.e. total pore pressure minus the hydrostatic pressure) and it can be seen that in the suspension region, the pore pressure is equal to the total vertical stress, so that the sediment is effectively fluid supported. In the region of intermediate density, the pore pressures are consistently lower than the total stress, indicating that a frame-work of soil has begun to exist, with non-zero effective stresses. The magnitudes of the effective stresses are, at this stage, extremely small but their development marks the transfer from a suspension with non-Newtonian fluid behaviour to a consolidating soil whose behaviour may be characterised by the parameters and models of soil mechanics. This point is emphasised: the difference between the so-called inter- mediate region and a traditional soil is one of degree only, not of fundamental behaviour. However, the description "intermediate" is retained in order to emphasise the comparative smallness of the effect- ive stress values, while "consolidating mud" or "consolidating sediment" are used as synonyms.

Figures 1 and 2 can be used to illustrate the determination of a correlation between void ratio e or specific volume v and the effective vertical stress σ_v' throughout the region of intermediate density. The specific volume is the total volume occupied by unit volume of solid material, related to the void ratio by v = 1 + e. The specific volume has the advantage of being directly comparable to the height of the soil layer in these one-dimensional experiments. Thus, the effective vertical stress may be obtained at some height from Fig. 2 and the corresponding specific volume calculated from the density value at the same height from Fig. 1. As consolidation proceeds, the effective stresses increase, extending the range of the relationship. The purely suspension phase disappears within the first few hours if no further sediment is added. Therefore, no further account is taken of the suspension, and experiments are assumed to start at the stage when effective stresses (albeit small) exist everywhere throughout the sediment in the settling column. (In Fig. 1 for example, this occurs shortly after the $6\frac{1}{2}$ hour profile).

At the sediment/water interface, the effective stress in the consolidating sediment must, by definition, be zero. It might be expected that a unique value of specific volume would be associated with this constant zero effective stress, but experimental results published by Been (1) and Been and Sills (3) showed that this void ratio decreases with time, as the whole sediment layer consolidates, from an initial value dependent upon the initial concentration of the sediment slurry. In these experiments, the void ratio at the surface of the intermediate region, at zero effective stress, could change from around 12 to around 5 over a period of a few months.

It therefore appeared that, at very low effective stress, the assumption of a unique specific volume corresponding to a particular effective stress was not justifiable. Nevertheless, Been showed that, on the time scale of these experiments, of the order of a couple of months, it was possible to identify some specific soil parameters which could be used in an idealised model to predict with surprising success the consolidation in the settling columns. The model followed the large strain, self weight analysis initially proposed by Gibson, England and Hussey (4) and developed by Lee (5) and Lee and Sills (6). It was then modified by Been to sidestep the assumption of the original model that the specific volume at the soil surface would be constant.

Been's experiments were all undertaken with slurry introduced quickly into the column with initial densities varying from 1.04 Mg/m^3 to 1.22 Mg/m^3. The higher densities were already greater than those associated with the development of effective stress and the density profiles showed that the subsequent consolidation behaviour (at least

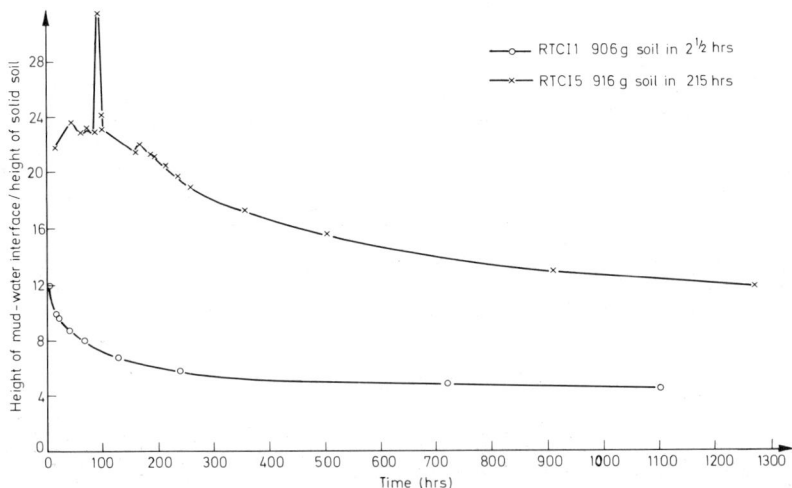

Fig.3. Normalised surface settlement for two experiments with different rates of gradual deposition.

in the early stages) was different from those tests where the intermediate region developed from a suspension.

      In a further set of experiments, sediment was introduced into the
columns at a controlled rate, (Sills & Thomas (8)). It is clear from
this work that the density established in the intermediate region will
be very dependent on the concentration of the overlying suspension.
Thus a slow rate of introduction of sediment (providing a dilute
suspension) allows a less dense structure to be established and maintained than that from a faster deposition rate. This is illustrated
in Fig. 3, showing the surface settlement in two experiments using very
similar total masses of sediment. In RTCI1, 906g of sediment was
introduced over a period of 2½ hours, while in RTCI5, 916g was
deposited in 215 hours. The settlement curves show that a greater
thickness of layer is associated with RTCI5, at the slower rate of
deposition. The settlement has not finished after 1200 hours, so it
is not possible to predict what the long-term comparison will be.

      In summary, then, it has been established that the consolidation
of a very soft soil is strongly dependent on the initial density and/or
deposition rates, where no distinction has been made between primary and
secondary consolidation. The purpose of this paper is to investigate
whether time (affecting secondary consolidation or creep) plays a
significant role in these very soft soils. Now, therefore, "consolidation" will be used to describe the volume changes caused by water flow
from the soil under the influence of excess pore pressures alone.

4.   EXPERIMENTAL CONDITIONS

In order to study the effects of time separately from the
consolidation process, individual experiments must be continued well
beyond the point at which consolidation has virtually ceased. There
will therefore be advantage in shortening the consolidation stage of an
experiment, which could be done either by accelerating the drainage
and/or reducing the drainage path. The provision of a drained boundary
at the base of the settling column, with a boundary pore water pressure
equal to the hydrostatic pressure, should therefore provide a significant improvement. Another possibility is to introduce sediment at
initially higher densities than suspension densities. Then, at the
start of an experiment, effective stresses will already exist, pore
pressures will already be less than total stresses and the drainage
path length for a given effective stress range will be substantially
less than the layer thickness for the same mass of sediment deposited
from a suspension.

The density profile of such a layer will be different from that
of a similar mass of sediment deposited from a dilute slurry, and this
may affect the early consolidation. Nevertheless it should still prove
possible to measure volumetric strains occurring at known effective
stresses after the dissipation of excess pore pressures.

The experiments therefore fall into three categories. The first
follows the pattern described for the earlier work, in which soil
slurries, ranging in initial density from 1.09 Mg/m^3 to 1.25 Mg/m^3 are
introduced into 100mm diameter settling columns up to 3.5m high with
surface drainage only. Where appropriate, the earlier results are
included. The second category contains experiments in which base
drainage is provided by connection to a stand pipe containing water at
the same level as the water in the settling column. The third category
consists of short column experiments, with initial densities ranging
from 1.25 Mg/m^3 to 1.42 Mg/m^3 and initial heights around 200mm. In all
of these three categories, density, pore pressure and total stresses
were measured. The sediment used is referred to as Combwich 6, and is
an estuarine silty clay from the River Parrett in Somerset. It has a
LL of 63, PI of 29 and particle size analysis by sedimentation shows
that around 30% of the individual particles are of clay size or smaller.
For all except one experiment, (using sea water), the sediment was mixed
with tap water, in which it flocculates. The behaviour of the soil in
the columns - and also in the field - is governed by the flocs and not
by individual particles.

5.   DESCRIPTION OF EXPERIMENTAL RESULTS

Several experiments were undertaken to evaluate the merits of
allowing drainage to occur at the base of the column. Fig. 4 compares
surface settlement profiles for tests DE1a, DE1b at initial densities
near 1.02 Mg/m^3. Heights are divided by equivalent height of dry
solids to allow for the small difference in initial density between the
two. Initially at this density a suspension phase is present; once
settlement of this to the consolidating soil phase is complete there is
a marked decrease in settlement rate. Little difference is seen
between the two experimental curves. Using theory for large strain

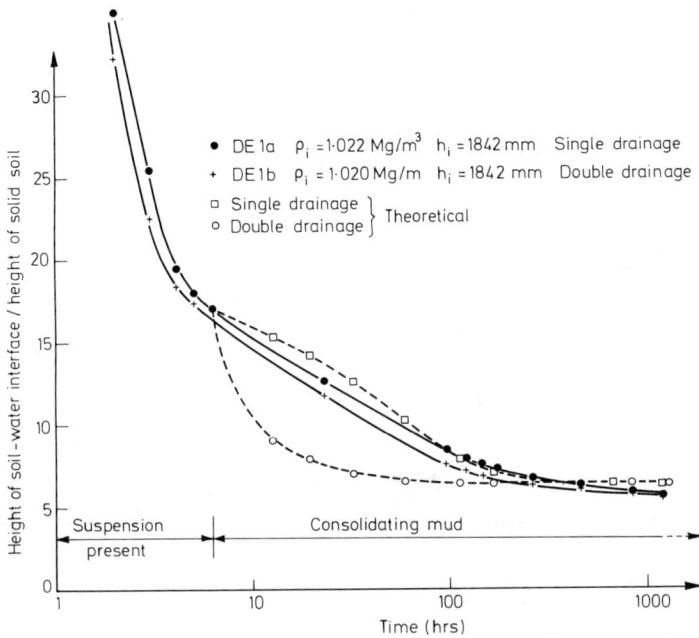

Fig.4. Comparison of single and double drainage conditions. Experimental and theoretical profiles of normalised surface settlement.

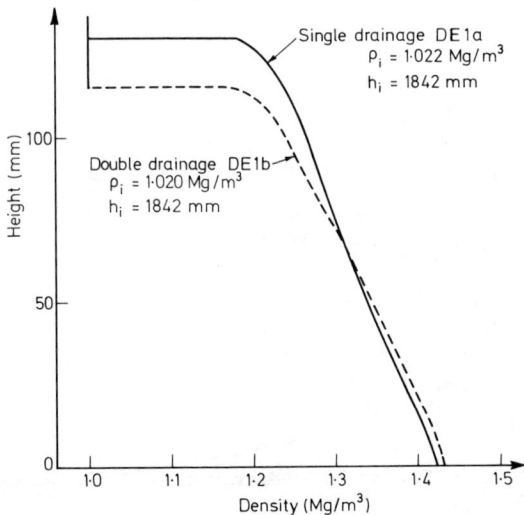

Fig.5. Density profiles for DE 1a, 1b after 940 hours.

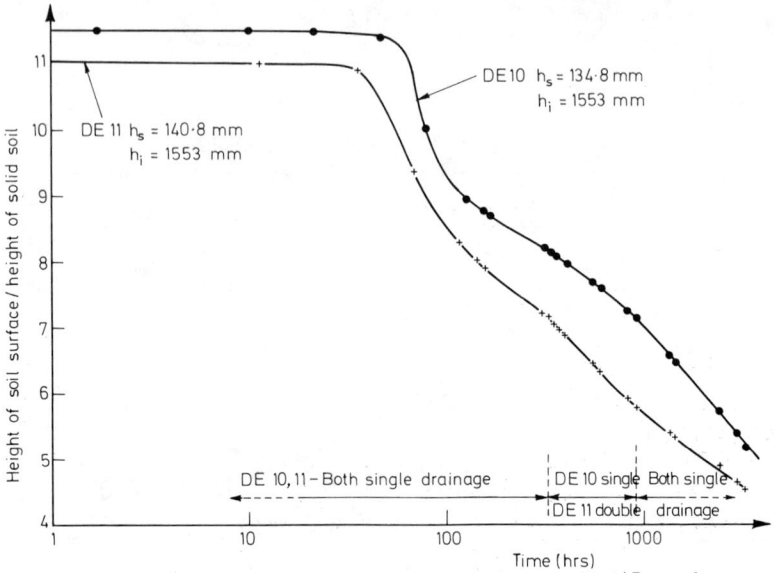

Fig.6. Normalised surface settlement for DE 10, ρ_i = 1.147, and DE 11, ρ_i = 1.153 Mg/m³.

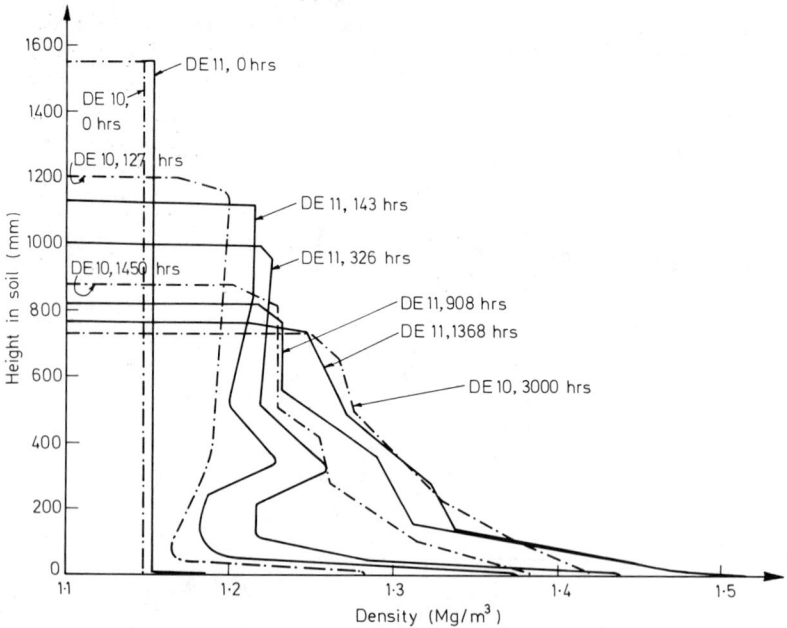

Fig.7. Typical density profiles for DE 10, DE 11.

consolidation with double drainage (Lee (5)) and applying modifications used by Been (1), as described earlier, with parameters obtained at the end of the suspension phase, theoretical settlement lines are plotted. The theoretical and experimental lines for single (surface) drainage are very similar, while the double drainage theory preducts a considerably faster rate of settlement than that which occurs. Density profiles for each test after 960 hours obtained by X-ray attenuation and shown in Fig. 5 show that the density profiles at that time are nearly identical for both drainage conditions. The difference in height (130:115mm) is accounted for by the slight difference in initial density and the ratio of heights is nearly equal to the ratio of solids mass present in each sediment; normalised settlement curves coincide at this time. Thus, there is little improvement in consolidation time shown by using double drainage conditions.

Effects of single and double drainage are compared in Figs. 6 and 7 for slurries at higher initial density, DE10 and DE11. During the first 40 hours, little settlement occurs; the soil, although of low density, appears relatively stable. Magnitudes of effective stress during this time are described later. The settlement rate then increases sharply as the structure collapses and slows again to a period where the height decreases almost linearly on a logarithmic time plot.

To avoid the possibility of altering this initial settlement period, DE11 was not permitted to drain from the base until this stable settlement rate had been reached. After 326 hours the base drain was opened to the hydrostatic level in DE11 and left open until a possible leak required that it be closed again at 907 hours. Again no significant increase in settlement rate can be detected. The consolidation rate in the column with double drainage was being slowed considerably by a comparatively dense soil layer existing near the base, as seen in Fig. 7. At 908 hours the void ratio at the base is about one-third that at the surface; the correspondingly lower permeability causes the drainage path to remain in an upward direction through most of the soil column.

The most simple technique for introducing sediment into the settling columns is rapid pumping of a well mixed slurry at uniform density. To compare results from this method with other possible field situations, experiments RTCI2 and RTCI3 were begun with a gradual, steady rate of input of a dense slurry into the top of a column already filled with water, at 7.2 g/hr and 7.9 g/hr of solid material respectively for 105 and 70 hours. The total solids masses are comparable to those for KB1 and KB10 which were deposited rapidly at densities close to 1.02 Mg/m^3. Fig. 8 shows normalised settlement height changes with time after completion of deposition. This time scale may be somewhat arbitrary since it does not take into account settlement due to structural changes during gradual deposition. Corresponding density profiles are shown in Fig. 9. A very large difference is seen in behaviour between the two types of deposition. The sediment input at constant density settles rapidly from a suspension into the consolidating mud phase and then slows in settling rate. The gradually deposited sediment, although at a density which would previously have been considered a suspension and free of effective stress, appears to be nearly stable for 10 to 20 hours and then settles very slowly.

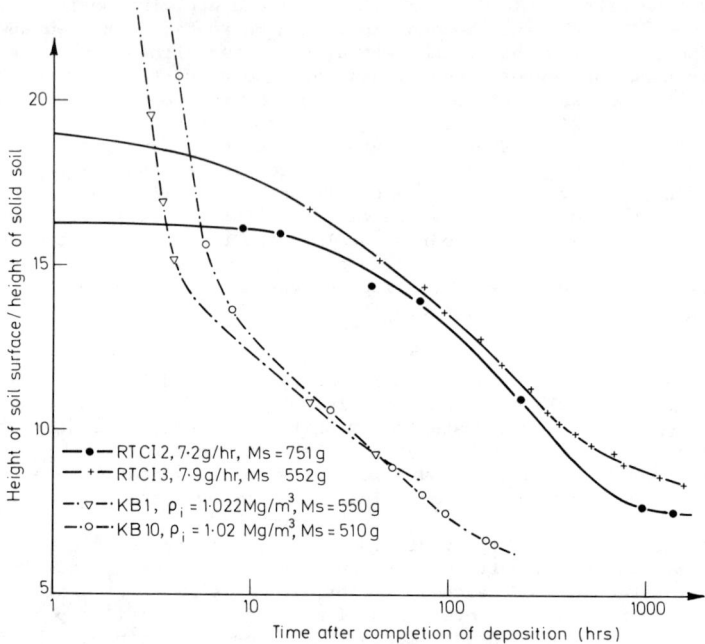

Fig.8. Comparison of instantaneous and gradual deposition – normalised surface settlement for four experiments.

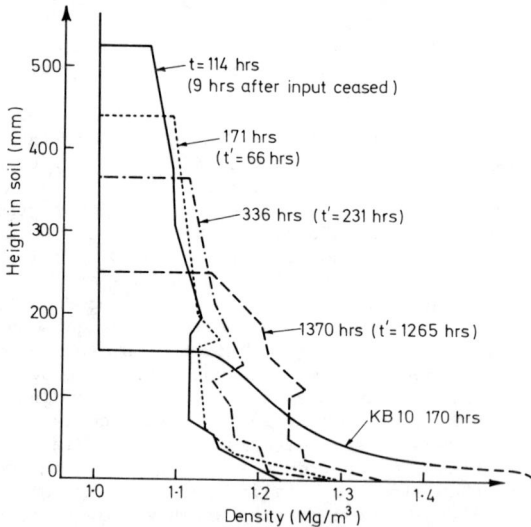

Fig.9. Density profiles for RTCI2, gradual deposition over 105 hours and KB10, instant deposition at 1.02 Mg/m³ after 170 hours.

Around five hours after completion of deposition, heights and average densities (represented by h/h_s) are similar for both methods of input: however, the gradually deposited sediments take ten times as long thereafter to decrease in thickness by a factor of two. Comparison of density profiles for RTCI2 and KB10 at 170 hours shows much lower values for the sediment deposited gradually and similar densities to those in KB10 at 170 hours are not reached by RTCI2 until 1370 hours. Towards the base the density in KB10 is much higher, indicating a structure less stable in an open form.

Changes in density measured at the surface of settling mud, corresponding to zero effective stress, are shown in Fig. 10 for a number of experiments with different initial and boundary conditions. Generally the surface density is clearly defined. Where it is not, the intersection point of the density profile extended from its slope lower in the soil with the surface height is used. Given the diffi- culty of obtaining very accurate density measurements within several millimetres of an interface with abrupt density change, this method seems suitable. Previous results have shown that, for a slurry placed quickly into the column, if the initial density is below 1.13 Mg/m^3 (specific volume greater than 13), then the slurry is a fluid-supported suspension. This settles rapidly to form a consolidating layer, in which effective stresses exist. This behaviour is seen clearly in Fig. 10. For initial specific volumes below about 13, there is no suspension phase present and the surface density increases slowly with time. At initial values greater than 13, the specific volume increases initially at the surface, probably due to segregation of large particles which fall through the flocculated structure to the base. When the falling suspension surface meets the denser layer beneath, a sharp fall in density occurs to a specific volume depending on the suspension density but in no cases lower than 13, i.e. the value which determined whether or not a suspension phase would be present. Following this a more gradual decrease in specific volume occurs on a much longer time scale following similar trends to those seen in sediments which never existed in a suspension phase. This value of specific volume between 12 and 14 (density from 1.14 to 1.12 Mg/m^3) is a critical point for slurries deposited instantaneously at uniform density. Below this range the slurry will behave as a consolidating soil (although with variable behaviour as discussed later) while above this range there is no definite structure present and the system is highly unstable; segre- gation and large deformations occur. Slurries with initial specific volume below 10 show less change in surface density during the first 100 hours although this rate generally increases after longer periods. Lines are also shown for saltwater deposition (RTSW2) and double drainage conditions (DE11); for both these variations the trends are seen to be similar to those for fresh water, singly drained conditions at the same initial density (KB15).

A very different result from those described above occurs in the case of gradual deposition (RTCI2). Despite the extremely high specific volume (33) at the end of the depositional period, there is no evidence either of significant segregation or the presence of a suspen- sion phase. The surface density increases only very slowly with time, showing no sudden change, and even after 500 to 1000 hours is still at a density which in an instantly dumped slurry would suggest that a

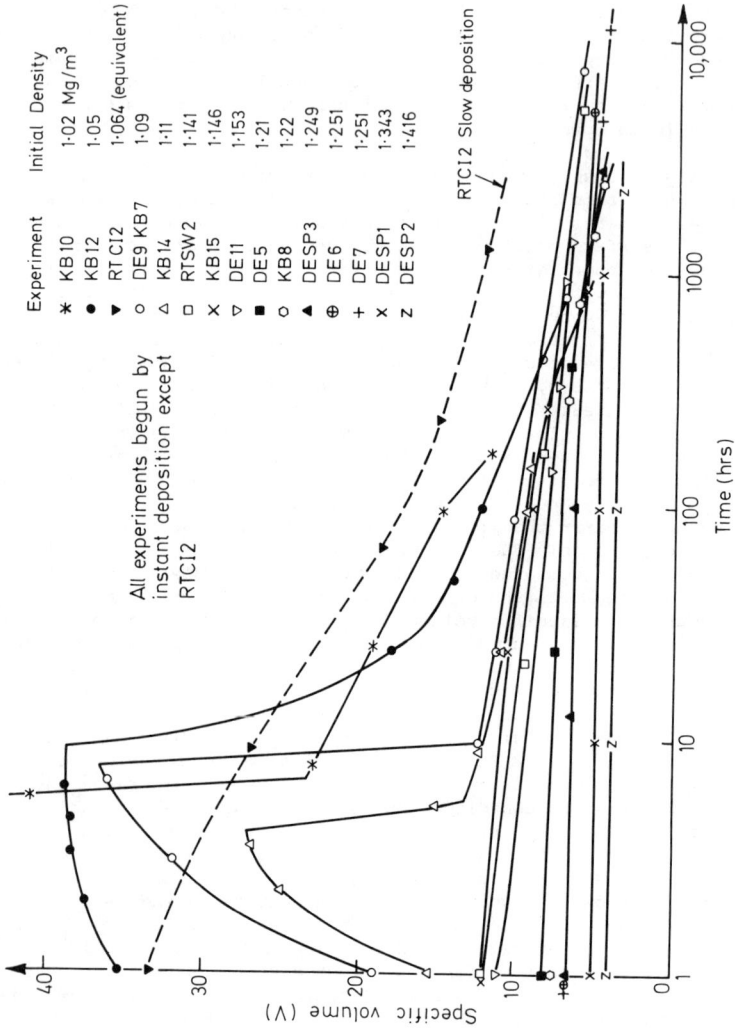

Fig.10. Surface density changes in settling sediments.

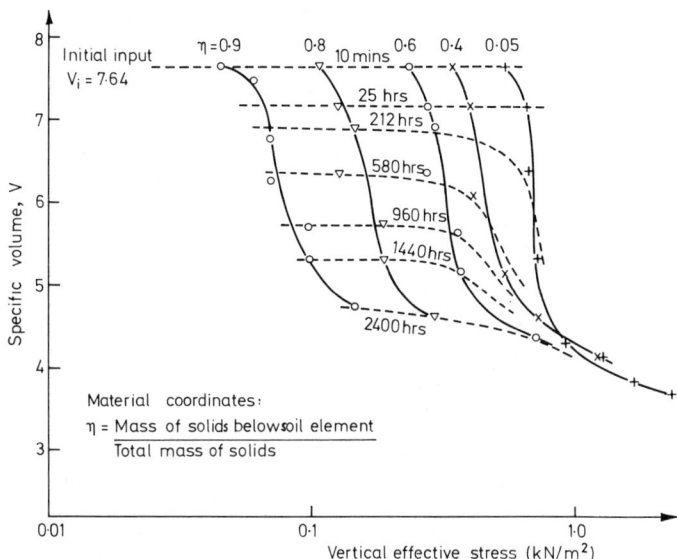

Fig.11. Consolidation paths followed by different soil elements in KB 8, initial density $\rho_i = 1.22$ Mg/m³, $h_i = 1.748$m. Dotted lines represent all points in soil at stated times.

suspension might be present. It is also of interest that for some suspensions with very low initial densities the rapid volume change at the end of the suspension phase can occur to a specific volume closer to that in the gradually deposited sediment after about 10 hours. For these sediments, however, the subsequent rate of specific volume decrease is much faster. A clear indication is provided by Fig. 10 that the characteristics of deposition, including initial density, height, rate of deposition and the effect that these and the boundary drainage conditions have on the subsequent stress history are factors of major importance in determining the consolidation state of a sediment.

Figs. 11 to 14 show consolidation curves, represented by plots of specific volume against effective vertical stress, for a number of experiments. Fig. 11 is for KB8, deposited at initial density 1.22 Mg/m³ which is considerably denser than the critical value discussed with reference to Fig. 10. Solid lines plotted are those followed by individual elements in the soil, defined by material co-ordinate :

$$\eta = \frac{\text{mass of solids below the element}}{\text{total mass of solids present}} .$$

Very different behaviour is seen from that found in experiments at lower initial densities. Due to the comparatively high initial density,

effective stresses are present in the slurry at input. Ten minutes
after input, the slurry density is nearly uniform throughout the
column, at a specific volume of around 7.6. Measurements of the total
vertical stress and the pore water pressure show that, at this stage,
the effective stresses range from zero at the surface to about 0.06
kN/m^2 near the base of the column. Therefore, a change in effective
stress from zero to 0.06 kN/m^2 does not necessarily cause significant
density or specific volume changes. The subsequent behaviour shows
comparatively large changes in specific volume with time, with only
small increases in effective stress. This increase of density takes
place over 1000 hours and more. Thus, any given value of effective
stress may apparently be associated with a range of specific volumes,
demonstrating time-dependent volume change, or creep. Paths followed
by different soil elements are nearly parallel and vertical in $v - \sigma_v'$
space and there is no sign of a unique consolidation curve for the soil
in this state. Dotted lines connect all points in the soil at any
given time and clearly show the dominant effect played by time during
this stage. This phenomenon can be explained by a collapse in the soil
structure; rather than particle or floc spacing decreasing causing
interparticle forces to increase, the flocs break down or rearrange so
that a smaller stress increase results. This process continues for
several thousand hours until the density becomes sufficient for effective stresses again to increase. The subsequent consolidation behaviour
is both time and stress dependent with the stress level playing an
increasingly important role as densities increase. Once primary consolidation is complete, however, (excess pore water pressures dissipated)
the soil will continue to increase in density under purely time-
dependent creep at constant effective stress.

     The consolidation behaviour at zero effective stress has been
shown to depend upon the stress history (method of deposition, density
and drainage path lengths of initial slurry, for example). Comparable
results for positive effective stresses are presented in Figs. 12 and 13.
These show density profiles with depth (Fig. 12) and positions in
consolidation space $(v:\sigma_v')$ (Fig. 13) for three experiments with the same
initial density (1.25 Mg/m^3) but different initial heights (DE6 1748mm,
DE7 1195mm, SP3 200mm) and two experiments with similar solids mass but
different initial density (DE7 3.9kg, 1.25 Mg/m^3 and DE9 4.1 kg, 1.09
Mg/m^3). Only DE9 begins in the suspension phase. DE6 and 7 are doubly
drained, although, as has previously been shown, this does not have a
large effect on consolidation behaviour due to early formation of a
less permeable base layer. Comparison of DE7 and DE9 in Fig. 12 shows
that segregation in the latter has had a marked effect, causing base
densities to be very high while the surface region is at a lower density
than that at a similar time in DE7. Profiles of DE6 and DE7 are very
similar in shape, but higher densities are attained in DE7 due to the
shorter drainage path. Dissipation of excess pore pressures in DE7 was
virtually complete by 4320 hours but a further density increase occurs
between 4320 and 11000 hours at constant effective stress. This
increase is nearly uniform throughout the height of the sediment.
Sample SP3, although consolidated for less time than either DE6 or DE7,
has reached higher densities. Again this is due to the drainage path
being much shorter; the increase in effective stresses due to dissipation of excess pore pressure is faster and causes creep at higher
effective stresses to be of longer total duration. The effects of these

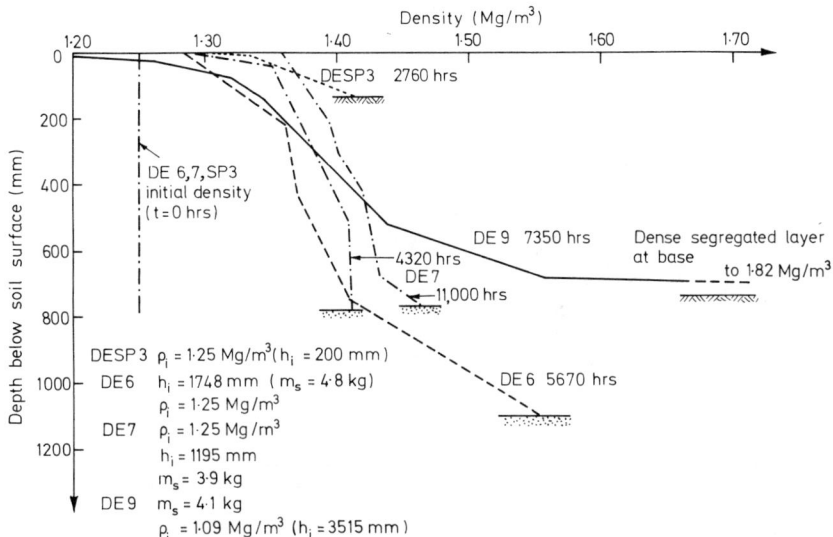

Fig. 12. Density profiles, DE 6, 7, 9, SP3 showing effects of varying initial conditions and subsequent stress history differences.

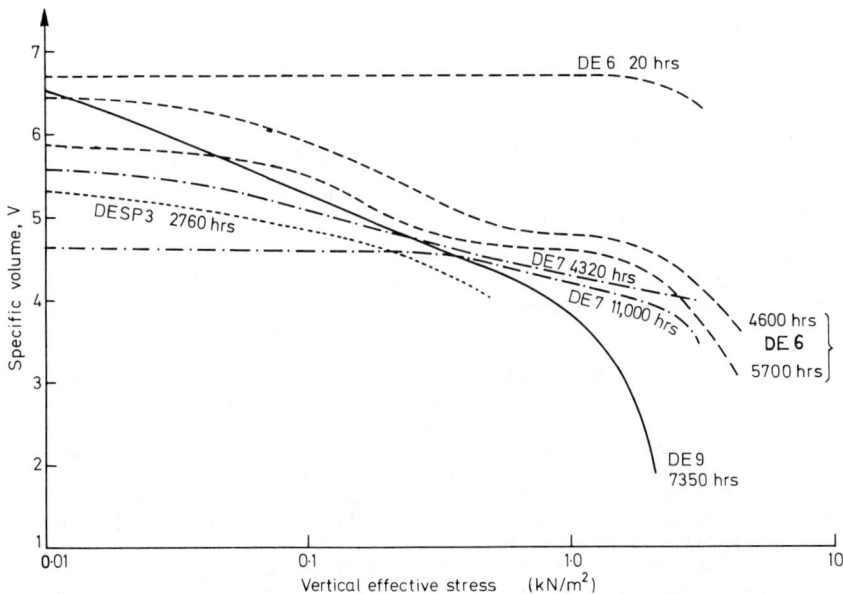

Fig. 13. Consolidation states for DE 6, 7, 9, SP3 at various times after deposition. Lines connect all points in soil at stated times.

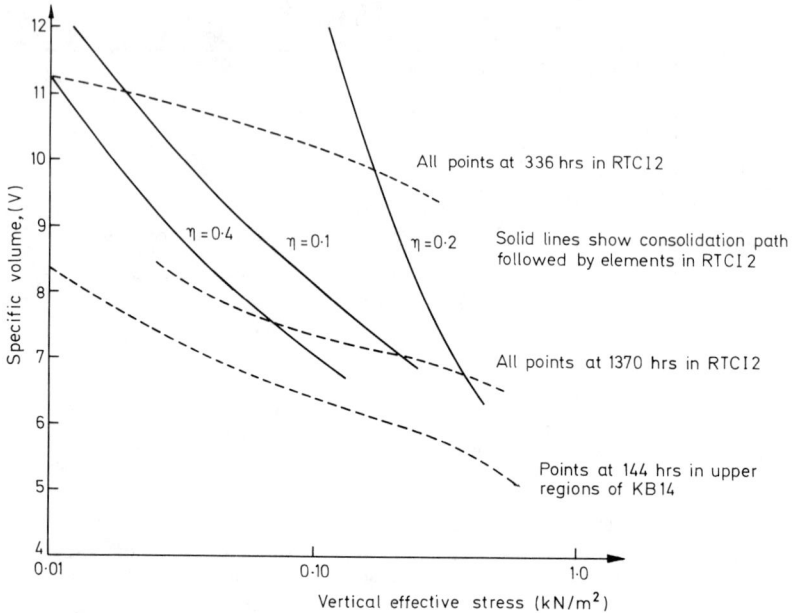

Fig.14. Comparison of consolidation paths followed by different elements in RTCI 2, gradual deposition at 7.2 g/hr for 105 hrs, with consolidation state in KB14, instant deposition at density $\rho_i = 1.11$ Mg/m³, after 144 hrs.

density differences are also seen in Fig. 13 where lines connecting all points in each sample are compared. As in Fig. 11 there is no unique consolidation line; the tendency is seen for specific volume to decrease at constant stress. The effects due to varying initial conditions are also clear; lines for different experiments at comparable times after deposition do not coincide and are not parallel.

Consolidation lines for 3 elements in RTCI2 (gradual input) are shown in Fig. 14. The first point of interest occurs when comparison is made with Fig. 11. Lines for soil elements near the base ($\eta = 0.1$ and $\eta = 0.2$) are interchanged in the gradually input sediment. The soil at $\eta = 0.1$ has developed higher densities, for any recorded effective stress, than the soil above at $\eta = 0.2$. Similar behaviour was not seen in any of the columns with instantaneous input. This might be expected, however, as the sediment deposited during the early stages of gradual deposition would initially have had a very short drainage path so that despite the low stress levels, time dependent density increase could be considerable. The other feature of interest is that specific volumes at given stress or times are much higher than for other tests. This was also a feature of the observations made on surface density changes (Fig. 10). Although there is no other experiment with initial conditions comparable of height and density or solid mass, experiment KB14 did have an initial density 1.11 Mg/m³ similar to that in RTCI2 at the end of deposition. KB14 had 1170g solid mass and

and RTCI2 750g, so the upper two thirds of KB14 should be adequate for qualitative comparison. The line connecting all points in KB14 at 144 hours is shown and is well below that reached in the slow input experiment after 10 times that period.

6.   DISCUSSION OF RESULTS

The experiments have been carried out under a variety of conditions. The first of these to be examined was the provision of double drainage by allowing flow from the base of the columns. The efficiency of such a drainage boundary will depend on the local permeability. However, the permeability is strongly dependent on the porosity and large changes occur in the porosity with only small changes in effective stress. Thus the permeability can be substantially reduced where the effective stresses are highest, as shown in Fig. 7. Providing a base drainage facility therefore does not necessarily have a significant effect on the rate of consolidation.

The effect of the rate of input on the behaviour of the sediment as illustrated by Fig. 8 is perhaps the most dramatic single feature. The slow deposition rates have allowed the framework to be stable in a much more open condition than is possible for the case of fast deposition. It is not possible at this stage to identify with certainty the process that is responsible for this difference but some speculation may be allowed. A significant difference between the extremes of fast and slow deposition is the length of time that sediment flocs rest on the surface of the consolidating soil before being buried by further sediment. As flocs land on the surface, they will initially establish an open framework: it appears that if sufficient time elapses before the loading is increased by more sediment, then this open framework can be maintained. If the loading is increased faster, then the framework collapses under it. This suggests that some bonding or strength development with time is possible. Fig. 9 shows that substantial density increase then occurs for the slowly deposited sediment, but at a much slower rate than for a sediment deposited instantaneously. It is not possible to predict whether creep could ultimately eliminate the distinction between the slowly and quickly deposited sediments.

The data shown in Fig. 10 refer only to sediment behaviour very close to the surface of the bed, where the pore pressures are hydrostatic and there is therefore no primary consolidation occuring. The effective stresses are presumably close to zero and the results may be used to infer something about the creep process and its dependence on specific volume. The conclusions will strictly be true only for constant, very small effective stress but may nevertheless be indicative of more general behaviour. Five experiments (KB10, 12, 14, DE9 and KB7) show the existence of a suspension stage. Of these, the two most dilute (KB10, 12) subsequently show some of the characteristics associated with the slow rate of deposition: that is, effective stresses develop for an open framework, at specific volumes otherwise associated with the suspension phase. It may be that the low initial density has allowed the surface layers sufficient time to develop some resistance. The magnitude of creep that may be expected to occur appears to be less as the initial specific volume reduces, and the

identifiable onset of creep appears to be later. Thus for initial specific volumes around 7 and 8, creep is very small for the first 400 hours
or so, while for specific volumes of 5, it is at least 1000 hours before
it begins to be measurable.

In most field situations, volume changes will be caused by changes
in the effective stress as well as by creep. Some of the characteristics of these combined processes are illustrated in Figs. 11, 12 and
13. Fig.11 has shown that the changes in specific volume that would
have occurred due to effective stress increases would have been very
small over a time period of 2400 hours. However, the implication of
Fig.11 is that even though the effective stresses are still small, the
creep component of the volume change has become significant. Figs.12
and 13 confirm the complexity of the inter-action between effective
stress driven volume change and creep. Despite the identity of the
sediment used in the experiments, it is difficult to identify common
features in the different curves. This could be explained if the amount
of creep were dependent on the actual value of effective stress, since
the different curves represent different stress histories and, in
particular, different amounts of time spent at the various effective
stress values throughout the column. Perhaps the most certain conclusion is that with the passage of time and after the dissipation of
excess pore pressures, specific volume values continue to decrease at
all effective stress levels, as shown by DE6 and 7. This general conclusion would hold also for slow deposition of sediment.

One consequence of this significant time-dependent reduction in
specific volume is the difficulty of identifying a virgin consolidation
line. If a soil has experienced a reduction in volume through creep,
then it will probably respond, at least initially, to a load increment
causing consolidation in a stiffer manner than if the creep had not
occurred. This will cause the soil to be apparently over consolidated.

7. CONCLUSIONS.

Self weight consolidation of very soft sediments is a complex
phenomenon governed by diverse physical variables and processes. Consolidation paths (specific volume - effective stress) followed by
elements at different heights in the same soil mass differ greatly and
the stress history undergone by any element will also affect the consolidation path followed at a later stage under increased loading or due to
passage of time alone. Factors influencing stress history include
drainage pathlengths, pore water salinity, deposition rate, initial
density and time.

Several distinct regions of behaviour have been identified during
the settling and consolidation process. If the input slurry has a
density less than about 1.12 Mg/m³ it will exist initially as a suspension with no effective stresses present. A denser layer, in which
effective stresses are present, will accumulate rapidly upwards from
the base but will not necessarily behave in the same manner as material
input initially at that density, due to segregation, reorientation and
flocculation which have been allowed to occur freely during the suspen-

sion phase and due to other time dependent processes of volume change
which have differed between the two situations. If the soil is input at
a density higher than 1.14 Mg/m^3 there is often an almost immediate
increase in effective stress without significant density change. Volume
reduction in either case at later stages is governed by the superposition of two effects: the decrease in specific volume due to effective
stress increase as pore pressures dissipate and the decrease in specific
volume with time alone at near constant effective stress. At any
instant, the relative importance of these two effects, controlled by the
factors described above, will determine the instantaneous direction in
specific volume - effective stress consolidation space in which an element will move. A large sediment depth, even at high specific volumes,
will often cause time dependent compression to dominate at almost all
stages of a test, due to the very slow rate of pore pressure dissipation. Allowing base drainage may not greatly accelerate consolidation
since a localised rapid effective stress increase will increase densities and decrease permeabilities. A gradual, steady rate of deposition
causes a pronounced increase in resistance to structural compression and
effective stress change dominates early volume change, with a relatively
stable, low density structure remaining even after a long period of
consolidation.

Although varying widely with the different experimentally controlled boundary and input conditions, the time dependent creep
measured in these tests is of sufficient magnitude, even on timescales
of several weeks or months, that it may well provide an adequate
explanation for the apparent overconsolidation reported from a number of
sources where tests have been carried out on sediments taken from near
the seabed surface. Corresponding to apparent overconsolidation due to
creep a significant increase in soil sensitivity, as obtained from
strength measurements, has been found. These results will be reported
elsewhere.

The experiments reported here have identified fundamental behaviour, both qualitative and quantitative, on one particular soil. There
is a need for further measurements with different sediments and stress
histories, including further loading and unloading after creep has
occurred, both in the laboratory and in the field.

REFERENCES

1. Been K. 1980: D.Phil. thesis - Stress strain behaviour of a cohesive
soil deposited under water.
2. Been K. 1981: Non-destructive soil bulk density measurement using
X-ray attenuation. OUEL 1384/81; Geotechnical Testing Jnl,Dec.1981
pp.169-176.
3. Been K. & Sills G.C.1981:Self-weight consolidation of soft soils: an
experimental and theoretical study. Geotechnique XXXI 4 pp.519-535.
4. Gibson R.E., England G.L. & Hussey M.J.L.1967: The theory of one-
dimensional consolidation of saturated clays: 1. Finite non-linear consolidation of thin homogeneous layers. Geotechnique Vol.17,pp.261-73.
5. Lee K.1979: D.Phil. thesis - An analytical and experimental study of
large strain soil consolidation.

6. Lee K. and Sills G.C.1981: The consolidation of a soil stratum
including self weight effects and large strains. Internat. Jnl. for
Numerical & Analytical Methods in Geomechanics,5,pp.405-428.
7. Sills G.C. and Been K.1981: Escape of pore fluid from consolidating
sediment. OUEL 1378/81. Transfer processes in cohesive sediment systems,
Plenum Press.
8. Sills G.C. and Thomas R.C.1983: Settlement and consolidation in the
laboratory of steadily deposited sediment. IUTAM conference: seabed
mechanics. Newcastle.

Expt. No.	Soil Mass kg	Initial Density Mg/m³	Initial Height mm	Drainage S-single D-double	Duration hrs	Comments
DE1a	0.46	1.022	1842	S	1180	Double drainage
DE1b	0.41	1.020	1842	D	1180	comparisons
DE5	4.75	1.210	1748	S,D,S	518	
DE6	5.69	1.251	1748	D	5670	
DE7	3.89	1.251	1197	D	11000	
DE9	4.19	1.094	3510	S	7350	
DE10	2.93	1.147	1553	S	3000	
DE11	3.06	1.153	1553	S,D,S	3000	
DESP1	0.87	1.343	195	S	Up	18 short columns
DESP2	1.045	1.416	193	S	to	in each testing
DESP3	0.642	1.249	199	S	4000	series
RTCI1	0.906	372g/hr, 2.5 hrs		S	1100	
RTCI2	0.751	7.15g/hr,105 hrs		S	1368	Continuous sediment
RTCI3	0.552	7.9g/hr, 70 hrs		S	1608	input experiments
RTCI5	0.916	4.24g/hr,215 hrs		S	1272	
RTSW2	2.655	1.141	1711	S	9000	Salt water 1.024 Mg/m³
KB1	0.55	1.022	1929	S	43	
KB7	2.04	1.09	1742	S	79	
KB8	5.00	1.22	1748	S	2400	
KB10	0.51	1.02	1895	S	170	
KB12	1.19	1.05	1802	S	315	
KB14	1.17	1.11	816	S	144	
KB15	1.23	1.146	643	S	818	

TABLE 1   EXPERIMENTAL DETAILS

INSTRUMENTATION PERFORMANCE IN SOFT CLAY SOILS
(A CASE HISTORY)

James E. Laier (M.ASCE)* and William H. Brenner**

INTRODUCTION

In 1978, IDEAL BASIC INDUSTRIES, INC. initiated construction of a
large cement manufacturing plant located south of Mobile, Alabama.
As shown on Figure 1, the plant site abuts the Theodore Ship Channel
which leads to Mobile Bay and the Gulf of Mexico. Plant components
include numerous process and service structures along with materials handling and storage facilities. A crane unloads limestone
to barges moored at the dock and a conveyor transports the limestone to a mechanical Stacker/Reclaimer which places the material
onto primary and reserve stockpiles. Plant construction was completed in 1981 and the facility is currently in operation.

The cement plant site is typically underlain by a 60 foot (18.3
meters) thick deposit of very soft to firm plastic, deltaic clay.
Consequently, initial foundation design concepts included precast
concrete piles to support all major plant components. Soon after
construction started, however, the design team realized that
millions of dollars could be saved by supporting the Reserve Limestone Storage Area (RLSA) on shallow foundations. Because deltaic
clays underlying the RLSA were too weak to safely support the
desired 55 foot (16.8 meters) high limestone stockpile, provisions
were made to improve in-situ soil strength characteristics by surcharging the RLSA using stage loading techniques. Vertical wick
drains were installed to accelerate consolidation.

Prior to loading the RLSA, in-situ soil displacement and stability
characteristics were predicted by Dr. Charles C. Ladd of Massachusetts Institute of Technology using laboratory testing techniques
to evaluate stress history, consolidation characteristics and
stress-strain-strength properties of foundation clays (1).

* Principal Engineer, Southern Earth Sciences, Inc., Mobile,
  Alabama

** Principal Geologist, Southern Earth Sciences, Inc., Mobile,
   Alabama.

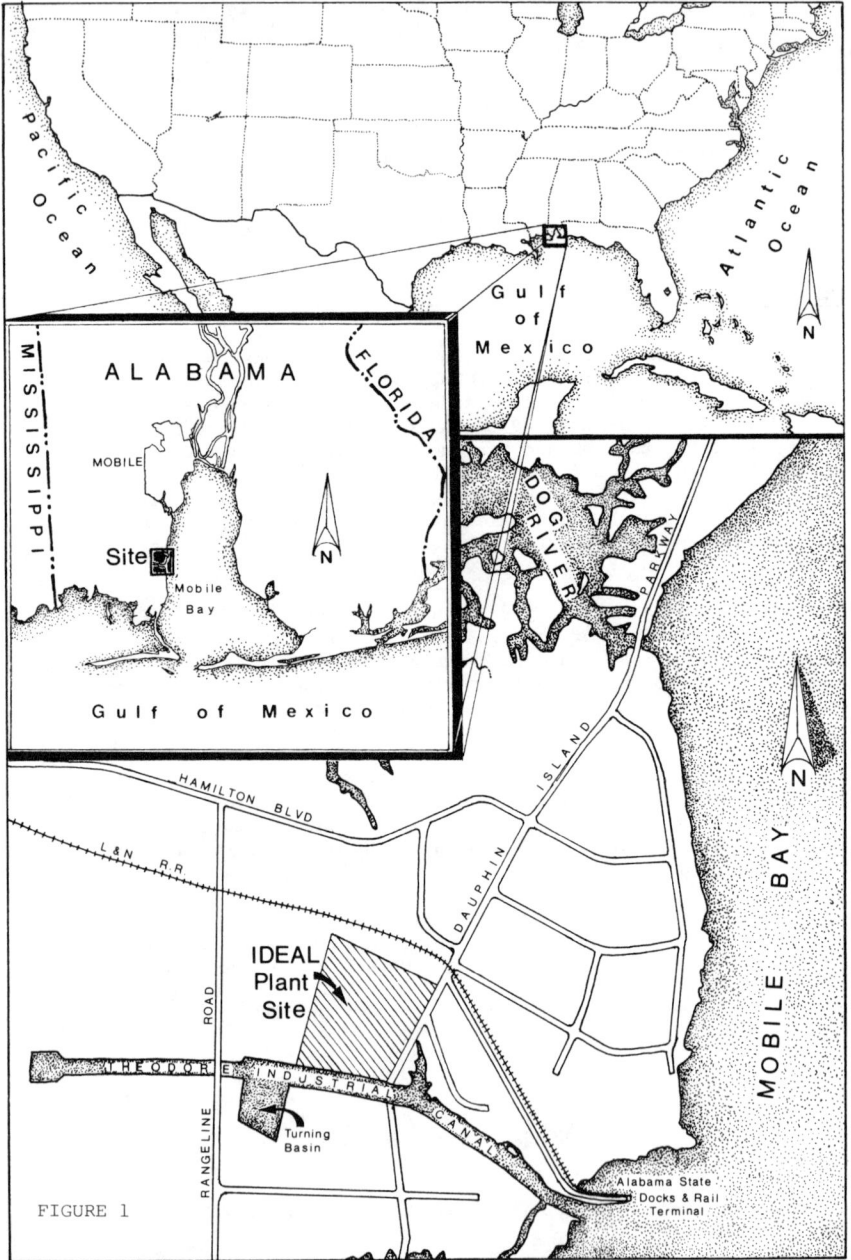

FIGURE 1

Based on Dr. Ladd's work (1), loading limits were established and
initial construction of the stockpile, to a height of 30 feet (9.1
meters), was accomplished using stage loading techniques. Stockpile performance was carefully monitored and field performance
data were systematically compared with predicted performance characteristics to assess stability and in-situ rate of consolidation
needed to plan subsequent loading schedules.

This paper is the first in a series of papers intended to describe
the comprehensive field monitoring system used to concurrently
measure large vertical strains, lateral soil deformations and pore
pressures within a large consolidating mass of cohesive soil.
Special attention is given the overall concept of instrumentation
for this project, including hardware installation and typical
problems associated with instrumentation performance. Selected
comparisons are made between typical instrument readings to
illustrate performance characteristics and general conclusions are
developed regarding the suitability and reliability of project
instrumentation. Subsequent papers will be devoted to specific
instruments with comprehensive discussion pertaining to the
suitability and reliability of each instrument for this project.

SITE AND SOIL CONDITIONS

The RLSA is bordered by a stacker/reclaimer on the west, by a dock
on the south and by wetlands on the north and east. Ground surface
contours within the proposed RLSA footprint are shown on Figure 2,
as they existed prior to loading. Typical ground surface elevations
ranged from 15.5 feet (4.7 meters) to 18.5 (5.6 meters) Mean Sea
Level. The site sloped gently toward the north and east and the
surface was typically covered with clay/sand (SC) fill as the site
had been used for a storage and assembly area during the pile
driving phase of the project.

Figure 3 illustrates a typical soil profile along the cross-section
through the centerline of the stockpile. Generally, a layer of
sandy soil extends from ground surface to about elevation +6.0 feet
(1.8 meters) MSL. This sandy top layer is underlain by a deposit
of soft to firm clay and silty clay down to about elevation -52
feet (-15.8 meters) to -64 feet (-19.5 meters) MSL. This clay
deposit contains shell, pieces of wood and traces and lenses of
sand and silt. The clay is characteristic of the deltaic deposition which took place in Mobile Bay during the Holocene Epoch
(2). Odeometer tests show that these clays are preconsolidated
as illustrated on the stress history profile shown on Figure 4.
Tests show the clay deposit to be precompressed by about 1.0 to
2.0 KSF (50 to 100 N/M^2). A dense sand layer, which lies beneath
the clay deposit, represents the top of an essentially incompressible Pleistocene deposit that extends to great depths.

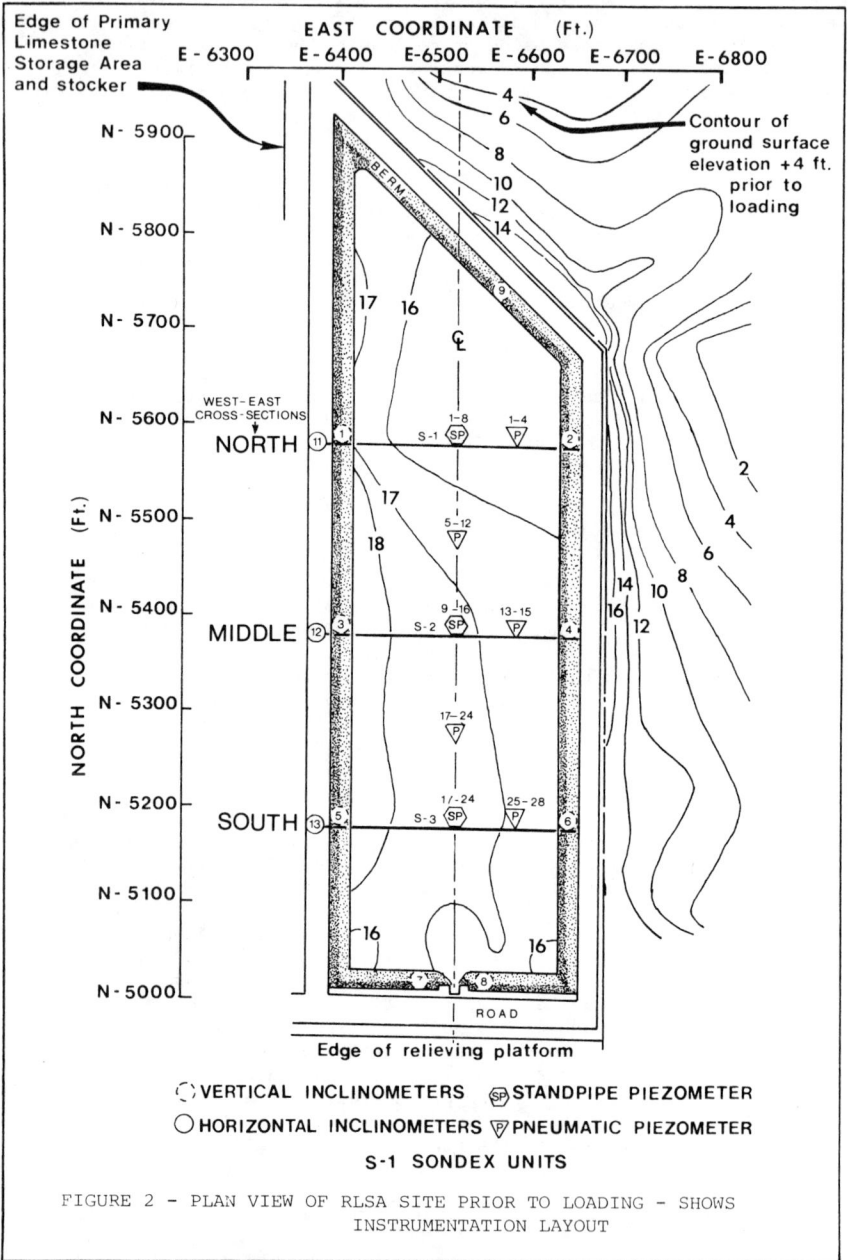

FIGURE 2 - PLAN VIEW OF RLSA SITE PRIOR TO LOADING - SHOWS INSTRUMENTATION LAYOUT

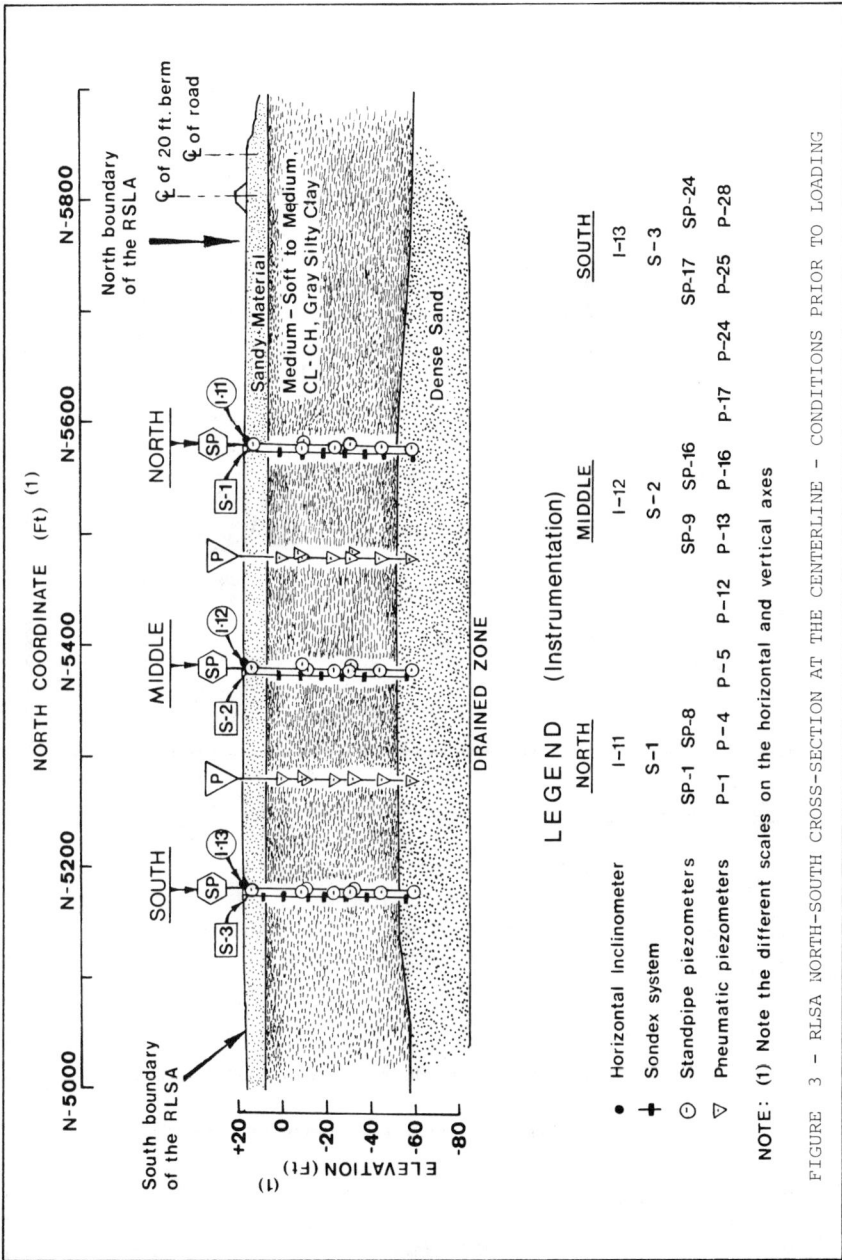

FIGURE 3 – RLSA NORTH-SOUTH CROSS-SECTION AT THE CENTERLINE – CONDITIONS PRIOR TO LOADING

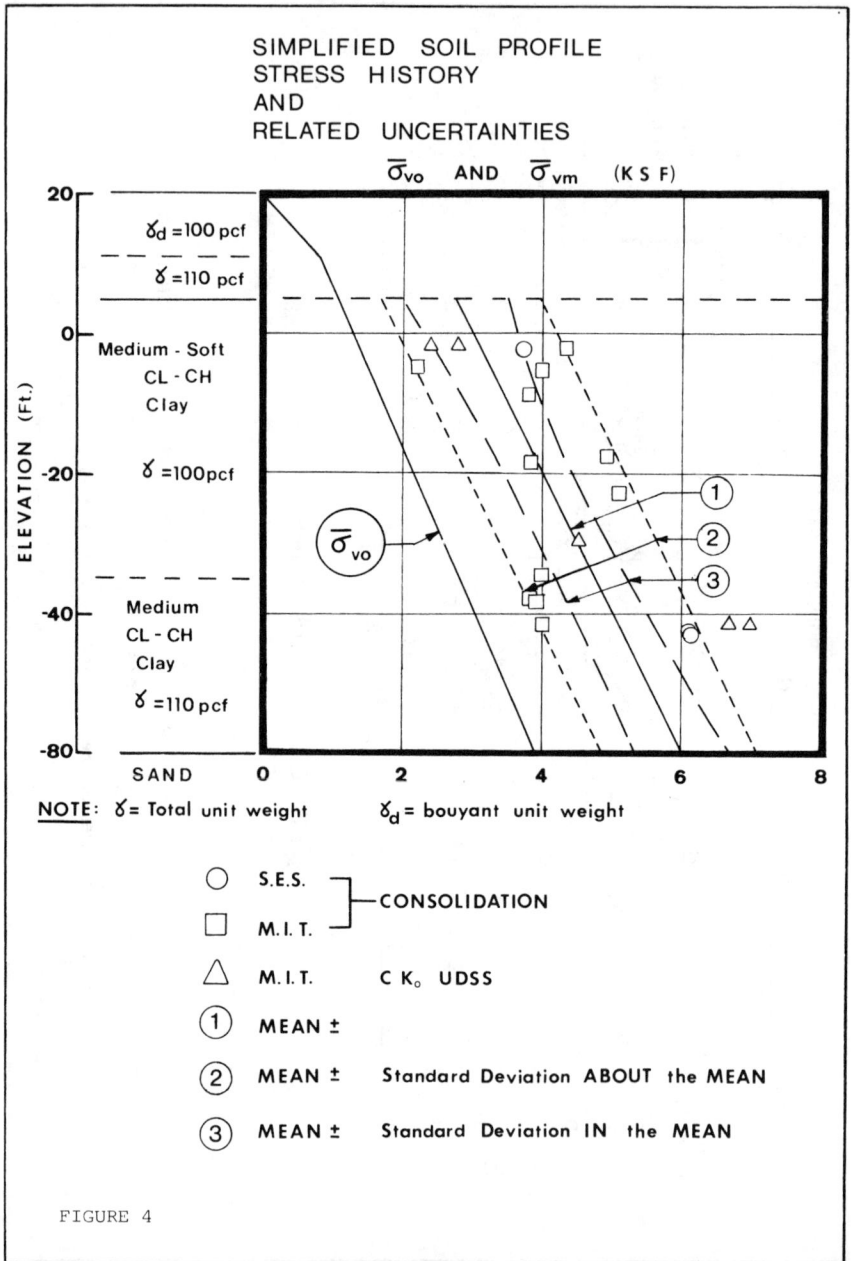

SIMPLIFIED SOIL PROFILE
STRESS HISTORY
AND
RELATED UNCERTAINTIES

FIGURE 4

INSTRUMENTATION

The IDEAL BASIC INDUSTRIES design team concluded that comprehensive
field monitoring would be necessary in gauging design assumptions,
safeguarding against catastrophic slope stability failures during
loading of the RLSA stockpile and for developing timely field consolidation data necessary for the accurate planning to stockpile
loading schedules.

Instrumentation: Based on a detailed analysis of soils data, Dr.
Ladd and the authors developed the following comprehensive instrumentation program as illustrated on Figure 2:

a. Vertical Inclinometers: Nine (9) vertical inclinometer casings,
    each about 90 feet (27 meters) long, were installed to measure
    horizontal soil displacements beneath the RLSA stockpile.
    Periodically, an inclinometer sensor* was lowered into each
    casing to electronically measure deviations from the vertical.

b. Horizontal Inclinometers: Three (3) horizontal inclinometers
    each about 250 feet (75 meters) long were placed across the
    base of the stockpile (before filling from about elevations
    15.5 feet (4.7 meters) to 18.5 feet (5.6 meters)) to measure
    settlement of the ground surface across the stockpile.
    Horizontal inclinometers ** were identical to vertical in-
    clinometers, except that the inclinometer sensor was designed
    to operate in the horizontal plane, from west to east, across
    the RLSA.

c. Sondex Units: Three (3) SONDEX*** units were installed to
    measure consolidation throughout the mass of soft to firm
    deltaic clay beneath the RLSA.

d. Pore Pressures Instrumentation: Two (2) types of piezometers
    were used to monitor pore pressures at the RLSA. Eight (8)
    GEONOR**** Open Standpipe Piezometers were positioned near
    each of three SONDEX units, along the centerline at the NORTH,
    MIDDLE and SOUTH cross-sections. In addition to the Standpipe
    Piezometers, eight (8) SINCO***** Pneumatic Piezometers were
    located along the centerline of the stockpile, midway between
    the NORTH and MIDDLE cross-sections and midway between the
    MIDDLE and SOUTH cross-sections. Also, four (4) SINCO
    Pneumatic Piezometers were located 65 feet (20 meters) east of
    the stockpile centerline at the NORTH, MIDDLE and SOUTH cross-
    sections.

*   Slope Indicator Inclinometer Sensor Model 50325 EM
**   Slope Indicator Inclinometer Sensor Model 50329 EMV
***   Slope Indicator Sondex Unit, Model 508XX and 50819
****   GEONOR Standpipe Piezometer, Model M-206
*****   SINCO Pneumatic Piezometer, Model 51481X

HARDWARE INSTALLATION PROCEDURES

The following section outlines general procedures used to install
principal pieces of instrumentation equipment. A detailed equipment list, including installation check lists, may be obtained
from Southern Earth Sciences, Inc.

Procedures for Installing Vertical Inclinometer Casing: Installation
generally consisted of preparing and inspecting sufficient sections
of inclinometer casing for installation within a 90 foot (27 meters)
deep boring (typically 16 sections of casing) and installing a
quick-connect grout plug into the bottom end of one length of casing.
As each drill hole was completed, preassembled sections of inclinometer casing (with the quick-connect grout plug at the bottom end)
were inserted into the cased hole until the entire string of 16
pieces were completely assembled within the drill hole. During
assembly, the sealed inclinometer casing was gradually filled with
water to overcome buoyancy effect of groundwater and special care
was used to maintain parallel alignment of internal casing grooves
necessary for proper probe operation.

With the inclinometer casing fully assembled, within the borehole,
the inclinometer sensor was placed into the string to ensure
correct groove alignment in both tracks. Following the assembly
check, a grout pipe was inserted into the inclinometer casing and
engaged to the quick-connect coupling at the bottom of the casing
and grout was pumped through the grout pipe, filling the annulus
between the inclinometer casing and the sides of the borehole,
until grout overflowed at the top of the drill casing. Finally,
the steel casing was extracted from the ground. Figure 5 shows a
sketch of a typical inclinometer assembly.

Procedures for Installing Horizontal Inclinometer Casing: Installation of horizontal inclinometer casing consisted of selecting
the EAST-WEST cross-section location and placing an assembled
string of casing (with grooves properly aligned) into a two foot
(0.6 meters) deep trench and backfilling around the in-place
casing with natural soils.

Procedure for Installing SONDEX Casing: Installation of SONDEX
casing consisted of preassembling about 100 feet (30.06 meters) of
1.25-inch (3.18 cm) PVC belled pipe inside a 100 foot (30.06 meter)
length of 2-inch (5.08 cm) length of flexible SONDEX tubing.
Female ends of the belled pipe were designated as the downward ends
of the string. Next, eight extension anchors were placed over the
flexible tubing and positioned approximately 10 feet (3.06 meters)
apart, starting 2.0 feet (0.6 meters) from the bottom of the casing.
The SONDEX probe was then inserted into the belled pipe to verify
that readings were possible at each sensing ring (anchor) location
before sensing rings were secured in place. After the instrument
check, a bottom assembly check valve was placed over the bottom
of the flexible tube/PVC pipe assembly.

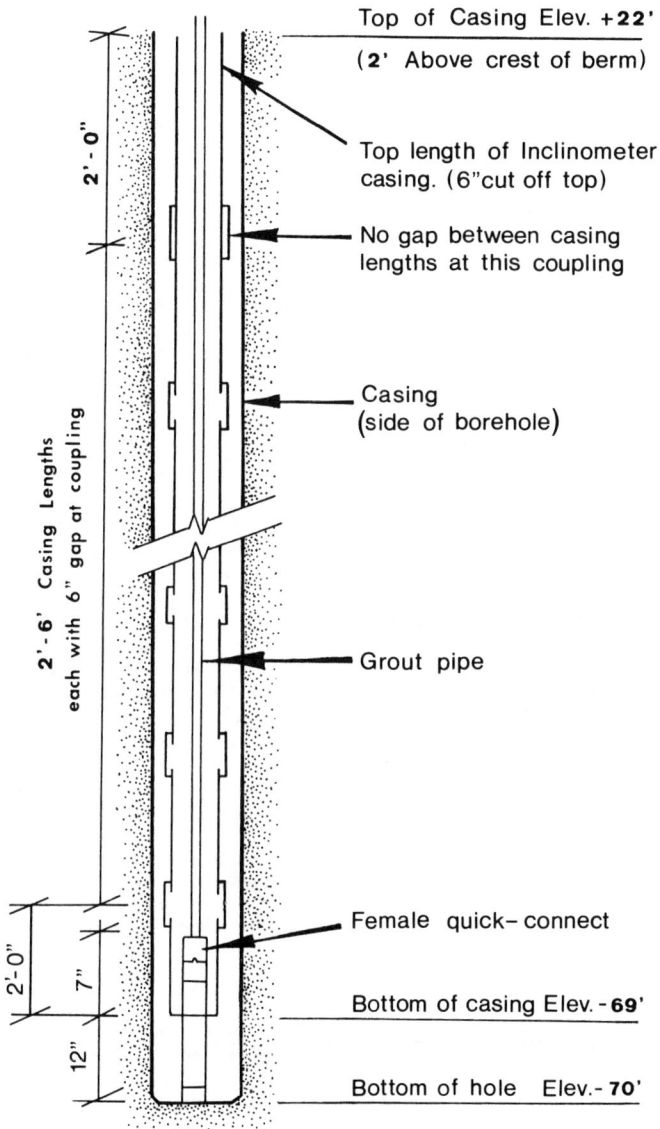

Top of Casing Elev. **+22'**

(**2'** Above crest of berm)

Top length of Inclinometer casing. (6"cut off top)

No gap between casing lengths at this coupling

Casing (side of borehole)

Grout pipe

Female quick–connect

Bottom of casing Elev.-**69'**

Bottom of hole   Elev.-**70'**

2'-0"

2'-6' Casing Lengths each with 6" gap at coupling

2'-0"

7"

12"

FIGURE 5 - SKETCH OF VERTICAL INCLINOMETER CASING

Following assembly of the casing, a cased borehole was drilled 15
feet (4.6 meters) into the deep sand stratum using "HW" casing.
Using a stainless steel cable support system, the SONDEX tubing was
gradually lowered into the cased borehole until the bottom of the
assembly of the SONDEX casing rested on the bottom of the drill hole.
When necessary, water was placed inside of the belled pipe assembly
to overcome buoyancy of the groundwater and the annular space between the flexible tubing and the belled pipe was filled with oil.
Finally, the tubing was cut so that the belled pipe extended about
2.0 feet (0.6 meters) above the top of flexible tubing. With the
SONDEX tubing in place, the "HW" casing was raised to about 5.0
feet (1.6 meters) above the lowest anchor ring so that the ring could
be expanded and set into place. This process was repeated until
all of the anchor rings had been set, making continuous measurements and keeping records of where the expansion anchors were located during the installation process. Finally, grout was pumped
around the in-place SONDEX tubing system through an internal grout
pipe similar to that used in installation of vertical inclinometer
casing, to set the casing in place and provisions were made to
attach the top, reference flange to a plywood platform, so that
during stockpile construction a system of risers could be used to
keep track of reference elevations and to permit long term utilization. Figre 6 shows a sketch of our typical SONDEX installation.

Procedures for Installing Pneumatic Piezometers: General procedures
for installing pneumatic piezometers in the soft, compressible clay
consisted of drilling and casing a borehole to within 3 feet (1
meter) of the desired piezometer elevation. Next, piezometers were
placed at the end of a string of "B" Drill Rod, lowered into the
cased hole, and pushed into place using "B" rod. With the piezometer in place, the "B" Rod and casing were withdrawn from the borehole while a grout slurry was pumped around the piezometer tubing
within the borehole. The pneumatic tubing was then placed in a
shallow trench extending to the toe of the stockpile. General procedures for installing pneumatic piezometers in sand were similar
to procedures for installing standpipe piezometers as discussed
below.

After installation of all piezometers within a cluster, the pneumatic
tubing was placed in a trench extending laterally to the toe of the
stockpile. Here the tubings were brought together in a common protective terminal. The trench was then backfilled with excavated
materials.

Procedures for Installing Standpipe Piezometers: General procedures
for installing Standpipe Piezometers consisted of connecting a continuous length of polyethylene tubing to the GENOR M-206 piezometer and sliding a 10 foot (3.06 meters) section of "E" Rod over
the tubing down to the GENOR M-206 piezometer body. Next, a borehole was drilled and cased at the approximate mid-point between
ALIDRAINS to 15 feet (4.6 meters) above the proposed piezometer
elevation and the piezometers were pushed into place using a system
of 1 inch (2.54 cm) diameter pipes placed outside of the "E" Rod.

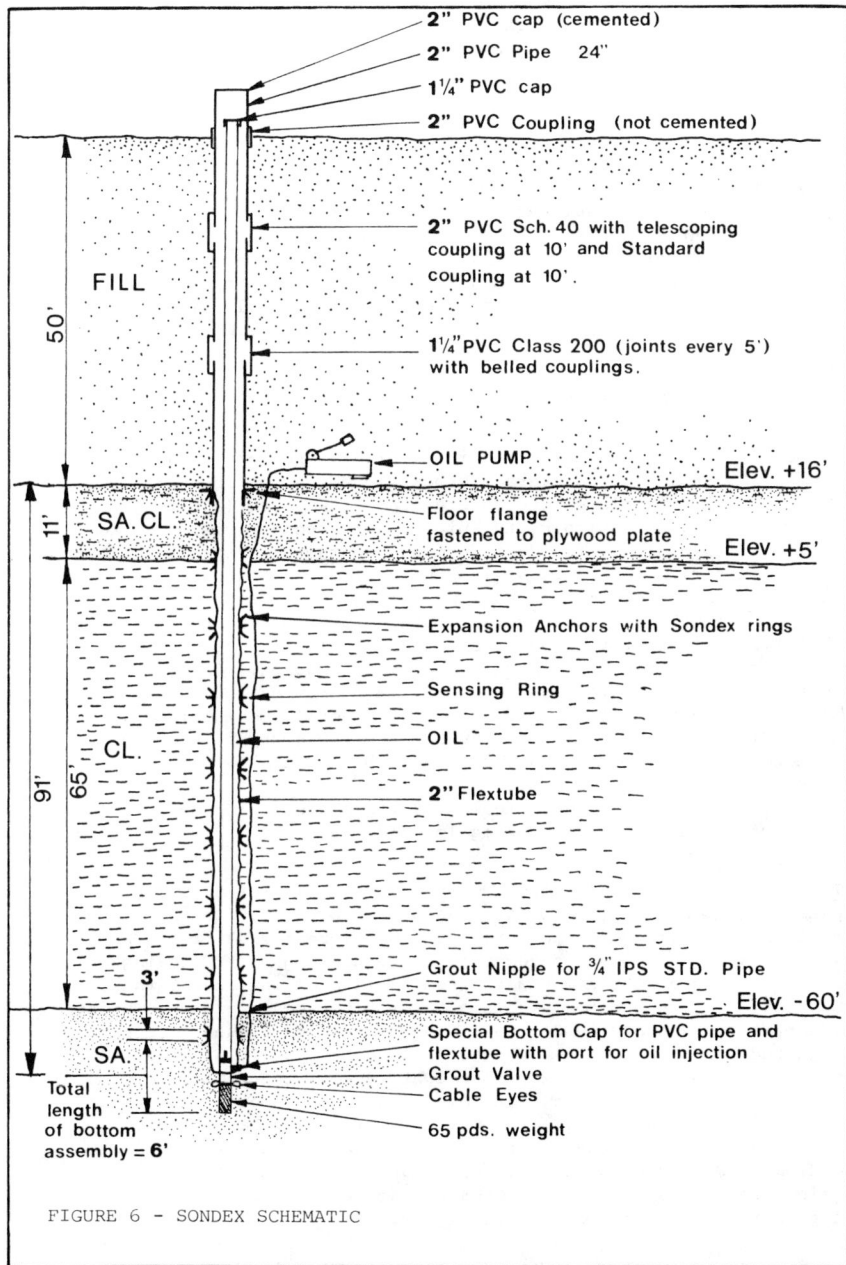

FIGURE 6 - SONDEX SCHEMATIC

A positive head of water was applied to the piezometer to prevent
tip clogging while pushing piezometer. The in-place piezometer
elevation was defined as the mid-height of the porous piezometer
filter.

Following installation of each piezometer, the casing was pulled
and the open hole grouted to ground surface to maintain continuity
within the soil mass. Next, water was placed into the polyethylene
tubing until its level was about 6 feet (1.8 meters) above ground
surface. Using the capillary reader, measurements were made periodically until the water level stabilized within the system and
plots of time versus water level were used in estimating in-place
soil permeability. Provisions were made to measure the elevations
to the top of each piezometer pipe and to periodically extend each
pipe as fill was placed onto the stockpile.

SELECTED INSTRUMENT READINGS

Stage one loading was initiated in May 1981, and completed in
September 1981, with an average stockpile height of about 30 feet
(9.1 meters). Settlement to date averaged two feet (0.6 meters).
Unfortunately, non-uniform stockpile loading conditions made it
difficult to evaluate field performance characteristics of the RLSA
stockpile during this initial loading period. Although Dr. Ladd
requested that stockpile elevations be maintained within a one foot
(0.3 meter) tolerance during first stage loading, Figure 7 shows
that fill heights varied by almost ten feet (3.06 meters) across
the stockpile during the first 145 days of loading. After se-
lective fill redistribution to achieve a fairly uniform stockpile
elevation, centerline fill heights were generally maintained at
about 31 feet (9.5 meters) at the NORTH cross-section, 29.0 feet
(8 meters) at the MIDDLE and 27.5 feet (8.4 meters) at the SOUTH
cross-section. These elevation differences mainly reflect variations in original ground surface elevations along with measurable
settlements that occurred during loading.

Typical Vertical Displacement Measurements: Figures 8, 9, and 10
present typical settlement data based on SONDEX systems and
Horizontal Inclinometer performance. Settlement data from SONDEX
systems were corrected to account for meandering of PVC casings within the fill and settlement data from horizontal inclinometers were
corrected to account for settlements from the west reference points
(starting in January 1982). All figures show increased settlement
magnitudes from north to south across the stockpile, at least partially due to differential fill heights across the stockpile. As
shown on Figure 8, surface settlements at the centerline increased
between the end of the loading period until early March 1982.
Figure 8 also shows that surface settlements measured using SONDEX
equipment were consistently larger than those settlements measured
using the horizontal inclinometer system prior to correcting the
latter for reference point settlements.

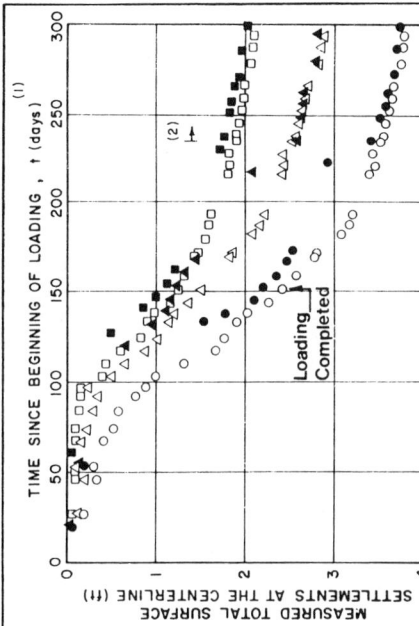

Location	Sondex system		Horizontal inclinometer	
	Symbol	Instrument No.	Symbol	Instrument No.
North	○	S-1	●	I-11
Middle	△	S-2	▲	I-12
South	□	S-3	■	I-13

NOTES: (1) Beginning of loading on May 13, 1981
(2) Settlement data for horizontal inclinometers corrected for reference point settlements (0.30 ft at the North and 0.32 ft at the Middle)

FIGURE 8 – MEASURED TOTAL SURFACE SETTLEMENTS AT THE CENTERLINE VERSUS TIME DURING CONSTRUCTION TO A FILL HEIGHT OF ABOUT 30.0 FEET (9.1 METERS)

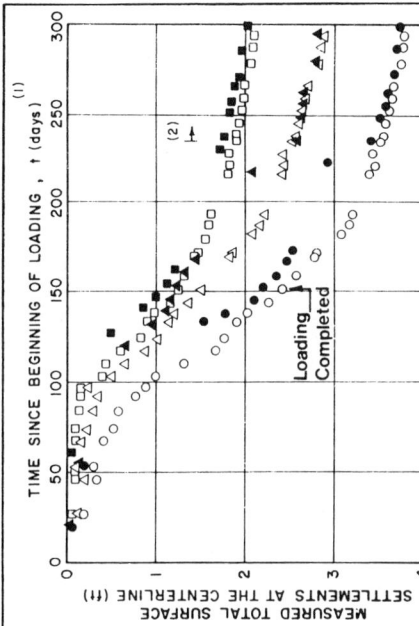

Cross-section			Ground surface elevation measured	Initial ground surface settlement from
North	Middle	South		
○	△	□	at the stockpile centerline	Sondex system
●	▲	■		Horizontal inclinometer
○	△	□	65 ft East of the stockpile centerline	Horizontal inclinometer

NOTES: (1) Corrected for initial ground surface settlement
(2) Beginning of loading on May 13, 1981

FIGURE 7 – ACTUAL FILL HEIGHT VERSUS TIME DURING CONSTRUCTION TO A FILL HEIGHT OF ABOUT 30.0 FEET (9.1 METERS)

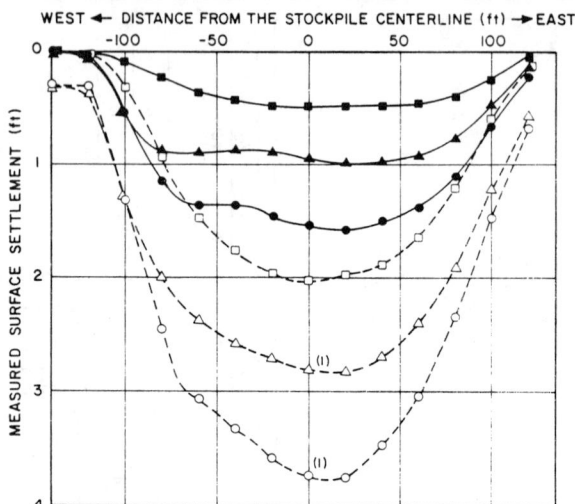

FIGURE 9 - TYPICAL MEASURED SURFACE SETTLEMENTS ACROSS THE RLSA
DURING STOCKPILE CONSTRUCTION TO A FILL HEIGHT OF
ABOUT 30.0 FEET (9.1 METERS)

TABLE 1
SELECTED INITIAL PIEZOMETRIC WATER ELEVATIONS

PNEUMATIC PIEZOMETERS				STANDPIPE PIEZOMETERS			
LOCATION	"P" No.	TIP ELEV. (ft)	PWE (ft)	LOCATION	"SP" No.	TIP ELEV. (ft)	PWE (ft)
	13	14.6	15.5		9	14.3	14.3
65' East	14	-10.2	8.5		10	-11.4	8.0
of C.L.	15	-22.9	9.0	CENTER	11	- 9.1	7.5
	16	-30.9	10.0		12	-23.3	9.0
					13	-30.1	9.0
					14	-30.6	8.0

NOTE: "PWE" = PIEZOMETRIC WATER ELEVATION

FIGURE 11 – TYPICAL MEASURED HORIZONTAL DISPLACEMENTS ALONG THE WEST AND EAST TOES OF STOCKPILE CONSTRUCTION TO A HEIGHT OF ABOUT 30.0 FEET (9.1 METERS)

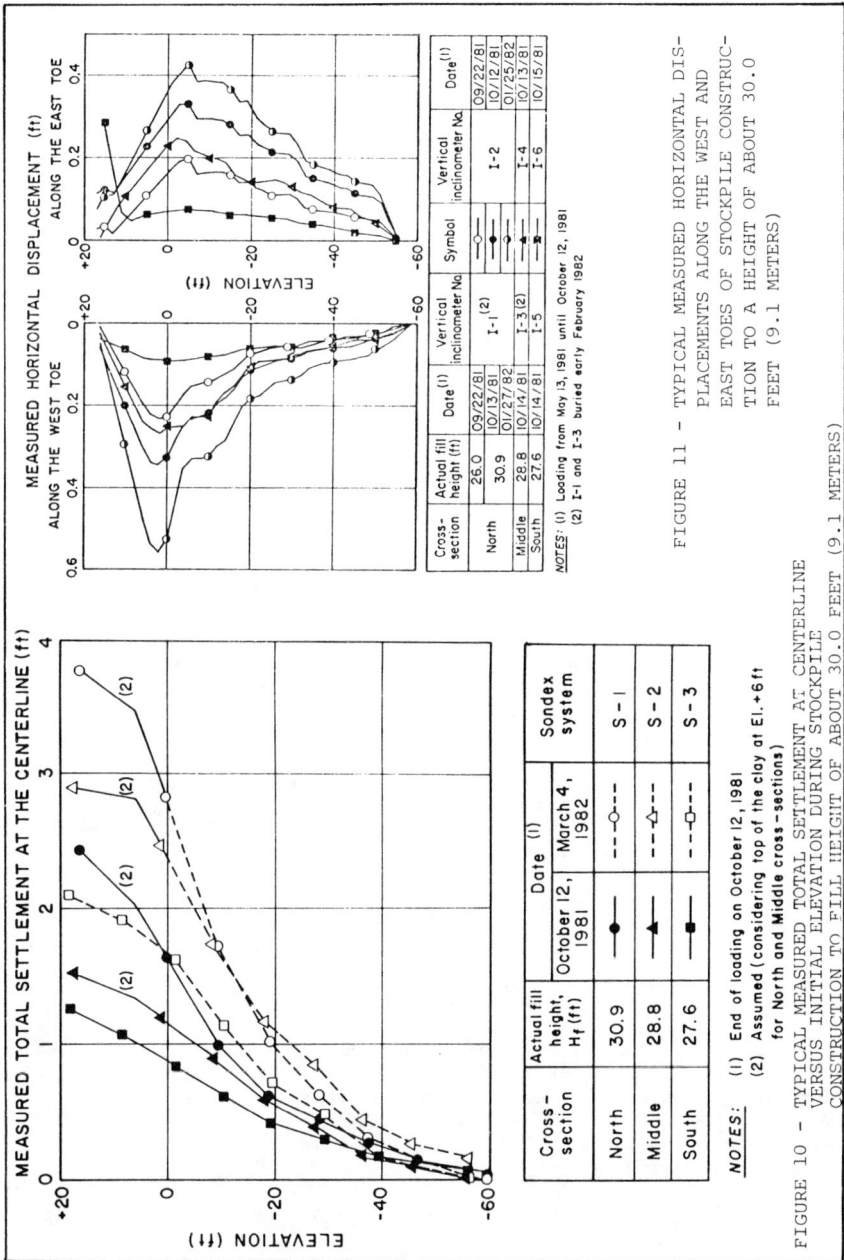

FIGURE 10 – TYPICAL MEASURED TOTAL SETTLEMENT AT CENTERLINE VERSUS INITIAL ELEVATION DURING STOCKPILE CONSTRUCTION TO FILL HEIGHT OF ABOUT 30.0 FEET (9.1 METERS)

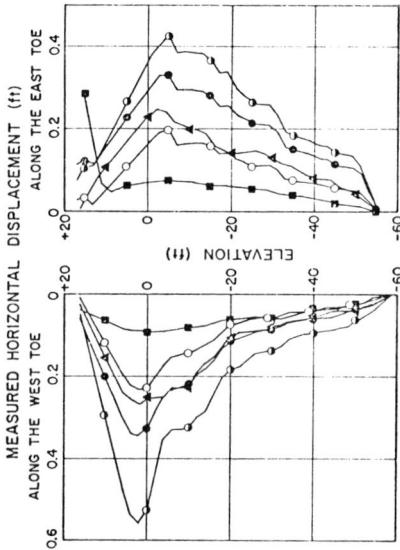

Typical Horizontal Displacement Measurements: Figure 11 represents
a plot of horizontal displacements along the east and west toes of
the stockpile at the end of the loading period for the NORTH, MIDDLE
and SOUTH cross-sections. It also plots displacement of the NORTH
cross-section during loading (late September 1981) and about three
months after completion of loading (late January 1982). As with the
SONDEX readings, horizontal displacements were larger at the north
end of the stockpile compared to the southern end. Plots for the
NORTH cross-section show that horizontal displacements continued to
increase, even after loading was completed.

Typical Measured Pore Pressures: Figures 12 and 13 present plots
of excess pore pressure (Δu) measured in feet of water, versus
time for the MIDDLE cross-section at the centerline and 65 feet east
of the centerline for standpipe and pneumatic piezometers, respectively. Initial piezometric water elevations selected by Dr.
Ladd (1) and presented on Table 1, were used to compute Δu. All
pneumatic piezometer readings were corrected for settlement of the
piezometer tips.

All pore pressure data collected during the loading period were
considered generally erratic when compared with predicted data. At
least three factors may have contributed to this erratic behavior:

1)   Non-uniform loading conditions

2)   Slow response time for standpipe piezometers

3)   Many piezometers were probably located off-center
     between the ALIDRAINS (which were spaced on a 5.0
     foot (1.52 meters) triangular pattern) and hence
     were affected by ALIDRAIN performance.

The basic reason for installing such a large number of piezometers
(52 in all) was a result of our inability to install piezometers
midway between ALIDRAINS. We assumed that only about 33% of the
piezometers would indicate response applicable of the mid-drain
condition. Also, we expected to lose 30% of our piezometers to
malfunctions and/or damage sustained as a result of consolidation
related problems.

SUITABILITY AND RELIABILITY OF INSTRUMENTATION

In general, horizontal and vertical inclinometer systems worked
well and remained operable three years after installation. Unfortunately, SONDEX and piezometer systems were not as reliable
and a high percentage of instruments failed within one to two
years after installation. The following section briefly summarizes
our experience with each type of equipment.

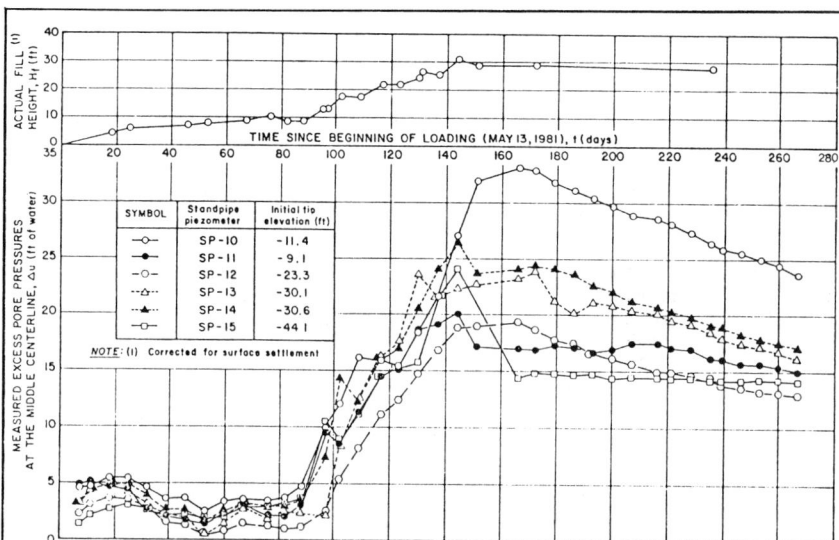

FIGURE 12 - MEASURED EXCESS PORE PRESSURES VERSUS TIME DURING
STOCKPILE CONSTRUCTION - CENTERLINE, MIDDLE CROSS-
SECTION

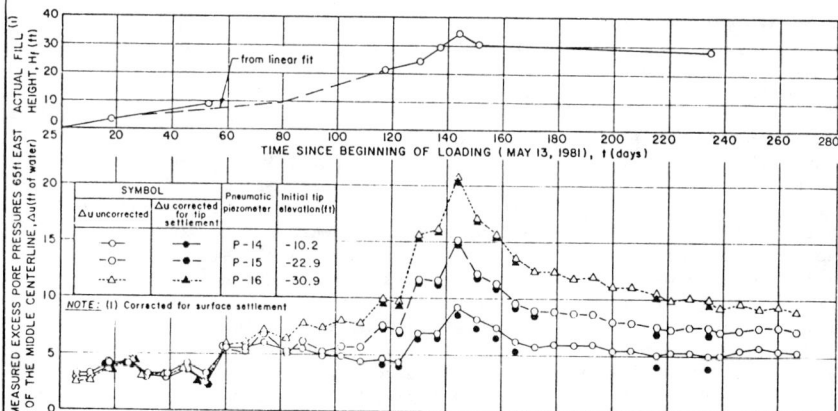

FIGURE 13 - MEASURED EXCESS PORE PRESSURES VERSUS TIME DURING
STOCKPILE CONSTRUCTION - 65.0 FEET EAST OF THE
CENTERLINE, MIDDLE CROSS-SECTION

Horizontal and Vertical Inclinometers: The horizontal and vertical inclinometer systems have worked exceptionally well and all
are still operable. The use of telescopic couplings to accommodate
large settlements resulted in additional benefits. The telescopic
couplings made repair a simple task of excavating five foot sections,
unscrewing eight button-head socket screws and replacing section
damaged during loading and/or reclaiming of the RLSA. In our
judgment, consideration should be given to telescopic inclinometer
casing even when settlement is minimal because of the ease with
which repairs can be made.

Sondex Units: Failure of the single Sondex unit during the first
year of operation was, in our judgment, directly related to the
inability of slip couplings and casing to tolerate settlements
and/or lateral movements during periods of limestone removal from
the RLSA. Plant operations which required reclaiming of the limestone to original ground elevations, encroached onto SONDEX locations, inducing lateral movements within the soil-limestone mass.
Considering the close tolerance existing between the Sondex probe
and the casing during testing, slight bending of the casing made
the system inoperable resulting in failure of the system.

Piezometers: During the late stages of loading, an excessive
amount of material (almost 50 feet, (15.2 meters)) was placed on
the north end of the RLSA. This resulted in a near massive
failure with large soil movements occurring very rapidly. Once
this situation was discovered, immediate steps were taken to remove excess material. However, during this period of soil movement, 25% of the pneumatic piezometers failed (all located in this
area). The most likely explanation for their failure was crimping
or shearing of the pneumatic tubing leads. In future installations,
where large soil movements are anticipated, consideration should
be given to placing the pneumatic leads in protective conduit or
extend them up through the stockpile or surcharge.

Six months after initial stockpile loading, approximately four
feet (1.2 meters) of settlement had occurred and 46% of the
pneumatic piezometers and 12% of the standpipe piezometers had
failed. After one year of loading, 4.5 feet (1.4 meters) of settlement had occurred, 75% of the pneumatic piezometers and 100% of
the standpipe piezometers had failed. As previously stated,
failure of the pneumatic piezometers was probably related to
crimping or shearing of the tubing which ran laterally through the
Alidrain sand blanket to protective terminals. Failure of the
standpipe piezometer is, in our opinion, a function of time.
Reading the standpipe piezometers required applying tension to the
.37 inch (0.94 cm) polyethylene tubing. With time, the Swagelock
unions slipped and the systems failed. It should be noted that
even though all of the standpipe piezometers failed within the
first year of operation, we were able to repair 60%. Because this
created a constant maintenance problem, we ultimately decided to
replace all standpipe piezometers with pneumatic piezometers.

ACKNOWLEDGEMENTS

The authors gratefully acknowledge Mr. Robert L. Matoush and Mr. James Hutchenson of Ideal Basic Industries, Inc. for their assistance in obtaining project design data and their continued support and encouragement during preparation of this paper. Appreciation is also due to Dr. Charles C. Ladd and Ms. Laure Noiray for permitting us to use information contained in Ms. Noiray's Masters Thesis prepared for Massachusetts Institute of Technology. The authors are also grateful to Mr. John Dunnicliff, Geotechnical Instrumentation Consultant, for his valuable assistance in development of our field instrumentation program. We would also like to express our appreciation to Mr. Bruce Wilson and Mr. Ben Jordan for drafting the figures for this paper and to Mrs. Brenda Brantley for typing this manuscript.

REFERENCES

1. Ladd, C.C., "Unpublished Interim Reports No. 1, No. 2, No. 3 and No. 4 to Brown & Root, Inc. and Ideal Basic Industries, Inc.", 1979, 1980 and 1981

2. Reed, P.C., Geology of Mobile County, Alabama, Alabama Geological Survey

3. Noiray, Laure, Predicted and Measured Performance of a Soft Clay Foundation Under Stage Loading, Thesis presented in partial fulfillment of the requirement of the Degree of Master in Civil Engineering, Massachusetts Institute of Technology, June, 1982

4. Navdocks DM-7 (1971), "Design Manual-Soil Mechanics, Foundations and Earth Structures", Naval Facilities Engineering Command, Department of the Navy, Washington, D.C.

Sedimentation and Self Weight Consolidation of Dredge Spoil

T. W. Lin[1] and R. A. Lohnes[2] M. ASCE

Abstract

Settlement test results on dredge spoil slurries suggest that self weight consolidation begins when an interface forms in the settlement column and the slurry reaches a critical concentration as a result of the solid particles becoming locked into a three dimensional lattice. The rate of settlement for a specific material is characterized by a coefficient of consolidation and a graphical method for determining the coefficient is outlined.

Introduction

The study of the settling behavior of dredged materials has become a concern of soil engineers because of the need for designing efficient dredge spoil disposal sites.

The flocculent settling phenomenon was first described by Coe and Clevenger (2) who described the settling behavior of ore pulp by the concept of solids discharging capacity. Kynch (6) derived a mathematical expression for the settling of suspensions assuming that the settling rate of particles depends on the local solids concentration. Based on the Kynch theory, Talmage and Fitch (10) showed that the concentration versus fall velocity relationship of particles, as required in thickener design, can be obtained from a single settling test.

Fitch (3) later classified the settling behavior of suspensions into four categories according to the degree of interparticle cohesiveness and solid concentration: 1) discrete settling, 2) flocculent settling, 3) zone settling, and 4) consolidation. Several researchers (3,4,8) found that only in low concentration suspensions did the material settle as individual flocks. In zone settling the slurries are at high concentrations, and the suspended solids agglomerate and settle as a coherent mass with no jockeying for position by the particles. If all the settling particles are locked into a three-dimensional lattice when zone settling starts, it is reasonable to conclude that consolidation process also begins at this moment. The lattice structure implies effective stresses between the solid

---

[1]Geotechnical Engineer, Miller Consulting Engineers, Riverton, Wyoming 82501

[2]Professor of Civil Engineering, Iowa State University, Ames, Iowa 50011

particles and suggests that zone settling can be described and analyzed
according to self weight consolidation theory.

Self Weight Consolidation Theory

Large strains are generally associated with the consolidation of
slurries. Therefore, conventional one-dimensional consolidation theory
is not adequate for describing the settling behavior. Gibson et al.
(5) formulated the general equation governing large strain consolidation. Later, Lee and Sills (7) obtained an analytical solution to
the problem in which consolidation is caused solely by the material's
own weight. The solution indicates that a slurry layer, having an
uniform initial concentration, c_i, or uniform void ratio, e_i, will end
with a linear, final void ratio distribution from bottom to top after
100% consolidation; and the void ratio, e, at any time, t, is:

$$e(z,t) = e_i - \beta \left[z_1 - z - 2z \sum_{ln} \frac{\cos(m\pi z/z_1)}{m^2\pi^2} \exp\left(-\frac{C_F m^2 \pi^2 t}{z_1^2}\right)\right]$$

$$(1)$$

The corresponding pore pressure, u, distribution in the slurry is:

$$u(z,t) = 2(\rho_s - \rho_f) \cdot z_1 \sum_n \frac{(-1)^n \sin[m\pi(1 - z/z_1)]}{m^2\pi^2}$$

$$\times \exp\left(-\frac{C_F m^2 \pi^2 t}{z_1^2}\right) \qquad (2)$$

where

ρ_s = unit weight of solids

ρ_f = unit weight of fluid

z = material coordinate, which labels only the solids

z_1 = actual material height

β = slope of final void ratio distribution

e_i = initial void ratio corresponding to c_i

C_F = coefficient of consolidation

n = 0,1,2...

m = 1/2(2n + 1)

Slurry height at any time, $h(t)$, can be expressed in terms of material coordinate z by integrating Eq. (1) from 0 to the total material height z_1; that is

$$h(t) = \int_0^{z_1} [1 + e(z,t)]dz \tag{3}$$

According to Eq. (1), the void ratio at the slurry surface, where $z = z_1$, remains unchanged during consolidation. However, Been and Sills[1] (1) found that the void ratio at the top of the slurry decreases to a value of e_o at 100% consolidation. In order to accommodate the theory to this real situation, Been and Sills assumed that the void ratio difference is the result of an imaginary overburden layer (Fig. 1a). The solution is still applicable if the actual material height z_1 is replaced by the modified material height z_o, where

$$z_o = z_1 + (e_i - e_o)/\beta \tag{4}$$

Equation (1) then becomes:

$$e(y,T') = e_i - \beta z_o \left[1 - y - 2\sum_n \frac{\cos(m\pi y)}{m^2\pi^2} \exp(-m^2\pi^2 T') \right] \tag{5}$$

and Eq. (2) turns out to be:

$$u(y,T') = 2(\rho_s - \rho_f)z_o \sum_n \frac{\cos(m\pi y)}{m^2\pi^2} \exp(-m^2\pi^2 T') \tag{6}$$

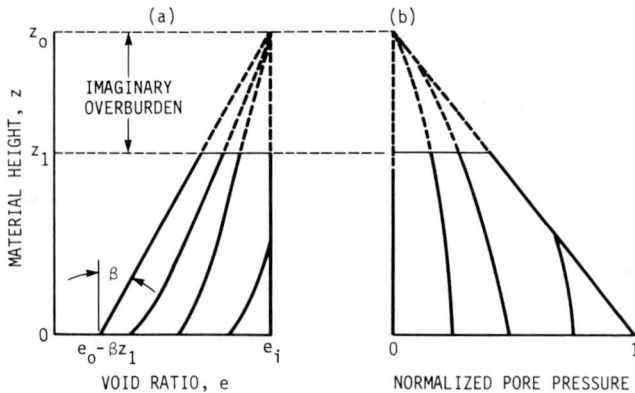

Figure 1. Void ratio and pore pressure distributions according to Been and Sills modification of self weight consolidation theory.

where $y = z/z_o$, and $T' = (C_F t)/z_o^2$. Equations (5) and (6) are shown in Figs. 1a and 1b respectively and are valid only for $0 < z < z_1$.

Equation (6) and Fig. 1b show that the excess pore pressure on the slurry surface is $(\rho_s - \rho_f)(z_o - z_1)$ at the beginning of consolidation but reduces to zero after 100% primary consolidation. In actual settling tests, however, no excess pore pressure exists on the surface at any time. Hence, Been and Sills further modified Eq. (6) to be:

$$u_1(z,T') = u(z,T') - u(z_1,T')$$

where $0 \leq z \leq z_1$, or as:

$$u_1(y,T') = 2(\rho_s - \rho_f)z_o \sum_n \frac{\exp(-m^2\pi^2 T')}{m^2\pi^2} (\cos m\pi y - \cos m\pi r)$$

(7)

in which, $r = z_1/z_o$, $0 \leq y \leq r$, and $u_1(y,T')$ is the excess pore pressure distribution in real soil. Based on Eq. (7), the degree of consolidation for the modified case, $S_m(T')$ can be expressed as:

$$S_m(T') = \frac{\int_0^r u_1(y,0)dy - \int_0^r u_1(y,T')dy}{\int_0^r u_1(y,0)dy}$$

$$= \frac{\sum_n \left\{ \left[\frac{\sin(m\pi r)}{m^3\pi^3} - \frac{r\cos(m\pi r)}{m^2\pi^2} \right] \cdot [1 - \exp(-m^2\pi^2 T')] \right\}}{\sum_n \left[\frac{\sin(m\pi r)}{m^3\pi^3} - \frac{r\cos(m\pi r)}{m^2\pi^2} \right]}$$

(8)

$S_m(T')$ varies with time factor, T', as well as r. Figure 2 shows the plots of $S_m(T')$ versus $\sqrt{T'}$ relationship for different r values.

Coefficient of Consolidation, C_F

The coefficient of consolidation is a parameter which relates the theoretical time factor to real time and the modified material height, and thus applies the consolidation theory to predict the behavior of a specific soil layer.

The coefficient of consolidation, C_F, is calculated according to

$$C_F = \frac{T' z_o^2}{t}$$

(9)

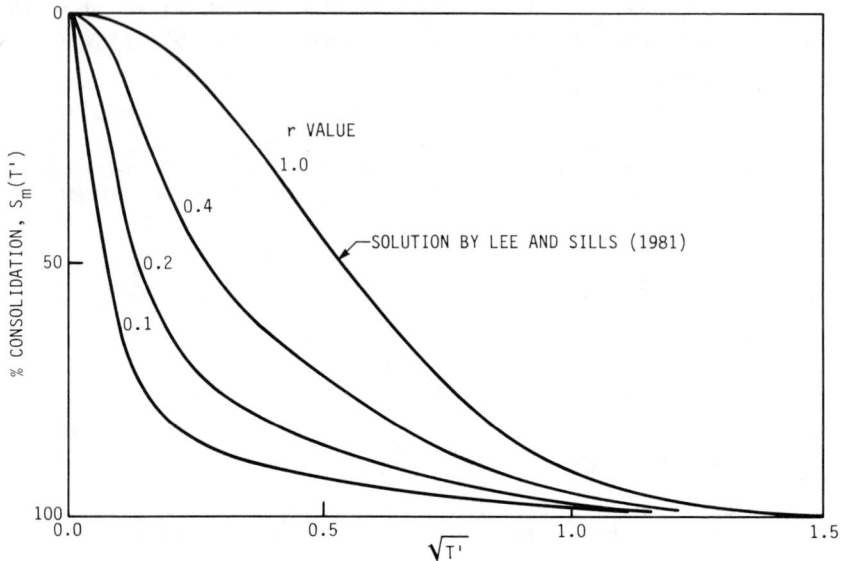

Figure 2.   The $S_m(T')$ vs. T' plots for different r values.

Recalling Eq. (4), z_o can be expressed as:

$$z_o = z_1 + (e_i - e_o)/\beta \qquad (4)$$

The quantities z_1 and e_i are related to the initial test conditions and are easily determined. The parameters e_o and β, however, are used to describe the final state of the slurry and require more analysis.

    Been and Sills plotted e_o and β versus the initial concentration, c_i, of the slurry, but they do not clearly define when self weight consolidation actually begins. Our interpretation is that self weight consolidation begins at the time the interface forms in the settlement column. At this moment, the particles are locked into a flexible, three dimensional lattice and a critical concentration, c_c, is reached. Thus it seems more appropriate to relate e_o and β to c_c. Slurries prepared at various concentrations may have their initial concentrations above or below the critical concentration for that particular material. All the settling column tests with initial concentration, c_i, lower than the critical concentration will start the consolidation process when the critical concentration, c_c, is reached and thus result in only one set of e_o and β. Any test with c_i higher than c_c will produce a set of e_o and β values corresponding to its initial concentration, because consolidation begins immediately after the test starts. Conceptually, there exists a limiting maximum concentration, c_m, at which a normally consolidated soil stratum is formed and no self weight consolidation occurs. Thus, if $c_i > c_m$, the uniform initial concentration profile is maintained because no self weight consolidation occurs, or $\beta = 0$ and $e_o = e_i$. Between the two limits c_m and c_c, the values e_o and β are assumed to be functions of c_i only. Many functions satisfying the conditions at the limiting concentration, c_m, were evaluated, but a logarithmic function gave the best fit. Thus, models for describing the variation of e_o and β between c_m and c_c are assumed to be:

$$e_o = e_m + k_2 \ln(c_m/c_i)$$

$$\beta = k_1 \ln(c_m/c_i) \qquad c_c \leq c_i \leq c_m \qquad (10)$$

where

e_m = void ratio corresponding to the limiting concentration and can be related to c_m by:

$$e_m = (G_s \cdot \gamma_w/c_m) - 1$$

k_1 = proportional constant having a unit of 1/length in order to satisfy the dimension equality in Eq. (4)

k_2 = dimensionless constant

G_s = specific gravity of solids

γ_w = unit weight of water

Theoretically, H_{100} is related to e_o and β as:

$$H_{100} = \int_0^{z_1} [1 + e(z,\infty)]dz = (1 + e_o)z_1 - \frac{1}{2}\beta z_1^2 \qquad (11)$$

Experimentally, the quantity H_{100} for a zone settling test can be taken as the slurry height corresponding to the later horizontal portion of the slurry height versus $\sqrt{t}$ curve. In order to solve the three unknown quantities c_m, k_1, and k_2 in the model, at least three zone settling tests on the same material should be performed. After determining these constants, the e_o and β values for any specific tests are obtained by Eq. (10), and the corresponding z_o is calculated from Eq. (4). C_F can then be calculated from Eq. (9).

Settling Tests

The settling behavior of dredge spoil slurries was studied by performing settling tests in a 140-mm (5.5-in.) inside diameter and 1.83-m (72-in.)-high plexiglass column with lake bottom sediment samples collected from the proposed dredging site at Lake Panorama,

Iowa. The sediment has the following properties: grain-size distribution, 37% clay and 63% silt; natural water content, 78.9%; in-situ
unit weight, 8.64 kN/m^3 (55pcf); organic content 5.6% by weight;
specific gravity, 2.74; liquid limit, 59.4; plasticity index, 24.7;
soil classification, MH. The settling tests follow procedures proposed
by researchers at the Waterways Experiment Station (9). The sediment
sample was mixed thoroughly with tap water to form a uniform slurry of
desired concentration. Then, the slurry was pumped into the settling
column, while air was supplied from the bottom of the column to prevent
the settlement of rapidly falling particles during filling. After
filling, the air supply was stopped, and the settling test began. The
settling behavior of the suspension was observed carefully, and when a
sharp interface between the sediment laden water and the clear supernatant water formed, its height was recorded. After that, the slurry
height was observed at regular intervals and the interface height
versus time curves were plotted. The above procedure was repeated for
slurries of different initial concentration. In order to obtain the
concentration profiles at different times and to study the influence of
sample extraction on the settling behavior of the slurry, another
series of settling tests were performed but with sample withdrawal at
regular time intervals through sample ports along the length of the
column. Hypodermic needles were used to extract the slurry samples.

The tests and experimental conditions are listed in Table 1.
Tests designated by "N" were performed without sample withdrawal,

Table 1. List of test conditions.

Type of Test	Experiment	Initial Concentration g/l	Initial Height m
Without sample extraction	N-1	75.3	1.797
	N-2	101.0	1.797
	N-3	147.0	1.797
	N-4	191.5	1.803
	N-5	226.5	1.810
With sample extraction	W-1	76.0	1.791
	W-2	99.1	1.791
	W-3	125.0	1.791
	W-4	190.0	1.791

Note: 1 m = 3.28 ft

whereas tests designated by "W" are those in which samples were extracted.

Test Results

Observations

At the beginning of the test, the suspension starts to settle and the upper portion becomes less concentrated than the initial concentration, c_i. The supernatant water above the falling suspension is turbid partly because of finer particles still suspended and partly because of the upward moving particle flux. A zone of darker water exists, but no sharp interface is apparent. Subsequently, the jostling particle flux quiets down, the supernatant water clears up, and a sharp interface forms. From then on, all the particles appear to be locked into a three-dimensional lattice and settle as a mass and consolidation begins.

Figure 3 compares the settling behaviors of tests N-1, N-3, and N-5, in which the observed slurry height, H, is plotted versus square root of time, $\sqrt{t}$. In general, the lower the initial concentration, the faster the settling rate. If the test starts with a high c_i, as in test N-5, the interface immediately forms, and the settling curve shows an early convex upward portion. In the low c_i test (N-1), however, a period of time has elapsed before the interface forms, and the early convex portion is flattened. After the early settling period, all curves show a convex upward transition period followed by a nearly linear portion, in turn followed by a concave upward portion, and

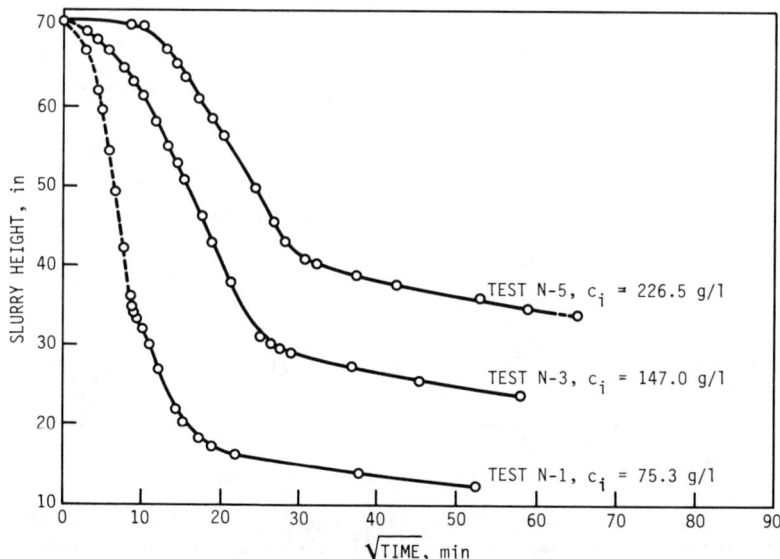

Figure 3. Comparison of settling behaviors of slurry at different initial concentrations (1 in. = 0.0254 m).

finally a very flat curve. The shapes of the experimental settlement
curves (Fig. 3) are similar to the shapes of the theoretical self
weight consolidation curves (Fig. 2).

Critical Concentration, c_c

The settling test results reveal that when the initial concentration, c_i, is higher than 147 g/l, the interface forms immediately after
the test starts (N-4 and N-5). If self weight consolidation starts at
the moment the interface is formed, a critical concentration, c_c, must
exist in the settling process when the slurry begins to consolidate.
The critical concentration should depend on the nature of the material
and the settling environment, especially electrolyte concentration of
the suspending medium.

The test starts with an initial concentration, c_i, and initial
height, H_i; therefore the average critical concentration of the slurry
is $c_i H_i / H_c$ when the interface forms and the slurry height is H_c.
Table 2 summarizes the calculated values of critical concentration for
N-1, N-2, and N-3. The three calculated c_c values are similar and
average about 148 g/l, which is close to the lowest concentration
slurry in which the interface formed immediately.

Table 2. Calculation of critical concentration c_c.

Test	c_i g/l	H_i m	H_c m	c_c g/l
N-1	75.3	1.797	0.888	152.4
N-2	101.0	1.797	1.255	144.7
N-3	147.0	1.797	1.768	149.4
N-4	191.5	1.803		
N-5	226.5	1.810		

Note: Critical concentration is calculated according to $c_c = c_i H_i / H_c$,
       1 m = 3.28 ft

Effect of Sampling on Settling Behavior

Tests N-4 and W-4, both with initial concentration above the critical concentration, are compared in Fig. 4. The interface forms immediately after the test starts, and the slurry in W-4, the test with
sample extraction, shows a much higher settling rate than that in the
test without sample withdrawal, test N-4.

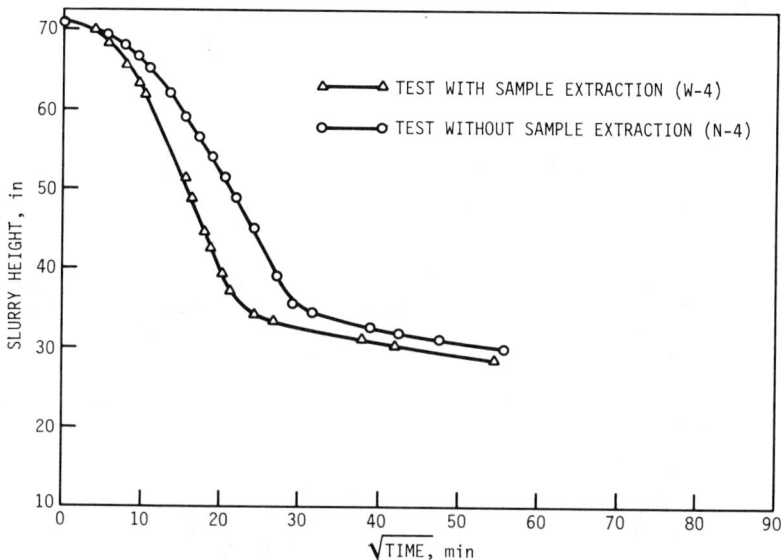

Figure 4. Comparison of tests with and without sample extraction
(c_i = 190 g/l, 1 in. = 0.0254 m).

This phenomenon can be explained by the consolidation mechanism. After the interface has formed, the system is in self weight consolidation, where excess hydrostatic pore pressure has developed. The insertion of hypodermic needles creates passageways for pore fluid to escape and aids the dissipation of excess pore pressure, thereby accelerating the consolidation rate. Tests on comparable samples below the critical concentration indicated that sample withdrawal did not affect the settling behavior until the interface formed. Before the interface forms, particles are in sedimentation, and no excess pore pressure exists; thus, the sampling procedures have little influence on settling behavior prior to interface formation.

Curve Fitting Method for Determining C_F

The degree of consolidation, $S_m(T')$, for the modified self weight consolidation theory is given by Eq. (8) and can be simplified as:

$$S_m(T') = \frac{\sum_n \left\{ \left[\frac{\sin Mr}{M^3} - \frac{r \cos Mr}{M^2} \right] \cdot [1 - \exp M^2 T')] \right\}}{\sum_n \left[\frac{\sin Mr}{M^3} - \frac{r \cos Mr}{M^2} \right]} \quad (12)$$

where

$$M = m\pi = \frac{1}{2}(2n + 1)\pi$$

$$n = 0, 1, 2 \ldots$$

Thus, $S_m(T')$ is a function of both time factor, T', and the ratio r. For a specific test, z_1 and z_0 and therefore r, can be obtained by the approach discussed in the previous section. Knowing the value of r, a theoretical $S_m(T')$ versus $\sqrt{T'}$ curve can be plotted.

The $\sqrt{t}$-fitting method can be employed to relate the theoretical curve to the zone settling curve for the specific test. To illustrate this method, the results of test N-3 are used as an example. The observed slurry height for this test is plotted against $\sqrt{t}$ in Fig. 5.

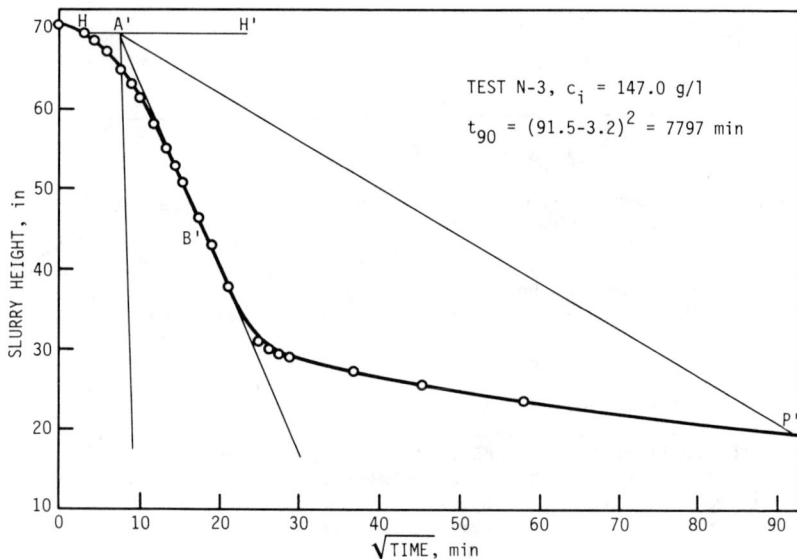

Figure 5.   t-fitting method applied on the settling curve of test N-3 (r = 0.22, 1 in. = 0.0254 m).

Because r value for this test is about 0.22, the theoretical $S_m(T')$ versus $\sqrt{T'}$ curve for r = 0.22 is constructed from Eq. (12) and shown in Fig. 6. The following procedures are performed to obtain both T' and t for 90% self weight consolidation of the slurry:

1.   On the theoretical curve (e.g., Fig. 6) locate the inflection point, B, and draw a tangent line through B to intersect the 0% horizontal line at point A.

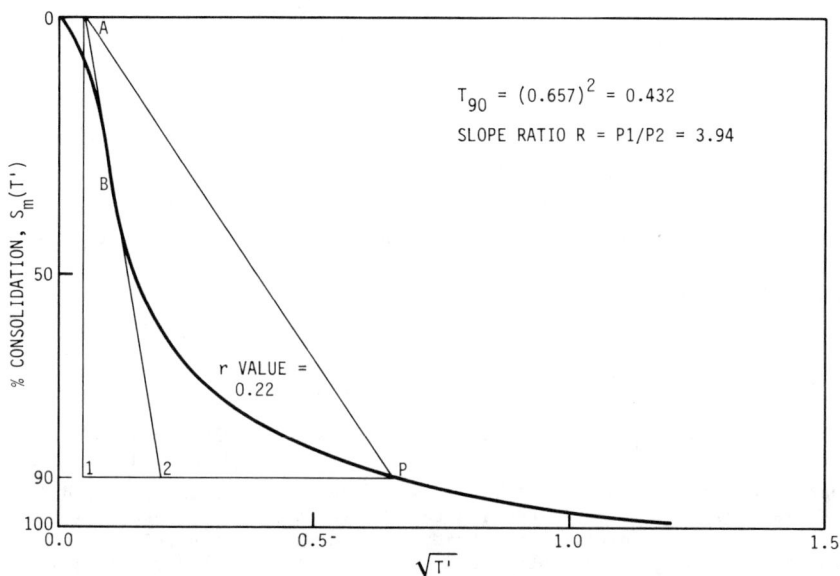

Figure 6.   t-fitting method applied on the theoretical $S_m(T)$ vs. T curve with r = 0.22.

2. Find the 90% consolidation point, P, on the theoretical curve. The time factor for 90% consolidation is obtained by reading the T'-coordinate for point P. For r = 0.22,

$$T'_{90} = (0.657)^2 = 0.432$$

3. Connect points A and P and estimate the ratio, R, of the slope of line AP to that of the tangent line AB. In this example, R = 3.94.

4. On the experimental zone settling curve (e.g., Fig. 5) locate the starting point of self weight consolidation, H, (i.e., the height at which the interface forms) and draw a horizontal line HH' through it. Line HH' is the experimental 0% consolidation line.

5. Draw a tangent line through the inflection point, B', of experimental settling curve to intersect line HH' at point A'.

6. From point A' on the experimental curve, draw a straight line A'P', having a slope R times the tangent line A'B' to intersect the settling curve at point P'. Experimental point P' corresponds to the theoretical time factor when the slurry reaches 90% self weight consolidation, and the corresponding time is t_{90}. In the low c_i test, the time required for slurry

to reach the critical concentration should be subtracted from the obtained t_{90} to yield a modified one. In other words, the origin of the height versus $\sqrt{t}$ plot needs to be adjusted to eliminate the time for sedimentation. In this example, the modified t_{90} = 7797 min.

After z_o, T'_{90}, and t_{90} are determined, the coefficient of consolidation, C_F, for the first test can be calculated from Eq. (9). The value of C_F should vary with initial concentration as well as with the nature of the material and electrolyte concentration of the suspending medium.

Application of Curve Fitting Method
of Determining C_F to Test Results

The results of the five settling tests, N-1 to N-5, are used for calculation of C_F according to the previously described method. Before that method can be applied it is necessary to determine the modified material height, z_o, corresponding to each test. Because tests N-1, N-2, and N-3 have initial concentrations lower than the critical concentration, these three tests yield only one set of e_o and β values corresponding to c_c of 148 g/l. Three simultaneous equations can be set up by substituting experimental values of H_{100} and z_1 for each of the three tests in Eq. (11). Only two of these equations are needed to solve for e_o and β; therefore the third equation can be used as a check. The equations for tests N-1 and N-2 yield e_o = 5.3 and β = 0.90 in.$^{-1}$. Substituting these values in the equation for test N-3 results in a calculated value of H_{100} = 0.439 m. The observed value of H_{100} for test N-3 is 0.478 m, so the values are in fairly good agreement.

Two additional experimental equations can be generated by substituting the previously determined values of e_o and β in Eq. (10). By combining Eqs. (10) and (11) and substituting the results of test N-5 into it, a third equation results. The simultaneous solution of these three equations allows the computation of the limiting concentration, c_m, and the constants k_1 and k_2. The values obtained are: c_m = 360 g/l, k_1 = 0.0259 m^{-1}, and k_2 = -1.50. The observed value of H_{100} from test N-4 has not been used in the calculations so far and thus may be used as a check. Using the calculated values of c_m, k_1, and k_2 in the combination of Eqs. (10) and (11), the calculated value of H_{100} is 0.638 m, which compares favorably with an observed value of 0.660.

The values of e_o, β, z_o, and r for each test are shown in Table 3. Note that r increases as the initial concentration, c_i, increases.

Figure 2 shows the theoretical relationship between the degree of consolidation and the square root of the time factor for various r

Table 3. Summary of the conditions and results of zone settling tests.

Test	c_i g/l	H_i m	z_1^a mm	H_{100} m
N-1	75.3	1.797	49.39	0.267
N-2	101.0	1.797	66.24	0.338
N-3	147.0	1.797	96.41	0.478
N-4	191.5	1.803	126.03	0.660
N-5	226.5	1.810	149.58	0.833

Test	e_i	e_o	β mm^{-1}	z_o mm	r
N-1	17.5	5.3	21.0	391	0.13
N-2	17.5	5.3	21.0	409	0.16
N-3	17.5	5.3	21.0	439	0.22
N-4	13.3	5.7	16.2	427	0.30
N-5	11.1	5.9	11.9	429	0.35

Note: z_1 is calculated using G_s = 2.74, 1 m = 3.28 ft, 1 mm = 0.039 in.

values. Note that as the r value decreases, the initial convex upward portion of the curve flattens and becomes more nearly vertical. Figure 3 shows the experimental curves for three different initial concentrations which represent three different r values. These experimental curves show the same trend as the theoretical curves. The calculated r value allows the determination of T'_{90}, which with z_o is used in Eq. (9) to calculate C_F for the five tests. The values for t_{90}, T'_{90}, and C_F are shown in Table 4. The test which is nearest to the critical concentration, N-3, has the lowest coefficient of consolidation, and there is a trend for increasing values of C_F with the initial concentrations both above and below c_c. This is reasonable because as the initial concentrations decrease below c_c the permeability of the suspensions increases, whereas with concentrations above the critical value the unit weights and thus the effective stress increase with increasing concentration. Higher permeabilities and higher effective stresses increase consolidation rates.

Table 4.   Calculation of the coefficient of consolidation, C_F.

Test	c_i g/l	z_o	t_{90} min	T'_{90}	C_F cm^2/min
N-1	75.3	391	497	0.215	0.665
N-2	101.0	409	2052	0.298	0.245
N-3	147.0	439	7797	0.432	0.110
N-4	191.5	427	7744	0.545	0.129
N-5	226.5	429	5730	0.608	0.194

Note: 1 cm^2/min = 0.155 in.^{2}/min.

Conclusions

The following conclusions are drawn from this study:

1. During the settlement of a suspension of solids in water, an interface will form when the suspension reaches a critical concentration. The formation of the interface marks the beginning of self weight consolidation.

2. The settlement rate is characterized by a coefficient of consolidation, which can be determined graphically from the time versus interface height curve.

3. The coefficient of consolidation has its minimum value at an initial concentration which is close to the critical concentration.

4. Sample extraction during the consolidation process increases the consolidation rate.

Acknowledgments

This research was supported by funds provided by the Central Iowa Power Cooperative and their support is gratefully acknowledged.

Appendix I: References

1. Been, K., and Sills, G. C., "Self Weight Consolidation of Soft Soils: An Experimental and Theoretical Study," _Geotechnique_, Vol. 31, No. 4, 1981, pp. 519-535.

2. Coe, H. S., and Clevenger, G. H., "Methods for Determining the Capacities of Slime-Settling Tanks," _Transactions_, American Institute of Mining Engineers, Vol. 55, No. 9, 1916, pp. 356-384.

3. Fitch, B., "Sedimentation Process Fundamentals," Transactions,
    American Institute of Mining Engineers, Vol. 223, 1962, pp. 129-
    137.

4. Gaudin, A. M., and Fuerstenau, M. C., "Experimental and Mathe-
    matical Model of Thickening," Transactions, American Institute of
    Mining Engineers, Vol. 223, 1962, pp. 122-129.

5. Gibson, R. E., England, G. L., and Hussey, M. J. L., "The Theory
    of One-Dimensional Consolidation of Saturated Clays,"
    Geotechnique, Vol. 17, No. 3, 1967, pp. 261-273.

6. Kynch, C. J., "A Theory of Sedimentation," Faraday Society
    Transactions, Vol. 48, 1952, pp. 116-176.

7. Lee, K., and Sills, G. C., "The Consolidation of Soil Stratum,
    Including Self-Weight Effects and Large Strains," International
    Journal for Numerical and Analytical Methods in Geomechanics,
    Vol. 5, 1981, pp. 405-428.

8. Michaels, A. S., and Bolger, J. C., "Settling Rates and Sediment
    Volumes of Flocculated Kaolin Suspensions," Industrial and
    Engineering Chemistry Fundamentals, Vol. 1, 1962, pp. 24-33.

9. Palermo, M. R., Montgomery R. L., and Poindexter, M. E., "Guide-
    lines for Dredging, Operating, and Managing Dredged Material
    Containment Areas," U.S. Army Engineer Waterways Experiment
    Station, Corps of Engineers, Vicksburg, Miss., Technical Report
    DS-78-10, 1978.

10. Talmage, W. P., and Fitch, E. B., "Determining Thickener Unit
    Areas," Industrial and Engineering Chemistry, Vol. 47, No. 1,
    1955, pp. 38-41.

Appendix II: Notation

c_c critical concentration of slurry when self weight consoli-
        dation begins

c_i initial concentration of slurry

c_m maximum concentration of slurry above which self weight con-
        solidates will not occur

C_F coefficient of consolidation

e void ratio at any time

e_i initial void ratio corresponding to c_i

e_o final void ratio at slurry surface

e_m void ratio corresponding to c_m

G_s specific gravity of solids

H_c	height of interface between consolidation slurry and supernatant water
H_i	initial slurry height in test column
H_{100}	slurry height at end of consolidation
k_1, k_2	empirical constants
r	z_1/z_o
$S_m(T')$	percent of consolidation
t	time
T'	theoretical time factor
u	pore pressure at any time
u_1	pore pressure in real soil
z	material coordinate, which labels solids
z_1	actual material height
z_o	modified material height
β	slope of void ratio distribution
γ_w	unit weight of water
ρ_s	unit weight of solids
ρ_f	unit weight of fluid

Investigation of Settlement of Kobe Port Island

Masato Mikasa* and Naotoshi Takada**

1. Introduction

Kobe Port Island is a reclaimed land that occupies an area of 463 ha a few hundreds meters off Kobe city, which was constructed on 11 to 13 m deep seabed in the period of 1966-1980 (6). It is an enormous fill work exceeding 15 m in height in average on a soft alluvial clay stratum 8 to 16 m thick. The vast fill material of disintegrated granite was taken from the Rokko mountains just behind Kobe city, carried by belt-conveyers and pusher barges and dumped into the reclamation site. Since a large amount and long term settlement of the island was considered unavoidable, settlement analysis was conducted before the start of the project mainly for the alluvial clay stratum. The actual settlement of the island, however, exceeded the predicted value markedly when the fill grew up to show its top surface above the sea level.

A new investigation of settlement of the island, therefore, was planned in 1975 to clarify the behavior of both the alluvial and the diluvial strata under the big fill load by means of several newly developed in-situ measuring apparatuses and Mikasa's consolidation theory (1).

In the field survey, a pull-type settlement gauge was newly developed and set at each boundary of the layers down to 55 m below the sea level, where appeared the top of the diluvial clay layer that was considered to be responsible for the unexpected settlement. The settlement gauge was also set at several depths in the alluvial clay layer to investigate its detailed settlement behavior. The distribution of pore water pressure in the alluvial clay layer was also measured with newly devised simple pore pressure gauges. A new type cone penetrometer, moreover, was developed and used to measure the in-situ strength of the alluvial clay, because the undisturbed sampling of soft clay which was in an early stage of consolidation under big fill load was found quite difficult. The unit weight of the fill was measured not only by the ordinary material-replacing method at the fill surface, but also by a large pressure cell buried under a newly placed fill.

The consolidation process of the alluvial clay layer was analyzed by Mikasa's consolidation theory that takes into account the change of the volume compressibility and permeability during consolidation together with the effect of finite strain and selfweight of clay.

The results of analysis could well explain the observed consolidation behavior of alluvial clay stratum, while the behavior of the diluvial strata was not analyzed by the consolidation theory at all, because they apparently showed such substantial settlement that might not be explained by any

*Professor, Civil Engineering Department, Osaka City Univ., Sugimoto Sumiyoshi-ku Osaka Japan
**Associate Professor, Civil Engineering Department, Osaka City Univ., Sugimoto Sumiyoshi-ku Osaka Japan

available theory. Our settlement observation was discontinued at the end of 1976 when a new fill was placed at the site and the ground condition was utterly altered. Later in another survey, the diluvial layers below the scope of the present survey, deeper than 90.7 m, were found to be settling considerably, and the whole state of affairs became clearer (6).

2. Soil Profile

At the present investigation site shown in Fig.1, the filling began in 1968, and the fill surface reached the level K.P.+2 m* in 1976. The data of four past borings shown in Fig.2 were available near the present investigation site, No.14. No.4 and No.8 of these are the borings before the fill work started. The alluvial clay layer, which is mostly responsible for the settlement of the island, does not show the same thickness in these five borings, and shows the least at the present investigation site. According to the settlement calculation shown later (Fig.17), however, a considerable amount of settlement had already taken place at the site at the date of boring. Therefore, the surface of the alluvial clay layer is not considered to have been so undulated as is seen in Fig.2. Below the alluvial clay layer comes a diluvial sandy layer 30 m thick that contains several thin clayey and silty layers with low N-values, which is underlaid by a diluvial clay layer about 24 m thick that was suspected to be the cause of the abrupt settlement. The primary properties (2) that is the properties as soil material, such as liquid limit, plastic limit and specific gravity, of the two clay layers were uniform enough to be analyzed as homogeneous** layers.

* K.P. (Kobe Peil) ± 0 m is the datum line in Kobe district.
** See the note in page 10.

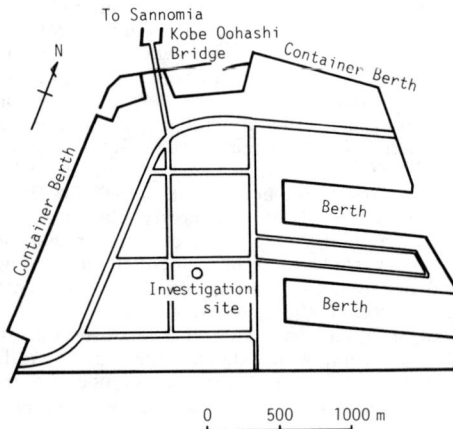

Fig.1 Investigation site

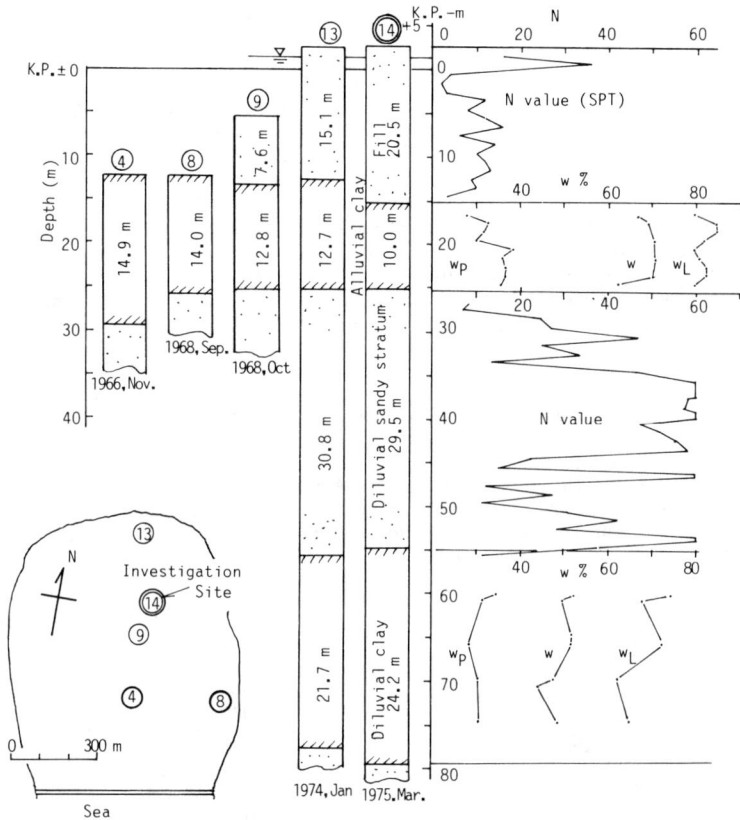

Fig.2   Soil profiles

3. Field Survey

3.1 Apparatuses Newly Developed for the Present Field Survey

A new pull-type settlement gauge (Fig.3) was developed by the authors to meet the requirements in this survey. It has a pair of wings pinned at the end of a prismatic bar, which makes an anchor and is hung by a vinyl-coated stainless steel cable that is passed over pulleys set on the ground surface and tensioned by a weight as shown in Fig.3(b). In setting, the anchor is temporarily attached to the end of the boring rod with a left-handed screw, and inserted into a pre-bored hole to a desired depth. Then it is pulled up so as to let the wings open by the aid of springs and penetrate into the bore hole wall.   After ascertaining the fixing of the anchor, the rod is released and the cable is tensioned.   This settlement gauge is very light, or rather it has a small negative weight caused by the tension of hanging cable.   Therefore it is very easy to install it even in the soft clay or in a deep stratum.   Another advantage of this gauge is that two or three sets can be installed at different depths together in one borehole.

Pore water pressure in the alluvial clay layer was measured by a simple pressure gauge shown in Fig.4, which was also developed by the authors. A flexible vinyl tube enclosed in the piezometer point cell that is saturated with water is filled with spindle oil and connected to a copper pipe leading to a bouldon gauge set on the ground surface. Since the spindle oil is lighter than water, the density being 0.9 t/m^3, the pressure measured by the bouldon gauge on the ground is expected to be positive even after the excess pore pressure in the clay layer dissipated.

The density of the fill was measured 1) by means of ordinary material-replacement method at the fill surface and 2) by means of vertical pressure measurement in the fill in the course of filling. For the former method artificial light weight aggregate round in shape and uniform in size was conveniently used as the standard density material to fill test holes as big as 0.13 m^3. For the latter method pressure cells of 50 cm diameter and 2.5 cm thickness were prepared. They had flexible faces and were filled with oil which was to be led to a bouldon gauge on the ground surface to read directly the total vertical pressure at the point caused by the surcharge fill weight.

As to the strength of the alluvial clay at that time, undisturbed samples were taken by the ordinary thin wall tube with stationary piston, and a series of shear test was performed. But the results showed an unconceivable scatter of the unconfined compressive strength along the

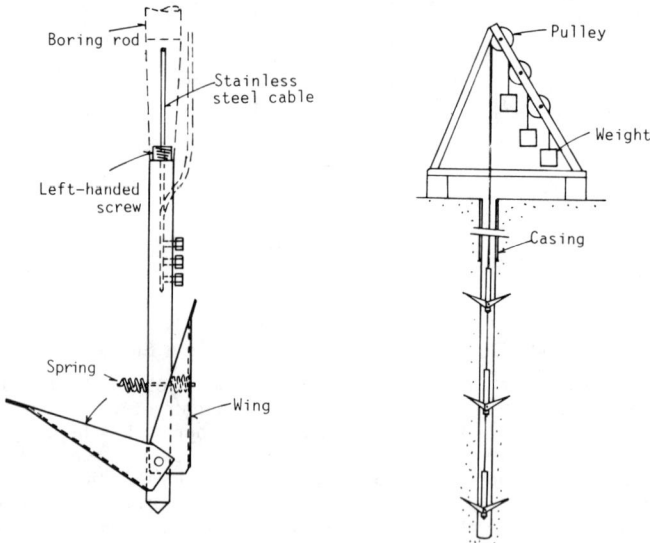

(a) Anchor   (b) Triple installation in a bore hole

Fig.3   Pull-type settlement gauge

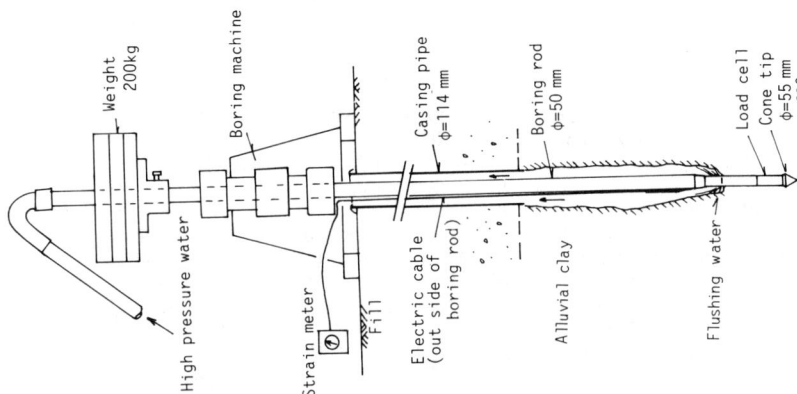

Fig.5 Cone penetrometer

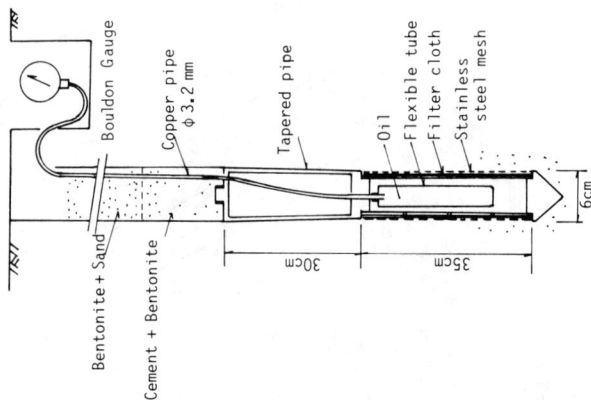

Fig.4 Pore pressure gauge

depth (Fig.10), the reason for which was considered to be attributed to the sampling disturbance caused by the heavy fill load over 20 tf/m^2 that had not consolidated the alluvial clay sufficiently yet. Thus as an alternative method, in-situ cone penetrometer was planned.

The cone penetrometer of conventional type, however, was not appropriate here because the alluvial clay was so deep ranging -18 m to -28 m below the ground surface ; it was considered difficult to measure the point resistance with it at desired accuracy, because the skin friction and buoyancy of the long driving rod would be too big. Thus again a new type cone penetrometer was developed by the authors, which is illustrated in Fig.5. The cone of 55 mm diameter and 60° apex angle is connected to a load cell, so that the penetrating resistance can be measured directly at the point. To make clearance around the driving rod as the passage for the electrical cable and also to reduce skin friction of the rod, high pressure water was flushed from the slits of the connecting pipe. The surrounding clay was eroded by flushed water and became slurry, and then the slurry flowed upward keeping the necessary clearance between the rod and the ground. There was a concern that the high pressure water might increase the pore water pressure around the cone tip. But eventually it did not affect the undrained penetrating resistance at all.

3.2 Results of Field Survey

Fig.6 shows time-settlement curves during the survey. The settlement rate of the ground level was about 40 cm/year. The undulation of the curves is chiefly due to measuring error, since the referring origin was chosen at a distant point from the site. Fig.7 shows time-consolidation

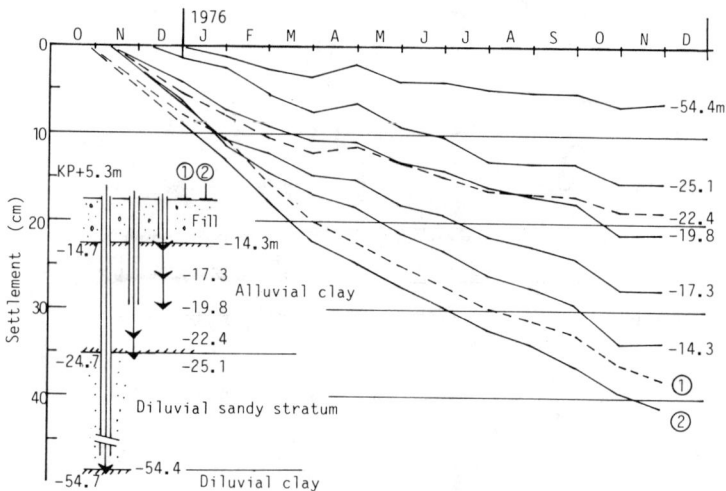

Fig.6   Record of settlement gauge

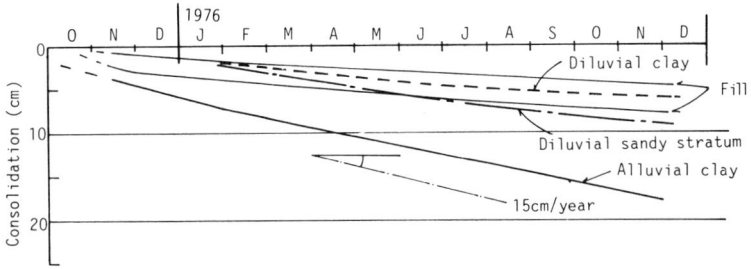

Fig.7 Time–consolidation curves of each stratum

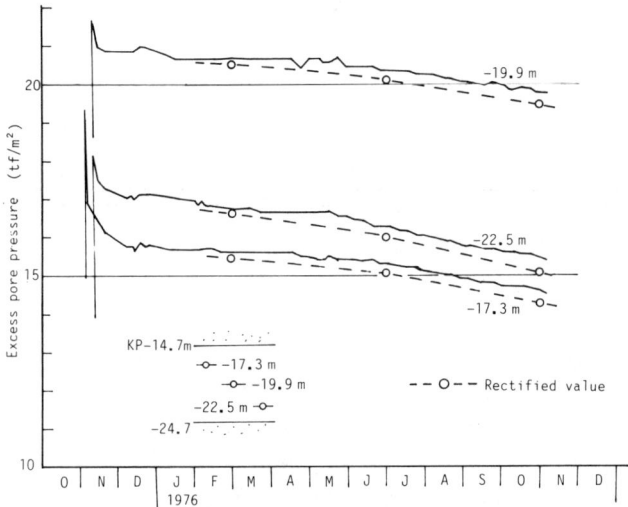

Fig.8 Observed pore pressure in alluvial clay layer

(or compression) relation of each layer obtained from the smoothed time-settlement curves of Fig.6. The alluvial clay layer naturally showed the highest consolidation rate, about 14 cm/year. It should be noted, however, that the diluvial sandy stratum as well as the deeper diluvial strata also showed considerable consolidation, though they were in a dense or overconsolidated state and were not considered, at the biginning of the project, to consolidate so much under the fill weight.

Fig.8 is the pore pressure record in the alluvial clay layer. In the figure, broken lines indicate the rectified pressures considering the ground settlement that increases the reading pressure of bouldon gauge. High pore pressure and low dissipating rate at the middle point of the layer are to be noted.

Fig.9 is the record of effective pressure increase against fill height measured by four pressure cells. The cells were set horizontally on the

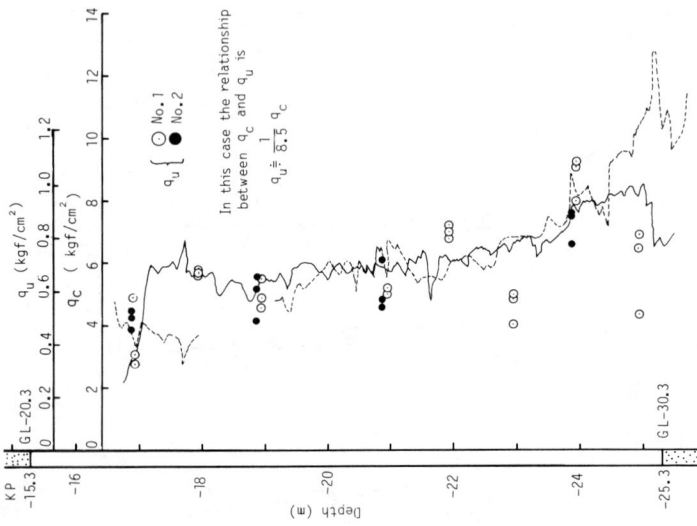

Fig.10   Cone penetration resistance and unconfined
          compressive strength

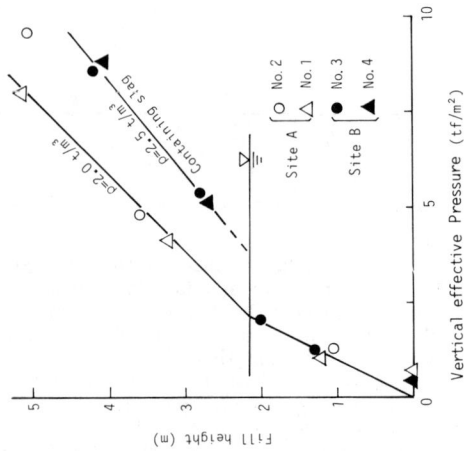

Fig.9   Relation of vertical pressure
         and fill height

fill surface 2m below the water level then. After their setting, addtional fill was placed upon them carefully by bucket crane and bulldozor. At site A a straight line bending at the water level was obtained, and the density of the fill material was reasonably calculated from the slope of the line, whereas at site B an exceptionally large density was measured above the water level due to slag which happened to be contained in the fill material. Considering the pressure data at site A together with the results by material-replacement tests, the average density of the fill material for the consolidation analysis was determined as $\gamma' = 1.13$ tf/m^3 and $\gamma_t = 2.16$ tf/m^3 (saturated with sea water of $\gamma_w = 1.03$ tf/m^3).

The distribution of cone penetration resistance q_c along the depth is shown in Fig.10 together with the unconfined compressive strength q_u of the "undisturbed sample". Though the clay layer has no sandy seams and is very uniform, the values of q_u scatter markedly, while the penetrating resistance q_c shows much less fluctuation and distributes on a concave line except near the upper boundary, which represents a strength isochrone at an early stage of consolidation under the fill weight. Two scales for q_u and q_c in the figure relate as $q_c = 8.5 q_u$. Compared with the relation $q_c = 5 q_u$, which is a well established standard in Japan (5), q_u value is apparently underestimated in this case showing the sample disturbance of the "undisturbed" sample.

4. Consolidation Analysis

Here we present a consolidation analysis for the alluvial clay layer. The consolidation (compression) of the sandy fill and the diluvial sandy stratum, though they are important constituents of the total settlement as shown in Fig.7, could never be analyzed by any "consolidation theory". Moreover, the consolidation of the diluvial clay stratum ranging from -55 to -79 m was never estimated to be so large as shown in Fig.7, because the total consolidation pressure p = 65 to 84 tf/m^2 under the fill weight did not exceed the consolidation yield stress (pseud-preconsolidation stress) of the diluvial clay p_y = 70 tf/m^2 and 90 tf/m^2 at the upper and lower surface, respectively. Therefore, the whole settlement of diluvial layers was considered to be utterly beyond the scope of consolidation analysis. (Afterwards in a new survey the "consolidation of diluvial clay" shown in Fig.7 was found to be due mostly to the compression of the layers deeper than -90.7 m.)

4.1 Governing Equation and its Finite Difference Form

The consolidation equation to be used for the present analysis was derived by Mikasa (1) as follows :

$$\frac{\partial \zeta}{\partial t} = c_v \zeta^2 \left[\frac{\partial^2 \zeta}{\partial z_0^2} - \frac{d}{d\zeta} (m_v \gamma') \frac{\partial \zeta}{\partial z_0} \right] \tag{1}$$

where t is time. ζ is the consolidation ratio (=f_0/f, f(=1+e) is the volume ratio), γ' is the submerged unit weight, and z_0 is the original coordinate, or the coordinate in the original state* in which the clay is assumed to have a certain uniform volume ratio f_0 throughout the layer, c_v is coefficient of consolidation and m_v is volume compressibility. z_0 is measured positively in downward direction. This equation is free from the assumptions employed in Terzaghi's consolidation theory that 1) volume compressibility m_v, 2) permeability k, 3) thickness of the clay layer and 4) the consolidation pressure

* The original state is not the equivalent of the initial state; see section 4.3.

are all constant during the consolidation period and that 5) the selfweight of clay does not affect the consolidation process. The assumptions used in deriving Eq.(1) are 1) clay is homogeneous*, 2) clay is saturated, 3) one-dimensional consolidation, 4) soil particles and water are incompressible. 5) Darcy's law is applicable, 6) f-log p and f-log k relations are not time-dependent, 7) c_v is constant. Assuming a linear f-log p relation, $d(m_v \gamma')/d\zeta$ in Eq.(1) can be written as

$$\frac{d}{d\zeta}(m_v \gamma') = -(1 - \frac{0.8686Cc}{f}) \frac{Gs - Gw}{f_o p} \gamma_w \qquad (2)$$

where G_s and G_w are the specific gravity of soil grains and sea water, respectively. Transforming Eq.(1) into a finite difference equation, we get

$$\Delta\zeta_{z_0} = \frac{c_v \Delta t \zeta_{z_0}^2}{(\Delta z_0)^2} [(\zeta_{z_0 + \Delta z_0} - 2\zeta_{z_0} + \zeta_{z_0 - \Delta z_0})$$

$$- \frac{\Delta z_0}{2} \frac{d}{d\zeta}(m_v \gamma')(\zeta_{z_0 + \Delta z_0} - \zeta_{z_0 - \Delta z_0})] \qquad (3)$$

where $\Delta\zeta_{z_0}$ is the increment of $\Delta\zeta$ corresponding to the time increment Δt at the depth z_0 in the original coordinate. Referring to Fig.11, the consolidation ratio at time $t + \Delta t$ is then

$$\zeta_{z_0, t + \Delta t} = \zeta_{z_0, t} + \Delta\zeta_{z_0} \qquad (4)$$

The boundary condition at a pervious layer in Mikasa's theory is given by the magnitude of compression strain of the clay element adjacent to the boundary, and this is the case for the upper and lower boundary of the present alluvial clay layer. Since the fill height increased stepwise in this case, the consolidation analysis was performed as such a stepwise increasing boundary condition problem.

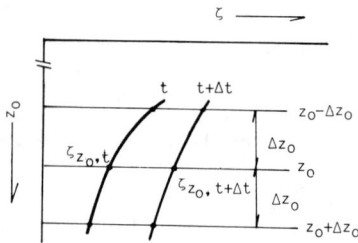

Fig.11   Schematic illustration of isochrone

4.2 Consolidation Parameters

The consolidation process at site No.14 after the date of boring could have been calculated using the data of borehole No.14 itself, only if reliable undisturbed samples had been available. But the unconfined compressive strength shown in Fig.10 showed that the "undisturbed sample" in this survey had suffered a significant disturbance. Therefore, the settlement calculation

* "Homogeneous" means here "the same sort of soil" but not "the same state of soil"

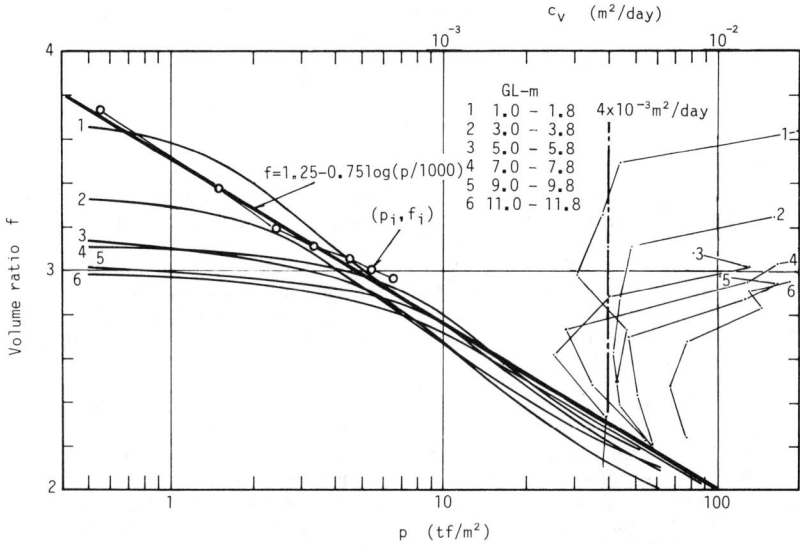

Fig.12   f–log p and f–log c_v relations (bore hole No.8)

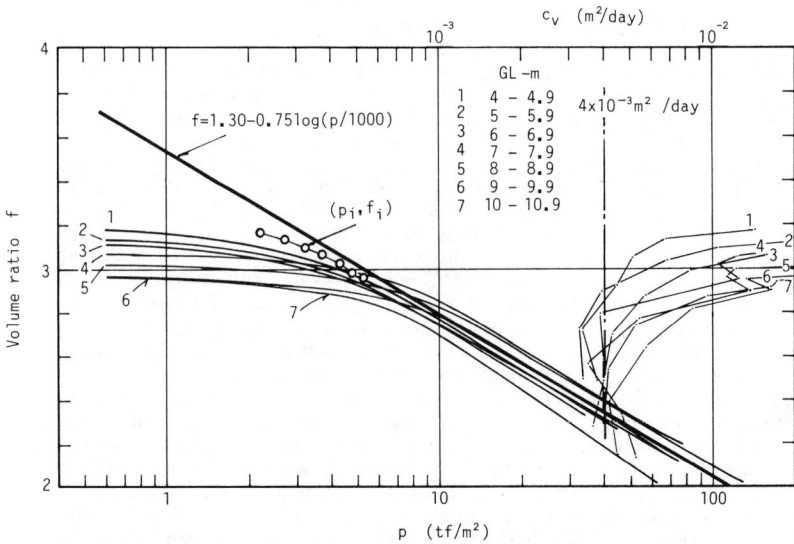

Fig.13   f–log p and f–log c_v relations (bore hole No.4)

was planned to start from the beginning of the reclamation work, using the data before the seabed clay was subjected to any overburden pressure. However, since we had no soil investigation data obtained just at site No.14 before the fill work started, the consolidation parameters and the volume ratio distribution along the depth at that time, which is necessary as the initial condition of the alluvial clay in the settlement analysis, were substituted by the data of the nearest bore hole No.8 and No.4 shown in Fig.2 about 350 m and 430 m off, respectively. Only the depth of the clay, however, was chosen as that of site No.14 according to the procedure shown in the next section to make a most reliable analysis of the settlement at site No.14.

Figs.12 and 13 are the f-log p curves from bore hole No.8 and No.4, whose normally consolidated regions gather in rather narrow bands. The rows of small circles show the relations between volume ratio and effective overburden pressure in the layers, the latter being obtained by integrating submerged unit weight of soils (using Figs.14 and 15). In Fig.12 the circles are almost located on the extended line of the average f-log p relation in normally consolidated region indicating that the clays were in normally consolidated condition. In Fig.13, however, the location of circles suggests a slight overconsolidation of the clays from borehole No.4. In conclusion, the f-log p relations in normally consolidated regions were chosen as the thick lines in Figs.12 and 13, and c_v was determined as 4×10^{-3} m^2/day as the average within the concerning pressure range. (c_v values were obtained by curve rule fitting method and corrected by multiplying secondary consolidation ratio r (3), (4).) Eq.(1) that is based on the assumption of constant c_v, therefore, was used for the present analysis.

4.3 Original and Initial Conditions

In Figs.14 and 15 the water content and volume ratio distributions in sites No.8 and 4 against real coordinate z are shown. The volume ratio was calculated from the measured water content assuming full saturation and a unique G_S value of 2.66. The volume ratio distribution of borehole No.8, Fig.14, takes the shape at the stationary state after the selfweight consolidation of very soft clay ended, whereas that of borehole No.4, Fig. 15, distributes almost linearly along the depth except in the uppermost part near the original seabed where the data were not obtained. The values of volume ratio in the shallow part where data were lacking in both boreholes were assumed to make smooth curves similar to the theoretical ones of stationary state after selfweight consolidation (broken lines in Figs.18 and 20), which took f_0=4.6 for No.8 and f_0=4.0 for No.4 both at the surface.

Now we shall set the original state f_o and the original coordinate z_o to deal with the large strain consolidation. The original condition (state) is not the initial condition at the start of calculation, but an imaginary state in which the clay is homogeneous and has a certain volume ratio not less than any value in the initial condition. In the present case, the original state was chosen as the maximum volume ratio before the fill work started —— the volume ratio at the surface of the original seabed, which, by the drawing of f-z curves in Figs.14 and 15, were f_o=4.6 for No.8 and f_o=4.0 for No.4. Then the initial distribution of volume ratio obtained by the survey is replotted to the original coordinate z_o as shown in Figs.14 and 15 down to -19.17 m and -19.25 m, respectively. Detailed procedure for this is as follows : divide the clay layer into elements of Δz=1 m, then the thickness of each element in the original state is $\Delta z_0 = (f_0/f)\Delta z = (f_0/f) \times 1$ m $= \zeta$ m ; the original coordinate z_0 of each element is obtained by summing

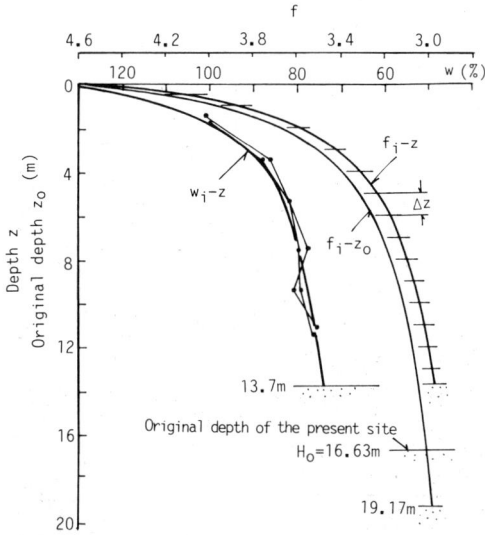

Fig 14   Initial water content, volume ratio and original
thickness H_O (bore hole No.8)

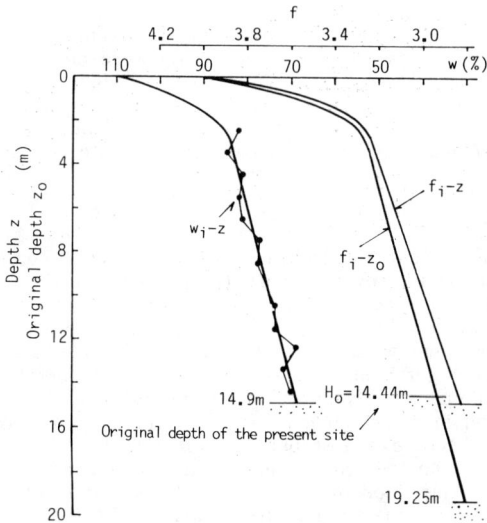

Fig.15   Initial water content, volume ratio and original
thickness H_O (bore hole No.4)

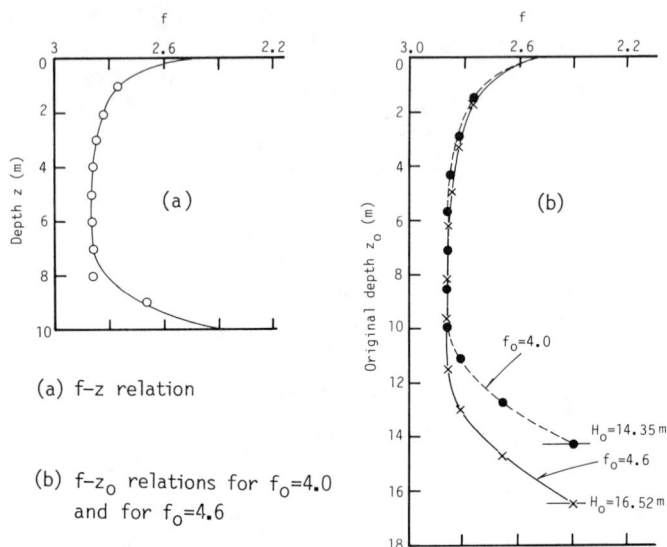

(a) f-z relation

(b) f-z_0 relations for f_0=4.0
     and for f_0=4.6

Fig.16 Volume ratio distribution against z and z_0
         at present investigation site

up Δz_0 from the surface to the depth.

Fig.16(a) shows the volume ratio distribution against the real depth z of the investigation site No.14 obtained by the present survey. The clay thickness being just 10.0 m at the time of boring, its original thicknesses in the presumed original states of f_0=4.6 and f_0=4.0 were calculated as H_0 =16.52 m and H_0=14.35 m, respectively, which, with a slight modification, were used as the original thicknesses of the assumed clay layers as shown in Figs.14 and 15. Fig.16(b) shows the volume ratio distributions of site No.14 replotted against the original coordinates for the two original states for reference.

4.4 Calculated Results

Two time-settlement curves calculated with the data of borehole No.8 and No.4 are shown in Fig.17, together with the progress of fill load. Since the observed fill height does not include the additional fill height to make up for the settlement of seabed, the fill load was rectified with the calculated settlement by the time as shown in the figure. In the early stage of consolidation, the time-settlement curve by the data of bore hole No.4 gives a smaller settlement rate than the other, chiefly because of its slightly overconsolidated initial condition. In the later stage of consolidation, the two curves become almost parallel owing to the similar consolidation characteristics of the two clay groups in normally consolidated region. Compared with the slope of 15 cm/year drawn for reference, the calculated settlement rate agrees satisfactorily with the observed one in Fig.7.

Figs.18 to 21 show the calculated isochrones of volume ratio f and effective stress p plotted against the original coordinate z_0. In the figures

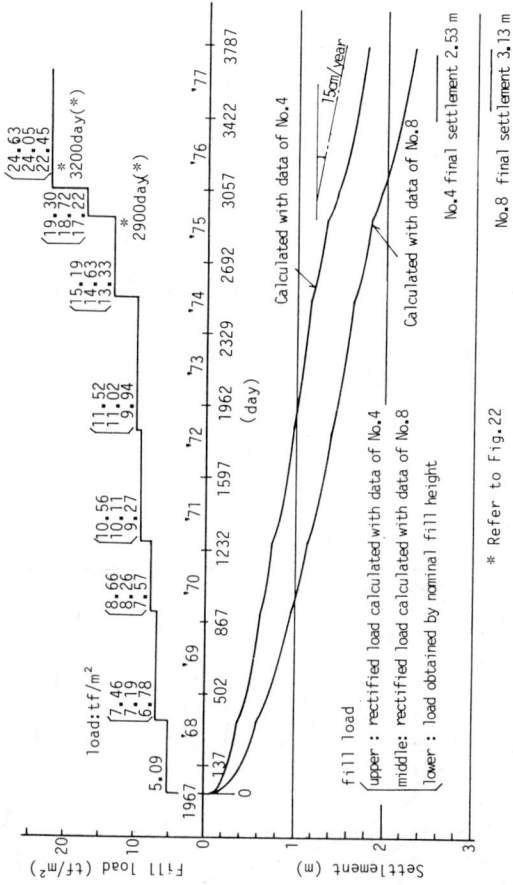

Fig.17 Calculated time-settlement curves of alluvial clay layer and fill load progress

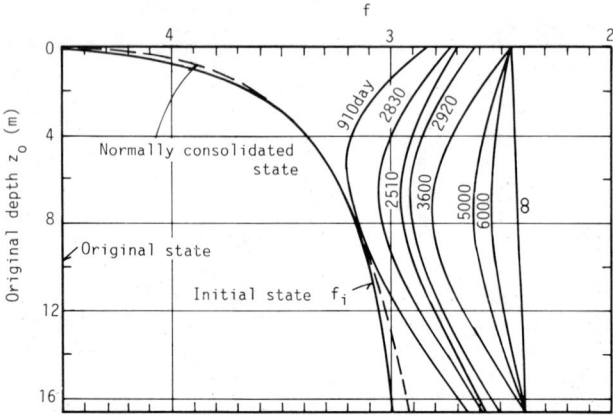

Fig.18   Calculated isochrone of volume ratio
(by the data from bore hole No.8)

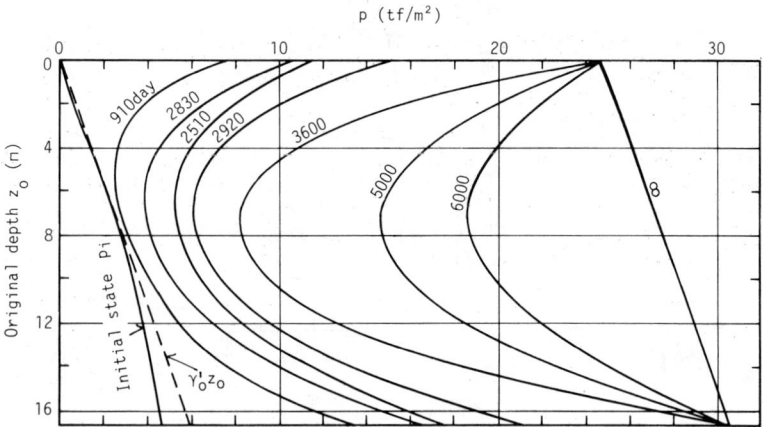

Fig.19   Calculated isochrone of effective stress
(by the data from bore hole No.8)

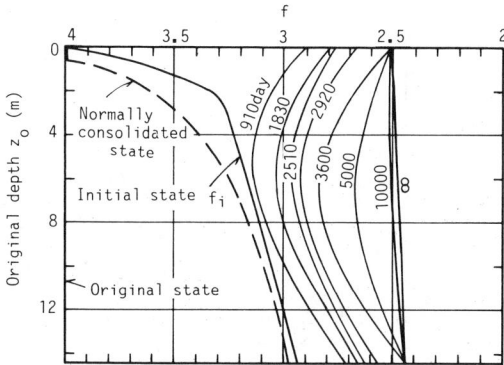

Fig.20   Calculated isochrone of volume ratio
         (by the data from bor hole No.4)

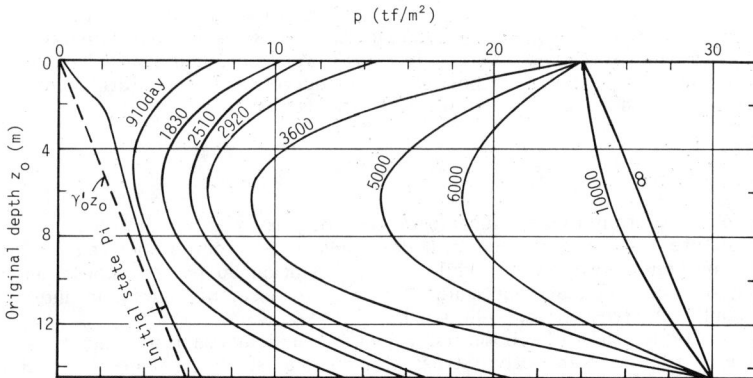

Fig.21   Calculated isochrone of effective stress
         (by the data from bore hole No.4)

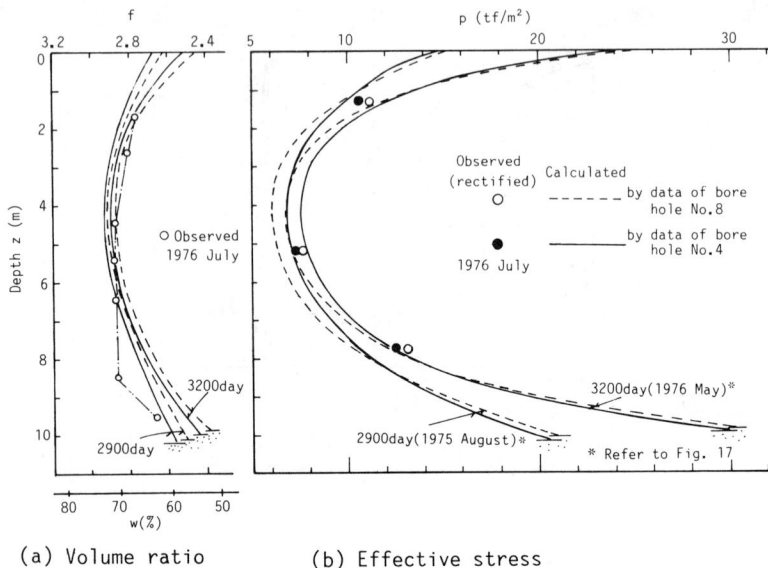

(a) Volume ratio   (b) Effective stress

Fig.22   Calculated and Observed isochrones

broken lines are drawn for reference which represent the state after self-weight consolidation. In this state the effective stress p increases linearly with the inclination of γ_0', and the volume ratio changes in proportion to log z_0. The time-settlement curves in Fig.17 were obtained from Figs.18 and 20 by integrating the area between f_i-z curve(the initial state) and f-z_0 curve at each time and dividing it by f_0 as follows :

$$ S = \int_0^{H_0} \frac{\Delta f}{f_0} dz_0 = \frac{1}{f_0} \int_0^{H_0} \Delta f\, dz_0 $$

We can observe from these figures that the progresses in f and p are largely different. For example, 3600 day lines in f-z_0 curve show about 70 to 75 percent consolidation, while in p-z_0 relation 50 percent consolidation is hardly achieved at that time. This difference, of course, is due to the nonlinear stress-strain relationship of the clay.

Figs.22(a),(b) show the observed values and calculated isochrones of both volume ratio and effective stress plotted against the real depth z* at the date of boring. The lines of August 1975 (2900 day) are drawn for reference to illustrate the difference between the states of alluvial clay layer before and after the fill surface reached the sea water level. The observed effective stress was obtained by subtracting measured pore pressure from the overburden total pressure.** Calculated isochrones of 3200-days (May 1976) well coincide with the observed ones. The observed volume ratio at the lower part of the clay, however, are a little larger than the calculation indicates, which is considered to be due to the inhomogenuity of the clay layer that is not taken into account in the analysis.

* Transforming z_0-f and z_0-p relations into z-f and z-p relations at any

time is just the reversed procedure of obtaining z_o-f relation described in section 4.3.

** Since the overburden total pressure depends on the fill height, which is rectified as described before, the total pressure has two different values by using the data from No.8 boring and No.4 boring because of the difference in the rectified settlements. Thus the observed effective stress has two values differing slightly.

5. Conclusions

Kobe Port Island is a big artificial island constructed on a very soft seabed clay layer. This report deals with an investigation on the settlement of this island conducted in 1975 - 1976 in the midst of its construction work. The main conclusions are as follows:

(1) Several in-situ measurement apparatuses were developed to meet the difficult field conditions. They are 1) pull-type settlement gauge, 2) double-cell type pore pressure meter, 3) pressure cell type density meter and 4)direct measuring cone penetrometer. All of these worked well and clarified the ground behavior under heavy fill weight.

(2) The settlement rate of the island during the survey was found to be about 40 cm/year and to be due not only to the consolidation of the soft alluvial clay layer, which had been roughly forcast before the reclamation work started, but also to the consolidation (compression) of sandy fill, diluvial sandy layer and diluvial clay layer, all of which had not been considered beforehand to consolidate so much under the proposed fill load. The last three occupied about 65 percent of the total settlement, and made the cause of the unexpected settlement observed in the course of construction almost clear.

(3) The settlement of the alluvial clay layer, the only settlement that could be analysed by any available consolidation theory at all, was analysed by Mikasa's consolidation equation that takes into account the change of k, m_v (c_v being constant), and consolidation pressure together with the effect of finite strain and selfweight. The c_v value was determined by the method proposed by Mikasa. The calculated results coincided well with the observed behaviors of the alluvial clay layer in settlement rate, stress condition and volume ratio distribution.

This big reclamation project may also be regarded as an enormous field one-dimensional consolidation experiment from the viewpoint of soil mechanics. The coincidence between the observation and the analysis of the consolidation of alluvial clay layer, we hope, may serve as a strong counteraction for the prevailing opinion that the consolidation theory is not reliable enough to meet the actual complicated soil behaviors and ground conditions.

The consolidation or compression of sandy fill and diluvial layers was not analyzed in our reseach work at all. At present, scrupulous field survey is considered to be the only effective approach for this problem.

Appendix.--References

1. Mikasa, M., "The Consolidation of Soft Clay -- A New Consolidation Theory and its Application," Kajima Shuppan-kai (in Japanese), 1963.
2. Mikasa, M., "Classification of Soil Properties and its Implications," Tsuchi-to-Kiso, Japan Society of SMFE, vol.12, No.4, pp.17-24, (in Japanese), 1964.
3. Mikasa, M., "Determination of Consolidation Parameters from Oedometer

Test," 19th Annual Convention of JSCE, III-7, (in Japanese), 1964.
4. Mikasa, M. and Ohnishi, H., "Soil Improvement by Dewatering in Osaka South Port, Geotechnical Aspects of Coastal Reclamation Projects in Japan," Proc. of 9th ISSMFE, Case History Volume, pp.639-664, 1981.
5. Muromachi, T., Relation Between Cone Penetrating Resistance and Uncon-fined Compressive Strength of Clayey Soils," Journal of JSCE, vol.42, No.10, pp.7-12, (in Japanese), 1957.
6. Nakakita, Y. and Watanabe, Y., "Soil Stabilization by Preloading in Kobe Port Island, Geotechnical Aspects of Coastal Reclamation Projects in Japan," Proc. of 9th ISSMFE, Case History Volume, pp.611-622, 1981.

DREDGED SLURRY STABILIZATION FOR SEAGIRT MARINE TERMINAL

W. Morin, F. ASCE;[1] T. Shafer, M. ASCE;[2] and K. Gangopadhyay, M. ASCE[3]

ABSTRACT

The Seagirt Marine Terminal is to be constructed on reclaimed land
created on dredged spoil from the construction of the I-95 Fort
McHenry Tunnel in Baltimore, Maryland. The first phase of development will be stabilization of the dredged spoil and underlying sediments by consolidation to support a design load of 600 psf (28kPa).
The estimated settlements within the dredged slurry area ranges up to
7 ft (2.1m). Three stabilization procedures were considered applicable; prefabricated wick drains; lime columns; and compaction grouting. Wick drains were selected to stabilize the slurry and underlying subsoil. Lime columns were considered an alternative procedure.
Compaction grouting was selected to stabilize the compressible stratum underlying the granular retaining dike. A test fill program was
designed to check the effectiveness of two types of wicks, the amount
of settlement and settlement time. The amount of settlement was
reasonably close to that calculated using Terzaghi's one-dimensional
consolidatin theory. Ninety-five percent consolidation can be
achieved with a wick spacing of 5 ft (1.5m) within 8 months. Both
types of wicks performed equally well.

INTRODUCTION

The Maryland Port Administration is conducting planning and design
studies for the Seagirt Marine Terminal located along the north shore
of the Patapsco River in the Canton area of Baltimore City, Maryland.
The 113 acre Terminal will consist of three container berths, constructed on reclaimed land created by the disposal of dredged spoil
and excess excavated material from the construction of the I-95 Fort
McHenry Tunnel Construction. It is unique in that it is the first
time in the U.S. that a dredged spoil disposal area has been slated
to become a marine terminal. The first phase of development will be
the stabilization of the dredged spoil and underlying harbor bottom
sediments, which involves consolidation of these strata to support a
design load of 600 psf (28 kPa) without excessive settlement. One of
the primary goals of the planning process was the determination of
the most appropriate and economical procedure for stabilizing the

[1]Vice President and Chief Geotechnical Engineer, STV/Lyon Associates, Inc., 7900 Westpark Drive, McLean, Va. 22102.

[2]Senior Project Engineer, STV/Lyon Associates, Inc., 21 Governor's
Court, Baltimore, Md. 21207.

[3]Vice President, EBA Engineering, Inc., 2116 Maryland Avenue, Baltimore, Md. 21218, formerly Project Engineer, Lyon Associates, Inc.

dredged slurry and the underlying subsoils. The condition of the
site as of January, 1984 as shown in Figure 1.

BACKGROUND OF PROJECT

The site was divided into three main portions: (a) 27-acres (10.9
hectares) in the east was developed to receive highly compressible
and chemically contaminated harbor bottom sediment called the Muck
Disposal Area; (b) 113-acres (45.7 hectares) in the west was developed to receive the remainder of the dredge spoil called the Spoil
Disposal Area; and (c) 6-acres (2.4 hectares) located between the
Muck Disposal Area and the Spoil Disposal Area was developed to treat
the effluent from the dredgings and to channelize the effluent to the
harbor called the Sedimentation Channel and Treatment Chamber. The
proposed 3-berth marine terminal will be constructed primarily over
the Spoil Disposal Area. There are no plans to develop the Muck Disposal Area in the near future. This paper is mainly concerned with
stabilization of the Spoil Disposal Area.

In order to contain the dredged spoil, an earth embankment was constructed on the north and west sides of the site, while a cellular
cofferdam containment structure was constructed on the south and east
side of the site along the Patapsco River and Colgate Creek respectively. The total length of the containment structure is 5,600
linear feet (1,707m) and it consists of seventy-six, 62 feet (19m)
diameter circular cells and connecting arcs filled with sand and
gravel fill.

The dredging of the trench for the tunnel construction started in
June, 1981. A hydraulic suction cutterhead dredge was used for the
dredging and the dredged spoil was transported to the disposal site
via a 27 inch (69cm) diameter pipeline. The subsurface investigation
conducted along the alignment of the tunnel estimated the quantities
of various types of dredge material as follows:

        Muck Disposal Area
                Very soft harbor bottom   600,000 Yd.3
                Sediments   (460,000 m3)
        Spoil Disposal Area
                a. Organic clayey silt   650,000 Yd.3
                    and organic sandy silt   (500,000 m3)
                b. Sandy silt to gravelly sand   1,155,000 Yd.3
                    (884,000 m3)
                c. Stiff clay to stiff clayey silt 938,000 Yd.3
                                                     (718,000 m3)
                        TOTAL   3,343,000 Yd.3
                                                    (2,562,000 m3)

It was estimated that the volume of dredged material requiring disposal, considering the bulking factor, would be 5 million Yd^3 (3.8
million m^3).

A relatively large proportion of sandy material was deposited along
the cofferdam in the Spoil Disposal Area, during the initial stages

Figure I. SEAGIRT MARINE TERMINAL

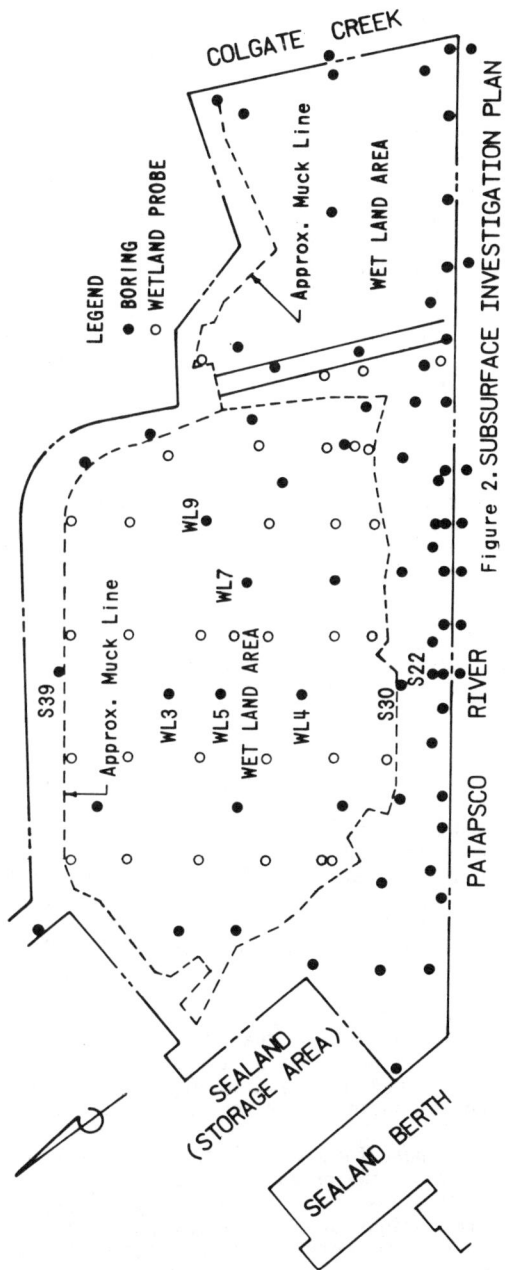

Figure 2. SUBSURFACE INVESTIGATION PLAN

LEGEND
● BORING
○ WETLAND PROBE

COLGATE CREEK

Approx. Muck Line

WET LAND AREA

Approx. Muck Line

WET LAND AREA

S39
WL9
WL7
WL5
WL3
WL4
S30
S22

PATAPSCO RIVER

SEALAND (STORAGE AREA)

SEALAND BERTH

of the dredging operation. A berm with a width of 100 feet (30m) was
created adjacent and parallel to the containment structure. Beyond
the berm a series of dikes were constructed by bulldozing the dredge
materials to a height of 10 feet (3m) above the elevation of the cofferdam and the adjoining berm to increase the capacity of the containment area.

The material disposed within the embankments consisted mainly of silt
size material. The action of the high-capacity cutterhead dredge completely disintegrated the soil matrix, producing a true slurry. Unsuccessful attempts were made to accelerate the sedimentation of the
dredged spoil by adding various types of synthetic flocculating
agents.

The uppermost soil strata in the area enclosed by the embankments,
where the largest portion of the dredged spoil was placed, consists
of very soft dredged spoil, underlain by compressible organic silt
and clay. Dewatering by progressive trenching was performed in this
area during the summer of 1983 to facilitate crust formation.

The south side of the site behind the cofferdam structure, forms
another distinct area. The uppermost soil strata in this area consists of sand and gravel with pockets of silt and clay. The sand and
gravel stratum is underlain by the compressible organic silt and
clay. The uppermost soil stratum underlying the site is a thin cover
of marine and estuarine, very soft grey silt and clay with some sand,
approximately 40 feet thick. The Patapsco River and its tributaries
are some of these channels in which the sediments were deposited in
varying thicknesses. At the Seagirt site, these sediments are approximately 40 feet thick.

SITE CONDITIONS

The field investigation included conventional boring and sampling,
cone penetration tests, vane shear tests and special field testing
and sampling. The special field tests were performed with a probe
rod with plates and attached load cells which were used to determine
the penetration resistance in in the very soft dredged slurry. A
special sampling device was constructed to collect the samples of
very soft dredged slurry. Locations of borings and probes are shown
in Figure 2. The boundary of the Wet Land Area indicates the extent
of slurry deposition.

Laboratory testing was performed on disturbed and undisturbed samples
collected during the field investigation which included classification tests, triaxial tests, one dimensional consolidation tests, and
unconfined compression tests. Additionally, large diameter slurry
consolidation tests were performed on the special samples.

The subsurface stratigraphy is shown in Figure 3, a section perpendicular to the original shoreline.

The various strata are described more or less from the surface downward. The strata are variable due to their mode of deposition.

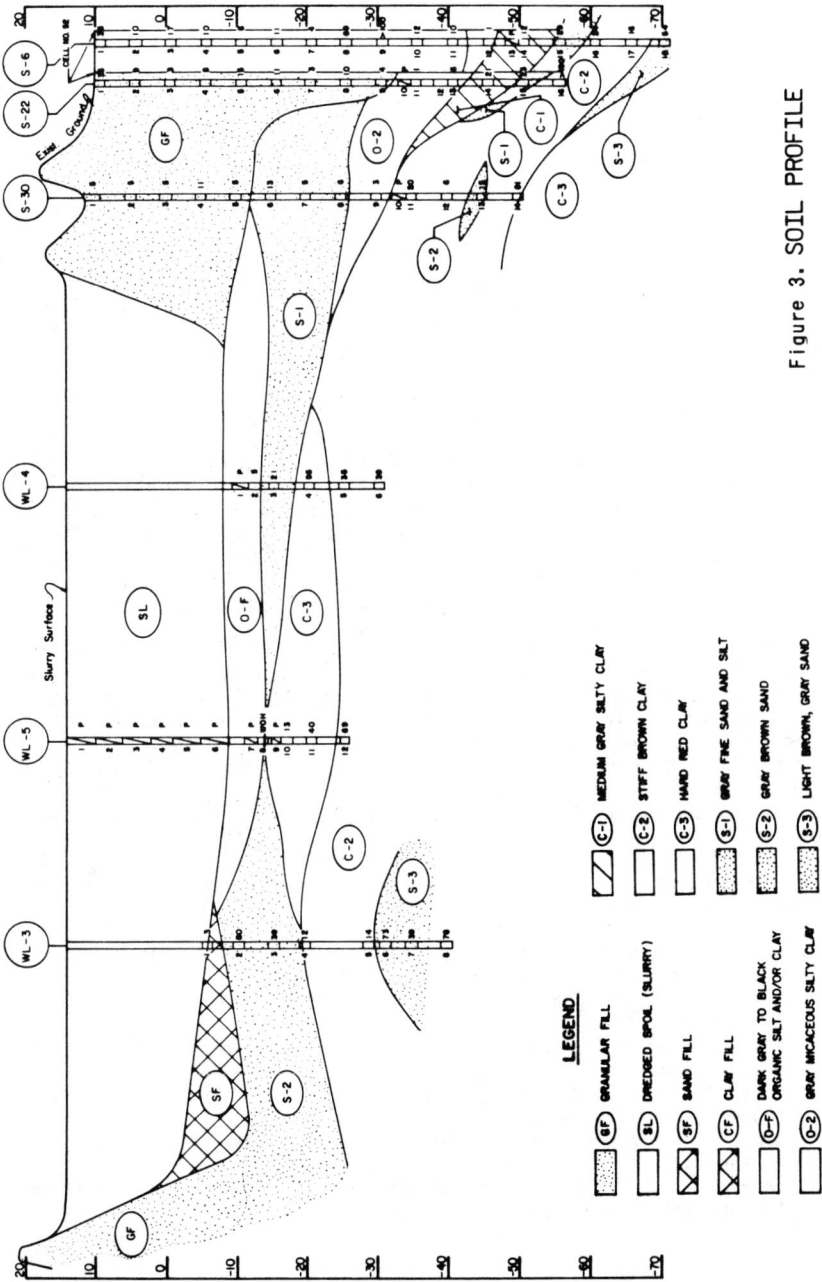

Figure 3. SOIL PROFILE

LEGEND

GF	GRANULAR FILL
SL	DREDGED SPOIL (SLURRY)
SF	SAND FILL
CF	CLAY FILL
O-F	DARK GRAY TO BLACK ORGANIC SILT AND/OR CLAY
O-2	GRAY MICACEOUS SILTY CLAY

C-1	MEDIUM GRAY SILTY CLAY
C-2	STIFF BROWN CLAY
C-3	HARD RED CLAY
S-1	GRAY FINE SAND AND SILT
S-2	GRAY BROWN SAND
S-3	LIGHT BROWN, GRAY SAND

Geologically recent sediments have been disturbed, intruded or mixed, with dumping during the past several decades of extraneous fill and organic materials. The addition and partial removal of fill during the tunnel construction has further altered the strata in the western part of the site.

Granular Fill (GF): The granular fill is a variable mixture of gravel and sand with a little silt and clay. This material was used to construct the spoil retaining dikes and as backfill behind the containment structure. The in-situ relative density varies from loose to dense.

Slurry (SL): The dredged material covers most of the area designated for spoil disposal to an initial surface elevation of about +14.0 ft (4.3m). The slurry consists predominantly of silt mixed with variable amounts of clay and fine sand and it contains a significant amount of organic matter. The consistency varies from a viscous liquid to very soft. Natural moisture contents range from 50 to over 200 percent, and are generally higher than the liquid limit.

O-F Layer: The O-F layer is composed of soft, dark gray to black clayey silt to silty clay, with some sand. This material has a strong organic odor and exhibits a loss on ignition ranging from about 5 to 10 percent. Natural moisture contents range from 50 to over 100 percent, varying inversely with the percentage of sand. This deposit represents the uppermost layer of harbor bottom sediments and covers most of the shoreward side of the project area. It is a mixture of geologically recent sediments and a variety of recent fill materials.

O-2 Layer: The O-2 layer is composed of soft to medium, gray to dark gray silty clay to clayey silt, with some fine sand. The natural moisture content ranges from 50 to 100 percent. The deposit is also mixed with recent fill materials. It is present behind the cellular cofferdam at elevations between -20.0 ft (6.1m) and -40.0 ft (12.2m) extending northward to about 300 ft (91m) from the original shoreline.

The remaining strata are not significantly compressible. These are identified on the profile as:

S-1: loose to medium gray fine sand with some silt; underlies O-F in the western half of the spoil area.
C-1: medium to stiff gray silty clay; underlies S-1 in the western half of the site.
C-2: stiff, brown silty clay to clayey silt; occurs in western half of the site.
C-3: very stiff to hard red or gray silty clay of the Potomac Group; found at lower depths.
S-2: dense to very dense gray-brown sand; found only in western part of the site.
S-3: very dense light brown sand is the lowest stratum intercepted in the western part of the site;

SOIL PROPERTIES AND ESTIMATED SETTLEMENTS

There are three compressible strata that will experience significant consolidation. The first of these is the dredged slurry which covers most of the interior of the Spoil Area, originally averaging 22 feet in thickness. The other two are the 0-F and 0-2 layers which underly the dredged slurry. The 0-F layer is largely restricted to the shore-ward two-thirds of the site, while the 0-2 layer is generally found in the outward third, under the granular fill behind the cellular cofferdam.

The mean properties of the slurry have been grouped according to depth in Table 1 for both probe samples and samples recovered from borings. Overall mean values are also provided. The moisture content, compression index, and void ratio all decrease with depth, while the dry density increases. This would be expected, as some degree of self-weight consolidation has occurred.

A summary of the mean values of the consolidation properties for the three compressible strata is given in Table 2. There were considerable variations in the test data as indicated by the standard deviation of the mean values. The coefficient of consolidation of the 0-2 and 0-F are similar and slightly higher than the value for the slurry, although the 0-F stratum is under-consolidated, while the 0-2 is slightly over-consolidated. Typical e log p curves for the slurry 0-F and 0-2 strata are shown in Figures 4 through 6.

The estimated settlement under the design load of 600 psf (28 kPa) within the dredged slurry area ranges from 3 ft (0.9m) to 7 ft (2.1m), based on laboratory consolidation tests and calculated by means of the Tenzaghi one-dimensional consolidated theory. The lower extremes are in the western part of the site where there is a reduced thickness or absence of the dredged slurry and 0-F strata. Estimated settlement within the remainder of the spoil disposal area, with the exception of the granular fill area, GF, is 6 ft (1.8m) to 7 ft (2.1m) and the estimated time for this settlement varies from 6 years to 8 years. In the western part of the site, the estimated time for settlement will be approximately 15 months.

The 0-2 stratum under the granular fill area, GF, is normally consolidated under its existing overburden pressure. Additional settlement of 3 in (8cm) to 6 in (15cm) will occur under the design load of 600 psf (28kPa).

STABILIZATION

Both the dredged slurry and the underlying subsoil will require stabilization through consolidation before the area can be developed into a marine terminal. There are several procedures for stabilizing soft soil which involve means of accelerating the settlement rate or reducing the amount of settlement or both.

Three procedures reviewed were considered applicable to existing conditions at Seagirt: (1) prefabricated wick drains; (2) lime columns;

TABLE 1

SUMMARY OF MEAN PROPERTIES OF SLURRY

(1)	% -200 Sieve (2)	Moisture Contents % (3)	Dry Density PCF (Kq/m^3) (4)	Liquid Limit % (5)	Plasticity Index (6)	Organic Content % (7)	Comp. Index (8)	Initial Void Ratio (9)
FROM PROBES DEPTH =								
0-10' (0-3m)	80	125	40.1 (642)	53	17	11.3		
10-20' (3-6m)	69	85	53.8 (862)	39	15	10.4		
>20' (>6m)	-	67	60.6 (746)					
Overall Mean	76	104	46.6	47	16	10.1		
FROM BORINGS DEPTH =								
0-10' (0-3m)		127	31.6 (506)	58	22	9.9	0.72	3.99
10-20' (3-6m)		63	62.5 (1001)	47	18	7.1	0.49	1.85
>20' (>6m)		33	92.8 (1487)				0.20	0.78
Overall Mean		84	54.7 (876)	51	19	8.3	0.58	2.61

TABLE 2

SUMMARY OF MEAN VALUES OF VARIOUS STRATA

(1)	Moisture Content % (2)	LL (3)	PI (4)	C_c (5)	P_c, tsf (6)	Cv in²/sec* (7)	e_o (8)
SL Mean	105	50	17	0.72	0.101 (0.99kg/cm²)	1.43×10^{-4} (9.15cm²/sec)	2.9431
Standard Deviation	37	16	8	0.25	0.112 (0.109kg/cm²)	1.25×10^{-4} (8.00cm²/sec)	1.0599
0-2 Mean	66	89	47	0.82	1.20	1.21×10^{-4} (7.74cm²/sec)	1.9713
Standard Deviation	16	8	5	0.29	0.15 (0.15kg/cm²)	0.98×10^{-4} (6.27cm²/sec)	0.1621
0-F Mean	100	70	25	0.86	0.54	0.99×10^{-4} (6.34cm²/sec)	2.5854
Standard Deviation	61	12	14	0.20	0.14 (0.14kg/cm²)	0.62×10^{-4} (3.97cm²/sec)	0.895

*In loading range of 1 to 2 tsf (48 to 96kPa).

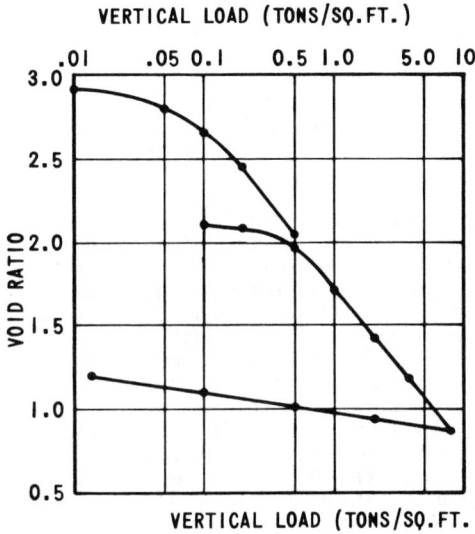

VERTICAL LOAD (TONS/SQ.FT.)

Figure. 4
BORING : WL-7
DEPTH : 16-20 FT.
STRATUM : SL

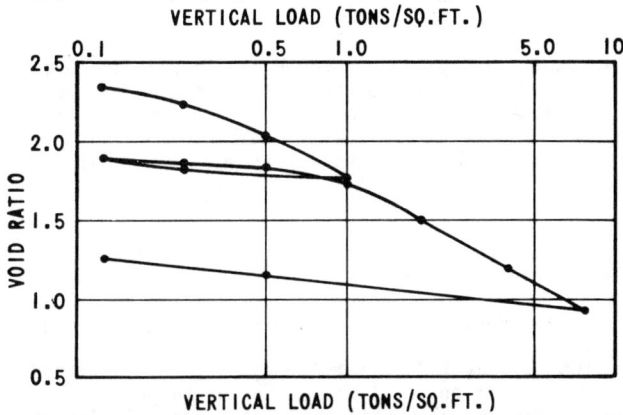

VERTICAL LOAD (TONS/SQ.FT.)

Figure. 5
BORING : WL-9
DEPTH : 24-26 FT.
STRATUM : 0-F

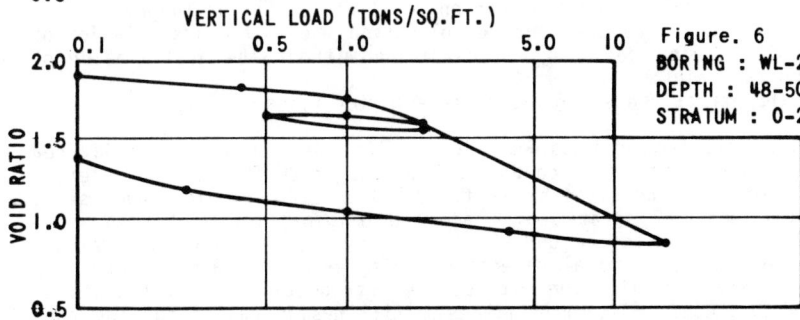

VERTICAL LOAD (TONS/SQ.FT.)

Figure. 6
BORING : WL-22 2
DEPTH : 48-50 FT.
STRATUM : 0-2

e - log p CURVES

(3) compaction grouting. The prefabricated wicks which were considered for this project are of two basic types. The first is a plastic ribbed or tufted wick, usually 4 in (10 cm) in width, encircled by a geotextile bonded filter fabric. The second is a needle-punched filter fabric with a hollow interior. Lime or cement stabilized columns have been used in soft clays (1) and were considered as an alternative procedure to wick drains, should the wicks not function satisfactorily.

Although simple surcharging would probably be effective in consolidating the 0-2 stratum underlying the granular fill, GF, for the design load of 600 psf (29kPa), this was not allowed out of concern for the cofferdam stability. Instead, compaction grouting was considered for the 0-2 stratum underlying the granular fill, GF. In this procedure, a stiff grout is injected into the stratum at sufficient pressure to displace the soil laterally, thereby causing excess hydrostatic pore pressures and enhancing the consolidation process. The grout can also be injected in such a manner as to form columns which extend through the compressible strata to the firmer underlying subsoils and, in this way, provide some vertical support. This reduces settlement and, if the volume of grout displacement equals the anticipated consolidation under the design load, total settlement should be negligible.

All of the procedures considered applicable would required field testing. Both types of wicks required field verification that they would not clog nor crimp nor otherwise cease to function under the large anticipated settlements. Lime columns have not been applied to such large stabilization projects and have hardly been applied at all in the U.S.

TEST FILL PROGRAM

A test fill program was designed to determine:

1. The amount of displacement settlement under the design loading.
2. The amount of primary consolidation actually achieved compared to that calculated from conventional consolidation theory.
3. The accelerated rate of consolidation due to the installation of wick drains compared to preloading without the installation of wick drains.
4. The effectiveness of two different wick types.

The test fill embankments were located in the proposed Berth II area where the existing soil condition consists of 20 ft (6.1m) of dredged spoil material underlain by 12 ft (3.7m) of soft organic clayey silt, the 0-F and 0-2 strata which are the original harbor bottom sediments. These soils are underlain by silty clay allowing only single drainage in the vertical direction; this results in a very long time for primary consolidation without the introduction of artificial internal drainage. The original test fill design called for the construction of three contiguous embankment sections, each to a total fill height of 11 ft (3.4m), which represents a surcharge load of

1100 psf (52kPa). Separate wick types were to be installed in the two end test sections. The two wick types selected for testing were the Mebradrain ribbed wick with encircling bonded filter fabric and the needle-punched Desol wick with hollow interior. The wicks were to be installed at 5 ft (1.5m) centers in a grid pattern beneath the central 25 ft (7.6m) x 25 ft (7.6m) portion of each end of the test embankment. The central test embankment was to be constructed without wicks and serve as a control section. Each of the three test sections was instrumented with settlement indicators, pneumatic-type piezometers, and inclinometers.

The proposed sequence of construction, which was designed to model that recommended for stabilization of the site, was as follows:

1. Place a bottom geotextile and 2 ft (0.6m) thick sand blanket on top of the existing crust over an area measuring 94 ft (29m) by 94 ft (29m).
2. Install wick drains in the two end sections.
3. Install geotechnical instrumentation.
4. Place a second layer of geotextile on top of the sand blanket in the three embankment areas.
5. Place 3 ft (0.9m) of granular fill to complete the berm.
6. Construct the three test fill embankments by placing two 3 ft (0.9m) lifts of granular fill.

It was anticipated, during the design of the test fills, that a relatively thick surface crust would have formed prior to test fill construction, as a result of the Site Dewatering Program implemented in the Spring of 1983. Although the Site Dewatering Program effected a noticable improvement in the site conditions, the thickness of the surface crust at the test fill location was only 4 in (10cm) to 6 in (15cm) when construction began in mid-August, 1983.

It proved to be impossible to construct three contiguous sections using light self-propelled equipment due to the instability at the site. The test fill layout was therefore revised to that shown in Figure 7. The sand fill and granular fill were placed by means of a crane stationed on the access road.

The rate of pore pressure dissipation, was very slow during the loading of the test sections. The rate of fill placement was slowed considerably due to concern for the stability of the test sections. Despite the slow placement rate, considerable lateral displacement of the underlying slurry occurred, particularly along the north and west edges of the fill sections. Due to the combination of slow fill placement rate and time constraints imposed by the overall project schedule the placement of fill material was halted in early December, 1983, after the two wick sections reached an overall height of 7.5 ft (2.3m), which represents an effective pressure of 600 psf (29kPa). Similar considerations caused the cessation of filling in the control test section after the 2 ft (0.6m) sand blanket had been placed. The rate of fill placement for the three test sections is shown in Figures 8, 9, and 10. Unfortunately, most of the instruments installed to measure settlement at various depths beneath the fill did

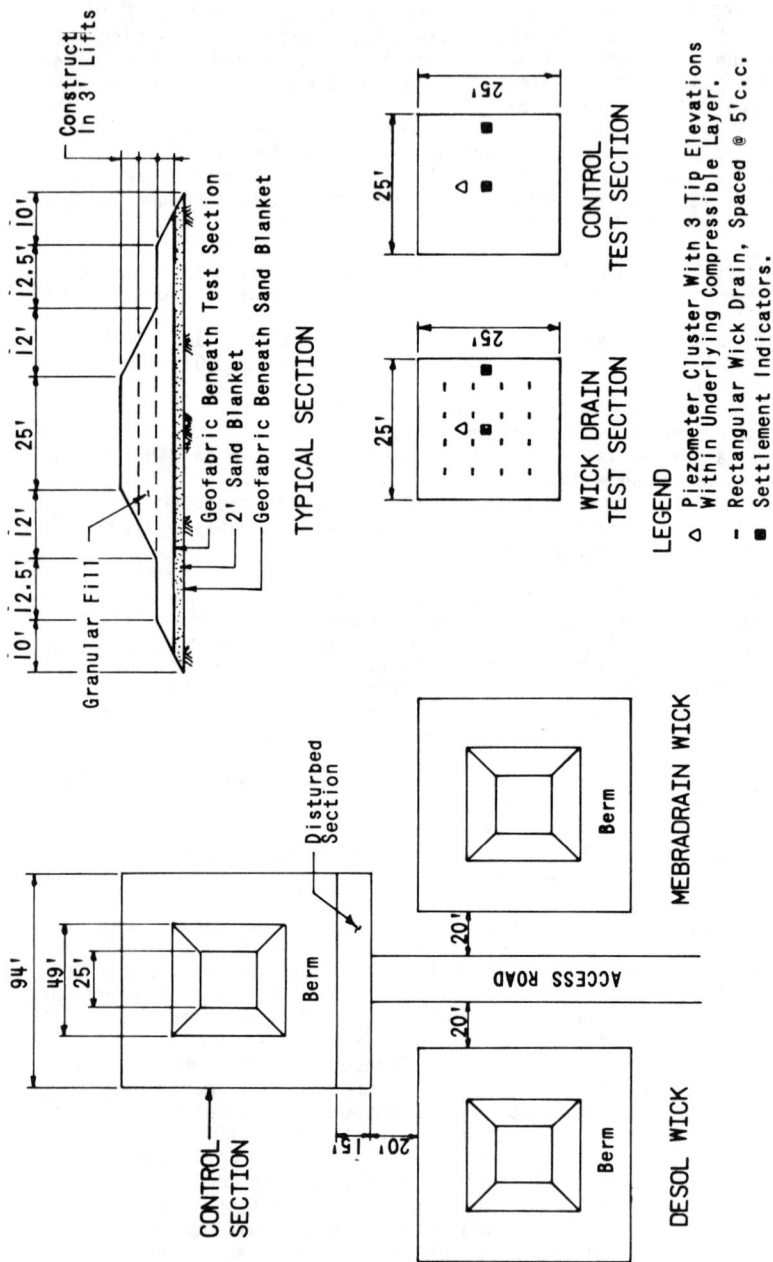

TYPICAL SECTION

Construct In 3' Lifts

Geofabric Beneath Test Section
2' Sand Blanket
Geofabric Beneath Sand Blanket

Granular Fill

CONTROL TEST SECTION

WICK DRAIN TEST SECTION

LEGEND

△ Piezometer Cluster With 3 Tip Elevations Within Underlying Compressible Layer.

- Rectangular Wick Drain, Spaced @ 5'c.c.

■ Settlement Indicators.

CONTROL SECTION

Disturbed Section

Berm

ACCESS ROAD

MEBRADRAIN WICK

DESOL WICK

Figure 7. TEST FILL EMBANKMENTS

not function properly. Most inclinometer casings could not be used after lateral displacements exceeded 2 ft (0.6m).

During the construction of the test fills, a boring was drilled in each of the two wick test sections. Undisturbed samples were collected in the slurry stratum (SL) and the underlying clayey silt strata (0-F and 0-2). One dimensional consolidation tests were performed on three samples in accordance with procedures outlined by Cargill (1983) along with classification and index tests on selected samples. The consolidation test results indicated that the slurry strata is underconsolidated but with a lower void ratio than found in the first phase of the field investiation. The 0-F and 0-2 strata are normally consolidated. The test results for the three undisturbed samples are given in Table 3.

TABLE 3
TEST RESULTS, SAMPLES FROM TEST FILLS BORINGS

STRATUM	SL	SL	0-F
Depth	5.5ft-7.5ft	9.0ft-11.0ft	29.0ft-31.0ft
	(1.7m-2.3m)	(2.7m-3.4m)	(8.8m-9.4m)
Moisture Content, %	68	53	186
Liquid Limit	65	47	170
Compression Index (C_c)	0.44	0.32	0.59
Initial Void Ratio (e_0)	2.07	1.41	1.99
Preconsolidation			
Pressure (P_c)	0.02tsf	0.05tsf	0.40tsf
	(1.9kPa)	(4.8kPa)	(38.1kPa)
Coefficient of Consolidation			
(C_v) at 1 tsf load			
(48kPa) in ft^2/day	.063	.078	.024
	(6.8x10^{-4}	(8.4x10^{-4}	(2.6x10^{-4}
	cm^2/sec)	cm^2/sec)	cm^2/sec)

The borings show that prior to the start of the test fill program, the Desol test section was underlain by 23 ft (7m) of slurry and 12 ft (3.6m) of 0-F and 0-2, while these respective strata thicknesses were 16 ft (4.9m) and 11.5 ft (3.5m) under the Mebradrain test section. These variations in the subsoils had not been expected. The test fill boring logs are shown in Figure 11.

The theoretical primary consolidation settlements of the compressible strata under the final height of the test fills were calculated using values for C_c, e_0, and P_c obtained from the consolidation test results. These are:

Test Section	Total Fill Height	Primary Consolidation Settlement
Desol Wick	7.5 ft (2.3m)	4.3 ft (1.2m)
Mebradrain Wick	7.5 ft (2.3m)	3.0 ft (0.9m)
Control	2.0 ft (0.6m)	2.6 ft (0.8m)

ANALYSIS OF TEST FILL DATA

The time settlements curves are shown in Figures 8, 9, and 10. The time-settlement curves for the wick test sections show a "flattening out" of the settlement rate after a period of 120 days in the Mebradrain test section and 125 days in the Desol test section. Although 95% of the ultimate primary consolidation has not yet been reached in any of the test sections, the time-settlement curves for the wick drain sections can be extrapolated to yield this value. Extrapolation yields 4.25 ft (1.3m) for the Desol test section and 3.5 ft (1.1m) for the Mebradrain test section. The time-settlement curve for the control section remains more or less linear, precluding extrapolation at this time (April, 1984).

The overall settlement recorded is the sum of the settlement due to primary consolidation and that due to lateral shear displacement of the slurry. The amount due to lateral displacement was calculated from the volume of material laterally displaced as determined from the inclinometer deflections. In the cases where the inclinometers ceased to function after experiencing lateral deflections in excess of 2 ft (0.6m), deflections were extrapolated from plots of lateral deflection vs. fill height. The volume displaced horizontally was equated to a volume of material under the test section displaced vertically. The shape of the vertical displacement volume can be inferred from a comparison of inclinometer deflections at the center of the cross-section to the inclinometer deflections along the berm. The estimated displacement settlement at the center of the test section was 0.6 ft (0.2m) for the Mebradrain section and 0.8 ft (0.3m) for the Desol section.

The amount of primary consolidation was obtained by subtracting the lateral displacement settlement from the total settlement. The primary consolidation is 3.4 ft (1.0m) for the Desol section and 2.8 ft (0.6m) for the Mebradrain section. The ratio of actual primary consolidation achieved to that predicted from theory is:

$$\text{Desol Section:} \quad \frac{\text{Actual}}{\text{Theoretical}} = \frac{3.4}{4.3} = 0.8$$

$$\text{Mebradrain Section:} \quad \frac{\text{Actual}}{\text{Theoretical}} = \frac{2.8}{3.0} = 0.9$$

The extrapolations of the time-settlement curves for the two wick sections show that 95% of the total primary consolidation will be achieved within 8 months of beginning the test sections. The actual rate of consolidation was determined by selecting points on the time settlement curve after loading was completed and using the relation:

$$T = \frac{C_v t}{H^2}$$

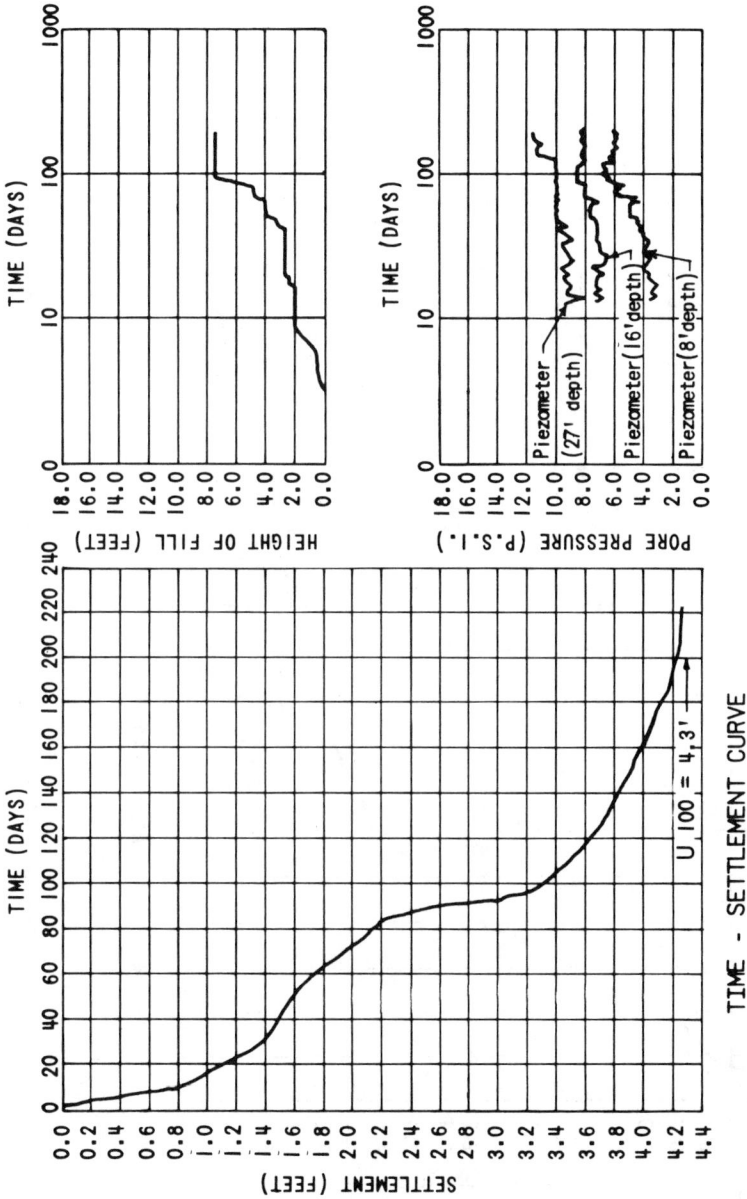

Figure 8. TEST FILL DATA (DESOL WICK TEST SECTION)

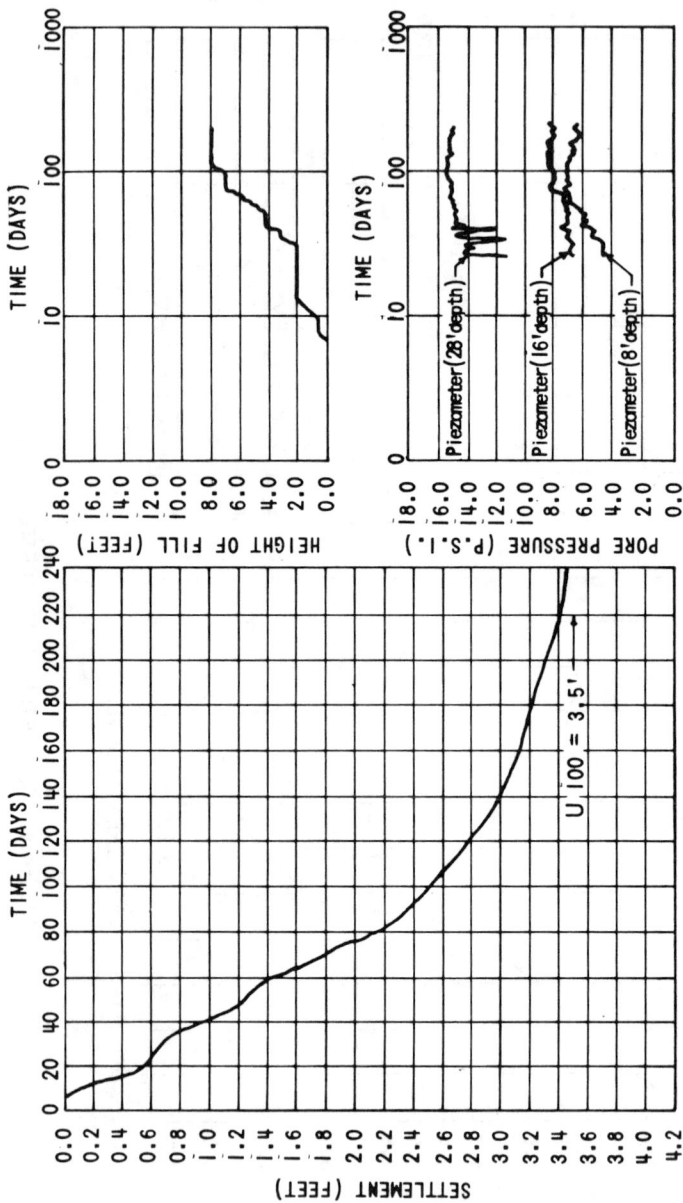

Figure 9. TEST FILL DATA (MEBRADRAIN WICK TEST SECTION)

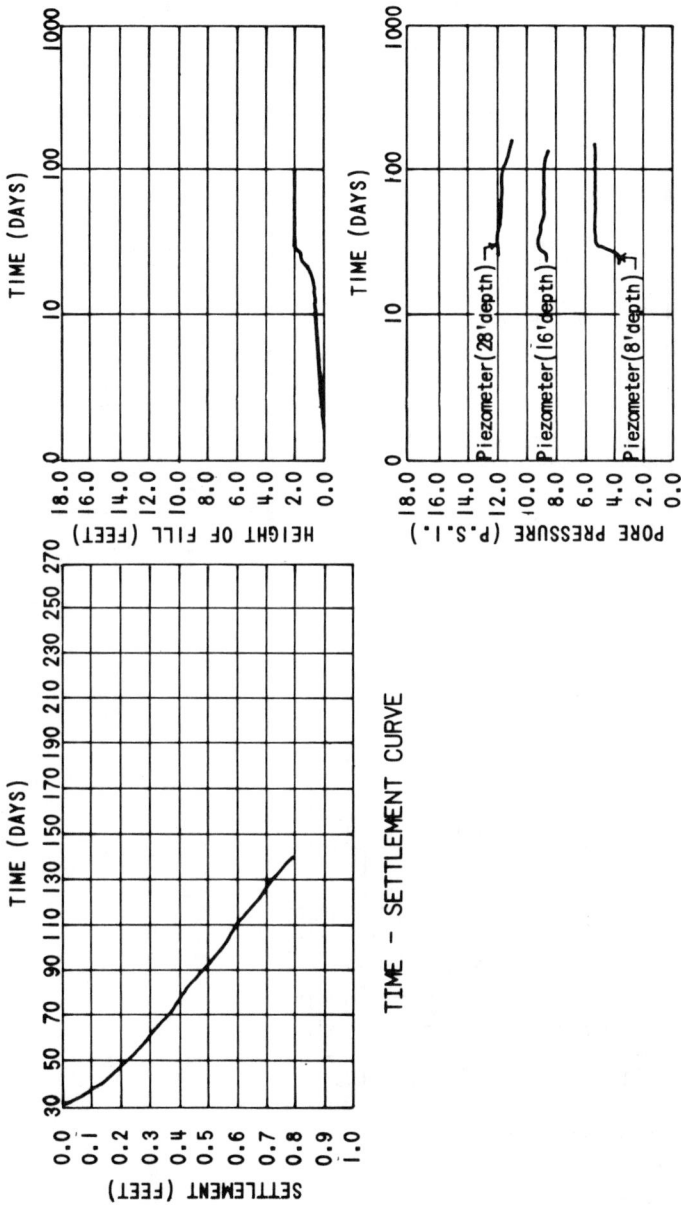

Figure 10. TEST FILL DATA (CONTROL TEST SECTION)

TEST FILL
DESOL WICK DRAIN

TEST FILL
MEBRADRAIN WICK DRAIN

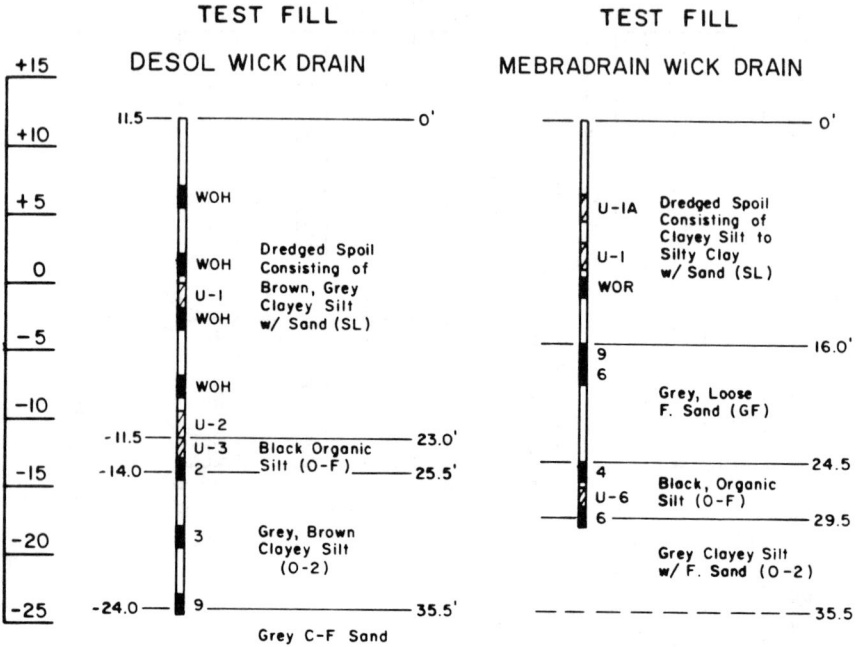

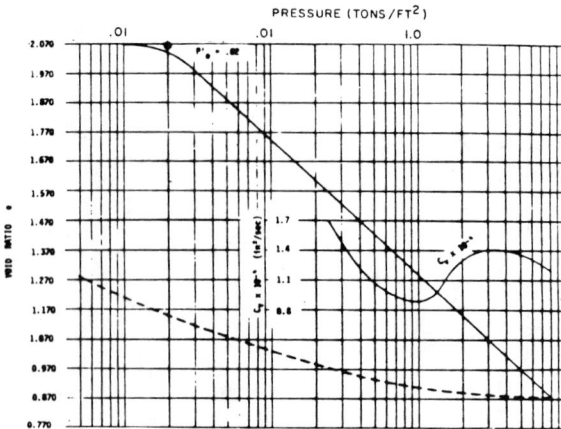

MEBRADRAIN WICK
TEST FILL I
SAMPLE IA
5.5'-7.5'
SL STRATUM

Figure II. TEST FILL BORING LOGS

Points were selected for a certain percent consolidation for a time period, t. The percent consolidation was used to obtain a value for T. The distance of the drainage path, H, is the radius of the de-watered cylinder. For a five foot square wick spacing, the drainage path, H = 2.8 ft (0.9m). The coefficients of consolidation calcu-lated by this method were 0.06 ft^2/day (6.4 x 10^{-4} cm^2/sec) for both wick sections.

The installation of wick drains accelerated the rate of consolidation considerably. The theoretical times required to achieve 90% consoli-dation without the presence of wicks were 16 years for the Desol wick test section and 3 years for the Mebradrain wick test section.

These C_v values compare well with those obtained in the vertical direction from the laboratory consolidation test and also from the large strain slurry consolidation tests performed at Northwestern Uni-versity. Further details on the latter tests are the subject of a separate paper.

A more valid assessment of the value of the wicks is possible by com-paring settlement rates between the control test fill and the test fills with wicks. Evaluation of average rates of settlement for each test fill section for periods when fill height was approximately 2 ft (0.6m) indicates that the test fills with wicks settled 3.7 times fas-ter than the control fill.

SITE STABILIZATION DESIGN

The test fill results were used in the design of the site stabiliza-tion contract for Berths I and II. This contract, which is scheduled to start in September, 1984, will include the placement of geotextile and a sand blanket, followed by the installation of wick drains. The Berth I area will then be surcharged under a load of 1,200 psf (57kPa) until early 1986 when the surplus surcharge will be rolled over to the Berth II area.

The rate of consolidation determined in the test fill program was used to calculate the required wick spacing (3). The wicks in Berth I will be spaced 6.0 ft (1.8m) on centers while the wicks in Berth II will be spaced 7.5 ft (2.25m) on centers. The wider wick spacing in Berth II is due to the longer period available for the soil surcharge in this area to remain in place.

CONCLUSIONS

Dewatering by progressive trenching was performed in the dredged slur-ry area during the summer of 1983 to facilitate crust formation. While this was only partially successful in that only 4 in (10cm) to 6 in (15cm) of crust was formed in the central area of the slurry, positive drainage was established and lowering of the water table is slowly occurring.

Settlement due to self-weight consolidation in the slurry area between completion of dredging operations, in April, 1982 and April,
1984, has varied between 6 in (15cm) and 1.5 ft (0.5m) over the site.
The water table during the last 6 months, after positive drainage
has been established, has dropped from elevation 10.5 ft (3.2m) to 10
ft (3.0m) as of April, 1984.

Test fills constructed on the slurry have indicated the following:

1. For a surcharge loading of 7.5 ft (2.3m) of fill, (825 psf,
    39kPa) and an average slurry thickness of 20 ft (6.1m), the set-
    tlements due to shear displacement were 0.6 ft (20cm) and 0.8 ft
    (26cm).

2. The ratio of actual to theoretical primary consolidation was 0.8
    for the Desol wick section and 0.9 for the Mebradrain wick sec-
    tion. The Terzaghi one-dimensional consolidation theory pro-
    vided reasonable predictions of primary consolidation for large
    strain settlement at the Seagirt site.

3. The installation of the wick drains accelerated the rate of con-
    solidation. Extrapolation of the wick drain test section time-
    settlement curves show that 95% consolidation will be achieved
    within 8 months from the start of loading. The average rates of
    settlement for the test sections with wicks have been 3.7 times
    more rapid than the control test section without wicks.

4. The time-settlement curves for the Mebradrain and Desol wick
    drains were fairly similar, yielding equivalent values for the
    coefficient of consolidation. Thus, it can be concluded that
    the two wick types performed equally well.

REFERENCES

1.   Broms, B. B. and Bowman, P., "Lime Columns. A New Foundation
Method," Journal of Geotechnical Engineering Division, ASCE, Vol.
105, No. GT 4, 1979.

2.   Cargill, K. W., "Procedures for Prediction of Consolidation in
Soft Fine-Grained Dredged Material," Technical Report D-83-1,
U.S. Army Waterways Experiment Station, 1983.

3.   Hansbo, S., "Consolidation of Clay by Band-Shaped Prefabricated
Drains," Ground Engineering, Vol. 11, No. 5, 1979.

ACKNOWLEDGEMENTS

STV/Lyon Associates, Inc. performed this investigation under a contract with the Maryland Port Administration. Mr. Gregory Halpin, Ad-
ministrator,   Mr. Robert Nelson, Director of Engineering, and Mr.
Neal Hasson, Chief of Design, were most cooperative in accommodating
various changes in the stabilization study as information was revealed during the investigation and as field conditions changed. The
Interstate Division of Baltimore City, the contracting agency for
Fort McHenry Tunnel, allowed access to the site and were of great
assistance in providing records of all prior investigations.   Dr.
Raymond Krizek was a consultant through all phases of the investigation, conducted the slurry consolidation tests, and contributed significantly to the overall program. STS Consultants, Ltd. were a subcon-
tractor on part of the investigation, particularly the probe testing
and sampling, and Mr. Barry Christopher, of STS, was of particular
assistance in evaluating geotextiles for the test fills.

Prediction of Viable Tailings Disposal Methods

J.D. Scott* and G.J. Cymerman**

Abstract

   A sand-fines-water diagram is proposed as a convenient tool to
analyze and describe numerous properties and behaviour of slurries
composed of coarse and fine particulate matter. The diagram can be
used to predict and plan disposal methods for slurried waste and
dredged material.

   Syncrude Canada Ltd. is considering a number of scenarios for
changing the nature of its tailings to a more amenable material
which consolidates more rapidly and can be disposed of in a smaller
storage area. The properties and behaviour of this tailings are
used as an example for developing a sand-fines-water diagram.

   At the Syncrude Canada Ltd. Oil Sands Plant the tailings stream
amounts to 130 x 10^6 tonnes of extraction tails every year. This
slurry, at a solids content of 50 to 55 percent, is a segregating
mix and therefore on deposition separates into a sand deposit and a
fines stream composed of silt and clay and some lost bitumen. This
fines stream settles out in the pond to form a high void ratio soil
matrix which consolidates slowly.

   As a result, approximately 14 x $10^6 m^3$ of liquid sludge is
formed every year and must be stored for many years at considerable
cost. The area required for the tailings pond to store the waste
sand and sludge is approximately 12 km^2. Environmental and economic concerns exist about the future reclamation of the tailings
pond and burial of mineable grade oil sand beneath the tailings
disposal area.

   Boundaries between sedimentation and consolidation states,
segregating and non-segregating mixtures, and fluid and solid states
are shown. Experimental programs to determine these boundaries are
included. The options being investigated by Syncrude Canada Ltd.
for new tailings disposal methods are discussed.

*AOSTRA Professor, Department of Civil Engineering, University of
Alberta, Edmonton, Alberta, T5G 2G7.

**Associate, Development - Mine and Extraction, Syncrude Canada Ltd.
10030 - 107 Street, Edmonton, Alberta, T5J 3E5.

1.0 Introduction

1.1 Tailings Disposal Problems

Waste disposal has always been a significant part of mineral processing. However, due to ever decreasing ore grades, and increasing environmental concerns, the costs of waste disposal per unit of valuable commodity are increasing at a fast pace.

It is estimated (2) that mineral and mining wastes in the U.S.A. amount to 1,300,000,000 tonnes per year, or 30 kg per capita per day. The problem is aggravated by the fact that almost all of this waste is in the form of slurry. It is becoming a rare occurrence that waste disposal into lakes, seas or rivers is permitted.

Moreover, reclamation of disposal sites after cessation of mining activities is often required.

Some coarse grained wastes lend themselves to relatively easy reclamation techniques involving gravity drainage, capping with topsoil, and revegetation.

However, in numerous cases, the presence of fine grained solids precludes this approach. This is the case of industries like:

- phosphate mining;
- heavy minerals in Australia and South Africa;
- oil sands in Canada;
- alumina, and
- some dredging operations.

The traditional approach to the disposal problem of fine grained waste and particularly waste clays has been to store the wastes in large impoundments surrounded by earth dams, taking advantage of terrain configuration where practical.

The disadvantages of such impoundments often include:

- deeper burial of potential ore bodies;
- high cost of construction and maintenance of the dams;
- slow rate of consolidation of fines, and
- lack of environmental appeal.

Ideally, clay waste should be solidified, placed in the mined out area and the surface reclaimed to the original state.

1.2 Tailings Disposal Techniques

Figure 1 illustrates simplified classification of common waste disposal techniques. The question mark relates to uncertainties regarding both:

- regulatory measures, and
- availability of alternative techniques.

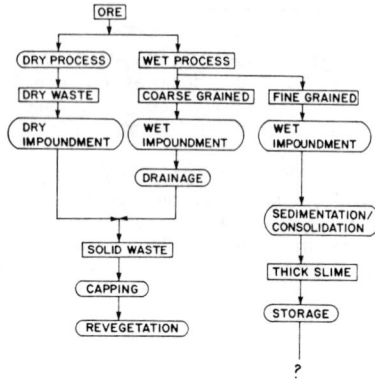

Figure 1.   Mineral Waste Disposal Chart

Considerable research and development effort continues in the area
of economical methods for solidification of fine grained wet waste.
Some of the possible approaches are illustrated in Figure 2.

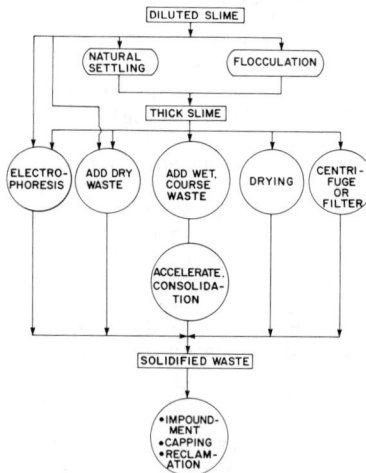

Figure 2.   Fine Grained Waste Solidification Chart

1.2.1   Electrophoresis

Solidification of fine grained slurries by electrophoresis has been demonstrated by Ritter (6). In an electrostatic field, clay particles deposit on an anode in a semisolid form. In this state, they can be impounded, capped and surface reclaimed.

However, the capital and energy costs involved in a large scale operation are prohibitive.

1.2.2   Centrifuging and Filtration

Several techniques were developed where fine grained slurry can be flocculated, and filtered or centrifuged to produce a semisolid soil (5).

Again, the capital and operating costs of these techniques on a large scale are prohibitive.

1.2.3   Drying

In desert areas, spreading fine grained slurries in thin layers results in satisfactory dewatering.

Unfortunately, many mining sites are located in climates that preclude this technique.

1.2.4   Dry Solid Waste Addition

In some applications, availability of dry solid waste, either coarse or fine, permits mixing of fine grained slurries with dry waste in such proportions that the product can be impounded and capped, or left for further consolidation until it becomes solid enough to be capped.

This application is obviously limited to those operations where dry solid waste is available in sufficient quantities.

1.2.5   Wet Coarse Waste Addition

There are at least three large industries that dispose of substantial quantities of coarse grained wet waste along with fine grained slurries. These are

- phosphate industry;
- heavy minerals - beach sand operations, and
- oil sand industry in Canada.

In each case it is advantageous to capture the fines in the voids of coarse grained material. At least three large scale operations

- Cooperative Farmers - Florida;
- Allied Eneaba - Western Australia, and
- Associated Minerals - Western Australia

successfully utilize this technique. If properly applied, the method offers a number of advantages:

- low cost;
- relatively fast land reclamation
- capture of fines in sand grain voids, and
- simplicity.

The major limitations in this case are:

- availability of coarse grained waste;
- ability to produce a nonsegregating mixture, and
- achievable rate of consolidation.

In order to successfully utilize this method, a good understanding of the system is necessary. This paper will describe a "Sand-Fines-Water Diagram" which has been developed as a planning tool for the evaluation of various waste disposal techniques involving mixtures of coarse and fine grained solids with water.

2.0 Sand-Fines-Water Diagram

The sand-fines-water diagram is proposed as a convenient tool to analyze and describe numerous properties and behaviour of slurries composed of coarse and fine mineral matter. The diagram can be used for predicting and planning disposal methods for most slurried waste or dredged material.

The determination of technically and economically viable disposal methods requires the combined input of metallurgical, chemical, hydraulic transport and geotechnical specialists. The properties of the slurries must be evaluated over a wide range from dilute slurries to solidified waste which can be safely capped and have the surface reclaimed for productive purposes.

To show the properties of a slurried material over such a wide range and include the change in properties when the ratio of fine to coarse material also changes, a ternary diagram (Figure 3) has been developed. Solids content (% of total mass) is shown on the right hand side and the percent of fines in the total dry solids is reported along the base of the diagram. These parameters have been found to be the most useful for all disciplines. For convenience in analyzing hydraulic transportation, the solids content by volume can

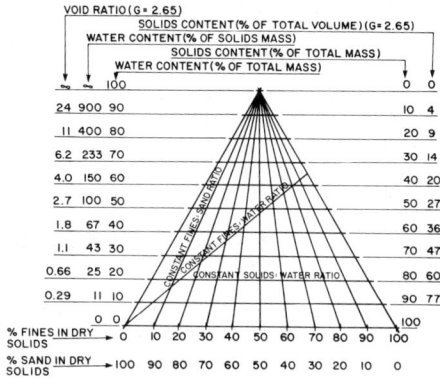

Figure 3.   Sand-Fines-Water Diagram

be included and for geotechnical engineering the water content and
void ratio can be added. If the fine and coarse materials have different specific gravities, however, the void ratio and volume solids
content will not plot quite horizontally.

A slurry with a specific amount of water (or fluid), fine solids and coarse solids (termed sand for convenience) will, of course,
plot as a point on the diagram. As the slurry consolidates it follows a constant fines:sand ratio path on the diagram.

As many properties of sand-fines-water mixes are controlled by
the fines-water ratio, these constant ratio paths are most useful in
predicting changes in properties during sedimentation and consolidation. Any line through the 100% sand apex is a constant fines-water
ratio path. The grain size split between fine and coarse material
is dependent on the grain size distribution of the mix and on the
grain shapes of the fine particles and these cause the waste or
dredged materials to segregate at a specific size. This size can be
determined by the segregation tests discussed later. For some slurries, the #200 sieve (74 microns) is an appropriate dividing size
although in the oil sands industry the #325 sieve (44 microns) has
been traditionally used. Recent testing at Syncrude indicates that
the 22 micron size may be more appropriate.

3.0   Properties of Slurried Materials

Geotechnical engineers are familiar with sedimentation and consolidation of fine grained materials and with their change in properties as their water content decreases from the liquid to plastic
to shrinkage limit.   Planning tailings disposal methods, however,
requires a broader understanding of the material characteristics.
All of these properties can be shown on the sand-fines-water diagram
and the Syncrude tailings are used as an example to show how the
slurry properties change with solids content and fines content.
Examples on how to use the diagram to plan and predict viable tailings disposal options are also given.

The Syncrude Canada Ltd. operation is an open pit oil sands
mine where 90×10^6 metric tonnes of oil sand are mined and processed annually.   The oil sand is a dense, cohesionless, uniform
quartz sand containing approximately 11% bitumen and 5% water (of
total mass) in the voids.   Approximately 85% of the mined solids are
sand and 15% are fines (finer than 44 microns).   Water is added in
the bitumen extraction plant to separate the bitumen.   The tailings
stream amounts to 130×10^6 metric tonnes every year and the tailings pond area required to store the waste materials is approximately 12 square kilometres.   The general layout of the Syncrude
site is shown in Figure 4.

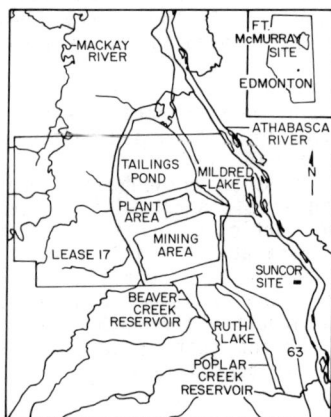

Figure 4.   Location of Syncrude Site on Lease

The tailings slurry pumped at a solids content of 50 to 55 percent segregates on deposition and the sand containing some fines is used to build the pond dykes and beaches while a fines stream composed of silt and clay and some lost bitumen flows into the pond. Approximately 44×10^6 m^3 of sand is deposited per year. The fines stream settles out and consolidates to a solids content of 20% fairly rapidly but the large volume of mining results in approximately 14×10^6 m^3 of liquid slimes or sludge forming every year. Over one-half of the water in the tailings is returned to the plant for reuse following clarification by natural sedimentation. Environmental and economic concerns exist about the future reclamation of the tailings pond and the burial of mineable grade oil sand beneath the tailings disposal area.

The grain size distributions of the dyke and beach sands and of the pond sludge are shown in Figure 5. Various mixes of these two materials have been tested to determine properties and locate the boundaries of changes in behaviour shown on the sand-fines-water diagram in Figure 6. Also shown in Figure 6 are the extraction tailings stream, the average sand deposit in the dykes and beaches, and the range of fine sludge in the pond. Figure 7 (1) is a depth profile in the sludge showing the depth and the state of consolidation by the summer of 1983.

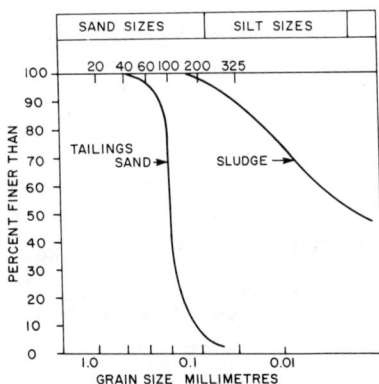

Figure 5. Grain Size Distribution

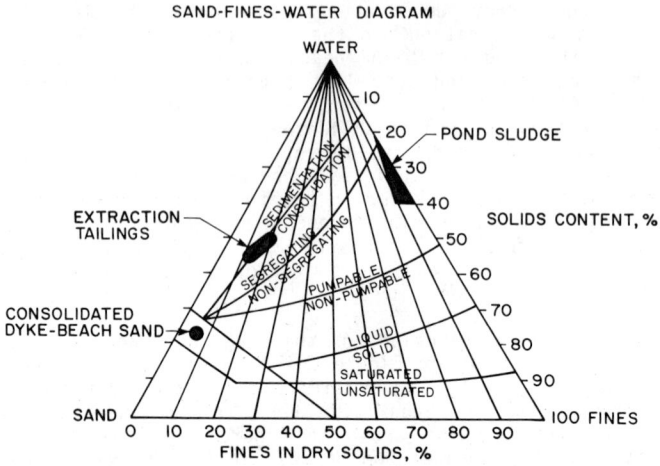

Figure 6. Boundaries for Syncrude Tailings

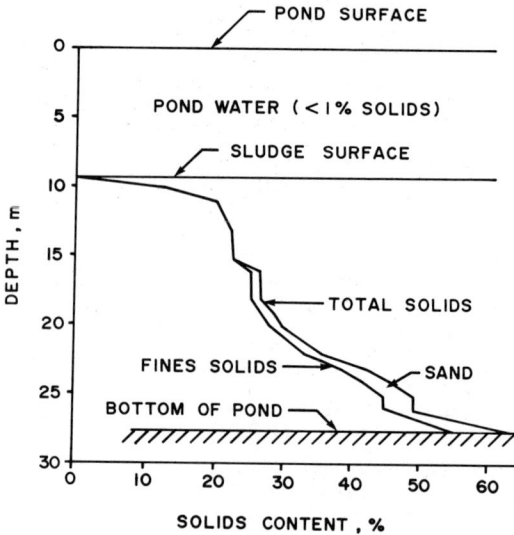

Figure 7. Sludge Solids in Tailings Pond

3.1 Sedimentation - Consolidation Boundary

This boundary shows at what solids contents effective stresses between the particles first occur. The sludge fines develop a structure between 12 and 16% solids depending on their gradation. It sediments to this value within days or several weeks after deposition and consolidates to approximately 20% solids in several months as shown in Figure 7. On the left hand side of Figure 6, where 100% sand is plotted, the sedimentation-consolidation boundary is the maximum void ratio of the sand determined by sedimentation through water.

3.2 Segregating - Nonsegregating Boundary

The existence of such a boundary has been recognized in the pipeline solids transport field for some time (4), (10) and has been variously described as settling - nonsettling behaviour or as a heterogeneous flow - homogeneous flow in pipelines. Determining the nonsegregating zone is also the basis for the thickened tailings disposal technique developed by Robinsky (7). Above this boundary the coarse particles will settle through the fines-water slurry and therefore segregate, while below the boundary no appreciable settling takes place regardless of the concentration of the coarse particles.

The tests developed to locate this boundary required the determination of a test procedure and a definition of segregation. Standpipe tubes of different heights containing various mixes were left for different lengths of times and the solids contents down the tube were then measured. A segregation index was defined as the change in solids content divided by the initial solids content and an average index for the tube of over 10% was considered to show segregation. To reduce end effects, a tube height of at least 90 cm was found necessary (Figure 8). It was found that if segregation was going to occur, it took place within several days but all tests were carried on for at least 30 days. Typical nonsegregating and segregating test results are shown in Figure 9.

Test results for the Syncrude sand-sludge material are shown in Figure 10. The boundary is fairly sharp and only a small change in solids content changes a mix from segregating to nonsegregating. As would be suspected, the boundary approximately follows a constant fines-water ratio. In this case a fines concentration of 28% solids appears necessary to hold the sand in suspension. Similar types of tube tests to measure segregation have been done by Sakamoto et al (9) where the change in density with depth was measured by an external nucleur source.

There is an indication (8) that nonsegregation in standing slurries occurs at lower solids contents than in pipelines or in freshly deposited beaches. This difference may be the result of thixotropy in the standing slurries in contrast to the constant shearing action in pipelines and during beach deposition.

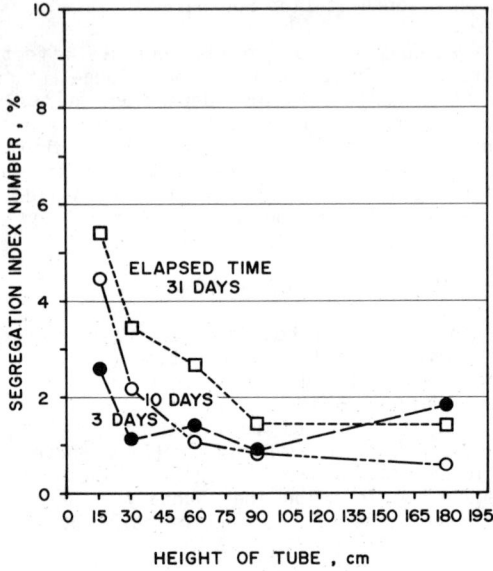

Figure 8. Segregation Index of 50% Fines with Tube Height

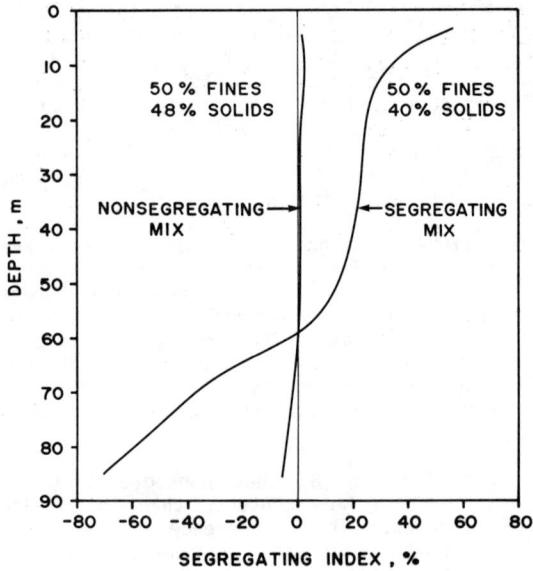

Figure 9. Segregating and Nonsegregating Mixes

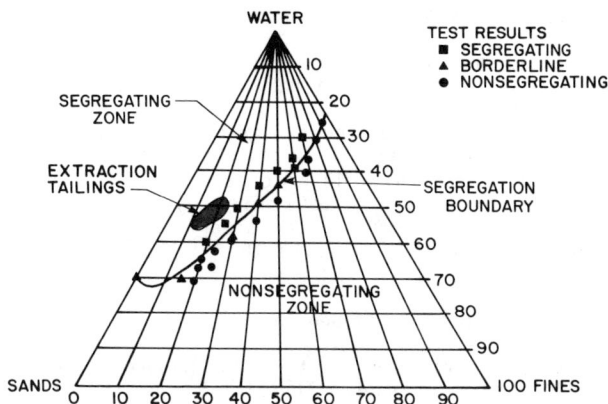

Figure 10.   Segregation of Sand-Sludge Mixes

Figure 9 also shows the tailings stream composition. It is a segregating mix and this characteristic is the cause of the fines segregation and sludge build up in the tailings pond. To prevent fines segregation it would be necessary to lower the water content of the tailings stream or to increase the fines content or to do both. Planning of alternate disposal schemes at Syncrude, in which segregation of fines does not occur, is based on this concept.

3.3 Pumpable - Nonpumpable Boundary

Slurries of very high solids contents can be pumped but only with high energy losses which may be excessively expensive. A pumpable boundary is therefore not a technical limit but is determined by the slurry properties and the economics of the disposal system.

Design parameters for a slurry pipeline are determined in a hydraulics laboratory where slurries of various concentrations within an appropriate operating range are pumped through a pipeline which can model the proposed field installation in pipeline size and essential topographic features. Measurements of velocity, head loss and slurry density are made.

Testing of slurries in a geotechnical laboratory can, however, provide data to enable the appropriate slurry compositions to be chosen for the more complex pumping tests. For the Syncrude tailings, contours of equal viscosity were plotted on the sand-fines-water diagram and coupling these with a limited number of pumping tests, the pumpable - nonpumpable boundary was estimated.

3.4 Liquid - Solid Boundary

The long term goal of a waste disposal program is to end with a waste material that has a consistency where it can be capped with more solid material and the waste area returned to productive use. At the Syncrude site the required consistency has been taken as being an undrained shear strength of 5 kPa. This slurry strength would allow a one metre thick layer of sand to be spread over the slurry mix. The boundary was determined by plotting contours of equal undrained shear strength on the sand-fines-water diagram. Of interest, the boundary lies close to the liquid limit line for the slurry mixes.

3.5 Sand Matrix - Fines Matrix Boundary

As the quantity of sand in the mix increases, the mix will change from a fines-water slurry containing sand grains distributed through it to a sand matrix where the sand grains are touching one another and the fines-water slurry is in the voids between the sand grains. The matrix boundary shows at what solids content and fines content this occurs. Along the matrix boundary, the sand structure is at its maximum void ratio.

Testing shows that this boundary is not sharp and as it is approached from the right, the viscosity, pumping energy, and shear strength all increase rapidly.

3.6 Saturated - Unsaturated Boundary

The slurry voids remain saturated with water until the saturated - unsaturated boundary is reached. For the high fines mixes this boundary is, of course, the shrinkage limit of the mix while in the sand matrix region it is the minimum void ratio of the beach sand after gravity drainage. The volumetric scales (void ratio and solids by volume) on the diagram are not valid in the unsaturated zone.

4.0 Syncrude Tails Disposal Case

4.1 Waste Materials at Syncrude

In order to illustrate the case of the Sand-Fines-Water diagram, a variety of waste materials available at the Syncrude operation in Alberta have been plotted in Figure 11.

4.1.1 Extraction Tails

The area marked "Extraction Tails" represents the range of compositions of total Extraction tails combined in a single stream. On disposal, this stream segregates immediately into a Fresh Beach and Thin Sludge also shown on the diagram.

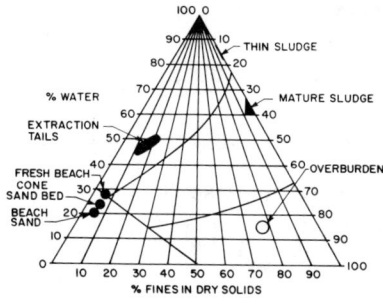

Figure 11. Waste Materials at Syncrude

4.1.2   Cone Sand Bed

Construction of four conical vessels each 16.8 m (55 feet) in diameter is underway. The primary function of these units will be to recover 2 to 3 million barrels of bitumen annually from the Extraction tails. Figure 12 shows the cross-section of the unit. Sand settles in the cone to form a contact bed. The composition of that material is shown on the diagram as Cone Bed Sand in Figure 11. Excess water with fines suspended in it will overflow the vessel as Thin Sludge.

4.1.3   Fresh Beach

Cone Bed Sand and Fresh Beach almost overlap on the diagram as the conditions of their formation are similar although the location is different. Fresh beach will gradually dewater due to gravity drainage. As a result, clear water will be released and the solids concentration in the beach will increase. The result of this process is Beach Sand.

4.1.4   Beach Sand

The composition of Beach Sand is shown in the bottom left corner of the SRW diagram.

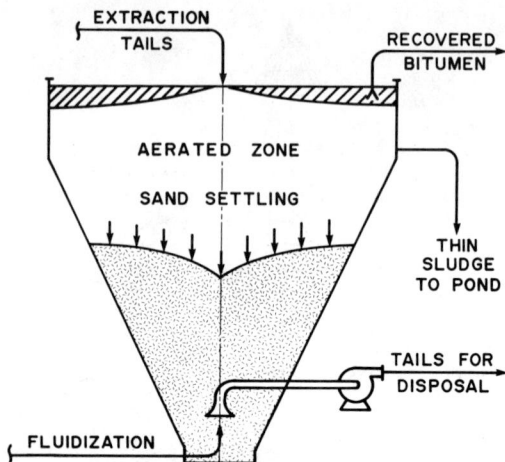

Figure 12. Tails Oil Recovery Cone

4.1.5   Thin Sludge

Thin Sludge is shown at the right side at the top of the diagram, it is also called Middlings. This stream is created everytime that settling of tails is allowed. It is a suspension of fine grained solids in water which enters the tailings pond.

Within a few weeks the fine solids settle to form a heavy, viscous sludge, releasing excess water, which is reclaimed and recycled to Extraction.

4.1.6   Mature Sludge

Extensive sampling of the Syncrude tailings pond shows that the concentration of the solids in the tailings pond sludge near the bottom exceeds 35% by mass. This is marked on the SFW diagram.

4.1.7   Overburden

Various facies of the overburden are marked on the SFW diagram as Pg, Pl$_2$, Kcw, Kc, and Kcb. An average blend of those facies would result in a composition shown in the left bottom corner, marked as O/B.

4.2   Discussion of Selected Schemes

Having all the available materials marked, a few schemes illustrating the use of the SFW diagram will be demonstrated.

4.2.1   Cone Sand/Sludge Mix

This scheme is illustrated in Figure 13. Mature sludge from the tailings pond is brought by pipeline and injected in the bottom of the Tails Oil Recovery cone thickener. There it mixes with Cone Sand Bed before it is withdrawn by a pump and pipelined to the final disposal area in the mined out pit. If the mixture does not segregate, sand should remain in suspension in the thick sludge.

However, due to a high sand to fines ratio of 5 to 1, a relatively high consolidation rate is expected. Figure 14 shows the rate of consolidation for this mixture calculated by finite strain consolidation theory by Carrier (1984). The excess water will drain, and will be recycled to the tailings pond.

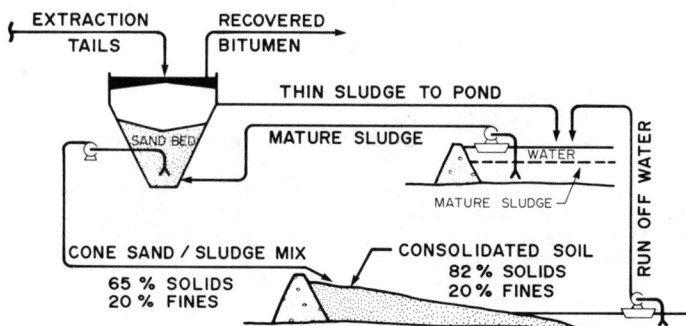

Figure 13.   Cone Sand/Sludge Mix

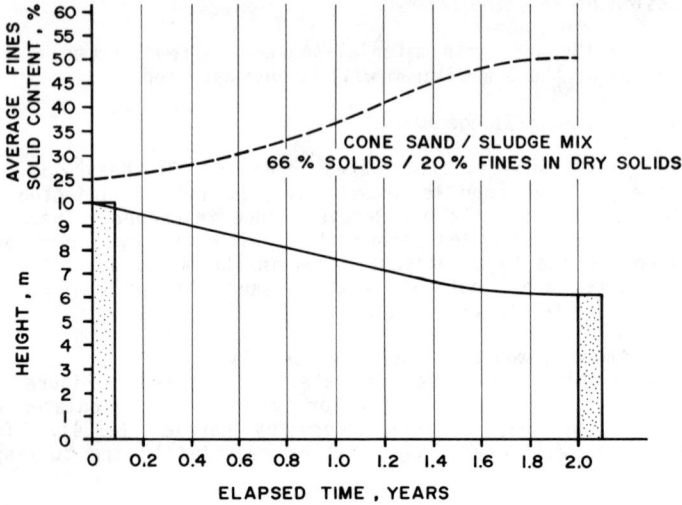

Figure 14.   Consolidation vs Time

After achieving sufficient strength of the deposit at perhaps 80% solids by weight, a sand cap will be placed on top of the mixture and reclamation of the area will start.

Among the advantages of this scheme are:

- relative simplicitiy,
- low cost, and
- maximized utilization of existing equipment.

The disadvantage relates to uncertainties regarding slurry behaviour upon discharge, due to proximity to the segregation line. In reality, the mixture may segregate on disposal releasing some run off sludge.

Figure 15 shows the representation of the Cone Sand/ Sludge Mix on the SFW diagram.

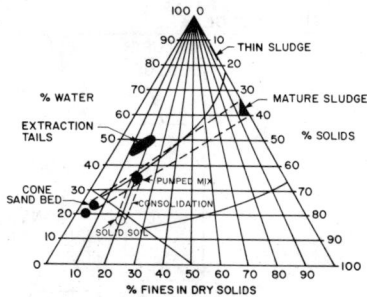

Figure 15.   Waste Materials at Syncrude - Cone Sand/Sludge Mix

4.2.3   Beach Sand/Sludge Mix

In this scheme, tailings are beached as usually, and allowed to drain by gravity to approximately 80% solids content by weight. Sand is then picked up by dry mining techniques, and mixed with mature sludge pipelined from the tailings pond. The scheme is illustrated in Figures 16a and 16b. Depending on the proportions of sand and sludge, a dry discharge or pipeline discharge can be arranged.

Considering that some 50 million tonnes of the material must be discharged annually, it is preferable to use pipelines rather than belt conveyors due to lower cost, although again, the rate of consolidation of the mix is of critical importance.

Figure 17 shows the case on the SFW diagram.

The obvious advantage of this scheme is that the solids concentration of the initial mix is higher and therefore, less consolidation is required to produce a stable deposit.

The disadvantage is the higher cost and increased complexity of the system.

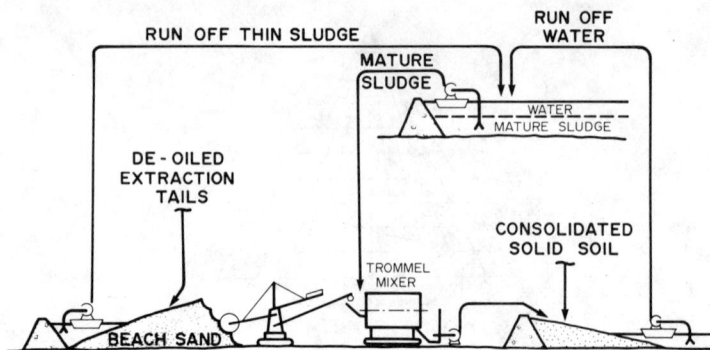

Figure 16a.   Beach Sand/Sludge Slurry Mix

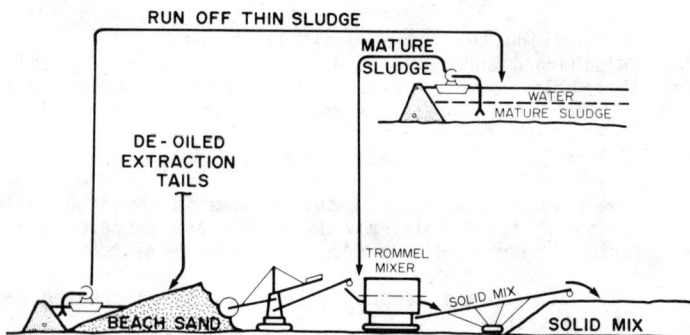

Figure 16b.   Beach Sand/Sludge Solid Mix

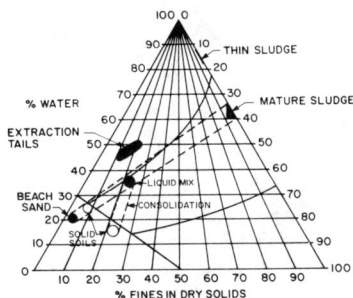

Figure 17. Waste Materials at Syncrude - Beach Sand/Sludge Mix

4.2.4   Sand/Sludge/Overburden Mix

        This example involves addition of solids to the mature
sludge, as shown on the SFW diagram in Figure 18. By doing so,
the solids concentration in the sludge can be artificially
increased to approximately 45% by weight. The sludge/O/B mixture can now be mixed with cone sand bed in the same manner as
in Beach Sand/Sludge mix. Figure 19 illustrates this case.

        The advantage of this technique is that the overall mix
is further away from the segregating boundary, and therefore,
the danger of segregation is more remote.

        The disadvantages are a higher cost and a greater com-
plexity.

5.0 Discussion and Conslusions

        A sand-fines-water diagram is proposed as a planning tool to
predict viable tailings disposal methods. Various properties and
characteristics of the slurry mixes can be shown in the diagram in a
convenient and consistent form.

        Although some of the properties of slurries, such as, sedimen-
tation, consolidation, liquid and solid limits are familiar to geotechnical engineers, other properties which are also needed to fully
characterize the behaviour of slurry mixes have not been fully

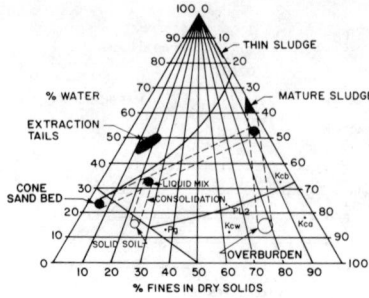

Figure 18.   Waste Materials at Syncrude - Sand/Sludge/Overburden Mix

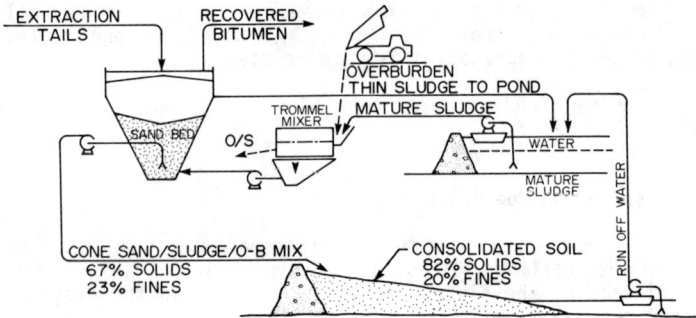

Figure 19.   Cone Sand/Sludge/Overburden Mix

appreciated in waste disposal design methods. Segregating - non-segregating mixes, pumpable - nonpumpable limits, and sand matrix - fine matrix zones are defined and test procedures to determine those boundaries are discussed.

The properties of the Syncrude Canada Ltd. tailings are shown as an example of how the sand-fines-water diagram can display the properties of various mixes and their relationship to one another.

The use of this diagram by Syncrude has guided the direction of research and testing on the tailings materials, and aided in economic studies on tailings disposal methods.

The promising methods developed to date include the use of a tails oil recovery cone thickener to provide cone sand which when mixed with mature sludge produces a nonsegregating mix which will consolidate faily rapidly. A large scale beach test of this mix is planned for the summer of 1984.

Another promising method is a sand-sludge-overburden mix which is in the early stages of testing. The disadvantage of increasing the volume of the tailings by adding overburden to it appears to be compensated for by the complementary interaction of the tailings sludge and fine grained overburden and the saving in cost of transporting the overburden to a disposal site.

The discussion of these possible modes of tailings disposal shows how fairly complex schemes can be evaluated after the appropriate boundaries have been determined and plotted on the sand-fines-water diagram. Such evaluations will then serve as a guide to further research and design studies.

APPENDIX I - REFERENCES

1. Bromwell Engineering, Inc., "Geotechnical Investigation of Mildred Lake Oil Sand Tailings Sludge Disposal", Report to Syncrude Canada Ltd., Dec., 1983, 104 p.

2. Bromwell, L.G., "Evaluation of Alternative Processes for Disposal of Fine-Grained Waste Materials", Seminar on Consolidation Behaviour of Fine Grained Waste Materials, Bromwell Engineering Inc., Oct. 1982, 26 p.

3. Carrier, III, W.D., "Analysis of the Tertiary Oil Recovery Vessel," Bromwell and Carrier Inc., Report to Syncrude Canada Ltd., February, 1984.

4. Charles, M. E. and Charles, R. A., "The Use of Heavy Media in the Pipeline Transport of Particulate Solids", Advances in Solid-Liquid Flow in Pipes and its Application, Pergamon Press, 1971, pp 187 - 197.

5. Liu, J. K., Lane, S. J., Cymbalisty, L. M. O., "Filtration of Hot Water Extraction Process Whole Tailings", U.S. Patent #4225433 granted to Petro-Canada Exploration Inc., Alberta Government, Ontario Energy Corp., Imperial Oil Ltd., Canada-Cities Service Ltd., Gulf Oil Canada Ltd., 30 Sept. 1980, 6 p.

6. Ritter, R. A., "Tailings Water Reclamation", Seminar on Advances in Petroleum Recovery and Upgrading Technology, AOSTRA, Calgary, May, 1981, 32 p.

7. Robinsky, E. I., "Tailing Disposal by the Thickened Discharge Method for Improved Economy and Environmental Control", Tailing Disposal Today, Vol. 2 Proc. of Second Int. Tailing Symposium, Denver, Colorado, Miller Freeman Publications, 1979, pp 75-95.

8. Robinsky, E. I., "Thickened Tailing Disposal Feasibility Study, Laboratory Tests for Determination of Thickened Tailing Properties, Progress Report No. 2", Report to Syncrude Canada Ltd., May, 1982, 41 p.

9. Sakamoto, M., Uchida, K. and Kamino, Y., "Transportation of Coarse Coal with Fine Particle - Water Slurry", Proceedings of Hydrotransport 8, 8th Int. Conf. on the Hydraulic Transport of Solids in Pipes, Johannesburg, South Africa. Aug. 25-27, 1982, pp 433-443.

10. Wasp, E. J., Kenny, J. P. and Gandhi, R. L., "Solid-Liquid Flow Slurry Pipeline Transportation", Trans Tech Publications Series on Bulk Materials Handling, 1977, 224 p.

WASTE PHOSPHATIC CLAY DISPOSAL IN MINE CUTS

Frank Somogyi[1], A.M., ASCE, W. David Carrier III[2], F.ASCE, James E. Lawver[3], and Jay F. Beckman[4]

ABSTRACT

A major research effort was undertaken to assess the applicability of waste clay disposal back into the mine cuts to the phosphate industry. The program included a large scale field test, extensive material characterization, and the development of predictive capabilities.

The field test involved the disposal of waste phosphatic clay into an abandoned mine cut approximately 40 feet deep, 300 feet wide and 1200 feet in length. Nearly 50,000 tons of clay at an initial solids content of 16% (void ratio of 14) were pumped into the test pit. The pit was monitored for almost two years, with periodic measurements of clay surface elevation, boundary piezometric levels, and solids content and pore pressure distribution. Material characterization included the measurement of compressibility and permeability by means of incremental loading and constant rate of deformation slurry consolidation tests, and insitu tests.

Preliminary predictions of consolidation behavior were made by formulating "engineering approximations" based on existing one-dimensional finite strain solutions. This method indicated that little lateral drainage could be expected, but could not incorporate the observed underseepage. A "quasi-two-dimensional" finite strain model was developed and found to produce results in close agreement with observed behavior.

INTRODUCTION

A possible method for the containment of fine-grained tailings associated with some strip mining operations is their disposal back into the mine cuts themselves. This method is illustrated in Figure 1 for a case in which the stripped overburden is piled in windrows, as is commonly done when draglines are employed. The technique obviously represents an economically attractive alternative to the conventional tailings impoundment within diked containment areas. In addition, the overburden rows provide easy access to the interior of the disposal area and thus may facilitate the placement of a soil cap after a desiccated crust has formed. Depending on their permeability and spacing, the overburden rows may also accelerate the consolidation of the fine-grained tailings by providing additional drainage.

[1] Assoc. Prof., Northwestern University, Evanston, IL
[2] Pres., Bromwell and Carrier, Inc., Lakeland, FL
[3] Tech. Mgr., IMC Corp., Bartow, FL
[4] Proj. Engr, Bromwell and Carrier, Inc., Lakeland, FL

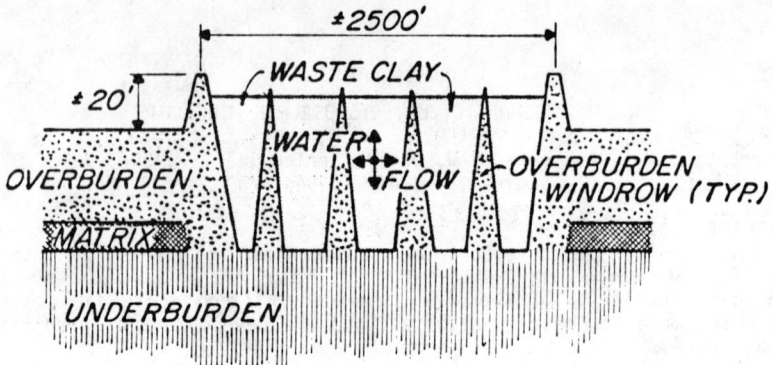

Figure 1. Typical Mine Disposal Area Resulting in Two-Dimensional Clay Consolidation.

As with any other tailings disposal strategy, the technical assessment of the above procedure requires accurate estimates of consolidation magnitude and rate during and after filling and subsequent capping. Because of the trench-like, or two-dimensional, geometry of the mine cuts, one-dimensional large strain consolidation theory may be inappropriate for the necessary estimates. A major research effort, including a full-scale field test, extensive tailings characterization and development of predictive capabilities, was undertaken to assess the applicability of this disposal alternative to the phosphate mining industry.

MATHEMATICAL MODELLING

Two approaches were adopted for modelling the consolidation behavior of the waste clay deposited in a mine cut. The first involved the rational extrapolation of one-dimensional large strain analyses to take into account two or three dimensional containment geometries. These solutions were intended for inexpensive preliminary estimates, and were termed "engineering approximations." The second approach consisted of the formulation and numerical solution of "quasi-two-dimensional" finite strain consolidation.

"Engineering Approximation"

An approximate technique for estimating the consolidation behavior of slurried mineral wastes deposited within essentially one-dimensional containment areas (i.e., a relatively uniform depth which is small compared to the areal extent) was presented by Carrier et al (1983). The basic equation is

$$h = a\left(\frac{W}{A}\right)^b t^{(c-b)} \tag{1}$$

where h = height of sediment
      W = dry weight of sediment solids
      A = size of disposal area
      t = time
   a,b,c = empirical coefficients curve-fit from a
           limited number of one-dimensional computer
           analyses

Equation (1) can be solved for W, producing

$$W = At^{(b-c)/b}\left(\frac{h}{a}\right)^{1/b} \tag{2}$$

which expresses the maximum amount (dry weight) of sediment that could
be placed in a disposal area of a given size and height within a
certain time period.

The consolidation behavior of material contained in irregular-
shaped disposal areas can be approximated by taking vertical slices
through the sediment and treating each resulting vertical strip (or
column) as a one-dimensional problem. In addition, the distribution
of dry solids was allowed to vary so that the overall surface remains
level. Carrier et al (1981) presented equations applicable to
"stepped," V-shaped and trapezoidal containment areas. In particular,
the "engineering approximation" for the self-weight consolidation of
sediment deposited in a trapezoidal area (shown in Figure 2) was given
by

$$W = t^{(b-c)/b}\left(\frac{h}{a}\right)^{1/b}\ w\ell + \frac{2mbh}{b+1}\left(w+\ell + \frac{4mbh}{2b+1}\right) \tag{3}$$

where w = base width
       ℓ = base length
       m = inside slope of disposal area

"Quasi-Two-Dimensional" Finite Strain Consolidation

The problem addressed was the prediction of the rate of quasi-
two-dimensional self-weight consolidation of a uniform initial solids
content slurry deposited into a rectangular cross section disposal
area with frictionless sides.

The term "quasi-two-dimensional" consolidation refers to an
idealization of the actual process, whereby the soil particles are
constrained to move in the vertical direction, while the pore fluid
may move both vertically and horizontally. Utilizing this
idealization, initially rectangular control volumes (or numerical
mesh) remain rectangular throughout the consolidation process. While
this avoids the need for, and mathematical complexities associated
with, multi-dimensional constitutive relations and failure criteria,
it precludes the ability to model lateral particle movements due to
lateral seepage forces, or the "self-levelling" flow that may occur
near the surface of the consolidating deposit.

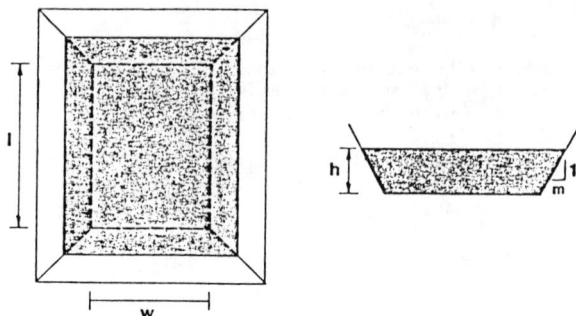

$$W = t^{\frac{b-c}{b}} \left(\frac{h}{a}\right)^{\frac{1}{b}} \left[wl + \frac{2mbh}{b+1}\left(w + l + \frac{4mbh}{2b+1}\right)\right]$$

Figure 2. "Engineering Approximation" for pseudo-Three-Dimensional
Consolidation in Trapezoidal Disposal Area.

The rectangular cross section was chosen for initial modelling
because of computational ease.   The assumption of frictionless
vertical boundaries was introduced to again avoid the need for multi-
dimensional constitutive relations and failure criteria.

The mathematical formulation paralleled the basic formulation
presented by Gibson, England, and Hussey (1967).   The present
formulation employs material coordinates in the vertical direction and
spatial coordinates in the horizontal direction, thus taking full
advantage of the idealization that particles are constrained to
vertical movement.

The equation describing the consolidation process is then given
by:

$$\frac{\partial e}{\partial t} + \frac{\partial}{\partial \zeta} \frac{-k_v}{\gamma_w(1 + e)} \frac{\partial u}{\partial \zeta} + (1 + e) \frac{\partial}{\partial x} \frac{-k_h}{\gamma_w} \frac{\partial u}{\partial x} = 0 \qquad (4)$$

where e   = void ratio
       k_v = vertical permeability
       k_h = horizontal permeability
       u   = excess over hydrostatic pore water pressure
       t   = real time
       x   = spatial ("real") horizontal coordinate
       ζ   = material vertical coordinate given by

$$\zeta(a) = \int_0^a \frac{da'}{1 + e_0(a')}$$

e_0 being the initial void ratio

a being the initial ("real") vertical position of the point in question

It should be noted that in Equation (4), the term $(1 + e)$ appears in the denominator of the "diffusion coefficient" in the vertical direction (and must be differentiated) and in the numerator of the horizonal "diffusion coefficient".

After introducing the following relationships (as in Somogyi, 1980):

$$e = A\bar{\sigma}^{-B} \qquad\qquad (5a)$$

$$\frac{\partial\bar{\sigma}}{\partial t} = (G - 1)\delta_w \frac{d\Delta\zeta}{dt} - \frac{\partial u}{\partial t} \qquad\qquad (5b)$$

where $\bar{\sigma}$ = effective stress = $\sigma_b - u$

σ_b = buoyant stress = $(G-1)\,\gamma_w\zeta$

A,B = curve-fit constants
G   = specific gravity of solids
γ_w = unit weight of water

$\frac{d\Delta\zeta}{dt}$ = term related to filling rate

Equation (4) can then be written in terms of excess pore pressure, as:

$$\frac{\partial u}{\partial t} + S\left[\kappa(1+e)\frac{\partial^2 u}{\partial x^2} + K\frac{\partial^2 u}{\partial\zeta^2} + (1+e)\frac{\partial\kappa}{\partial x}\frac{\partial u}{\partial x} + \frac{\partial\kappa}{\partial\zeta}\frac{\partial u}{\partial\zeta}\right] = 0 \qquad (6)$$

where $S = \frac{\bar{\sigma}^\beta}{\alpha}$

$\alpha = AB\gamma_w$

$\beta = 1 - B$

$\kappa = k_h$

$K = \frac{k_v}{(1 + e)}$

Equation (6) is the general equation describing the process and is to be solved numerically. The equation assumes constant submergence of the consolidating deposit. Details of the mathematical derivation can be found in Bromwell Engineering (1982).

Assuming perfect drainage all around, the boundary conditions are zero excess pore pressure (u) at all boundaries. In addition, the initial excess pore pressure distribution is identical to the buoyant stress distribution.

The Alternating Direction Explicit Procedure (ADEP) was chosen for the numerical solution to Equation (6). This procedure combines the computational ease of explicit procedures and the inherent stability, at least for linear problems, of implicit techniques. Although it is not certain that the ADEP is unconditionally stable when applied to Equation (6), it can be conveniently checked by trial.

A parametric variation study was performed in order to assess the effect of disposal area width and horizontal permeability on the rate of consolidation. The basic problem involves a 10-foot height with uniform initial solids content of 16%. The material properties used are the same as for the "engineering approximation" example.

The width of the disposal area was varied from 1.0 feet to 50.0 feet. The results of these analyses are presented in Figure 3. As expected, the width of the disposal area has a much greater effect on settlement rate for narrow areas than wide ones. Doubling the width from 1.0 feet to 2.0 feet produces a 350% increase in the time to attain 50% consolidation (from 3.6 to 12.8 days) and a 370% increase in the time to 90% consolidation. However, doubling the width from 10.0 feet to 20.0 feet causes only a 160% increase (from 108 to 170 days) for 50% consolidation and a similar amount for 90%. Coincidentally, the time for 50% consolidation doubles (55 to 110 days) when the width is increased from 5 to 10 feet, but the time for 90% consolidation increases 260%.

The case with 50.0-foot width is also of interest because the aspect ratio (width:height) of 5:1 is quite similar to that in the test pit, which will be presented later. Since the test pit is more than three times the size of this case, its consolidation rate can be expected to be 5 to 10 times slower.

The effect of possible anisotropic permeability, with horizontal permeability greater than the vertical, was also investigated for the basic 10.0-foot by 10.0-foot area. The results are shown in Figure 4. The time for 50% consolidation decreases from 110 days to 80 days if the horizontal permeability is twice the vertical and 55 days (half) if the horizontal is four times the vertical.

FIELD TEST DESCRIPTION

Site and Instrumentation

The field test involved pumping waste phosphatic clay into an abandoned mine cut surrounded by windrows of overburden. The test pit

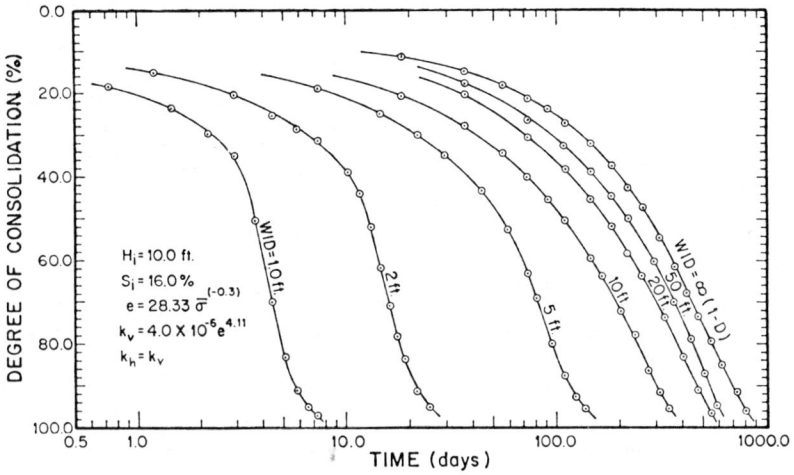

Figure 3. Effect of Area Width on Consolidation Rate.

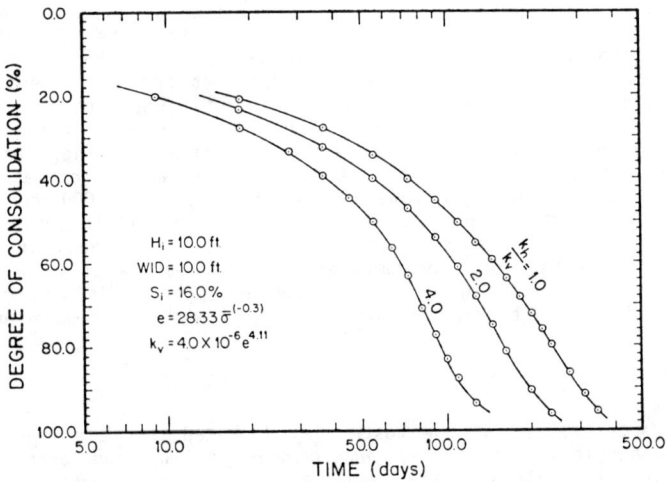

Figure 4.   Effect of Anisotropic Permeability on Consolidation Rate.

is approximately 270 feet wide and 1200 feet in length, corresponding to a surface area of 7.4 acres. The cross section of the pit is roughly trapezoidal in shape, with an average depth of about 40 feet. A plan view and three cross sections, indicating the pit geometry, topography and testing and instrument locations, is shown in Figures 5 and 6. The waste clay inlet pipe is located at the eastern end of the pit and a spillway to keep the water level inside the pit drained down to near the clay surface is located at the western end of the pit.

A grid of nine testing stations was established. All sampling, pore pressure measurements, and permeability tests in the clay were conducted at these stations. Permanent instrumentation included tide gauges in the pit and piezometers installed in the soil (overburden) surrounding the mine cut. Two tide gauges were installed in the test pit, one at each end of the pit. The tide gauges consisted of 24-inch diameter steel pipe embedded in the bottom of the pit in order to prevent movement. Four hydraulic and six pneumatic piezometers were installed in the soil surrounding the mine cut. These piezometers permitted continuous monitoring of the boundary pore pressures.

Filling History

The test pit was filled with gravity thickened waste clay dredged from a nearby tailings pond and discharged into the eastern end of the mine cut. Filling began in December, 1980 and continued intermittently for 220 days. The solids content (dry weight basis), measured during filling varied from 13.8 to 17.4 percent, with an average of 16.0 percent. The pit was filled to an average elevation of 160.3 by the end of April 1981. At this time, the difference in the elevation of the clay between the two ends of the pit was 0.7 feet, giving a slope of approximately 0.7 feet per 1,000 feet. The clay was then allowed to consolidate quiescently until mid-July 1981, at which time additional clay was pumped into the test pit, raising the average elevation to 161.7. After this final filling, the slope of the clay from one end of the pit to the other was approximately 1.2 feet per 1,000 feet. A total of approximately 50,000 tons (dry weight) of clay was pumped into the test pit. The filling history, as determined by measurements taken during pumping, is shown in Figure 7. The total weight of clay computed from the results of sampling the filled pit at the nine grid points shown in Figure 5 is also plotted in Figure 7 and can be seen to agree very well with the pumping data.

MATERIAL PROPERTIES

Characterization of the waste clay included the determination of index properties, mineralogy, and compressibility and permeability behavior. Laboratory tests were performed on composite samples recovered from the test pit at the end of April 1981.

Index Properties

Atterberg Limits, hydrometer and specific gravity tests were performed. The Liquid Limit was found to vary from 215 to 234, with

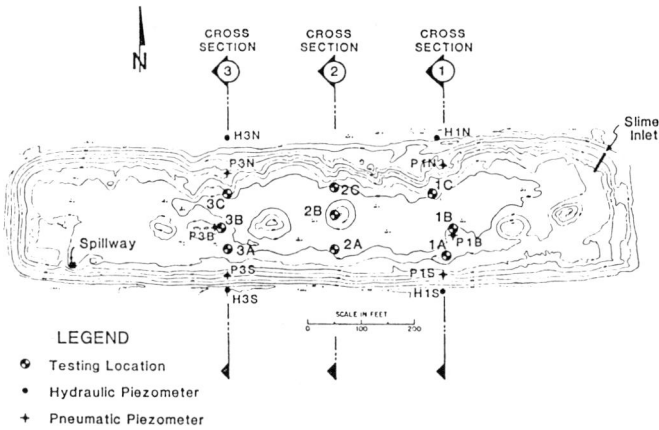

Figure 5. Plan View of Test Pit, Indicating Sampling and Testing Locations.

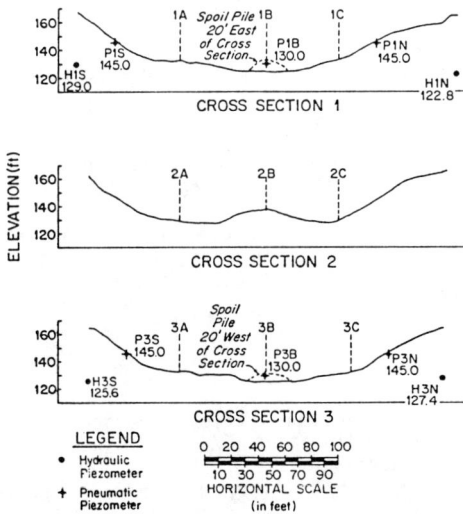

Figure 6. Cross Sections of Test Pit.

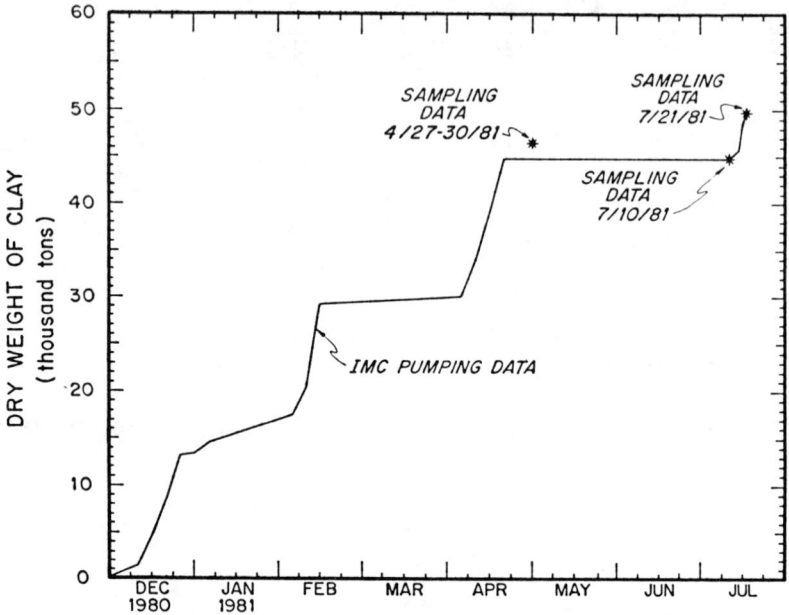

Figure 7. Filling History of Test Pit (Dry Weight of Clay vs. Time).

an average of 226.   The Plastic Limit, ranging between 51 and 57, averaged 54.   The resulting average Plasticity Index of 172 corresponds to an USCS designation of CH.   Hydrometer analyses indicated the clay (i.e., -2μ) fraction of the samples to be between 76 and 88 percent, with an average of 81.   From this, the computed activity varied between 2.04 and 2.28, and averaged 2.12.   Finally, the average specific gravity was found to be 2.71, ranging from 2.70 to 2.73.

Mineralogy

X-ray diffraction revealed little variation in mineralogy among the samples examined, with about 67 to 75 percent of each sample consisting of clay minerals.   The predominant clay mineral was found to be smectite, followed by illite, with minor amounts of kaolinite and attapulgite.

Compressibility

The compressibility, or relationship between void ratio and effective stress, of the waste clay was measured both in the laboratory and in the field.   Two types of slurry consolidation tests were performed in the laboratory on composite samples of the clay. They included a stress-controlled (or incremental loading) slurry consolidometer test and a constant rate of deformation slurry consolidometer (CRDSC) test. In the incremental loading test, the average equilibrium void ratio of the sample due to the applied stress is readily calculated, but the test is very time-consuming. The constant rate of deformation test, described by Znidarcic and Schiffman (1981), can be completed in much less time, but requires accurate boundary pore pressure measurements and correct analysis for estimating the void ratio and effective stress distributions within the sample during the test. The results of the two types of slurry consolidation test are presented in Figure 8(a), and show excellent agreement.

The void ratio-effective stress relationship can also be determined in the field from measurements of the solids contents and pore pressures within the consolidating deposit, as described by Somogyi et al (1981) and Carrier and Keshian (1979). The procedure basically involves computing the void ratio at a given depth from the solids content measured at that depth. The total stress is calculated using the solids contents of the overlying material. The effective stress is then computed as the difference between the total stress and the pore pressure measured at the given depth. The results of such field determinations are presented in Figure 8(b) and show very good agreement with the results of the CRDSC test. A power curve fit of the CRDSC data yielded the following compressibility relationship:

$$e = 23.0 \, (\bar{\sigma})^{-0.237} \tag{7}$$

where e = clay void ratio

$\bar{\sigma}$ = vertical effective stress (psf)

Permeability

The permeability of the waste clay at various void ratios was also determined in conjunction with both types of slurry consolidation tests. Constant head tests were performed after each load increment in the stress controlled test. During the permeability test, the externally applied load was adjusted to compensate for the effects of the seepage force. In the CRDSC test, the permeability can be calculated using either of two analytical techniques (viz. the "g-function" or "boundary" technique), as described by Znidarcic and Schiffman (1981). The resulting permeabilities are presented in Figure 9(a), and again show excellent agreement, even at the very high void ratios considered.

In situ permeability measurements were also made by performing rising head tests with a permeability probe. The test basically involves lowering a water filled tube connected to the probe a few centimeters below the equilibrium level and measuring the rate at which the water level rises. The permeability is then calculated using standard formulae. The results are presented in Figure 9(b) and

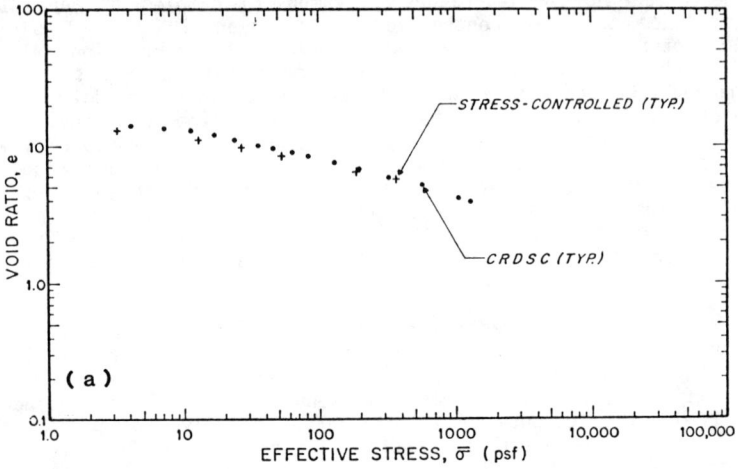

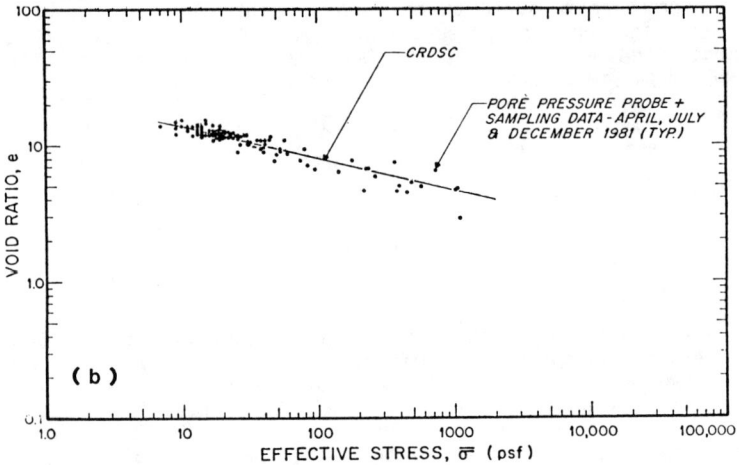

Figure 8.   Compressibility of Waste Clay.
   (a)   Stress-Controlled and CRDSC Results;
   (b)   Comparison Between In-situ Measurements and CRDSC Results.

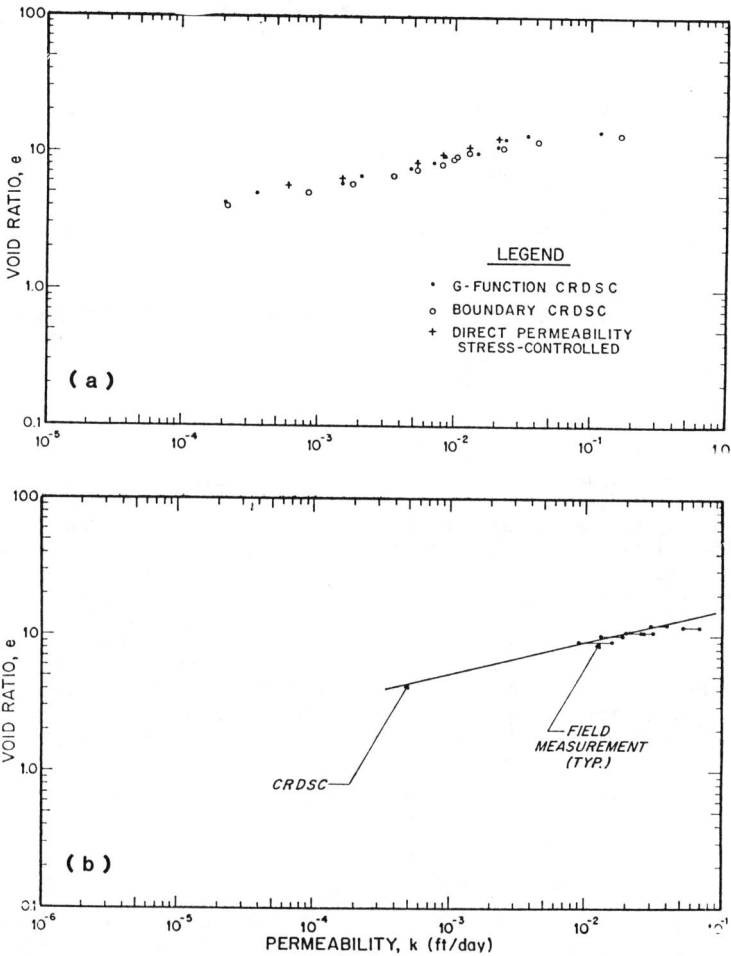

Figure 9. Permeability of Wast Clay.
(a) Comparison Between CRDSC and Constant Head Results;
(b) Comparison Between CRDSC and In-situ Results.

show very good agreement with CRDSC results, which can be expressed by the following power curve:

$$k = 1.03 \times 10^{-6} \ (e)^{4.19} \qquad\qquad (8)$$

where k = permeability (ft/da)
       e = clay void ratio

PREDICTED AND ACTUAL BEHAVIOR

"Engineering Approximations

Equations (1) and (3) were used to make preliminary estimates of the waste clay behavior in the test pit. Clay properties which had been measured in earlier studies conducted in the vicinity of the test pit were employed in these estimates. They are

$$e = 28.33 \ (\bar{\sigma})^{-0.30} \qquad (\bar{\sigma} \text{ in psf}) \qquad (9a)$$

$$k = 4.0 \times 10^{-6} (e)^{4.11} \qquad (k \text{ in ft/day}) \qquad (9b)$$

In addition, the dry weight of clay, W, was taken to be 46,500 tons in late April 1981 and 50,000 tons after the final filling. The maximum storage volume of the test pit is approximately 160 acre-ft and the average maximum depth is 34.4 feet. Therefore, the equivalent area of the "rectangularized" test pit, A, for use in Equation 1, is 4.66 acres. Finally, the idealized trapezoidal geometry of the test pit, for use in Equation (3), was taken to be a base width, w of 130 feet; a base length, l, of 1080 feet; and an inside slope, m, of 1.47. The resulting estimates are presented in Figure 10, and indicate virtually no effect of geometry. This suggests that one-dimensional analysis

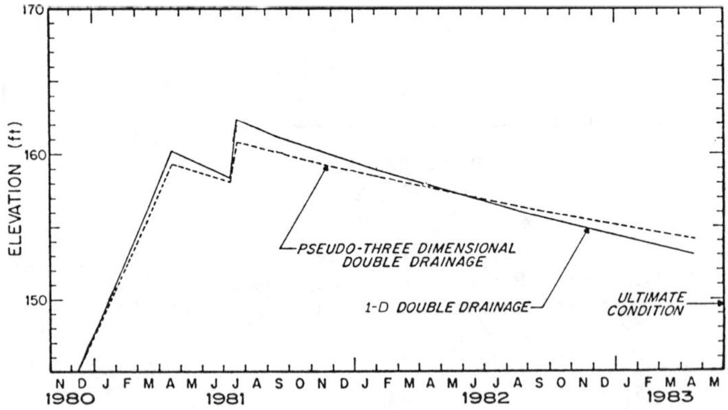

Figure 10. Preliminary Estimates of Waste Clay Behavior Using "Engineering Approximations".

may suffice and agrees with the "quasi-two-dimensional" results presented in Figure 3 for an aspect ratio of 5:1. Further use of the
"engineering approximations" was abandoned, for the above reason as
well as the present inability of the approximations to simulate the
effects of underseepage (which will be shown to significantly affect
behavior).

"Quasi-Two-Dimensional" Predictions

Numerical analyses were first performed to model the initial
filling period, which ended in April of 1981. The actual cross
section was idealized as a rectangle 34.4 feet high and 153.4 feet
wide. These equivalent dimensions represent an average cross-
sectional area of 5,275 square feet (equal to the actual). The
filling history was idealized as having occurred instantaneously, at a
uniform initial solids content of 16%. The measured clay properties,
as given by Equations (7) and (8) were used.

A one-dimensional analysis was conducted for a five year
period. The results are shown in Figure 11. The predicted
equilibrium conditions were a height of 20.5 feet, at an average
solids content of 25.1%. The results in Figure 11 indicate that the
1-D solution greatly underestimates the actual performance.

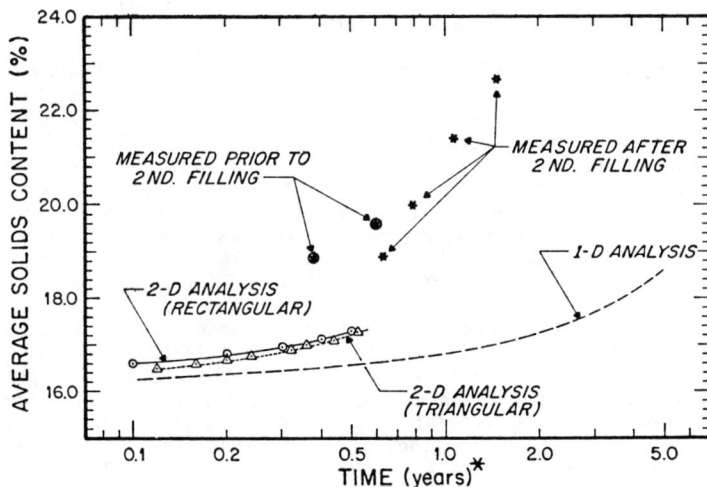

* Begin 12/10/80

Figure 11.   One and "Quasi-Two-Dimensional" Analyses of Test Pit
             Behavior (Seepage Not Considered).

Because of the expense and large storage requirement of the "quasi-two-demensional" solution, preliminary analyses were performed in an attempt to optimize grid size and time steps and maintain on-line processing capabilities. The results of the analysis conducted for 0.5 years are also shown in Figure 11. The change in predicted behavior is not dramatic, as anticipated based on Figure 3 for an aspect ratio of 5.

The obvious discrepancy between the analytical results and observed performance could be attributable to three physical phenomena. First, the horizontal permeability is greater than the vertical; second, the actual trapezoidal (or triangular) rather than rectangular cross section of the pit; and third, the existence of seepage out of the pit. Since no experimental data on the magnitude of anisotropy in permeability is available, and its existence has been questioned for pre-thickened clays, the first explanation will be avoided at present.Inspection of the actual cross sections of the test pit indicate it to be much more triangular (or trapezoidal) than rectangular. The computer program was modified to reflect this, and the results shown in Figure 11 indicate virtually no difference in the predicted behavior, at least for the half-year period analyzed. This is again probably because of the very shallow pit, relative to its width.The final possible explanation to be pursued was the existence of underseepage. Typical field measurements of solids content and pore pressures within and below the sediment are presented in Figures 12, 13 and 14 respectively, and clearly indicate the presence of underseepage. An average head differential between the sediment top and bottom of 13.4 feet was determined from the boundary pore pressure measurements.As a first attempt at modelling this phenomenon, a constant seepage head was imposed across the deposit. While this cannot physically occur for long-term conditions, it is reasonable for relatively short times (perhaps a few years), especially in light of the caking or sealing at the boundaries that is predicted and has been observed.The results obtained when a 13.4-foot seepage head was imposed across the deposit are plotted as squares on Figure 15. Excellent agreement with measured results can be seen, with the predicted solids contents being 19.1% and 19.7% at 0.4 and 0.5 years, as compared to the actual 18.8% and 19.6% at 0.38 and .60 years.

In order to check whether post-filling consolidation behavior is independent of filling history (for rapid filling), as has been proposed for the one-dimensional case, an attempt was made to predict the solids content at the end of the second (or final) filling period, by beginning with an appropriate height of 16% solids material at the start of the initial filling.The dimensions of the equivalent rectangle for this case are 43.3 feet by 183.0 feet. The increased width (over the previous 153.4 feet) is indicative of the sloping sides of the actual pit. The results, also plotted on Figure 15, indicate a predicted solids content of 19.1% at 0.6 years, as compared with the actual 18.9% at 0.63 years.

The analysis was continued for times after final filling by assuming as initial conditions a uniform deposit 35.9 feet high with solids content of 18.9%. These were the measured values on July 21,

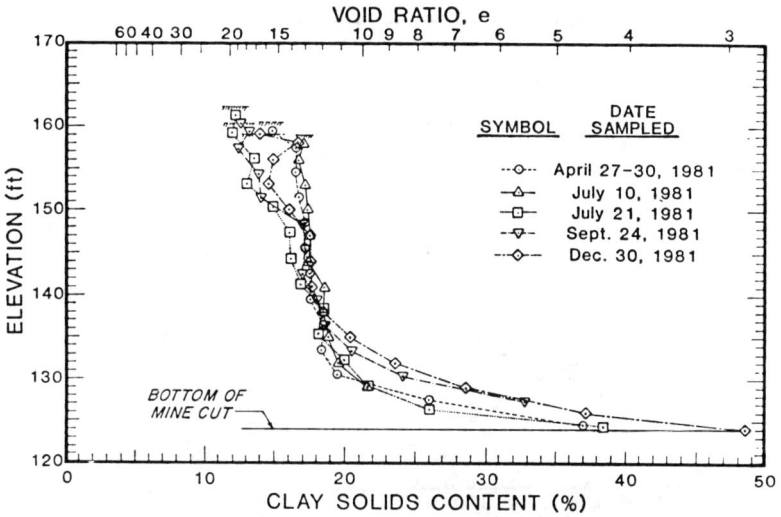

Figure 12. Solids Content Distribution at Test Location 1B.

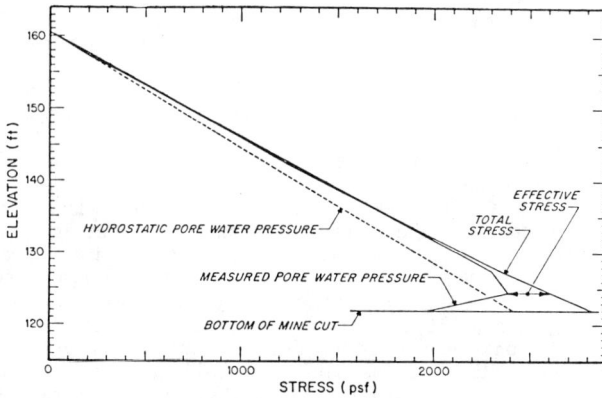

Figure 13. Geostatic Stresses at Test Location 1B on April 27-30, 1981.

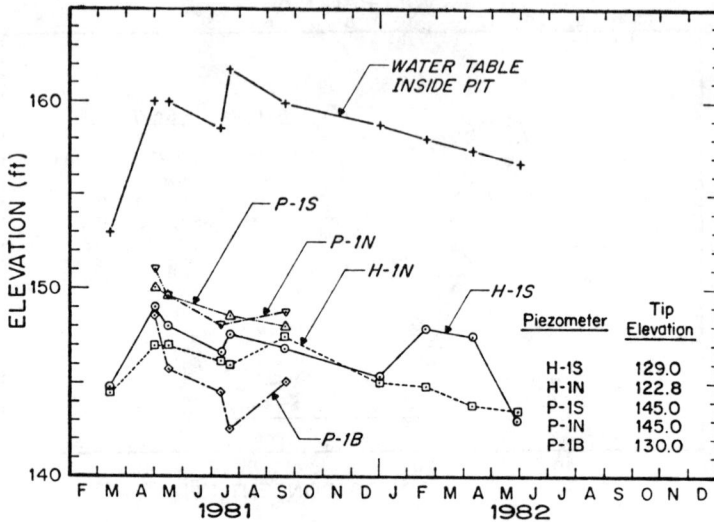

Figure 14. Boundary Piezometric Levels at Cross Section 1.

1981, after the second filling had been completed. The equivalent width of 183.0 feet was again used. The results are presented in Figure 16 and show very good agreement throughout the year following final filling, when all supernatant was decanted and crust development initiated.

CONCLUSIONS

The case history described herein demonstrates that the "quasi-two-dimensional" finite strain consolidation model can provide accurate estimates of full-scale behavior, if the actual boundary conditions are properly taken into account. Parametric variation studies using the model indicate that little improvement in consolidation rate occurs at aspect ratios (width:height) near or above about 5:1, unless lateral permeability is significantly higher than vertical permeability. Finally, the compressibility and permeability relations needed for finite strain analyses can be obtained from incremental loading slurry consolidation tests, constant rate of deformation slurry consolidation tests, or insitu sampling and pore pressure measurements.

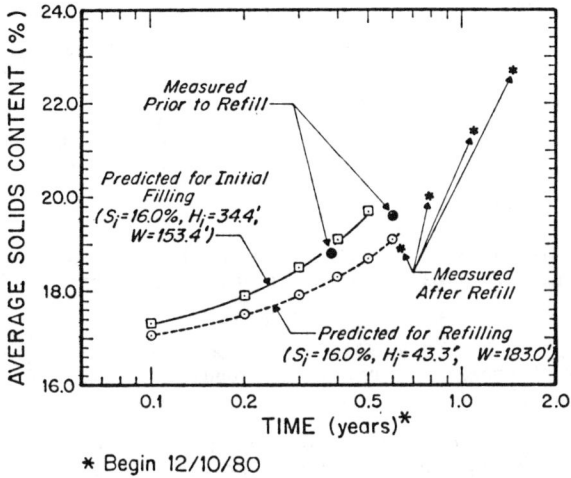

* Begin 12/10/80

Figure 15. Predicted Conditions in Test Pit at Completion of Initial and Second (Final) Filling Periods.

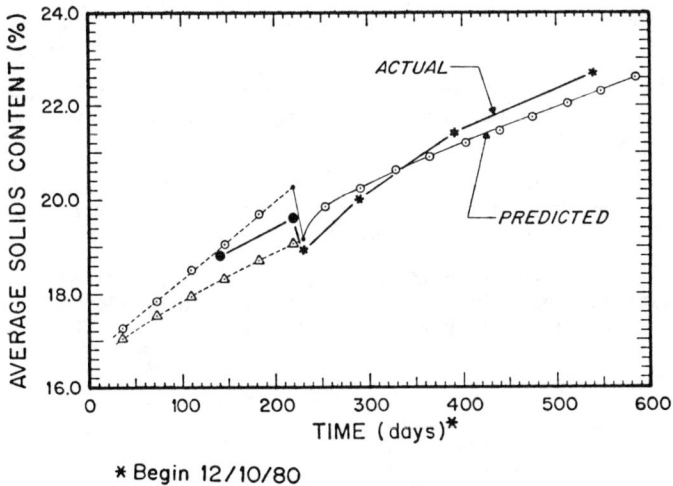

*Begin 12/10/80

Figure 16. Predicted Behavior in Test Pit During and After Filling.

ACKNOWLEDGEMENTS

The research program described herein was funded by the Florida Institute of Phosphate Research, under Contract No. 81-02-006. The field testing program employed IMC facilities and was conducted by personnel from Agrico, IMC and Mobil Chemicals. X-ray diffraction tests were performed at M.I.T., and CRDSC tests at the University of Colorado.

APPENDIX. - REFERENCES

Bromwell Engineering, Inc. (1982) "Waste Clay Disposal in Mine Cuts," Final Report to Florida Institute of Phosphate Reserch on Contract No. 81-02-006.

Carrier, W. D. III, Bromwell, L. G. and Somogyi, F (1983) "Design Capacity of Slurried Mineral Waste Ponds," Journal of Geotechnical Engineering, ASCE, Vol. 109, No. 5, pp. 699-716.

Carrier, W. D., Bromwell, L. G. and Somogyi, F. (1981) "Slurried Mineral Wastes: Physical Properties Pertinent to Disposal," presented at ASCE National Convention, October 1981, St. Louis.

Carrier, W. D. III and Keshian, B., Jr. (1979) "Measurement and Prediction of Consolidation of Dredged Material," presented at the Twelfth Annual Dredging Seminar, Houston, TX.

Gibson, R. E., England, G. L. and Hussey, M. H. L. (1967) "The Theory of One-Dimensional Consolidation of Saturated Clays, 1. Finite Non-Linear Consolidation of Thin Homogeneous Layers," Geotechnique, Vol. 17, pp. 261-273.

Somogyi, F. (1980) "Large Strain Consolidation of Fine-Grained Slurries," presented at Canadian Society for Civil Engineering 1980 Annual Conference, Winnipeg, Manitoba, Canada.

Somogyi, F., Keshian, B. Jr., and Bromwell, L. G. (1981) "Consolidation Behavior of Impounded Slurries," presented at Annual Spring Convention, ASCE, New York.

Znidarcic, D. and Schiffman, R. L. (1981) "Finite Strain Consolidation: Test Conditions," Technical Note, Journal of the Geotechnical Engineering Division, ASCE, Vol. 107, No. GT 5, pp. 684-688.

Centrifugal Modeling of Phosphatic
Clay Consolidation

D. G. Bloomquist[1], and F. C. Townsend[2],
A.M.   M.ASCE

ABSTRACT: Florida produces approximately 80 percent of the nation's
and 35 percent of the world's phosphate requirements. Unfortunately,
as a by-product of the beneficiation process, vast quantities of very
dilute clay slurry are produced, which are subsequently stored in
large retention ponds and left to settle over several years to an
average solids content near 20 percent.   One method of accelerating
the consolidation process is to surcharge the clay with a sand
tailings cap, which enhances consolidation and provides a stable
foundation for reclamation.

    Centrifugal model tests were performed on a phosphatic clay to
evaluate; (a) the scaling time exponent between prototype and
centrifugal tests and (b) computer predictions with the centrifugal
model results.

    The results of these tests indicate that self-weight
consolidation of these clays will achieve an average final solids
content of approximately 21-23 percent in 2.5 to 3 years.   Final
solids contents of 25 to 40 percent are possible if surcharging sand
caps are placed on the clays.   Unfortunately, the decreasing
permeabilities associated with consolidation cause increased time
spans of 15+ years in order to reach these solids contents.

    Current large strain consolidation computer programs are
formulated on effective stresses to predict consolidation behavior
and, thus, are inappropriate to predict sedimentation phenomena.
Conversely, centrifugal model tests using "modeling of models"
techniques can faithfully represent this prototype situation of
sedimentation advancing to consolidation.

INTRODUCTION

    Florida produces approximately 80 percent of the nation's and 35
percent of the world's phosphate requirements.   Unfortunately, as a
by-product of the beneficiation process, vast quantities of very
dilute clay slurry are produced, which are subsequently stored in

---

1. Assistant Engineer in Civil Engineering, University of
    Florida, Gainesville, Florida 32611.
2. Professor of Civil Engineering, University of Florida,
    Gainesville, Florida 32611.

large retention ponds and left to settle over several years to solids contents approaching 20 percent. However, once this self-weight consolidation phase is complete, the solids contents remain essentially constant at the 20-25 percent range. Unfortunately, solids contents of about 40 percent are desirable for reclamation. One method of accelerating the consolidation process is to increase the effective stress by surcharging the clay with a sand tailings cap, which enhances consolidation and provides a stable foundation for reclamation (Lawver, 1983). While capping waste clay ponds to enhance consolidation is attractive, the questions of immediate concern are; (a) What benefits are achieved regarding rate? (b) What is the final solids content achievable by placing a cap? and (c) How are these benefits influenced by the height of cap placed?

Currently, large strain consolidation numerical models (Somogyi, 1979; Cargill, 1982) are available for estimating the final solids content and rate of consolidation of waste clay ponds with and without a cap. Currently, the input parameters for these computer models are obtained from constant rate of deformation laboratory tests (Wissa, et al., 1983; Znidarcic, 1981), or available correlations for several waste clays (Carrier, 1983). However, field validation of these numerical models with and without a cap has not been accomplished. Thus, the unfortunate circumstance exists of striving to predict solids content as a function of time, yet having to wait many years to determine if the predictions are accurate. Fortunately, centrifugal model testing offers a viable alternative to full-scale field testing and can be used both to predict solids content as a function of time and to verify computer models. Thus, it is possible to use models, both centrifugal and numerical, to predict consolidation behavior of waste clays when subjected to various construction events and thereby furnish some information concerning the aforementioned questions.

OBJECTIVES

The following objectives for this research were:
a) To determine, using centrifugal modeling techniques, the consolidation characteristics (magnitude and rate) of various depth waste ponds, with and without a sand cap.
b) To compare the results of centrifugal model tests with finite strain consolidation predictions.

SCOPE OF WORK

To accomplish the desired objectives, three centrifugal model test series were performed; specifically:
1. Modeling of Models - model tests at 40, 60, and 80 g's replicating the same prototype were performed to determine time scaling relationships to permit calculations of prototype times from centrifugal model times.
2. Field Duplication - a model replicating the IMC tank test was performed to verify time scaling relationships.
3. Cap Tests - a variety of capping scenarios were performed to evaluate the concept of stage capping.

Numerical analyses consisted of two finite difference programs (Somogyi, 1979 and Cargill, 1982) which solve the Gibson (1967, 1980) equation. Input parameters were obtained by back-calculations from centrifugal tests or available data (Lawver, 1982). Subsequently, numerical and centrifugal model results were compared.

EQUIPMENT, MATERIALS AND PROCEDURES

UF Centrifuge Equipment

The UF geotechnical centrifuge has a 1-m radius and can accelerate 25 kg to 85 g's (2125 g-kg capacity). For the proposed centrifugal model tests, the waste clays are contained in a 14 cm diameter by 15.25 cm high plexiglas container. Modeling criteria to reduce radial acceleration gradients across the model recommend a model height of ± one-tenth the radius, and a width subtending an angle less than 15 percent. Accordingly, the 15.25 cm container is less than 20 cm high, and the 14 cm diameter subtends an angle of eight degrees. Previous tests (Bloomquist, 1982) evaluated side effects by performing tests in 5 cm, 10 cm, and 14 cm diameter containers at 60 g's. No difference in results was observed between the 10 cm and 14 cm diameter tests, suggesting that side effects are minimal for the 14 cm diameter container.

By performing the tests with approximately a 10 cm high waste clay column, sufficient freeboard height is available for protection against spillage and to provide for the sand cap. The plexiglas container is housed in a swinging aluminum bucket, with a vertical window that allows visual observations of the model. A set screw mounted to the aluminum housing unit allows for precise bucket orientation. A photo-electric pick-off and flash delay augment the system for photography. Figure 1 presents a schematic drawing of the centrifuge and photographic equipment.

A miniature accelerometer is mounted to the aluminum bucket to provide accurate monitoring of the acceleration, such that periodic adjustments (±0.5 g) can be made to maintain a constant g level.

Materials

The properties of the waste clay tested are listed below:

Clay	LL	PI	G_s	pH	CEC	Mineralogy of Clay Fraction
Kingsford	176	119	2.71	7.2	81.9	90% Mont, 5% Interlayer(14A) 2% KAO, 3% Qtz

Centrifugal Model Test Procedures

The sequence of events for performing the centrifugal model tests consisted of; (a) waste clay placement in centrifuge containers, (b) initial solids content determination, (c) cap placement (if performed), (d) acceleration to test acceleration, and (e) photographic monitoring of waste clay surface (to nearest .5 mm) with time.   Figure 2 presents an example of photographic monitoring of a waste clay surface.

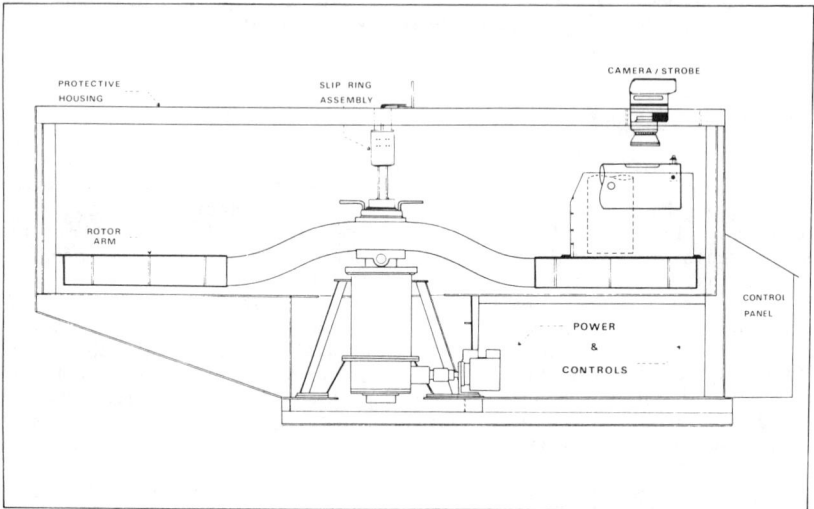

FIGURE 1:   Schematic of Centrifuge and Camera Set-up

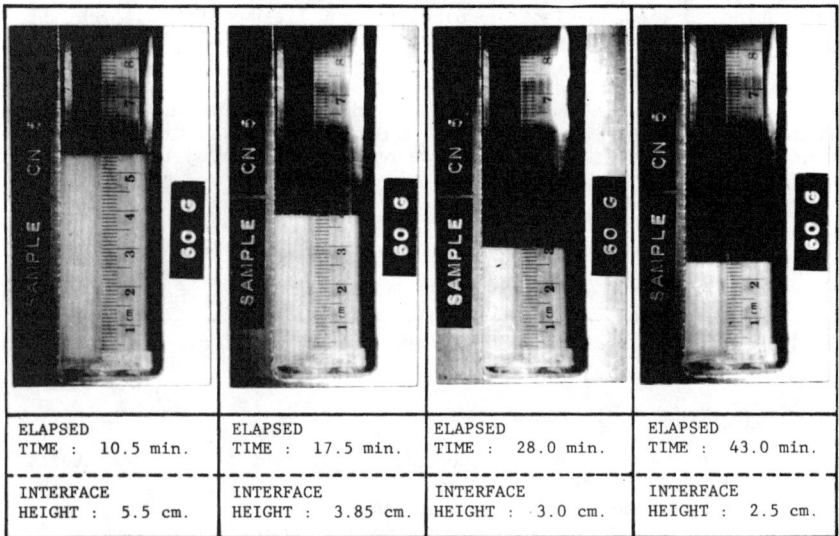

ELAPSED TIME : 10.5 min.	ELAPSED TIME : 17.5 min.	ELAPSED TIME : 28.0 min.	ELAPSED TIME : 43.0 min.
INTERFACE HEIGHT : 5.5 cm.	INTERFACE HEIGHT : 3.85 cm.	INTERFACE HEIGHT : 3.0 cm.	INTERFACE HEIGHT : 2.5 cm.

FIGURE 2:   Example of Photographic Monitoring

Scaling relationships pertinent to these models are: geometry (ℓ) ℓ_p/ℓ_m = N and elapsed time (t) tp/tm = N^x, where p and m indicate prototype and model respectively, N is the number of gravities, and x is a coefficient between 1.0 and 2.0 as determined by "modeling of models". Void ratio is calculated from solids content as e = (1-S)Gs/S, where S = solids content by weight.

TEST RESULTS AND DISCUSSION

Table 1 summarizes the results of ten centrifugal model tests on the Kingsford waste clays. Presented are the initial and final model conditions and the corresponding prototype conditions. Also presented are the duration of accelerations. The corresponding prototype elapsed times are calculated using procedures described in the Modeling of Models section.

TABLE 1: Summary of Centrifugal Model Tests on Kingsford Waste Clays

Test No. (1)	Initial Model Conditions Solids Cont. % (2)	Model Ht, cm (3)	Cap Ht, cm (4)	Accel. Level g's (5)	Dur. min. (6)	Final Model Cond. Solids Cont. % (7)	Model Ht, cm (8)	Prototype Conditions Initial Ht, m (9)	Cap (a) Ht, m (10)	Final Ht, m (11)	Elapsed Time, days (12)
MM-40	14.6	12.00	N/A	40	4200	21.3	7.90	4.8	N/A	3.14	1,100
MM-60	14.6	8.00	N/A	60	2340	21.3	5.2	4.8	N/A	3.11	1,137
MM-80	14.6	6.00	N/A	80	1405	21.3	3.9	4.8	N/A	3.11	1,082
SED-80	3.2	10.00	N/A	80	970	21.0	1.4	8.0	N/A	1.12	747
SSC-60-1	14.6	8.00	0.00	60	2700	21.1	3.90	4.8	0.00	2.34	1,312
	21.1	3.90	1.3	60	1300	26.5	3.00	2.34	0.78	1.8	3,250
SSC-60-2	21.1	3.10	2.1	60	1310	28.6	2.22	1.86	1.26	1.33	3,275
SSC-80-1	14.6	10.00	0.00	80	2460	24.0	5.70	8.0	0.00	4.56	1,894
	24.0	5.70	2.0	80	1350	30.4	4.27	4.56	1.60	3.42	6,000
	30.4	4.30	4.0	80	2940	40.0	3.00	3.44	3.2	2.4	13,067
SSC-40-1	14.6	12.00	0.00	40	5500	21.3	7.85	4.80	0.00	3.14	1,397
	21.3	7.90	2.00	40	3000	26.7	6.17	3.16	0.8	2.47	3,333
SSC-60-3	23.8	7.00	6.00	60	2350	37.0	4.10	4.20	3.6	2.46	5,875
Tank-80	12.60	7.90	N/A	80	2500	21.6	-	6.32	N/A	-	403

(a) All caps became submerged shortly after achieving prototype accelerations.

γ' cap = 714.4 Kg/m^3 (44.6 psf)

Sedimentation/Hindered Settlement/Consolidation (Model SED-80)

When a dilute (3-5% solids) clay slurry is pumped into a retention pond, a settling/sedimentation/consolidation process begins as illustrated in Figure 3 (Imai, 1981, Been and Sills, 1981). Initially, if the surface activity is minor, the particles settle at a more or less constant rate controlled by gravity in an approximately

Stokian fashion. Almost concurrently at the bottom a "consolidation" zone forms in which effective stresses are greater than zero and large strain consolidation theories incorporating self-weight are applicable. Above the "consolidation" zone exists a zone of "hindered settlement," where settlement is interferred with by neighboring particles. The "consolidation" zone gradually increases at the expense of the "hindered settlement" zone until only a single zone exists. Since excess pore pressures are continually dissipating as the two zones merge, at the time of merger, consolidation/sedimentation is completed and an equilibrium condition; i.e. final solids content, is reached.

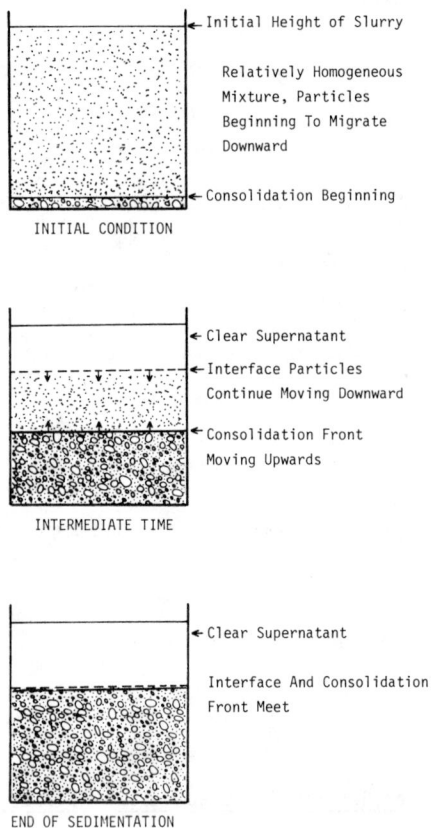

Initial Height of Slurry

Relatively Homogeneous
Mixture, Particles
Beginning To Migrate
Downward

Consolidation Beginning

INITIAL CONDITION

Clear Supernatant

Interface Particles
Continue Moving Downward

Consolidation Front
Moving Upwards

INTERMEDIATE TIME

Clear Supernatant

Interface And Consolidation
Front Meet

END OF SEDIMENTATION

FIGURE 3: Idealized Sedimentation/Consolidation Process

Figure 4 presents a log time-log solids content relationship for model SED-80 which replicates a 7.98 m (26.2 ft) deep pond of 3.2 percent initial solids which undergoes sedimentation/ consolidation for 747 days (based upon 1.6 scaling exponent). From this presentation, distinct slope changes defining regions of sedimentation, hindered settling, and self-weight consolidation are observed. For the Kingsford clay, Stokian sedimentation (based upon 3μ size clay) ends at 6.8 percent solids versus the 8.5 percent shown. Carrier, et al. (1983) have proposed an emperical relationship for estimating the void ratio at which sedimentation is complete as: e = 0.07 Gs (LL). For Kingsford clay, a solids content of 7.5 percent is calculated using this relationship.

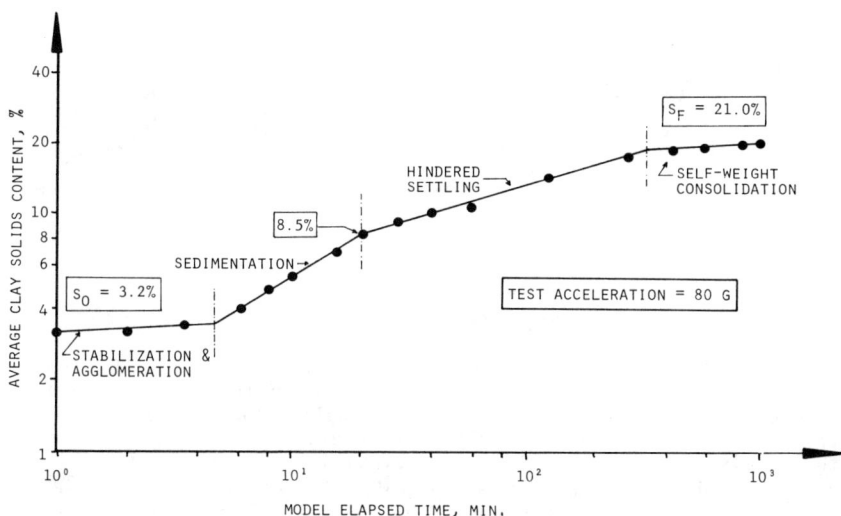

FIGURE 4:   Log Time vs Log Solids Content for Model SED-80

Modeling of Models

Modeling of models experiments on the Kingsford clay were performed at 40, 60, and 80 g's (Model MM-40, 60, 80) replicating a 4.78 m (15.7 ft) deep waste pond. These results are presented in Figure 5, with the scaling exponents shown in Figure 6. These scaling exponents were determined by noting particular average solids contents and the corresponding elapsed times to achieve them. The scaling exponent, x, was then determined for the different solids contents as, $t_A(N_a)^x = t_B(N_b)^x$ where t = time and N is the corresponding acceleration level.

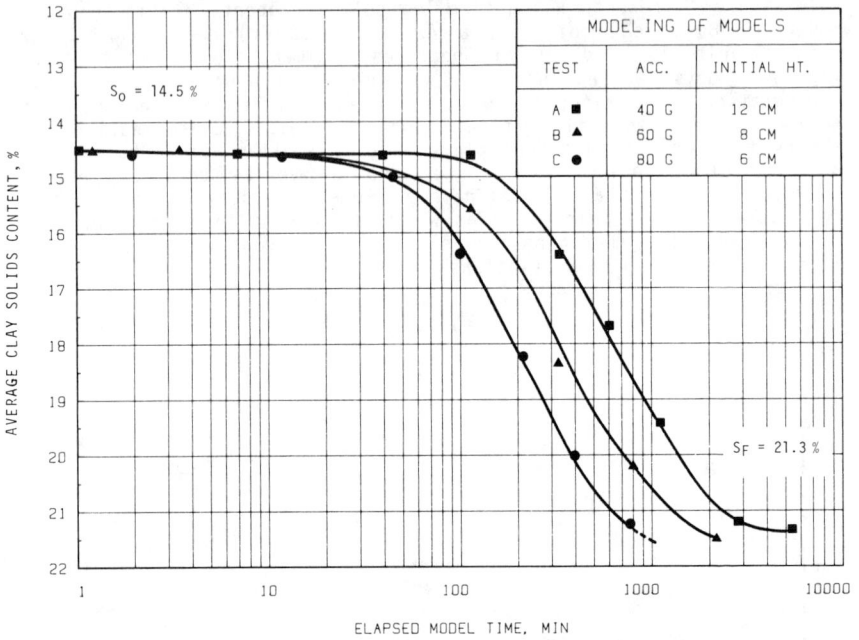

FIGURE 5: Modeling of Models Results

Theoretically, the time scaling exponent is 1.0 for sedimentation, and 2.0 for consolidation. For consolidation, other investigators, Croce et al. (1984), and Scully (1984) have determined an exponent of 2.0 from "modeling of models" experiments. However, the models of Croce et al. (1984) had an initial void ratio of 2.86 and sedimentation played no part. In the case of Scully's (1984) models the initial void ratio was 15.0 and his scaling exponent varied from 1.90 to 2.3. By contrast, the results presented in Figure 6 show that starting with an initial void ratio of 16.0 ($S_c \approx 14$), the scaling factor progresses from 1.6 to 2.0.* Since only the waste clay surface is monitored during the centrifugal test, these exponents represent the composite of the "hindered settlement" plus "consolidation" zones. As these two zones merge, consolidation predominates and the theoretical value of 2.0 is achieved. For these models, this achievement occurs at average solids contents greater than 20 percent (e=10.9).

---

* Accordingly, to calculate prototype times the appropriate scaling exponent for the various solids contents is used.

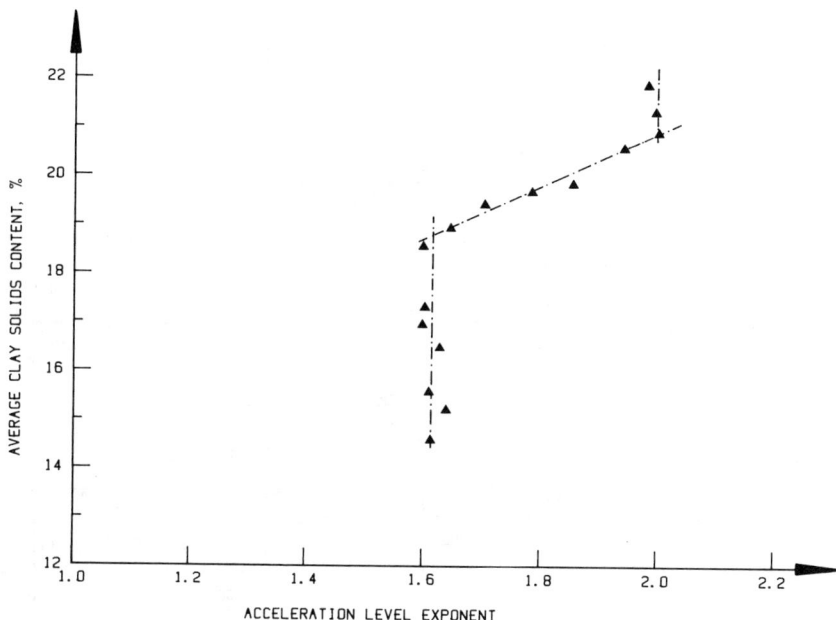

FIGURE 6: Time Scaling Exponent Versus Solids Content

We are aware that other investigators (Been and Sills 1981, and Yong, 1983) report that the threshold void ratio for the existence of effective stresses is approximately 4 to 6, which is lower than the value reported here.

Cap Enhanced Consolidation (Model Series SSC)

Examination of the modeling of models test series suggests that self-weight consolidation will only achieve a final average solids content of approximately 22 to 25 percent. Accordingly, capping the waste clays with a sand surcharge is a method for increasing the effective stresses to enhance consolidation, thereby providing greater storage volumes and less turnaround time between mining and reclamation. Table 1 summarizes the capped model results while Figure 7 presents typical results. Generally, capping was not attempted at solids contents less than 20 percent, as unreported models showed signs of surface distress when caps were placed on lower solids contents clays.

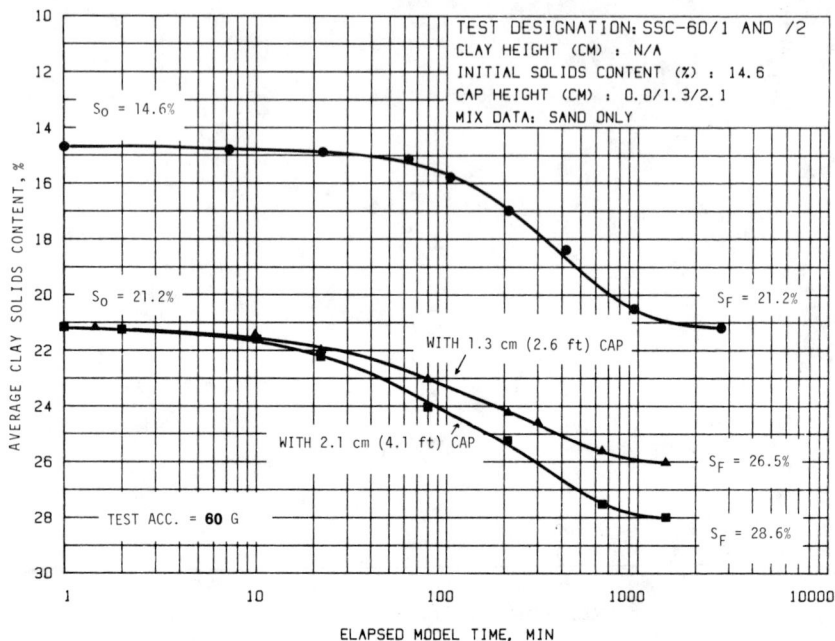

FIGURE 7:   Effect of Capping on the Final Solids Content

These results (see Table 1) show that solids contents up to 40 percent can be achieved by successive cap placement. The fear that the sand surcharges would fail as filters and become clogged thus stopping consolidation did not materialize as the caps became submerged as consolidation progressed.   Unfortunately, as one might anticipate as the solids content increased, permeability decreased, and consequently, timespans of 15+ years are required to achieve high solids contents.

LARGE STRAIN CONSOLIDATION MODELING

In 1967, Gibson, England, and Hussey extended Terzaghi's consolidation theory and provided a model which considers unlimited strain, the effects of self weight, and variation of permeability with void ratio, and a nonlinear void ratio/effective stress relationship. The governing differential equation is:

$$\left(\frac{\gamma_s}{\gamma_w} - 1\right) \frac{d}{de} \frac{k(e)}{(1+e)} \frac{\partial e}{\partial z} + \frac{\partial}{\partial z} \frac{k(e)}{\gamma_w(1+e)} \frac{d\sigma'}{de} \frac{\partial e}{\partial z} + \frac{\partial e}{\partial y} = 0 \qquad (1)$$

where
γ_s = the unit weight of solids
γ_w = the unit weight of water
e = the void ratio
k = the permeability
z = a material coordinate
σ' = the effective stress
t = the time

Somogyi (1979) developed several computer programs based on a manipulated form of equation (1). The revised governing equation used is:

$$\frac{\partial}{\partial z} - \left[\frac{k}{\gamma_w(1+e)} \frac{\partial u}{\partial z}\right] + \frac{de}{d\sigma'} \left[(SG-1)\gamma_w \frac{d(\Delta z)}{dt} - \frac{\partial u}{\partial t}\right] = 0$$

where
u = the excess over hydrostatic pore water pressure
SG = the specific gravity of solids with the remaining terms as previously defined

The relationship between void ratio and effective stress, and the variation in permeability with void ratio may be described by power functions of the form: (Carrier, et al. 1983)

$$e = A (\sigma')^{-B}$$

$$k = C e^D$$

where
e = the void ratio
σ' = the effective stress
A,B = constants
k = the permeability
C,D = constants

The constants are to be obtained from laboratory tests, primarily the slurry consolidometer or the constant rate of deformation slurry consolidometer.

Prototype Tank Test

In July 1977, International Minerals and Chemicals Inc. filled a 2.74 m x 4.27 m x 6.7 m metal container to a depth of 6.32 m with Kingsford waste clay having an initial solids content of 12.6 percent and allowed this material to self-weight consolidate for 403 days to 21.2 percent solids. This test provides a prototype with single drainage and vertical walls, which is ideal for centrifuge and computer comparisons. Accordingly, Figure 8 presents the comparisons between centrifuge model Tank-80 (Table 1) and Somogyi's (1979) program with this prototype. The elapsed time scaling exponent used for the centrifugal model comparison was 1.6 as shown previously in Figure 4. The numerical prediction was based upon values of A = 8.86 σ^1 (28.3 psf), B= -0.3, C = 1.4x10^{-9}cm/sec (4x10^{-6}ft/day), and D=4.11. These input parameters were obtained by laboratory data "tempered" with field observations (Lawver, 1982).

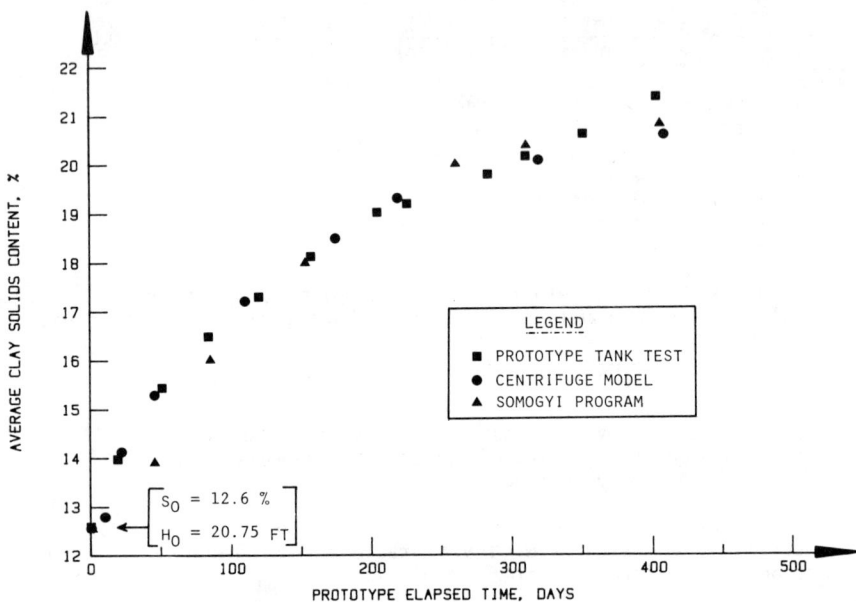

FIGURE 8: Comparisons of IMC Tank Test,
Centrifugal Model and Numerical Results

The results of these comparisons are quite encouraging, with the centrifugal model predicting final solids contents of 21.6 percent, the numerical prediction being 21.5 percent, and the field test (after 403 days), 21.2 percent.

In a similar effort, Lawver, et al. (1984) has successfully modeled a 7.62 m (25 ft) deep test pit of Kingsford waste clays using Somogyi's program. For this case, the input parameters were identical to those used for Figure 8.

Numerical and Centrifugal Model Limitations

One apparent limitation of the finite strain consolidation computer programs is that at low solids contents, sedimentation predominates and the effective stresses are zero, thus, rendering equation 1 inappropriate. But for what conditions is equation 1 appropriate? Been and Sills (1981) and Imai (1981) suggest the threshold void ratio for unique effective stress void ratio relationships is approximately 6 to 8, respectively. Nevertheless, to assess the effect of initial solids content on numerical predictions, centrifugal models with initial solids contents of five percent ($e_0 =$

51.5) and ten percent (e_0 = 24.4) were compared with numerical predictions as shown in Figures 9 and 10. For the centrifuge tests, a time scaling exponent of 1.0 was used until 8.5 percent solids and 1.6 thereafter. Obviously, numerical predictions during sedimentation deviate significantly from centrifugal test results (Fig. 9) until higher solids contents are achieved as shown in Figure 10. By comparison, Scully et al. (1984) obtained favorable comparisons between centrifuge models and finite strain consolidation predictions beginning at an initial void ratio of 15. The deviation between numerical and centrifugal predictions at 18 percent (Fig. 10) represents the change in time scaling exponent transversing from hindered settlement to consolidation. From these observations, numerical modeling of sedimentation and hindered settlement phases using input data from slurry consolidation tests can be erroneous. From the centrifugal model tests, these phases (hindered settlement/self-weight consolidation) cease when the time scaling exponent becomes 2.0 or approximately 20 percent solids content.

A similar limitation exists concerning centrifugal modeling in that the time exponent is unknown (it lies between 1.0 and 2.0). However, this drawback can easily be overcome by "modeling of models" techniques to determine the appropriate exponent value.

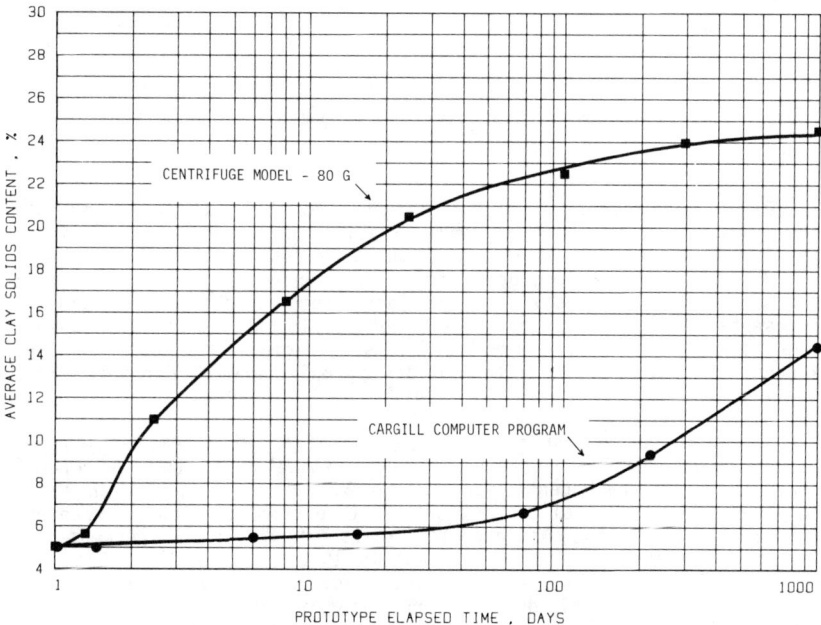

FIGURE 9: Comparison Between Centrifugal and Numerical Models at Low Solids Contents

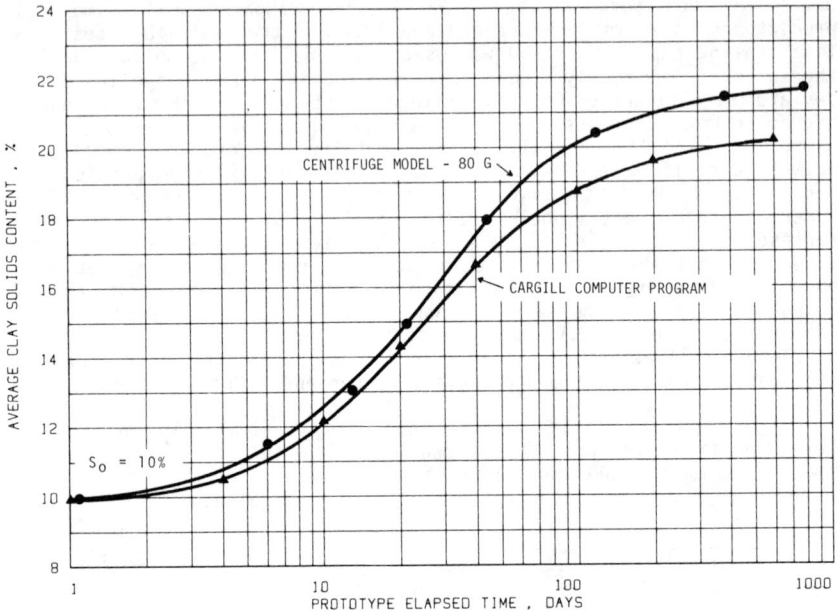

Figure 10:   Comparison Between Centrifugal and Numerical

CONCLUSIONS

Based upon centrifugal "modeling of models" experiments dilute clay suspensions pass through sedimentation to hindered settlement to consolidation phases. Accordingly, the time scaling exponent advances from 1.0 to 2.0 depending upon the solids content. Since centrifugal models observe the waste clay surface, this time scaling exponent represents a composite of "hindered settlement" and "self-weight consolidation."   Thus, for the Kingsford clay, the time scaling exponent advanced from 1.6 to 2.0 at approximately 20 percent solids (e = 10.8).

For these clays, self-weight consolidation will only achieve an average final solids content of approximately 21 to 23 percent (depending upon pond depth) in 2.5 to 3.0 years, after which time, increases in final solids contents are extremely slow.   However, average final solids contents in excess of 25 to 40 percent are possible if a surcharge of approximately 19.2 to 23.9 kPa (400 to 500 psf) is placed on top of the waste clays.   Unfortunately, the decreasing permeability due to densification of the clays in turn causes timespans of 15+ years to achieve these final solids contents.

Current finite strain consolidation programs are based upon effective stresses and do not consider sedimentation or hindered settling where effective stresses are essentially zero.   However, centrifugal model tests using "modeling of models" techniques will faithfully represent the prototype situation at these low solids contents.

ACKNOWLEDGEMENTS

The authors greatly appreciate the support of IMC, Bartow, Florida and assistance of Dr. J. E. Lawver, IMC, and Steve Olson, Agrico, Mulberry, Florida. The assistance of Messers David Israel and T. Carter, University of Florida, are also acknowledged.

REFERENCES

1. Bloomquist, D. (1982), "Centrifuge Modeling of Large Strain Consolidation Phenomena in Phosphatic Clay Retention Ponds," Ph.D Dissertation, University of Florida, Gainesville (ROD83-13610).

2. Been, K. and Sills, G. C. (1981), "Self-Weight Consolidation of Soft Soils: An Experimental and Theoretical Study," Geotechnique, Vol. 31, No. 4.

3. Cargill, K. W. (1982) "Consolidation of Soft Layers by Finite Strain Analysis" MP-GL-82-3, Geot. Lab, USAEWES, Vicksburg, Mississippi.

4. Croce, P., Pane, V., Znidarcic, D., Ko, H. Y., Olsen, H. W., and Schiffman, R. L. (1984), "Evaluation of Consolidation Theories by Centrifuge Modelling," Applications of Centrifuge Modelling to Geotechnical Design, University of Manchester, U.K.

5. Carrier, W. D., Bromwell, L. G., and Somogyi, F. (1983), "Design Capacity of Slurried Mineral Waste Ponds," ASCE Journal of Geotechnical Engineering, Volume 109, No. 5, May.

6. Gibson, R. E., England, G. L., and Husey, M. H. L. (1967) "The Theory of One-Dimensional Consolidation of Saturated Clays, 1. Finite Non-Linear Consolidation of Thin Homogeneous Layers," Geotechnique, Volume 17.

7. Gibson, R. E., Schiffman, R. L., and Cargill, K. (1980) "The Theory of One-Dimensional Consolidation of Saturated Clays, II. Finite Non-Linear Consolidation of Thick Homogenous Layers," Canadian Geotechnical Journal.

8. Imai, G. (1981), "Experimental Studies on Sedimentation Mechanism and Sediment Formation of Clay Minerals," Soils and Foundations Vol. 21, No. 1,March.

9. Lawver, J. E. (1982), "IMC-Agrico-Mobil Slime Consolidation and Land Reclamation Study," Progress Report #6, IMC, Bartow, Florida.

10. Lawver, J. E., and Carrier, W. D. (1983) "Mathematical and Centrifuge Modeling of Phosphatic Clay Disposal Systems," Presented at AIME meeting, March, Atlanta.

11. Lawver, J. E., Carrier, W. D., and Somogyi, F. (1984), "Waste Clay Disposal and Land Reclamation Techniques in the Florida Phosphate Industry," Presented at AIME meeting, February 1984, Los Angeles.

12. Scully, R. W., Schiffman, R. L., Olsen, H. W. and Ko, H. Y. (1984) "Validation of Consolidation Properties of Phosphatic Clay at Very High Void Ratios" ASCE Special Publication, Sedimentation/Consolidation Models, San Francisco.

13. Somogyi, F. (1979), "Analysis and Prediction of Phosphatic Clay Consolidation: Implementation Package," Bromwell & Carrier Engineering Inc., Lakeland, Florida or FIPR, Bartow, Florida.

14. Somogyi, F. (1980), "Large Strain Consolidation of Fine Drained Slurries," Presented at the May 20-30 Canadian Society for Civil Engineers held at Winnipeg, Manitoba.

15. Wissa, A. E. Z., Fuleihan, N. F., and Ingra, T. S. (1983) "Evaluation of Phosphatic Clay Disposal and Reclamation Methods, Volume 4: Consolidation Behavior of Phosphatic Clays," Research Project FIPR 80-002, Florida Institute of Phosphate Research, Bartow, Florida.

16. Yong, R. N., Siu, S. K., and Sheeran, D. E. (1983) "On the Stability and Settling of Suspended Solids in Settling Ponds. Part I. Piece-Wise Linear Consolidation Analysis of Sediment Layer," Canadian Geotechnical Journal, Vol. 20, No. 4, November.

17. Znidarcic, D. and Schiffman, R. L. (1981), "Finite Strain Consolidation: Test Conditions," Tech Note ASCE Journal Geot. Engineering Division, Volume 107, GT5.

APPLICATION OF FINITE STRAIN CONSOLIDATION THEORY FOR ENGINEERING DESIGN AND ENVIRONMENTAL PLANNING OF MINE TAILINGS IMPOUNDMENTS

by

Jack A. Caldwell[1]

Keith Ferguson, A.M., ASCE[2]

Robert L. Schiffman, M., ASCE[3]

Dirk van Zyl, A.M., ASCE[4]

ABSTRACT

A knowledge of the consolidation behavior of mine tailings, during and after deposition, is important in estimating final impoundment capacity, rate of pore fluid expulsion during and after deposition and post-deposition (reclamation) behavior. This paper describes three case histories where finite strain consolidation theory was used for the engineering design and environmental planning of tailings impoundments. These three cases are located in different climatic environments. Four different materials are considered: massive sulfide tailings remaining from copper and zinc extraction, and gold tailings from three different ore bodies.

[1]Division Head, Steffen Robertson and Kirsten, Denver
[2]Project Engineer, Steffen Robertson and Kirsten, Denver
[3]Professor, University of Colorado, Boulder
[4]Associate Professor, Colorado State University

INTRODUCTION

The rate at which mine tailings consolidates and the magnitude of consolidation settlement affect the design of the tailings impoundment at which they are deposited. The way the tailings consolidate must be understood in order to:

- Estimate the required impoundment capacity.

- Estimate the amount and the quality of seepage of water from the pores of the deposited tailings.

- Calculate the rate, magnitude and spatial variation of change of such tailings characteristics as density, hydraulic conductivity, compressibility, and strength.

- Estimate the amount of settlement of the top surface of the impoundment that will occur after reclamation of the facility.

This paper describes the use of finite strain consolidation theory in the design of tailings impoundments for three mines.

THEORY AND BACKGROUND

Two relatively recent papers (Gibson, England and Hussey, 1976; and Gibson, Schiffman and Cargill, 1981) describe the finite nonlinear strain theory for one-dimensional consolidation of saturated clay. They show that the classical theory of one-dimensional consolidation has three major limitations when applied to a soft material that is consolidating under its own weight and for which material is constantly being added to the deposit. The strains are assumed to be small; hence, the change in thickness of the consolidating layer is ignored. The self-weight of the deposit is inaccurately represented by a constant during the process of consolidation and is assumed to be uniform within the consolidating layer. The material properties a_v and k are assumed to be constant throughout the consolidation process. The first of these makes the use of the infinitesimal theory incorrect for settlement calculations of highly compressible materials. The amount of settlement in these deposits is an important parameter in design.

The main conditions considered in finite strain theory are:

- The magnitude of strains during consolidation is unrestricted; i.e., they may be finite or infinitesimal.

- The soil skeleton may deform in a linear or nonlinear manner; i.e., the void ratio and corresponding density of the soil skeleton is modelled as a function of the compressibility over the range of effective stresses.

- The coefficient of permeability is free to change during consolidation; i.e., is a function of the void ratio.

The theory of finite strain is given by the equation

$$\pm\left(\frac{\gamma_s}{\gamma_f} - 1\right)\frac{d}{de}\left[\frac{k(e)}{1 + e}\right]\frac{\partial e}{\partial z} + \frac{\partial}{\partial z}\left[\frac{k(e)}{\gamma_f(1+e)} \cdot \frac{\partial\sigma'}{\partial e} \cdot \frac{\partial e}{\partial z}\right]+\frac{\partial e}{\partial t} = 0$$

where z   = material coordinate

       e   = void ratio

       k(e) = function relating permeability to the void ratio

       σ' = effective stress

       γ_f = unit weight of fluid

       γ_s = unit weight of solids (specific gravity)

The use of plus or minus sign depends on the coordinate system (with or against gravity). All other theories of one-dimensional consolidation are special cases of the above theory as shown by Schiffman (1980).

The above equation is a complex relationship for which a general closed-form solution is not yet available. Numerical procedures and the software have been developed to solve the finite strain equations. The Program FSCON4 (Pane and Schiffman, 1981) solves the consolidation of a loaded layer using finite difference procedures. The Program FSLINE (Schiffman and Pane, 1984) solves the problem for a deposit whose volume of solids is increasing as a function of time. Input data to the program includes specific gravity, and the relationship between the effective stress and the void ratio, and between the void ratio and the hydraulic conductivity. This data is obtained by standard step loading laboratory testing or from the constant rate of deformation test (CRD) as described by Znidarčić (1982). Both surcharge loading and self-weight consolidation are considered in the analysis.

Case History No. 1: Gold and Copper/Zinc Mine, Northern Maine

The design of a new impoundment for a mine in Northern Maine included an analysis using finite strain consolidation theory to characterize the density, permeability, and consolidation of the tailings both during and after operation.

The project area has an annual precipitation of 1,000 mm, and an annual evaporation of 500 mm. Average summer temperatures range from $10°$ C to $16°$ C, while winter temperatures range from $-12°$ C to $-9°$ C. Figure 1(a) shows the site topography and the layout of the proposed impoundments. The bedrock underlying the valley is a sequence of weakly metamorphosed volcanic rocks that are predominantly andesitic and basaltic. The bedrock is overlain by medium-dense to dense glacial tills. Groundwater flow at the site generally occurs in the upper zone of the bedrock. Recharge is from inflow on the steep

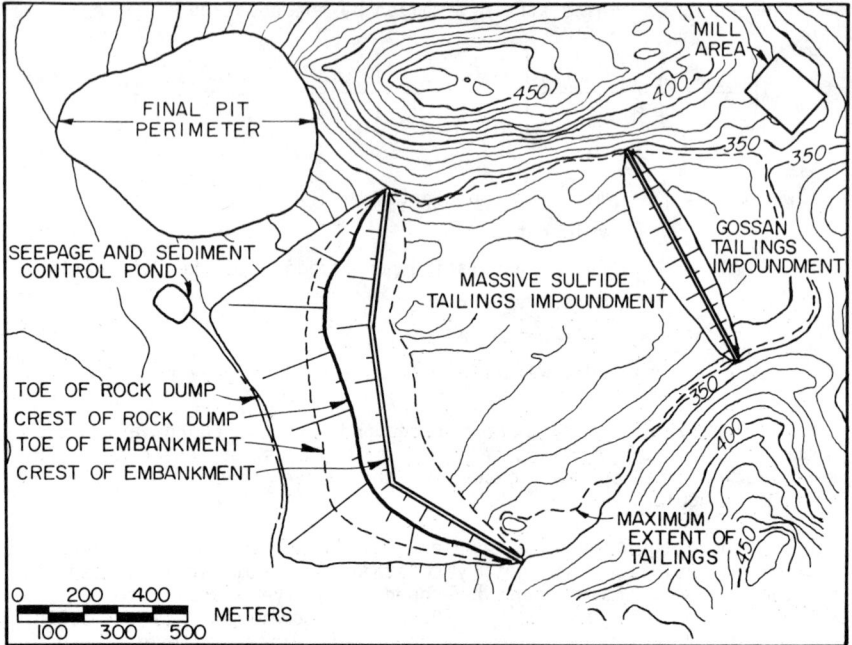

(a) GENERAL LAYOUT - TAILINGS IMPOUNDMENT

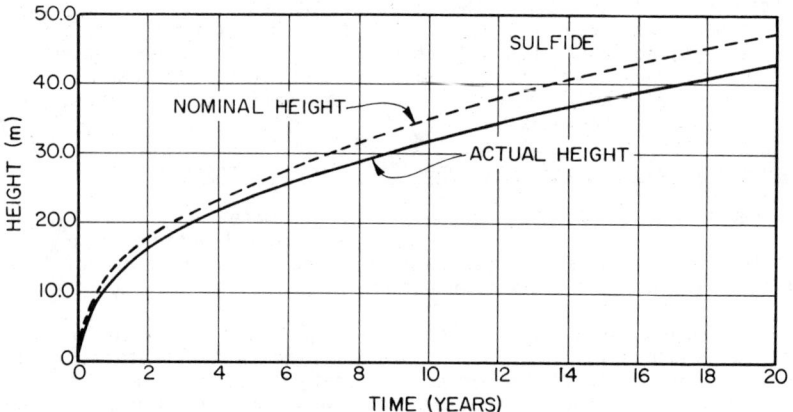

(b) HEIGHT-TIME CURVES - SULFIDE

FIGURE 1. SITE CHARACTERISTICS - CASE HISTORY NO. 1

bedrock controlled slopes above the impoundment; there is little, if any, recharge through the tills.

Two types of tailings will be produced and deposited at the impoundment. The gossan tailings (remaining after extraction of gold and silver) will be deposited first behind the upper embankment shown in Figure 1(a). Thereafter, the tailings from processing of the massive sulfide ore body, for extraction of copper and zinc, will be deposited behind the lower embankment.

The gossan tailings will be delivered to the impoundment as a slurry at 45% by mass of solids, and the massive sulfide tailings will be delivered as a slurry containing 30% solids. The tailings at both impoundments will be deposited below water. The specific gravity of the gossan tailings is 3.0 and of the sulfide is 4.2.

Figures 2 and 3 show the results of the laboratory tests and finite strain analyses of the gossan and sulfide tailings, respectively. It should be noted that both materials will be very finely ground (80% minus No. 400 sieve).

The effective stress-void ratio and hydraulic conductivity - void ratio results used in the analysis were based on a number of laboratory step loading/flow pump permeability and constant rate of deformation tests (Olsen, 1966 and Pane et al., 1983). These tests were performed in the laboratories of the University of Colorado, Boulder, under direct supervision of the third author. Slurry samples modelling field deposition conditions were used. Note the convex shape of effective stress void ratio curve for the gossan tailings at low stresses. The sulfide tailings behave like a typical, normally consolidated deposit.

Finite strain consolidation theory was used to investigate the progress of consolidation, the effect of consolidation on the final height of the deposited tailings, and also the distribution of the void ratio and the hydraulic conductivity with depth, following consolidation. The results for both the gossan and the massive sulfide tailings are shown in Figures 2 and 3.

For the gossan tailings deposited to a maximum depth of 14 m over 18 months, the predicted ultimate settlement is 3 m. Ninety-nine percent of the consolidation is calculated to occur in about 600 days. This indicates that under controlled filling the gossan will be almost fully consolidated by the time deposition has ceased, and that the potential for settlement in the long term of the reclamation cover is very small.

As a comparison, the Terzaghi theory would predict 90% consolidation to occur in 700 days when single drainage conditions and 14 m initial height are assumed for c_v = 0.196 m^2/day). This compares to 400 days obtained with finite strain theory or approximately a factor of 2.

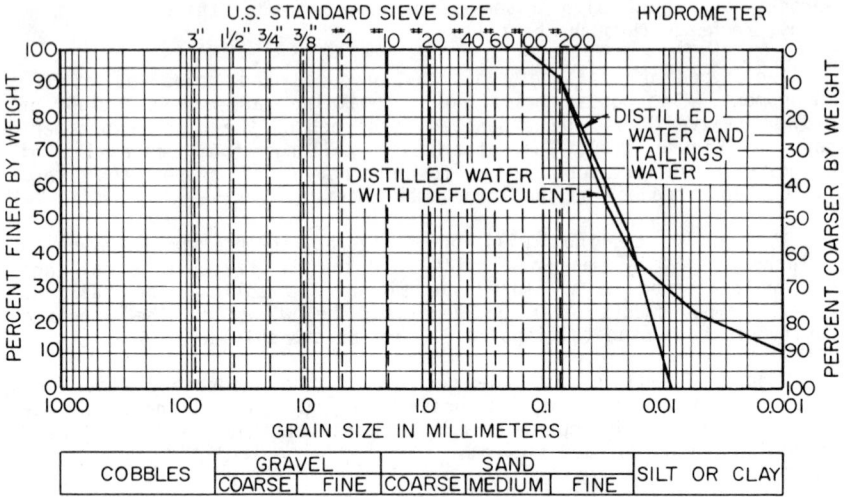

(a) PARTICLE SIZE DISTRIBUTION OF GOSSAN TAILINGS

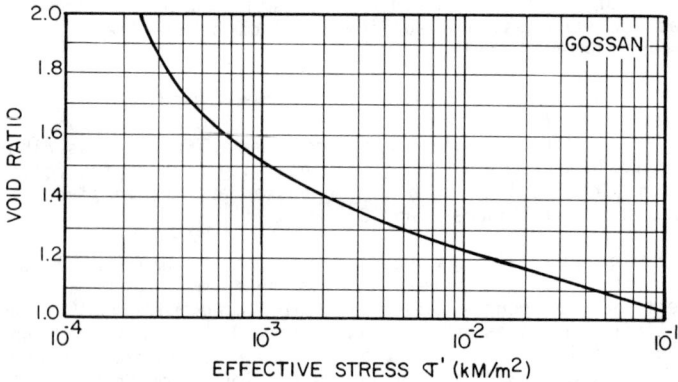

(b) VOID RATIO-EFFECTIVE STRESS

FIGURE 2. LABORATORY AND FINITE STRAIN ANALYSES
CASE HISTORY NO. 1, GOSSAN

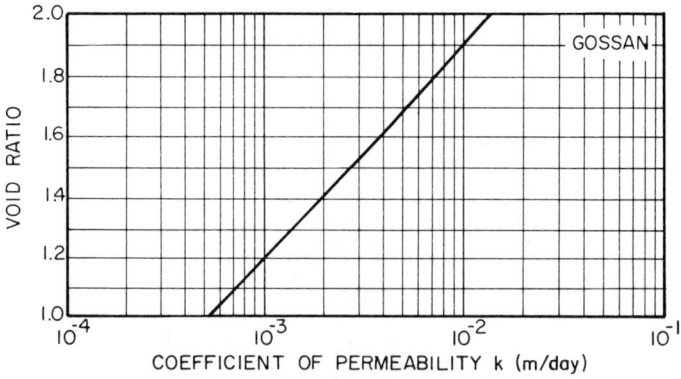

(c) VOID RATIO-HYDRAULIC CONDUCTIVITY

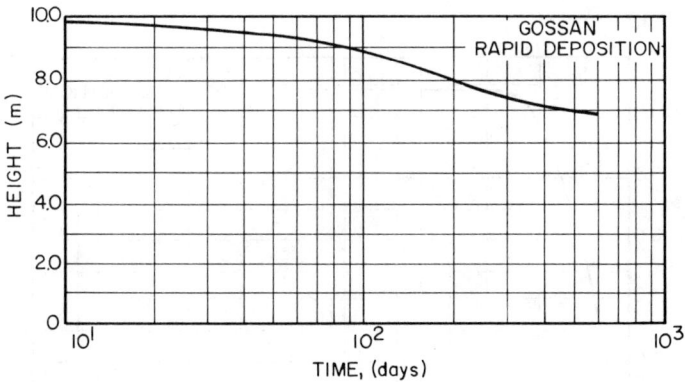

(d) POST-DEPOSITIONAL SETTLEMENT

FIGURE 2 - CON'T

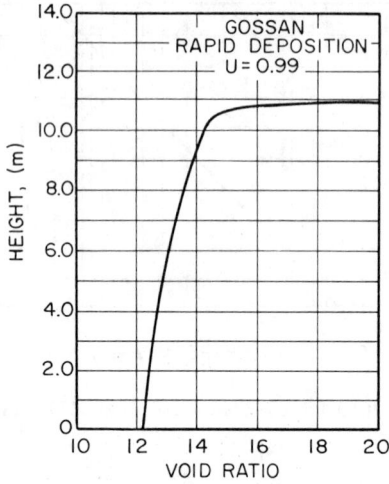

(e) HEIGHT-VOID RATIO

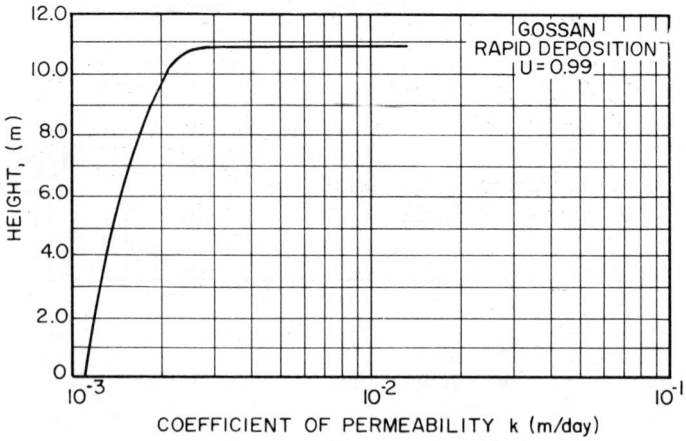

(f) HEIGHT-HYDRAULIC CONDUCTIVITY

FIGURE 2 - CON'T

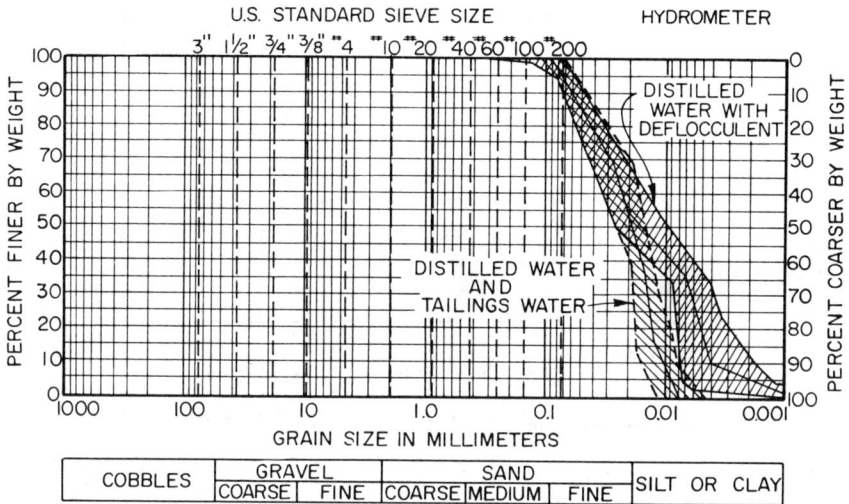

(a) PARTICLE SIZE DISTRIBUTION OF SULFIDE TAILINGS

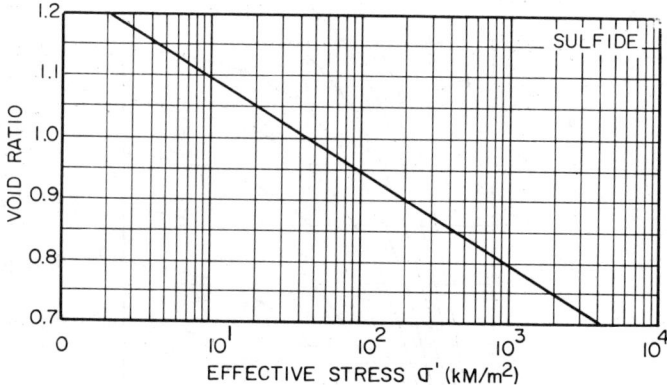

(b) VOID RATIO-EFFECTIVE STRESS RELATIONSHIP
FOR SULFIDE TAILINGS

FIGURE 3. LABORATORY AND FINITE STRAIN ANALYSES
CASE HISTORY NO. 1, SULFIDE

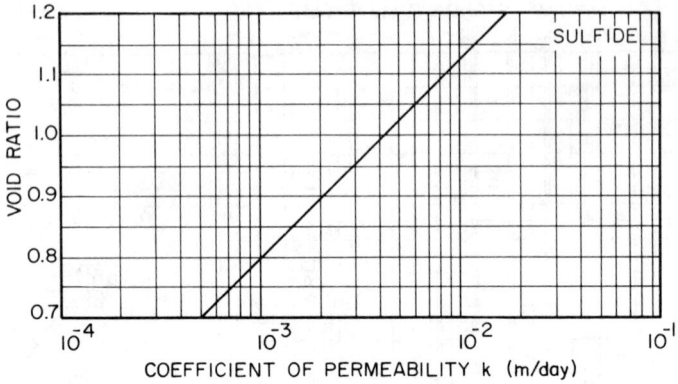

(c) VOID RATIO-HYDRAULIC CONDUCTIVITY

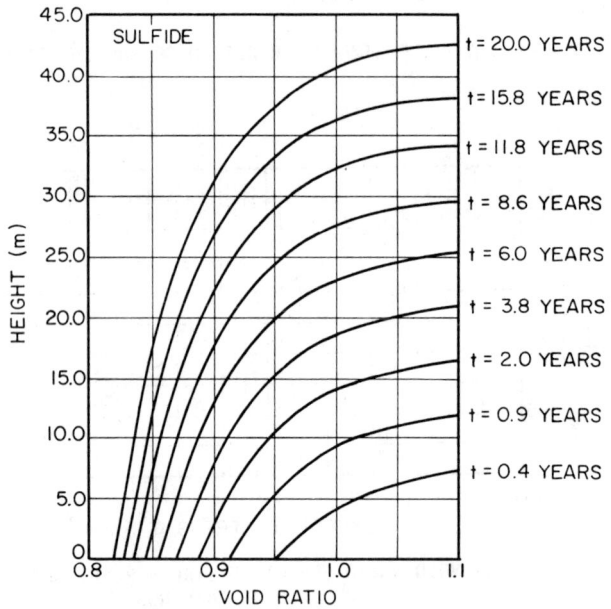

(d) HEIGHT-VOID RATIO AT DIFFERENT TIMES

FIGURE 3 - CON'T

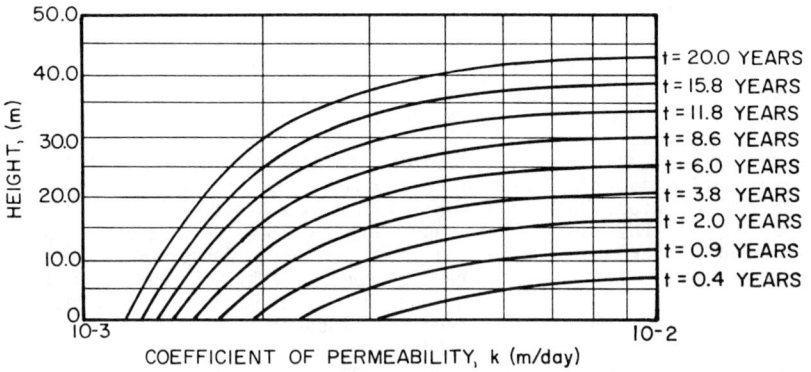

(e) HEIGHT-HYDRAULIC CONDUCTIVITY AT DIFFERENT TIMES

FIGURE 3 - CON'T

The void ratio and hydraulic conductivity profile of the gossan tailings confirm the presence of a thin, highly permeable, high-void ratio layer at the top of the deposit at the end of deposition.

A filling curve based on the impoundment geometry was used as the basis of the analysis of the behavior of the sulfide tailings [see Figure 1(b)]. For this relationship, based on the void ratio after deposition, a height of 47.15 m is reached at the end of twenty years. The finite strain theory analysis showed that after consolidation, the actual height would be 42.76 m. Therefore, a 4.39 m of consolidation settlement takes place during deposition. All primary consolidation is calculated to have occurred at the end of the 20-year deposition period. Note that the void ratio and hydraulic conductivity profiles for the sulfide tailings show a much more gradual change with depth than the gossan tailings. There is about half an order of magnitude change in hydraulic conductivity of the sulfide at the bottom of the impoundment during the construction period. This will influence seepage losses from the impoundment.

Case History No. 2: McLaughlin Gold Mine, Lake County, California

The consolidation characteristics of the tailings to be deposited at the McLaughlin tailings impoundment were evaluated in order to:

- Determine the required impoundment capacity,

- Provide preliminary information about the post-closure settlement of the tailings surface, and

- Consequently provide the preliminary design basis of reclamation plans submitted in permit applications.

The McLaughlin project will process 3,000 tons of gold-bearing ore per day for 24 years, and produce 26 million tons of tailings. The average annual precipitation at the site is 750 mm and the average annual evaporation is approximately 1,200 mm. The tailings will be deposited as a slurry with 40 percent by mass solids and consolidation will occur in a predominantly saturated environment until closure.

The impoundment is underlain by serpentinitic bedrock which has a very low hydraulic conductivity, 1.25×10^{-7} m/sec to 1.17×10^{-9} m/sec. Single drainage at the surface of the tailings will control the consolidation process.

Figure 4 shows the site layout, height-time curve (based on the production rate above), grain-size distribution of tailings, laboratory results for effective stress-void ratio and hydraulic conductivity - void ratio as well as the finite strain consolidation analysis results. These results indicate that the degree of consolidation at the end of impoundment filling will be 99 percent. The final average void ratio at closure will be approximately 1.52. A total of 26 feet of consolidation will have occurred for 119 feet of tailings depth. The actual height and consequently the embankment crest elevation was selected based upon a consolidated height of tailings of 93 feet.

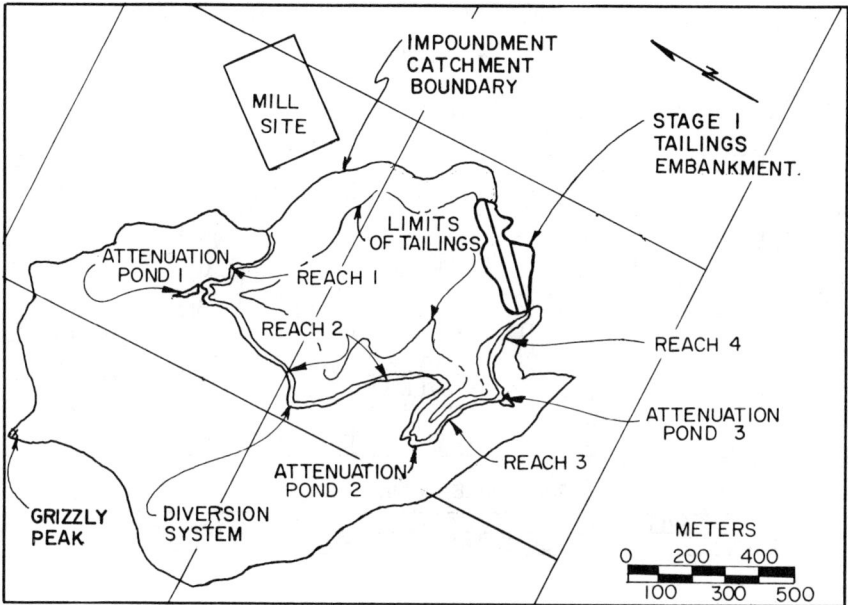

(a) SITE LAYOUT

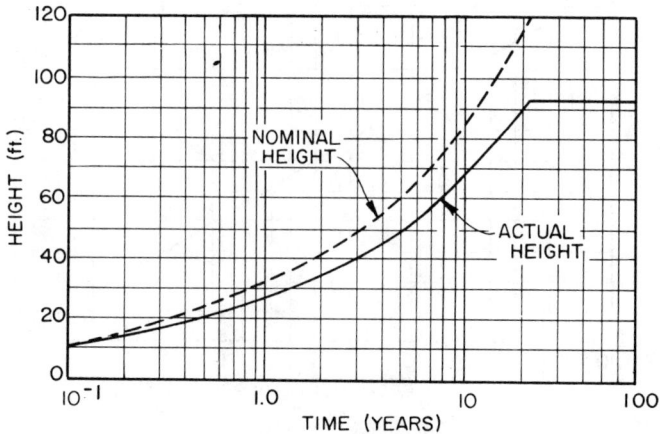

(b) HEIGHT-TIME CURVES

FIGURE 4. SITE CHARACTERISTICS, LABORATORY AND
FINITE STRAIN ANALYSES - McLAUGHLIN PROJECT

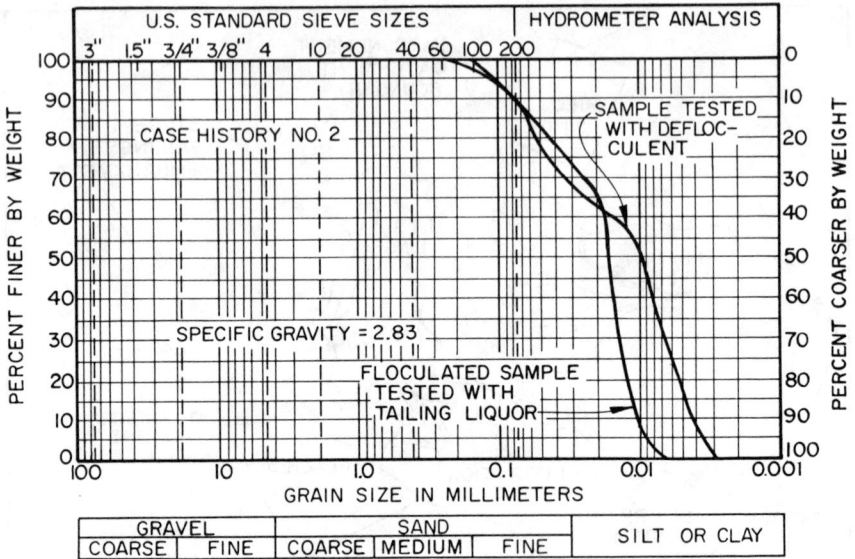

(c) GRAIN SIZE DISTRIBUTION

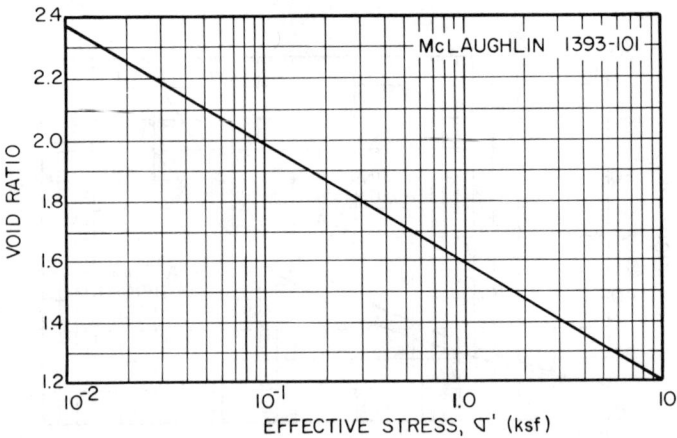

(d) VOID RATIO-EFFECTIVE STRESS

FIGURE 4. CON'T

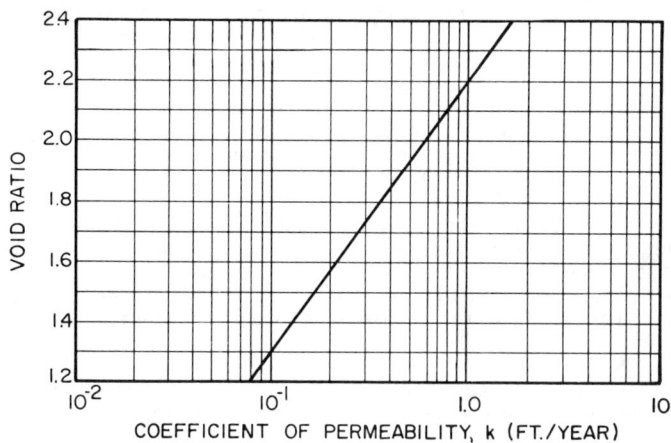

(e) VOID RATIO-HYDRAULIC CONDUCTIVITY

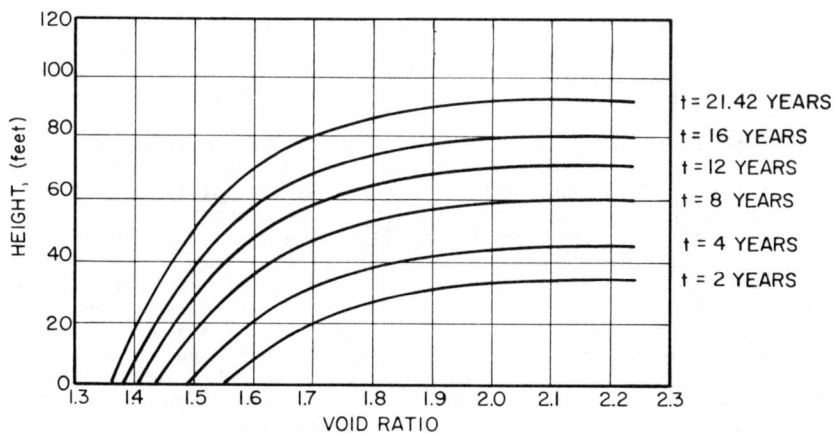

(f) HEIGHT-VOID RATIO AT VARIOUS TIMES

FIGURE 4. CON'T

The results had the following significant impacts on the designs of the tailings impoundment:

- The embankment height was selected for a tailings that is 99 percent consolidated. Classical consolidation theory predicts a lower degree of consolidation at the end of deposition and, consequently, a higher embankment would be required. Cost savings were therefore realized.

- The surface of the tailings post-closure will settle only due to the load imposed by the reclamation cover, and not as a result of further primary consolidation as is predicted by classical consolidation theory. Thus, the reclamation plan calls only for minimal surface contouring and a thin cover.

- The reclamation costs are thus lower than would be required if the design were based on classical consolidation theory.

Case History No. 3: Unidentified

Fine-grained tailings are to be deposited into an impoundment with very high rate of rise. The maximum impoundment depth is about 108 ft. The mine is located in a high rainfall, low evaporation climate where large volumes of surface runoff flows to the tailings pool. Tailings slurry contains some contaminants and is deposited at a pH of 10 to 12. Some local acid mine drainage with a pH of 4.5 to 5.5 flows to the impoundment. Depending on seasonal effects of surface runoff and groundwater discharge, the tailings pool pH could range from 6 to 11.

The major emphasis of this study was to estimate the rate of interstitial tailings water discharge to the tailings surface and foundation. This information was required for the design of water return facilities and to estimate water treatment requirements.

The pH level of the tailings slurry and depositional environment has a noticeable effect upon the engineering characteristics of the tailings. To illustrate, gradation characteristics are shown on Figure 5(a). The two flocculated tailings curves represent the variation for a pH range of 9.6 to 11.8. A standard grading curve using a deflocculant is also shown. Similarly, the effect of pH upon plasticity, settled density, and initial void ratio is given in Table 1. Figures 5(b) and (c) show the void ratio (d) as a function of the effective stress (σ') and hydraulic conductivity (k) as a function of void ratio. Testing included samples ranging from pH 7 to 10.2 (σ' vs e) and pH 7 to 9 (e vs k). These results indicate that pH has some effect on the effective stress-void ratio characteristics, but no effect on the void ratio - permeability characteristics.

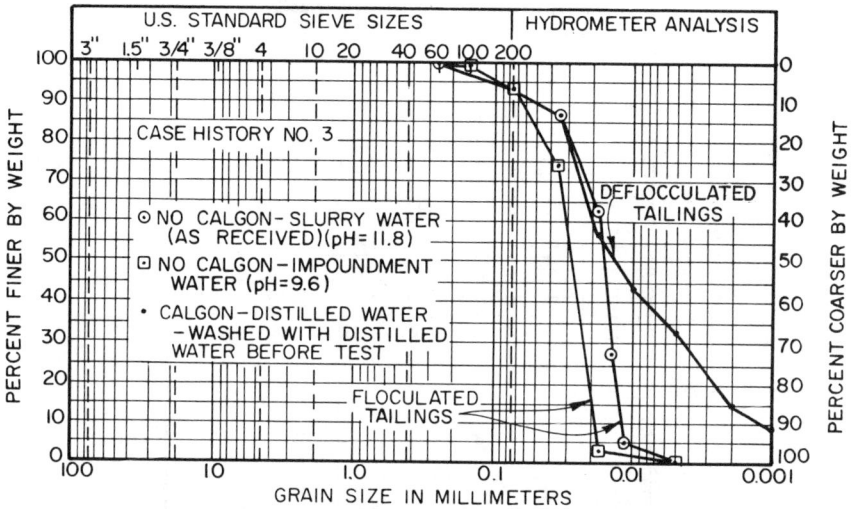

(a) GRAIN SIZE DISTRIBUTION

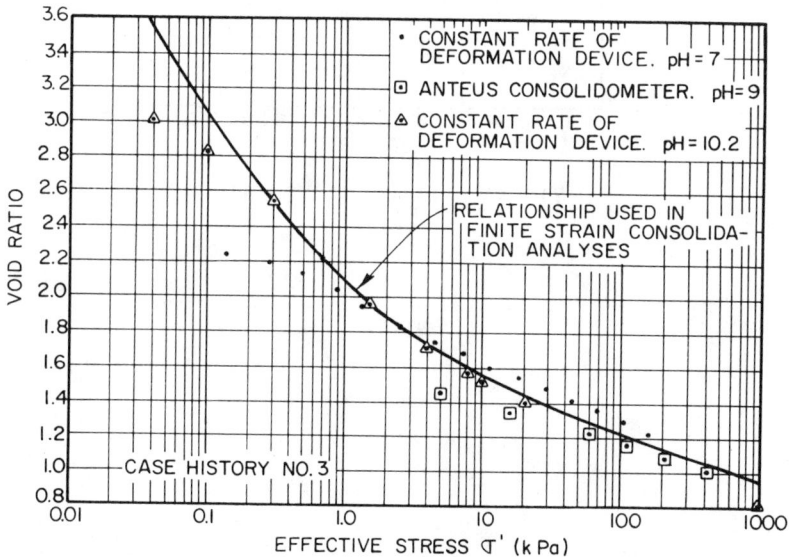

(b) VOID RATIO-EFFECTIVE STRESS

FIGURE 5. SITE CHARACTERISTICS, LABORATORY AND
FINITE STRAIN ANALYSES - CASE HISTORY NO. 3

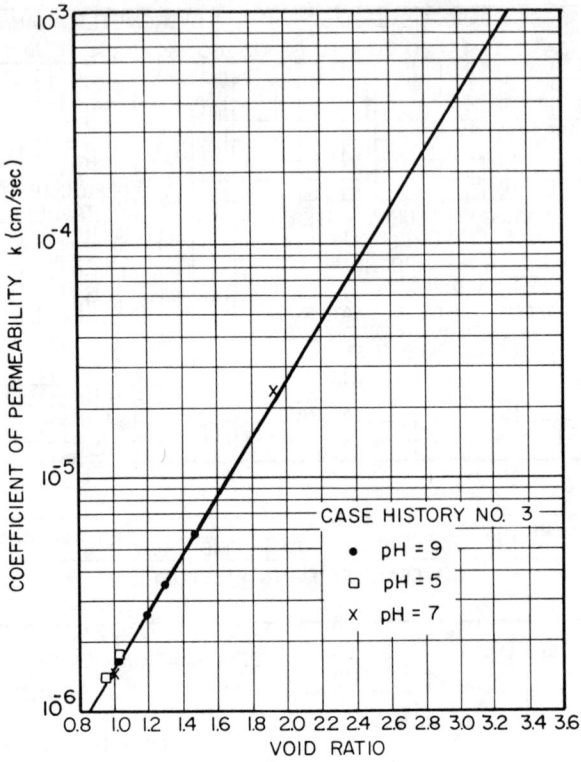

(c) VOID RATIO-HYDRAULIC CONDUCTIVITY

FIGURE 5. CON'T

TABLE 1 - pH Effect on Atterberg Limits and
Settled Density of Case 3 Tailings

pH	Liquid Limit LL (%)	Plastic Limit PL (%)	Plastic Index PI (%)*	Initial Settled Dry Density - γ_D(pcf)	Initial Void Ratio - e (%)
11.6	38.9	33.4	5.5	42.3	3.57
10.0	36.7	35.5	1.2	41.3	3.68
8.0	34.8	33.6	1.2	47.5	3.07
6.0	33.5	31.4	2.1	48.7	2.97
4.0	31.5	31.1	0.4	50.0	2.87

Of principal concern was the rate of tailings water discharge after reclamation of the impoundment. Results of the finite strain analyses showed that primary self-weight consolidation will occur very rapidly in the impoundment and will be essentially complete (degree of consolidation - U greater than 95 percent) at the time when deposition is stopped. Reclamation plans call for the placement of a cover four years after deposition is stopped, which will result in a surcharge loading of approximately 800 psf.

Results of depositional consolidation evaluations are shown in Figures 5(d) and (e) for column heights of 27 and 108 ft., respectively. For the first column, 90 percent of primary consolidation was predicted to occur by 0.15 years following impoundment closure. Ninety-one percent of primary consolidation was predicted to occur by 1.0 years following impoundment closure in the second column (representing the deepest portions of the impoundment). These results served as the initial conditions for surcharge (reclamation) loading evaluations.

Typical excess pore water pressures for various degrees of consolidation determined for the case of post-reclamation consolidation of a 27 ft. tailings profile are shown on Figure 6. The results were used to predict the volume of discharge to the impoundment surface and foundation (i.e., double drainage). The neutral excess pore water pressure axis illustrated on this figure was the key to these evaluations. The slope of the excess pore pressure curve at a given degree of consolidation represents the gradient that discharging waters will experience. The waters will flow in the direction of decreasing excess pore water pressure.

The predicted volume change above the neutral axis would be equivalent to the volume of water discharged to the impoundment surface. Likewise, volume change below the axis would be discharged

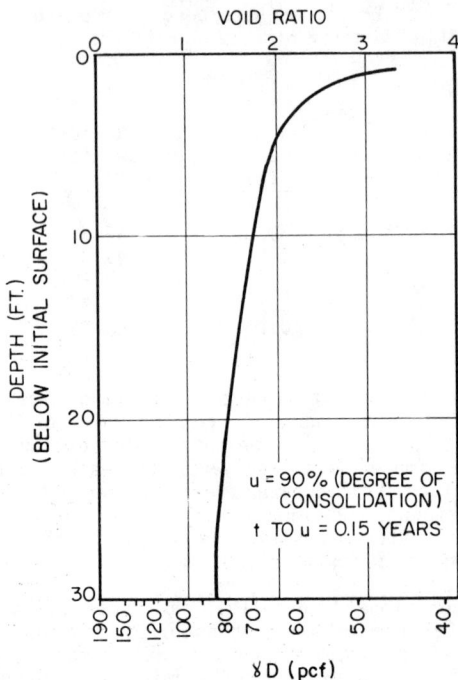

NOTES

1. VOID RATIO DISTRIBUTION SHOWN IS AT NEAR
    END OF PRIMARY SELF WEIGHT
    CONSOLIDATION FOR A 27' HIGH COLUMN

2. γD CALCULATIONS ASSUME SPECIFIC
    GRAVITY OF SOLIDS = 3.1

(d) HEIGHT-VOID RATIO FOR 27' COLUMN

FIGURE 5. CON'T

VOID RATIO

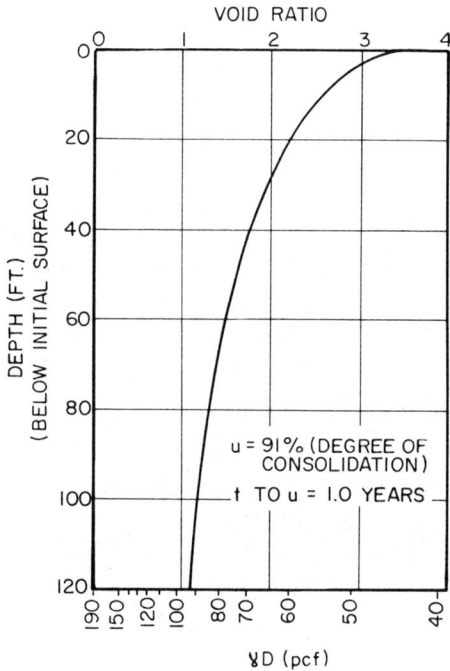

NOTES

1. VOID RATIO DISTRIBUTION SHOWN IS AT NEAR END OF PRIMARY SELF WEIGHT CONSOLIDATION FOR A 108' HIGH COLUMN

2. γD CALCULATIONS ASSUME SPECIFIC GRAVITY OF SOLIDS = 3.1

(e) HEIGHT-VOID RATIO 108' COLUMN

FIGURE 5. CON'T

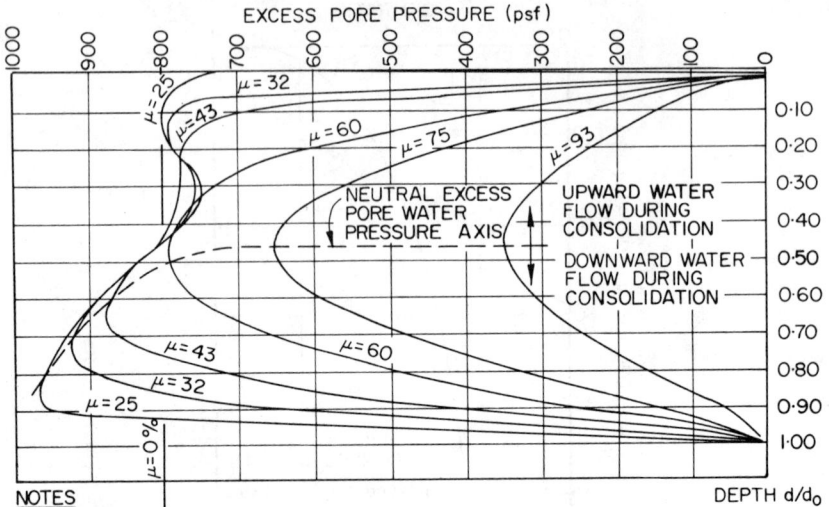

EXCESS PORE PRESSURE (psf)

NOTES

1. μ = DEGREE OF CONSOLIDATION (%)

2. FIGURE ILLUSTRATED IS FOR A 27'
    TAILINGS PROFILE SUBJECTED TO
    AN INCREMENTAL LOADING OF 800 psf

FIGURE 6. EXCESS PORE PRESSURE PROFILES FOLLOWING
           PLACEMENT OF RECLAMATION COVER

to the impoundment foundation. In order to complete an estimation
of consolidation discharge over an extended period, secondary consolidation was also considered. All secondary consolidation discharge
was assumed to flow to the impoundment foundation.

A summary of predicted consolidation discharge flows is given
on Figure 7. Finite strain results indicated completion of primary
consolidation at Year 0.8 following reclamation for the 27-ft. column,
and four years for the 108-ft. column. The results, therefore, include
estimations of combined primary and secondary consolidation between
Years 0.8 (Year 4.8 following closure) and 4 (Year 8 following
closure).

Comparison of Material Characteristics

The four tailings products referred to in the case histories
above were from different ore bodies and from different extraction
processes. Only the ore types and extraction processes for the gossan
tailings in Case History No. 1 and the gold tailings of Case History
No. 3 are similar. In order to compare the effective stress and
hydraulic conductivity versus void ratio results, the combined plots
of Figure 8 were prepared. It is interesting to note how well all
these results compare.

It has been observed that pH changes in the interstitial fluid
has the most significant effect on initial settled density (e.g.,
Table 1). The effects on void ratio - effective stress and void
ratio - hydraulic conductivity relationships are very small.

CONCLUSIONS

Three case histories have been presented where finite strain
consolidation theory was used to evaluate the consolidation behavior
of deposited mine tailings. The results were applied to the following
concerns:

(i)   Determining the required impoundment capacity and subse-
        quently the required embankment heights.

(ii)   The final engineering characteristics of the waste in
        order to evaluate post-impoundment closure performance
        and potential reclamation problems and alternatives.

(iii)   The rate of interstitial water discharge from the tailings
        during deposition and following closure in order to
        perform suitable water balance evaluations, and to design
        adequate water return and treatment facilities for impound-
        ment seepage discharge.

A comparison of the consolidation parameters for the tailings
investigated show remarkable similarity, despite variations in ore
type and mineral extraction processes.

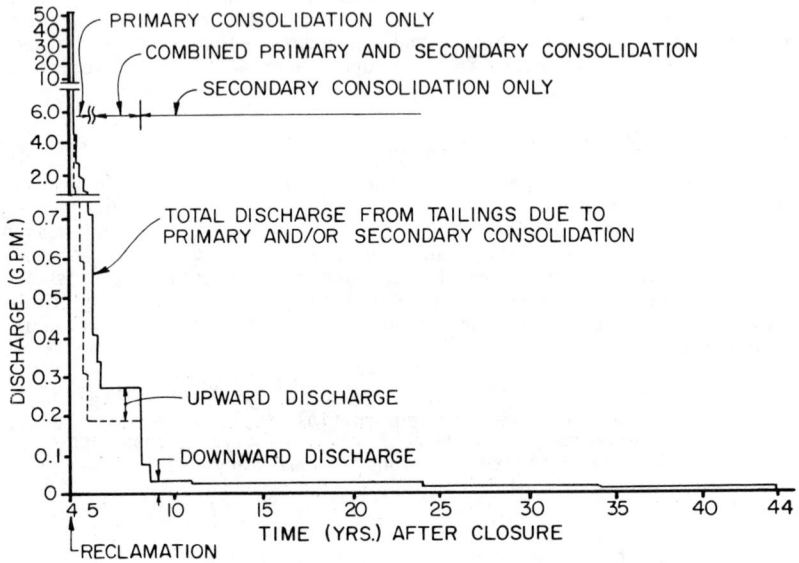

FIGURE 7. PREDICTED CONSOLIDATION DISCHARGE FLOWS

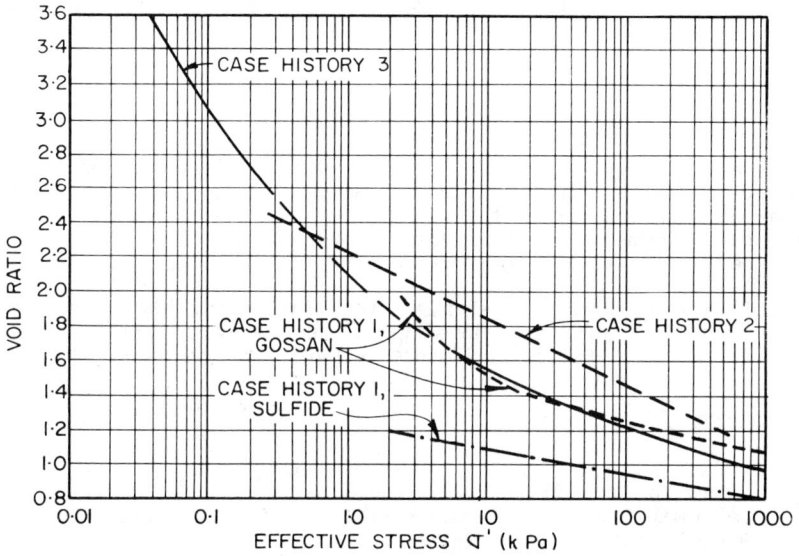

(a)

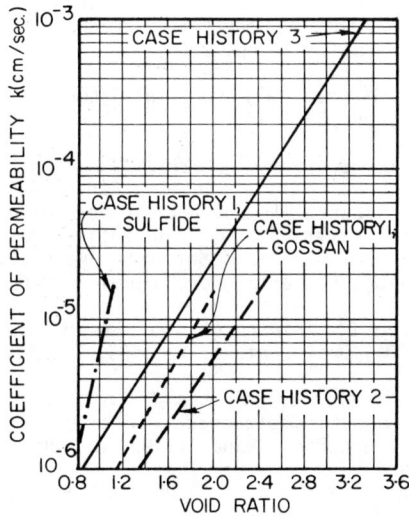

(b)

FIGURE 8. SUMMARY OF RESULTS

A number of simplified assumptions were made in these analyses. These include:

(i)   Laboratory test results represent field behavior, and

(ii)   Tailings deposition takes place under water.

The first assumption is not always correct because some spatial variation of tailings characteristics will be present due to the sedimentation behavior of the tailings. On the basis of water balance predictions, the second assumption is nearly correct as the tailings surfaces will be covered by water for most of the time.

It can be concluded that finite strain consolidation analysis is a useful tool for the investigation of the depositional behavior of mine tailings. Future field observations will have to be considered for validation of the approach.

REFERENCES

Gibson, R. E., England, G. L. and Hussey, M. J. L. (1967). "The Theory of One-Dimensional Consolidation of Saturated Clays, I. Finite Non-Linear Consolidation of Thick Homogeneous Layers." Geotechnique 17. No. 3, 261-273.

Gibson, R. E., R. L. Schiffman, and K. E. Cargill (1981), "The Theory of One-Dimensional Consolidation of Saturated Clays, II. Finite Non-Linear Consolidation of Thick Homogeneous Layers," Canadian Geotechnical Journal, 18, pp. 280-293.

Olsen, H. W. (1966), "Darcy's Law in Saturated Kaolinite," Water Resour. Res. 2, pp. 287-295.

Pane, V. and R. L. Schiffman (1981), "FSCON4-I, Version 1, Level A, One-Dimensional Finite Strain Consolidation of a Thick, Normally Consolidated Homogeneous Layer with Point Data Void Ratio-Effective Stress and Void Ratio-Permeability Relationships," Geotechnical Eng. Software Activity, Department of Civil Engineering, University of Colorado, Boulder, Colorado.

Pane, V., P. Croce, D. Znidarcic, H-Y Ko, H. W. Olsen, and R. L. Schiffman (1983), Geotechnique, Vol. 33, No. 1, pp. 67-71.

Schiffman, R. L. (1980), "Finite and Infinitesimal Strain Consolidation," Journal of Geotechnical Engineering Division, ASCE, 106, No. GT2, Proceedings Paper 15193, pp. 203-207.

Schiffman, R. L. and V. Pane (1984), "Nonliner Finite Strain Consolidation of Soft Marine Sediments," Proceedings of the 1983 IUTAM Conference on Seabed Mechanics, University of Newcastle-Upon-Tyne, Graham and Trotman, Ltd., U.K. (in press).

Znidarčić, Dobroslav (1982) "Laboratory Determination of Consolidation of Cohesive Soil," PhD Thesis, University of Colorado, Boulder, Colorado, 172 pp.

SUBJECT INDEX

Page number refers to first page of paper.

AUTHOR INDEX

Page number refers to first page of paper.